MEDITERRANEAN IDENTITIES - ENVIRONMENT, SOCIETY, CULTURE

Edited by **Borna Fuerst-Bjeliš**

Mediterranean Identities — Environment, Society, Culture
http://dx.doi.org/10.5772/66587
Edited by Borna Fuerst-Bjeliš

Contributors

Kerim Çiçek, Oğuzkan Cumhuriyet, Helena Trindade Lopes, Isabel Almeida, Irene Martín-Forés, António Canatário Duarte, Victor Mallia-Milanes, Nunziacarla Spanó, Emilio De Domenico, Michael Krom, Helen Powley, Phillippe Van Cappellen, Tugrul Varol, Mertol Ertuğrul, Halil Barış Özel, Conxita Royo, Jose Miguel Soriano, Fanny Alvaro, Sebastiano Sferlazza, Federico Guglielmo Maetzke, Donato Salvatore La Mela Veca, Marcello Miozzo, Ana Cristina Gonçalves, Adélia Sousa, Sandra Oliveira, Giovanni Laneve, Lorenzo Fusilli, Georgios Eftychidis, Luciano Lourenço, Adélia Nunes, Ana Sebastián-López, Fedra Alessandra Pizzato, Anna Maria Mannino, Paolo Balistreri, Alan Deidun, Gilkauris María Rojas Cortorreal, Francesc Navés Viñas, Julio Peña, Jaime Roset Calzada, Carlos Fernando López Ordóñez, Attilio Rigotti, Guadalupe Echeverría, Catalina Dussaillant, Emma McGee, Inés Urquiaga, Nicolás Velasco, Onur Gönülal, Cem Dalyan

Published by InTech
Janeza Trdine 9, 51000 Rijeka, Croatia

Publishing Process Manager Martina Usljebrka
Technical Editor SPi Global
Cover InTech Design team

First published November, 2017
Printed in Croatia
Legal deposit, Croatia: National and University Library in Zagreb

Additional hard copies can be obtained from orders@intechopen.com

Mediterranean Identities — Environment, Society, Culture, Edited by Borna Fuerst-Bjeliš
p. cm.
Print ISBN 978-953-51-3585-2
Online ISBN 978-953-51-3586-9

For EU product safety concerns:
IN TECH d.o.o., Prolaz Marije Krucifikse Kozulić 3, 51000 Rijeka, Croatia,
info@intechopen.com or visit our website at intechopen.com.

Meet the editor

Borna Fuerst-Bjeliš is a senior scientist and full time professor at the University of Zagreb, Croatia. She was awarded by the *Federico Grisogono* award for the scientific achievement and outstanding contribution to the development of geography in Croatia. Her field of expertise is environmental history and landscape change, spatially focused mainly on Mediterranean karst area, including the development of methodologies of research (historical cartography), borderland areas, spatial perceptions, and spatial (regional) identities. She is the coauthor of the book *Historical Geography of Croatia* (University of Split, Croatia) and the editor of the Croatian editions of *What is Environmental History* by D. Hughes (Disput) and *The Global Environmental History* by I. G. Simmons (Disput). She is the member of a number of international and national scientific organizations. Currently, she is the vice-president of the *European Society for Environmental History* and a member of the Steering Committee of the Specialty Commission of the International Geographical Union on *Marginalization, Globalization, and Regional and Local Responses.*

Contents

Preface

What is the Mediterranean? What is defining it? Perhaps no region in the world is so complex and meaningful. These kinds of questions do not cease to be asked by the scholars since the influential work on the Mediterranean civilization by Fernand Braudel from the last century. He opened quite new perspectives on the question of identity of the places, introducing the leading role of the environment, which he thought is embedded in the core of the civilizations. The perception of the Mediterranean leans equally on the nature, its climate, vegetation, and sea but also on the culture and history, lifestyle, and finally on the landscape that embraces both, the nature and culture. To approach the question of identity, it seems that we have to give importance to all of these. There is no Mediterranean identity, but Mediterranean identities. Although, at first glance, it seems that it is more than clear what the Mediterranean region is, everyone has a kind of an image of it; some authors argue that it is not even a region, because it lacks homogeneity that is essential for regions with expressed identity. In fact, it is true, but that is exactly what gives individuality to the Mediterranean. Mediterranean is not about the homogeneity and uniformity, but about the unity that comes from diversities, contacts, and interconnections. And this applies to marine and terrestrial species, waters, as well as for peoples, cultures, and lifestyles. The book is envisioned to embrace the environment, society, and culture of the Mediterranean in their multiple and unique interconnections over the millennia and to contribute to the better understanding of the essential human-environmental interrelations throughout the history of the Mediterranean. Environment is one of the most important parts of the story. Everything takes place in the environment. In the distant past, subsistence was conditioned by the environmental opportunities much more than today. However, the contemporary lifestyle and activities and even sociopolitical relations are still in some extent, of course, designed by the environmental opportunities or constraints.

The choice of 17 chapters of the book, written by a number of prominent scholars and scientists, clearly shows the necessity of the interdisciplinary approach to the Mediterranean identity issues. The book reflects some of the current thinking on the subject, based on a vast body of literature, but also opens new insights and perspectives. Each author approaches the issue in his own, original way complementing each other. Chapters have defined a number of current key issues that make the basic structure of the book.

The book consists of four sections. The first section on "Rethinking the Mediterranean" brings four chapters discussing the identity and uniqueness of the Mediterranean from different perspectives—cultural and natural. The authors Trindade Lopes and Almeida have given a strong accent on the Mediterranean as a contact area and its multicultural nature but have particularly emphasized its African and Asian roots, aiming to reclaim the importance of these other legacies (than Greek and Roman) to the construction of the cradle of the civilization. Mallia-Milanes discusses the questions of (dis)continuity of primacy of the Mediterranean in the international economy and exchange in the early new period in the context of the decline of the Ottoman Empire and the great discoveries. Powley, Van Cappellen, and Krom presented original new insights and a review of the unique relations between the water inflow, outflow, and biological productivity of the Mediterranean Sea. Comparing with other semiclosed and closed seas as well as the oceans, they clearly pointed to its uniqueness. On the other hand, Pizzato demonstrates the considerable performative effect of the Mediterranean as a symbolic space and source of meaning on Italian culture as well as in the process of Italian nation building between the nineteenth and the twentieth centuries. The second section addresses "Biodiversity: Nature at the Crossroads." As Mediterranean is considered a hotspot of biodiversity, this is the largest section, consisting of five chapters. Three chapters discuss marine ecosystems with different scope, and two others deal with terrestrial ecosystems. Mannino, Balistreri, and Deidun give comprehensive and overall review of marine biodiversity in relation to the changing climate and biological invasion, while Spanò and De Domenico address particularly the central Mediterranean biodiversity "crossroad" and Gönülal and Dalyan deep-water biodiversity. Martín-Forés questions the traditional approach of the negative impact of introduced plant species and establishes a new paradigm that changes traditional belief. She analyzes Mediterranean biomes on all the continents and the impact of main interchanges that occurred throughout the history on the landscape (change). The chapter by Çiçek and Cumhuriyet contains new contributions in terms of systematization of threatened species of Mediterranean amphibians and reptiles. The questions addressed are significant for understanding the human-environmental issues of the Mediterranean region. The third key issue is represented in the next section on "Risk and Hazards." Risks and hazards are the reality of the Mediterranean throughout its whole history. Particular attention in this book is given to the drought and fire relation as well as to the questions of wildfire prevention and recovery. Sferlazza, Maetzke, Miozzo, and La Mela Veca discuss the resilience of Mediterranean forests to climate change and present the results of the project LIFE ResilForMed in terms of developing the best management practices that may help Mediterranean forests to adapt to climate change and thereby safeguard their multifunctional benefits for future generations. Varol, Ertuğrul, and Özel question the methods of predicting the number of fires and burnt area, mainly through the drought indices. Gonçalves and Sousa point to the increasing trend toward larger fires in the Mediterranean due mainly to climatic and land-use changes. The case study in Portugal highlights the importance of vegetation regrowth in a short time after the fire to maintain both forest systems and soil conservation. Oliveira, Laneve, Fusilli, Eftychidis, Nunes, Lourenço, and Sebastián-López present the results of the PREFER project, as initiative and a kind of platform aiming to improve fire hazard mitigation measures and to promote comparable appraisals between different regions, based on a set of high-quality information and data sets. On the other hand, Duarte addresses another important Mediterranean issue of vulnerability of soil and water and erosivity problems in Mediterranean climatic conditions, pointing primarily to the exchange of high rainfall and drought periods. The last section addresses "Cultural Ecology and Mediterranean Lifestyle: Within and Beyond." The chapter by Rojas-Cortorreal, Navés Viñas, Peña, Roset, and López-Ordóñez gives new insights into the Mediterranean urbanism in relation to the environmental characteristics and the potential use of vegetation in designing the urban climatic comfort. In the review chapter by Royo, Soriano, and Alvaro, we can read about the wheat, one of the oldest domesticated crops at the very core of the Mediterranean diet with the immense significance for the Mediterranean culture and identity. The topic of Mediterranean diet and its widely known health benefits is further developed by Echeverría, Dussaillant, McGee, Urquiaga, Velasco, and Rigotti. *Mediterranean diet* is nowadays accepted as a generic term and as one of the pillars of the Mediterranean identity has spread far beyond the Mediterranean region itself.

The book has brought to the front the contemporary key issues of the Mediterranean region. Millennia-long human-environmental interrelations have built, shaped, and designed a kind of Mediterranean lifestyle, landscape, and distinct multifaceted identity. The book has also stressed the most serious concerns of the Mediterranean today—threats to biodiversity, risks, and hazards—mostly increasing wildfires, due to the climate, lifestyle, and land-use change and finally depletion of traditional Mediterranean practices and landscapes, as constituent parts of the Mediterranean heritage.

I would like to express my appreciation to all the contributors of this book especially to all the authors of the chapters. I am particularly indebted to the Publishing Process Manager, Ms. Martina Usljebrka and the InTech publishing team for their kind support and great efforts in bringing the book to fruition.

Prof. Borna Fuerst-Bjeliš, PhD
University of Zagreb, Croatia

Section 1

Rethinking the Mediterranean

Chapter 1

The Mediterranean: The Asian and African Roots of the Cradle of Civilization

Helena Trindade Lopes and Isabel Almeida

Additional information is available at the end of the chapter

http://dx.doi.org/10.5772/intechopen.69363

Abstract

In Antiquity, the regions encompassed by the Mediterranean Sea were extremely fertile allowing rapid prosperity. This wealth combined with the easy communication between banks contributed to a rich and successful transmission of knowledge, especially during the 1st millennium BC, which turned the Great Sea the core of Ancient History. Later, the Mediterranean civilization was acknowledged as the fundamental political, cultural, artistic and religious substratum for the construction of the so-called Western world. Yet, it was in Egypt and Mesopotamia, during the 4th and 3rd millennia BC that many of these foundations were first set. The Ancient Mediterranean world was not *just* influenced by its African and Asian neighbours but was in fact defined by a profound communion, at all levels, between these different regions. In the twenty-first century, however, many European countries still insist in portraying themselves as direct heirs of the combined Greco-Roman and Judeo-Christian traditions, disregarding their African and Asian roots. As a result of this flawed self-perception, a gap between Europe, Africa and Asia came to be, bearing deep consequences to the present. With this contribution, we aim to reclaim the importance of these other legacies to the construction of the cradle of the civilization.

Keywords: Mediterranean Sea, Egypt, Mesopotamia, transfer of knowledge, political and religious creations, cultural interactions

1. Introduction

The importance of the Mediterranean as fundamental for the construction of the so-called Western world, throughout centuries, is well established both in society and in academia.[1] When one looks into Ancient History, it becomes clear that the political, scientific, artistic and religious developments accomplished in the Mediterranean basin during the 1st millennium BC were, and still are, systematically acknowledged as structural for the composition of the European identity in particular and the Western world's in general. This notion can easily be attested by making a quick survey into the contemporaneous political, cultural and mass media discourses[2] [1]. Moreover, if one looks into the contents of the *curricula* currently taught in the compulsory education of the vast majority of European countries, one quickly realizes that regarding ancient past, the focus is on the events that took place in the 1st millennium BC Mediterranean world. The classical and the Judeo-Christian traditions are, thus, impregnated in the self-perception that the modern European citizen has.

Yet, the question about what and who forged an identity should not be addressed lightly neither should be simplified by taking into account just one or two contributes. Identity and its construction is one of the most complex and intricate subject matters not only on what concerns humanities and social sciences but also regarding the individual comprehension of one self. That is why we should always search further, both in time and space, to achieve a deeper and more concrete understanding of our roots.

Thus, this chapter represents a synergetic contribution for this wider goal, understanding the multiplicity of roots that contributed to the construction of the Western civilization by taking into account the interaction between Asia, Europe and Africa in Antiquity. The importance of the Mediterranean Sea will not be diminished. On the contrary, we hope to stress the prominence of the Great Sea, which witnessed the rise of so many cultural worlds, besides the classical ones, such as the Egyptian, the Mesopotamian, the Cyprian, the Cretan, the Anatolian and so on. We hope to stress the need to focus present-day attention on the understanding of the multiple features and traits that were already linked to this Sea in the past.

To achieve this purpose, we decided to structure our argument in several intertwined topics, such as the geographical, the political, the scientific, the cultural and The religious ones. But first, we should start by stressing what we already postulated above: for us, the world encompassed by the Mediterranean Sea is wider than its natural banks: the cradle of civilization was (and still is) the conjunction of African, Asian and European cultural backgrounds.

[1]When one talks about the importance of the Mediterranean in academia, it is only natural to recall the work of one of the great historians of the twentieth century, Fernand Braudel, who dedicated years of his research to the understanding of this Sea, throughout time.

[2]For instance, we can find in the UNESCO website page dedicated to the Acropolis, the following: «On this hill were born Democracy, Philosophy, Theatre, Freedom of Expression and Speech, which provide to this day the intellectual and spiritual foundation for the contemporary world and its values» [1].

http://dx.doi.org/10.5772/intechopen.69363

2. The geographical context, once and again…

The invention of writing is the moment traditionally evoked to stipulate the beginning of Antiquity. The first writing systems were developed in Mesopotamia and in Egypt,[3] by the end of the 4th millennium BC [2–4]. This fundamental invention, which revolutionized human thought,[4] as Bottéro referred to [5], did not happen by chance. On the contrary, the invention of writing was prompted by the progressive urbanization both territories were experiencing at that time, which entailed the rise of a hierarchized society and a centralized political power, led by the **en**[5] and by the pharaoh,[6] respectively [6–8].

To understand the urbanization process and its political developments, one must turn to the geographical characteristics of both Mesopotamia and Egypt since the historical context is always deeply influenced by the geographical one [9].

Mesopotamia is, first of all, a geographical concept, since its etymological root, from the Greek, means "between the rivers" [10]. Thus, in Antiquity, the eye of the Greek beholder understood the basilar importance of the rivers to the birth of this civilization: the Tigris and the Euphrates, and also their tributaries, such as the Khabur or the Diyala, among others[7] (**Figure 1**).

These multiple fluvial courses combined with a mild climate allowed for an exceptional environment, which prompted the establishment and growth of human communities: the presence of an abundant native fauna and flora; the existence of an abounding fertile land, favouring the development of agriculture and husbandry; and a quick contact between regions, since the main rivers were navigable. Moreover, the natural borders were extremely permeable, allowing for a simple and rapid communication with adjacent regions, and even beyond, thus motivating commercial and cultural exchanges. The Taurus and the Zagros mountains, in the north and east, the Syrian and Arabian deserts, in the west and southwest,

---

[3]The earliest records of writing in Mesopotamian and in Egypt are dated to *circa* 3200 BC. In the first case, there are several tablets with some inscriptions, which were found in the *Eanna* precint, at the site of Warka, ancient Uruk, dated to the Late Uruk period, level IV (3300–3100 BC). These inscriptions were focused in fixating basic information about the economic life of the city, namely in what concerned taxes and crops. As for the first records in the Egyptian writing, they consist of inscriptions in pottery dated to the Naqada III A-B period (3200–3000 BC), which were found in the tombs of the dynasty 0 (Necropolis B, Tomb U-j, Umm el-Qa'ab, Abidos), and also dated to the beginning of the dynasty 1.

[4]«L'écriture lui permet de transcender l'espace et la durée (…) le discours écrit seul peut fonder toute une tradition, non seulement dans l'ordre de la connaissance pure, du savoir, de la croyance, mais tout aussi bien dans l'ordre du goût et du plaisir de communiquer, disons dans l'ordre littéraire. Voilà porque j'ai dit que l'invention de l'écriture a révolutionné la pensée humaine (…)» [4].

[5]There is still a profound discussion among scholars regarding the titles of the Mesopotamia city-state leaders for the end of the 4th and during the 3rd millennia BC. Recognizing that there are more than one designation (**en**, **lugal**, *ensi*), we prefer to use the term **en** for the first centuries of Mesopotamian history.

[6]Contrary to the Mesopotamian **en**, **lugal** or *ensi*, which were considered human, though an exceptional one, the Egyptian pharaoh was considered divine.

[7]It must be underlyned that, throughout centuries, Mesopotamian inhabitants never referred to themselves as such. Instead, their own designation was defined by social or familiar kinship, political-ideological ties, religious or cultural affinity. However, the term which corresponds to the notion of "country" existed both in Sumerian (**kalam**) and in Akkadian (*mâtu*).

Figure 1. Mesopotamian territory and its main cities, courtesy of the Oriental Institute of Chicago.

and the Persian Gulf in the south allowed a myriad of contacts, direct and indirect, with Anatolia, the Iranian plateau, the Oriental Mediterranean Coast, Egypt, Oriental Africa and even the Indian Ocean.

Thus, several routes were developed very early in time. Some are attested as early as the 5th millennium BC, but it was during the 4th and 3rd millennia BC that the majority of them met an extraordinary growth. Through them, the Mesopotamian communities were able to acquire the goods they were lacking, such as lapis lazuli from the region of present-day Afghanistan, African gold from Egypt, copper from Cyprus, silver from Anatolia and so on. We should stress the importance of the Siro-Palestinian cities, such as Ugarit or Byblos, as mediators in these contacts established between Mesopotamia, Egypt and the rest of the Mediterranean world, diachronically [11]. The "in between the rivers" region was, hence, a platform that connected different worlds, cultures, goods and people.

With such geographical characteristics, not only was the territory particularly attractive for the establishment of human communities but it also reunited the necessary conditions for

http://dx.doi.org/10.5772/intechopen.69363

the development of the urbanization process. Since this geographical context was so rich and wide, the first cities, which arose in this region during the second half of the 4th millennium BC, did not feel the need to be unified in a centralized state. Every single one could survive and thrive on their own. So, the political model that we find in these ancient times in Mesopotamia, and which was maintained throughout the following centuries, was structured on city-states, governed by the **en**, **lugal** or *ensi*. This situation originated profound rivalries between each Mesopotamian urban centre. Throughout the 3rd millennium BC, the cities became systematically defiant of one another, in search for more glory and power. Yet, in the twenty-fourth century BC, the territory was finally unified by Sargon of Akkad, who set a new model of governance: the imperial one. Still, in the centuries to come, the logic of urban independence was so strong that every attempt of unifying the Mesopotamian city states did not last long [12].

Egyptian geographical context was likewise exceptional: the fortune of being crossed by the Nile [13], with its regular floods, which impelled Egyptians to keep a controlled calendar of the river rhythms very early in time, turned its banks extremely fertile. Again, this allowed the development of both agriculture and husbandry, making the area along the banks a natural place for the establishment of human communities. Soon, the urbanization process began, paralleling the one in Mesopotamia.[8]

This prosperity, however, was accompanied by a latent threat: the floods, though regular, were extreme and violent, forcing communities to work together, namely in the construction of channels and dikes. Moreover, Egypt presented natural borders that were well defined, which enclosed the territory. At the same time, these natural barriers protected its communities from the hostile exterior: the Nile falls, in the south, were an obstacle to who wanted to penetrate the Nilotic country from the Sub-Saharan territory, and the deserts in the east and west, sheltered Egypt from direct neighbours. In the north, the territory opened itself to the Great Sea, with the Delta region displaying increased levels of prosperity when compared to the south. The Mediterranean aroused the curiosity of the Egyptian mind, with the promise of new discoveries, adventures and exchanges with the ancient world (**Figure 2**).

Despite the existence of this northern challenge, the weight of the main geographical characteristics was heavier and presented them as determinant in the Egyptian choice for a political model: the monarchy, led by the pharaoh. One can say that contrary to the Mesopotamian geography, the Egyptian one encouraged political unification in a centralized state that embraced the whole territory since its origins. At the end of the 4th millennium BC, the southern Egyptian cities, like Hiérakonpolis, Abydos, Nagada and Tarkhan, searched the unification of the land in order to make the better of the geographical conditions. Hence, the pharaoh emerged as a charismatic divine figure which materialized the central and unified force of the Nile. From then on, there were three major political periods in the history of this civilization, where unification was the rule (Old, Middle and New Kingdoms). The moments of political fragmentation were short, in the *longue durée,* and always considered as a chaotic disruption to the primeval order [12].

[8]Though the urbanization process happens roughly at same time in both territories, it must be stressed that the early Egyptians city-states were smaller both in population density as well as in extension.

Figure 2. Egyptian territory and its the main cities, courtesy of the Oriental Institute of Chicago.

Due to the proximity and intimate contact with the Mediterranean world, both Egyptian and Mesopotamian political models spread along the Great Sea, naturally adapting themselves to those different contexts. Throughout centuries, we identify not only Phoenician and Greek city-states[9] [14, 15] but also Hellenistic monarchies in Anatolia and in Rome [16, 17]. And even though the imperial construction [18] appears in Mesopotamia and Egypt later in time (with the government of the Akkadian dynasty, between the twenty-fourth and twenty-second centuries BC and during the New Kingdom period, between the sixteenth and eleventh centuries BC), one must say that this political model, which was fully developed by the Persians, Alexander the Great and the Romans, also had its roots in these ancient civilizations.

The Mediterranean was the determinant point of connection for all these historical actors, promoting the circulation of people and goods, which led to the natural circulation of ideas, whether political, cultural or religious. We should note that the already above-mentioned contacts established since the 4th and 3rd millennia BC, between Egypt, Mesopotamia and the Siro-Palestinian powers, was a model systematically replicated throughout time, integrating new and different identities. During the 2nd millennium BC, Egyptians, Cretans and Aegeus developed an intense activity along the Great Sea. In the following millennium, it was time for the Phoenician and the Greek protagonism, the former exploring the Western Mediterranean, that is North Africa and the Iberian Peninsula, and the latter spreading its activity from the Black Sea to the Red Sea [19]. The colonization process was, thus, initiated with the establishment of their *emporia* [20, 21] which led to a natural rivalry between the two powers[10] [22], accompanied by the emergence of pirate activity (**Figure 3**).

On the other hand, the construction of great empires was set in motion. Carthage, the Phoenician colony in the North of Africa, started out by occupying a strategical place as the first port of the Western Mediterranean [19]. From the sixth century BC onwards, the Carthaginians invested in their military force in order to control the remaining of the Phoenician city states. A maritime empire was, thus, established. Simultaneously, the Greek victory over the Persians, in the year of 480 BC, in Salamis, allowed the establishment of Piraeus as the main port of the Eastern Mediterranean. Sometime later, during the second half of the fourth century BC, Alexander, the Great, built an extraordinary empire, linking Greece, Anatolia, Phoenicia, Egypt, the Syro-Palestinian coast, Mesopotamia, the Iranian plateau and reaching as far as the Indus valley.

Alexander, driven by the will to expand the Greek values further away, worked on the ancient and well-established perception that the Great Sea offered a link to the whole ancient world [23]. Despite the political division of his empire following his death, Alexander's goal persisted: the Hellenization of the Orient was a work in progress. Alexandria, in Egypt, became

[9]Though there is still a profound debate among scholars concerning the city-state model, what we intend to emphasize is the importance of geography in the development of this and other political models. The connection between geography and history is underlined in Ref. [9], for the Phoenician and Greek cases.

[10]It is interesting to note, however, how these two powers, though politically and commercially rivals combined their cultural actions regarding one of the most important writing tools of the present day Western World: the alphabet. Its invention by the Phoenicians, as Helena Trindade Lopes noted [22], «resulted from a long historical process of maturation, which manifested for the first time in the passage from the 2nd to the 1st millennia BC. This alphabet would spread eastwards and westwards. The Hebrews and the Aramaeans used it to transcribe their own language and the Greeks, around 800 BC, adapted it to the needs of a non-Semitic language and vocalized it. From that moment on, the "genetic" relationship between the Phoenician alphabet and the Greek alphabet was completed. The former went its own way and the latter, via the Etruscan and Roman world, gave origin to our modern alphabet».

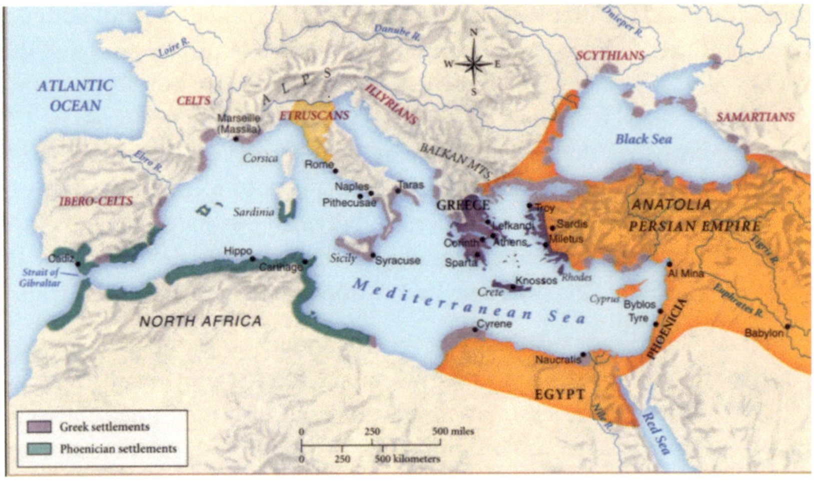

Figure 3. Mediterranean Greek and Phoenician colonies, courtesy of the Utah State University.

the cultural metropolis of the time, with its splendorous library and museum [24]. In the meantime, Cartagena was founded on the West, contributing to confirm the ethnic and cultural miscegenation of the Mediterranean.

Over this period, between the sixth and third centuries BC, Rome grew both in importance and in force within the Italian peninsula and Sicily, gaining effective control over the Greek colonies established there. In time, Roman power took over Greece, Spain and the Orient. With the government of Augustus, the Mediterranean finally met its destiny fully becoming the *Mare Nostrum*.[11]

3. Cultural interactions

3.1. The scientific knowledge that preceded *logos*

Throughout millennia, the Mediterranean stood as the privileged scenery for the rise and fall of several and distinct political projects. The Great Sea not only witnessed antagonisms and wars but also dialogs and interactions. Because of its singular features, which enabled the contact between Africa, Europe and Asia, it was possible for different people from diverse backgrounds to come together and share the task of building the ancient world, creating and negotiating ideas, beliefs, and artistic canons. The Mediterranean can, thus, be seen as a place where human creativity was combined and launched for the future.

On what concerns the transfer of knowledge, once and again, Egypt and Mesopotamia had a crucial embryonic role, developing several principles that prepared the ground for the emergence of a strong and structured scientific thought.

[11]It should be stressed that this expression was already in use during the Punic wars. However, as we hope it became clear, this notion was perceived much earlier in time, though not defined as such.

In Egypt, besides the incredible medical knowledge achieved [25], which were stimulated by the Nilotic fundamental cult of the dead, other breakthroughs were accomplished. The imperative need of establishing a calendar was felt by Egyptians not only in order to control the Nile but also to define the religious festivals. This led to remarkable advances regarding the knowledge about the firmament and the movement of the stars and planets. In fact, the beginning of the Egyptian year was defined by the presence of Sirius, *Sopdet*, in the sky, which happened in mid-July and was coetaneous with the annual flood of the Nile. To this first season, *Akhet*, two others would follow: *Peret*, which corresponded to the time of sowing, from November until mid-March, and *Chemu*, the harvest period, from mid-March until mid-July. Each season lasted for 4 months, comprising 30 days each. Annually, 5 epagomenic days were added, a period which was celebrated by the Egyptians as a moment where the mythical, remote past of their identity was remembered: the time when Horus fought his uncle Set for the divine right to occupy the Egyptian throne.

Likewise, the Mesopotamian sages also looked at the movements of the celestial bodies in order to set up a calendar for their religious and daily-life activities [26]. Astronomy was thus taking its first steps in the banks of the Nile, the Tigris and the Euphrates [27, 28].

Moreover, the Semites, whose presence in Mesopotamia is attested to the end of the 4th millennium BC, brought with them another particular interest that made their attention to be fixated in the sky: the search for divine messages, concealed by the deities in the natural world. One of their strongest beliefs was that deities painted nature with signs which bore their divine will, in order for communities to know it and more importantly, to act accordingly. This promoted the development of divination [29–31],[12] a phenomenon that, despite being completely immersed in the religious sphere, expressed a deductive thinking, which prefigured the scientific method [32, 33].

In fact, the Mesopotamian diviner, the *bārû*, was a highly qualified individual, extremely respected by the Mesopotamian society and, more importantly, by the royal court. As Caramelo stated, from the trivial to the most important decisions, Mesopotamians recurred to divination to obtain a protective, secured feeling on what concerned their actions [34]. These true sages spent years crafting the ability to scrutinize natural phenomena, studying and making advances regarding their predecessors' work. In order to decipher divine messages, whether in dreams, animal or birds behaviours, animal entrails, astronomical phenomena and so on, the *bārû* had to adopt a step-by-step process. This consisted of not only observing nature and gathering data but also refining, altering or expanding that same data (or the one previously collected). Consequently, the confirmation of the significances attributed to the divine sign was achieved.

On another level, the Rhind Papyr [35], for instance, displays valuable information regarding the development of the Nilotic mathematical thought. Egyptian mathematicians developed the application of fractions and equations [36], having a decimal and additive calculation system. This helped to solve many arithmetical problems, through the use of tablets. For instance, it was possible to calculate the surface of not only the square, the rectangle and the

[12]According to Seth Richardson [29], the first known Mesopotamian references regarding the divination techniques appear in the professions' list discovered in Tell Fara, ancient Šuruppak, and dated to *circa* 2600 BC.In the introduction of the same work, Amar Annus points out to the strong possibility that this technique was older than the written records which attest it, being known and transmitted by a previous oral tradition.

triangle but, curiously, also the surface of the circle. On the land "between the rivers", the circle was also analysed and Mesopotamian mathematicians were able to divide it in 360°. Plus, they also developed the sexagesimal system, which divided time in hours, minutes and seconds, a system in use to the present day [37, 38].

Naturally, the knowledge presented above helped the creation of several architectonic works being the paradigmatic pyramid one of the most emblematic ones in what concerned the land crossed by the Nile. The Egyptian pyramid [39] was a magnificent tomb, understood as a "house for eternity", with a geometrical shape and size that still today impresses anyone who has the fortune of beholding it.[13]

The earliest pyramidal complex known today is the step pyramid of Djoser. Corresponding to the first tomb completely built in stone, this complex is located in Sakkara and dated to the dynasty 3, *circa* 2670 BC. Its construction was supervised by Imhotep, the royal architect, whom in time was granted a higher place, a seat among deities [40]. It is important to stress the existence of this function in the Egyptian court, so early in time, since it shows the importance bestowed to artistic expressions by the Nilotic centralized power.

However, it was during the following dynasty, *circa* 2560–2540 BC, that the pyramidal shape achieved perfection with the construction of the Giza complex [41]. The three extraordinary pyramids of Khufu, Khafre and Menkaure still defy human comprehension, claiming their eternal seat, both in time and in space, just like it was intended by their constructors (**Figure 4**).

Figure 4. Khafre Pyramid and the Sphinx, picture by the author Helena Trindade Lopes.

[13]It must be pointed out that ancient Egypt produced other types of tombs, such as the mastabas and the *hypogea*.

The emblematic Mesopotamian building, though not a tomb, was also a sacred precinct, constructed "on high"—the ziggurat. This impressive monument was the highest structure in the Mesopotamian city, imposing its presence in the landscape. Unfortunately, most of the ziggurats constructed in Antiquity did not survive to the present day due in great measure to the perishable material used for its construction: the adobe. Yet, this singular monument is impregnated in the memory of the Western world, through the imagery of the Tower of Babel[14] [42], which was broadly spread by the Old Testament tradition. Just like the Egyptian pyramids, the ziggurat construction was in debt of the mathematical and astronomical breakthroughs accomplished by the Mesopotamian sages [43]. Hence, as in so many other aspects, the Mesopotamian and Egyptian scientific minds were running side by side, making their way until they flooded over the Great Sea. The advent of rational thought took place, as it is well known, in Greece, during the 1st millennium BC. But its roots are older and deeper, hidden behind the mantle of myth, a concept so wrongly disapproved by time.

3.2. The genesis of divine imaginary

Hence, besides scientific knowledge and artistic expressions, Egypt and Mesopotamia also shared with the Mediterranean world their religious experiences. Their imagery and metaphorical language was systematized during millennia, acquiring a primeval binding dimension between humans and deities.

One of the most striking aspects of both the Egyptian and the Mesopotamian mythical framework has to do with divine creation [44, 45] and its parallels with the monotheistic vision that arose on the Mediterranean shores. For the *homo religiosus* that dwelt in the margins of the great ancient rivers, the divine was naturally multiple in its manifestations but not so rarely it admitted an uncanny intimacy with singularity [46, 47].

The cosmogonic views illustrate this closeness very well. In Egypt, the creation of cosmos was structured around two fundamental archetypes: the solar one, which displays different demiurges—Atum and Amon—and is organized by successive generations of deities, just like human lineage, and the Memphite one, elaborated by the priests of the city protected by Ptah, who bestowed the creative power to the Verb [48, 49]:

> *Ptah, the creator god of Memphis, conceived the cosmos in its different manifestations in his heart and realized it through the creative and operative force of the word. The doctrine of the creator verb, usually recognized from the biblical text (Gen. 1) and situated in a particular historical, geographical and temporal context, actually dates back to a time and a place which was very different, the Nile Valley* [22].

Likewise, in Mesopotamia, two major cosmogonic views, the Sumerian and the Semite, expressed similar ideas. The first one revolves around the idea of a single deity, Namma [50], defined as the primeval mother, who engendered all the senior gods and goddesses, who, in

[14]Throughout centuries, this monument was the object of a tireless search by European travelers who crossed the region between the rivers. Regrettably, the monument was already lost, being visible only its negative. After the archaeological expedition on the site of Babylon, from 1899 onwards, several scholars presented hypothesis for the reconstruction of the ziggurat. Recently, Juan Luis Montero Fenollós and his team presented their perspective [42].

turn, gave birth to new and successive generations of divine beings.[15] For the Semitic tradition, however, we find a divine primeval couple, Tiamat and Apsu, whose power of creation resided on sexual intercourse. Interestingly, this act was in close association with the notion that without the act of naming nothing fully existed, since, as it is stated in the Babylonian epic of creation, *Enūma-eliš*:

> When skies above were not yet named / Nor earth below pronounced by name,
>
> Apsu, the first one, their begetter, / and maker Tiamat, who bore them all,
>
> Had mixed their waters together, / but had not formed pastures, or discovered reed-beds,
>
> When yet no gods were manifest / Nor names pronounced, nor destinies decree,
>
> Then gods were born within them. [51]

Moreover, these Sumerian and Semitic primeval deities were all understood as a matter, the primeval divine ocean,[16] thus claiming water as the primal element that existed in the genesis of the cosmos. Likewise, in Egypt we find Nun, the primeval watery matter, where the demiurge was asleep, waiting to begin its creative task.

This conceptualization about the origin of the cosmos was, without a doubt, one of the major contributions that Egypt and Mesopotamia shared with Western civilization: the creative power of name, word and water. Naturally, the contacts between the different people that crossed the Mediterranean world, in its wider sense, as we started out by stating, helped this religious dialogue and interaction. We should remember and stress the presence *in loco* of Hebrews, both in Egypt (Exodus) and in Mesopotamia (Babylon Exile), as they would be the main protagonists of the biblical narrative, which was definitely a product of their own Mediterranean interactions. The Old Testament, an extraordinary religious and literary work that in time would become basilar to the construction of the Western civilization, via the Judeo-Christian matrix, once and again was fashioned over more ancient and, specially, multiple roots.

Yet, the religious dialogue between the divine multiple of Egypt and Mesopotamia and the divine one of the Bible does not end here. On another creative perspective, the anthropogenic one, it is also important to stress how clay/dust was already present in these civilizations. On what concerns the Egyptian view, man was modelled after the "dust of the earth" by Khnum, the potter god, recalling, again, the Jewish tradition recognized in Gen. 2.7′ [22]. On their side, Mesopotamian accounts state that when the divine assembly decided humanity should

[15]«Namma, the primeval mother who gave birth to the senior gods» ETCSL 1.1.2, 16 [50].

[16]Though these primeval Mesopotamian deities were understood as the divinized ocean, Namma, the Sumerian deity, was considered feminine, while the Semitic pair, Tiamat and Apsu, corresponded to the feminine salty waters and the masculine sweet waters, respectively.

be created, a mother goddess, assisted by the wise Enki/Ea, mixed up clay to fashion the first humans.[17]

We could keep on tracing and presenting more examples that display the profound interactions between these Mediterranean religious imageries. But, perhaps the most striking is the one that refers to the monotheizing proposal that Amenhotep IV-Akhenaton presented to the world in the fourteenth century BC. This "heretical" pharaoh, as he would become known in the Egyptian tradition, developed and imposed a notion that established Aton, the solar disk, as the singular, true deity [52, 53].

> *The similarity between the Hymn to Aten, the programmatic text of the "monotheistic" reform of Amenhotep IV, and Psalm 104 is another example of these constant "coincidences" that ancient history consecrated, demonstrating that the human beings of the past, just as today's human beings do, travelled, and in doing so they took and brought with them ancestral values, derived from their dwellings, or singularities that they incorporated in their own traditions. The road, which was the Mediterranean, connected Egypt to the world and by doing so carried traces of this civilization to other spaces, other peoples, other cultures [22].*

We should note that this monotheizing process happened during the period in which Egypt opened itself to the Great Sea, that is the New Kingdom. The second half of the 2nd millennium BC was defined by a strategical game played in the oriental shores of the Mediterranean, with discords and negotiations between two main powers: Egypt, with their African roots, and the kingdom of Hatti, originally from Anatolia. The Siro-Palestinian city states and kingdoms, such as the famous Kadesh, were caught in this cross fire. As for Babylon and Assyria, who were the major Mesopotamian forces at the time, both were naturally observing and waiting for the result of the confrontation between Egypt and Hatti due to their own political and commercial expectations.

On the first half of the 1st millennium BC, Assyria would transpose the natural borders of Mesopotamia [54], engulfing not only the Oriental Mediterranean but also Northern Egypt (**Figure 5**).

And why is this important for our argument? Because while the Assyrian empire was being built, after several centuries of political, commercial and military expeditions and interactions, its patron deity, Aššur, became so prominent among all other Mesopotamian divine beings that it is considered one of the most striking cases of a monotheizing process in Mesopotamia [55, 56].

In conclusion, the religious systems of these two ancient civilizations inspired conceptions of different and multiple generations of *homo religiosus*. Their models about the creation of the cosmos and humankind were built upon the figure of the Demiurge and upon a tension between the divine *one* and the divine *multiple*. The Mediterranean waters, in time, spread these religious archetypes throughout the whole ancient world. Received, adapted and transformed, these ideas persisted until the present, echoing in the Judeo-Christian tradition an ancient and distant past.

[17]There are several Mesopotamian manuscripts that refer to the creation of humans. Though they present some differences, due to the main goals of each composition, in all the divine creative act revolves around the above mentioned motifs. *Vide* these antropogonic accounts in Ref. [51]: Tablet I of *Atrahasis*, and Tablet VI of *The Babylonian Epic of Creation*.

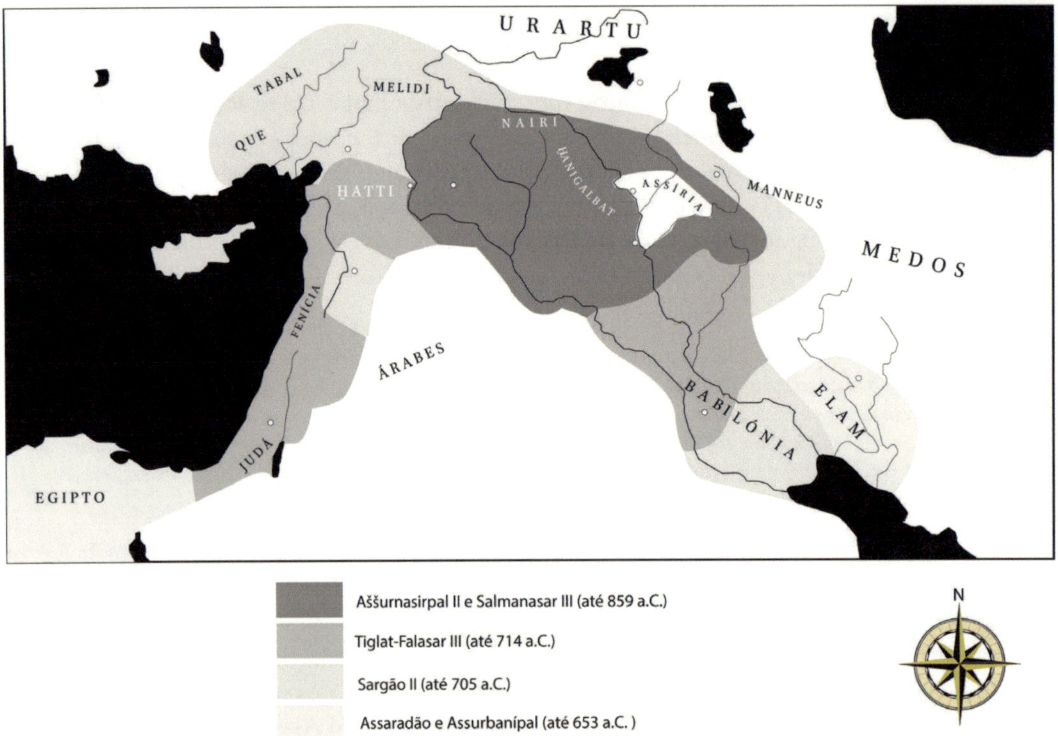

Figure 5. The Neo-Assyrian expansion map, between the ninth and seventh centuries BC, courtesy of Marcel Paiva do Monte.

4. The *boom* of cultural interaction

The cultural, scientific and religious vigour of these civilizations was transmitted, as already stated, to the Mediterranean world, especially the Greek one, which absorbed their traits with its natural depth. But because these first Greek academics recognized the reputation of the Mesopotamian and Egyptian knowledge, several Hellenic scholars travelled to these regions, or got in touch with their traditions through others, in order to learn and study these venerable and elderly sages.

That is why, during the fourth century BC, the astronomer, mathematician, physician and philosopher Eudoxus of Cnidus, who was contemporaneous of Plato, travelled to Heliopolis, the city of Re, the divine Sun of Egypt, to expand his astronomical knowledge. But Eudoxus was not the only Greek to visit and or to be influenced by Egypt and Mesopotamia: Thales of Miletus, Solon, Plato and Herodotus, to name just a few, are a small part of the highly esteemed scholarly Greek group which came across and absorbed these African and Asiatic cultural traits.

Egypt had a long tradition of transfer of knowledge within the temples. These complexes were not just the terrestrial dwellings of divine beings and, therefore, the central place for cultic activities. They also integrated the so-called "Houses of life", *per ankh*, which corresponded to scholarly centres, where knowledge was taught and scientific, literary and religious works were produced [57]. Some of these ancient Egyptian schools became very famous, within the Nilotic territory and beyond, such as the *per ankh* of Menphis, of Akhmim, of Abydos, of Coptos, of Esna, of Edfu and of Amarna, among others. The Teban school, for instance, was (and still is) known for the production of majestic works of art such as the scrolls of papyrus with passages of the "Book of the Dead". Simultaneously, the temples also had the *per medjat*, which literally means "House of the papyrus' scrolls", which preserved the texts written by these ancient sages. We could use here the term "library", not in the modern sense of a building or an institution but in the sense of a documental collection.

Mesopotamian schools, on their side, are attested since at least the late 3rd or beginning of the 2nd millennium BC. The **é.dub.ba** was the place where youngsters had their intensive formation years in order to become scribes. Despite the existence of older proofs, the extensive *corpora* retrieved in these ancient Mesopotamian schools are dated to the Old Babylonian period, *circa* eighteenth century BC onwards. These documents allow the understanding not only of the exercises of the scribe apprentices but also of the *curriculum* followed in the different **é.dub.ba** and moreover, the literary novelties that were developed there [58].

As for the collections of texts, we should stress the importance of the royal archives and libraries, where the Mesopotamian rulers kept not only their legal, social and diplomatic data but also literary, cultic and mythological works of art. Among the many libraries and archives recovered by the archaeologists in the sites of ancient Mesopotamia, perhaps the most famous is the library of the Assyrian king, Ashurbanipal, in his capital Nineveh [59]. During Ashurbanipal's government, who ruled between *circa* 668 and 627 BC, the Assyrian empire extended to its maximum length, incorporating multiple territories and cultural expressions as stated above (**Figure 5**).

Ashurbanipal was a very well-educated ruler, who was particularly interested in creating an up-to-date royal collection of texts regarding multiple matters. Modern researchers identified and translated several letters from scholars of many cities within the Assyrian empire, responding to the king's command: 'Write out all the scribal learning in the property of Nabû and send it to me! Complete the instruction'[18] [60]. From magic and ritual texts, to diplomatic and literary compositions, archaeologists found, between the 1850s and 1930s, over 32,000 cuneiform tablets in his library attesting, thus, the enormous effort of the king in order to collect and gather all the possible data of his time.

The Mesopotamian and Egyptian labour of collecting and preserving knowledge continued and developed in the centuries to come. The assembly of the museum and library of Alexandria constituted the culmination of these traditions [61]. From then on, scientific knowledge acquired a universal dimension as its growth was faster and its rhythm allowed taking a step further than its Egyptian and Mesopotamians predecessors: from empirical knowledge, it was possible to go forward to theoretical elaborations.

[18]Excerpt of the letter BM 45642 sent by the sages of Borsippa to the king Ashurbanipal.

The Ptolemaic rulers, just like Ashurbanipal, appealed to the sages of the whole ancient world in order for their library to reunite a "global" wisdom. Hence, from multiple regions of the Mediterranean, scholars arrived in Alexandria aiming to give their contribution for the concretization of the goal: Manetho an Egyptian from Sebennytos; Callimachus and Eratosthenes from Cyrene, Libya; Hecataeus from Abdera, Thrace; Apollonius of Rhodes; Aristophanes of Byzantium, Northern Greece; Herophilos of Cos, an Aegean island; Archimedes of Syracuse, Sicily; Plotinus and Horapollo, Egyptians of Asiut and Akhmim, respectively, among many others.

Soon, other centres would join Alexandria, like the ones in Antioch, Pergamum, Athens or Macedonia, building a true network of scientific cooperation, all committed to spread the scientific knowledge. This trend persisted beyond the Hellenic world, surviving in the Roman and Byzantine periods, as attested, for instance, by the Fayum portraits [62]. This artistic *corpus* was discovered in Egypt, in 1888, by the British archaeologist W.F. Petrie. The portraits, which are dated to the early centuries of the Christian era, were painted in wooden plaques or linen cloths by the Greeks, who established themselves in the Nilotic territory. Alone, these portraits manifest the amalgamation that Fayum was at the time, reuniting the ancient Egyptian sarcophagus tradition and funerary masks, with the Greek and Roman pictorial techniques and fashion trends. In conclusion, this *corpus* represents the diachronic fusion of the Mediterranean.

Later, from the seventh century AD onwards, the Arabic power met the incredible cultural and scientific patrimony contained within the Great Sea, expressed in many languages and indebted of many backgrounds. The Arabs would translate, study and publish many of these ancient works, contributing for the preservation and, more importantly, for the diffusion of this knowledge.

During the Medieval times, the European Christian monks would carry on this work, copying these manuscripts in Latin versions. The advent of the press, in the Renaissance, allowed for all this knowledge to spread his wings and fly out of the monasteries, reaching a broader audience that grew exponentially in the following centuries.

Consequently, the ancient past of Egypt and Mesopotamia resounds in our present-day civilization, persisting in the multiple layers of the cultural transmission that took place in the last millennia. However, these echoes were filtered by the monotheistic visions of Judaism, Christianity and Islam, at same time, they were chained by logic and reason. The ancient knowledge lost its magical and metaphorical essence. Likewise, modern Western societies lost the natural ability to dream and to be in full communion with the cosmos. The Western world lost the link to its Mesopotamians and Egyptians ancestors.

5. Recovering the Egyptian and Mesopotamian legacy

Who then, one could ask, could unveil the ancestral memory that the Classic and the monotheist traditions removed? The answer is not that difficult. It dwells in the free minds of the artists. They resisted logic and strict rules, they struggled to secured magic and dreams close-at-hand, they fought for giving back to society the ability to feel and aim higher.

One of the greatest contemporaneous painters, Anselm Kiefer [63, 64], presents the Mesopotamian and the Egyptian mythological and cultural framework as one of his major influences.[19] Kiefer understands the impact these African and Asian roots had in the Classical and Biblical world, as well as in the Cabala. Thus, through his works, Kiefer tries to reconcile all these multiple layers, teasing the viewer with innumerable references. Like so many others, his goal is to wake modern society, to make it question its logical chains and to free itself so it can be, once again, reconnected to nature, to the cosmic universe. For Kiefer, knowing and understanding these ancient roots, which were reunited in the Mediterranean, offers the possibility of freedom, to imagine, to feel, and to be.

In his 1996 painting *Man under a Pyramid* [65], displayed in the Tate Gallery, in London, Kieffer tried to show the union between man and cosmos, which is symbolized by the emblematic Egyptian structure. In its own words:

> *Là, dans ce tableau, c'est moi, mais ce n'est pas seulement moi, c'est un gisant, un archétype. Ci-gît. Je suis un home d'aujourd'hui qui a des souveniers précis des temps anciens [...] Je suis ici, composé de tous mes souveniers qui remontent jusqu'aux dinosaures et même plus loin. Le future est lié au passé, mais pas mélangé à lui [66].*

The pyramid, the "house for eternity", as it was imagined by the ancient Egyptians, stands out in Kiefer's work as a perfectly shaped symbol of the collective memory, where humans and cosmos are in a perfectly aligned communion.

To this case we could add other artistic contributions, within the fields of music, literature, cinema, dance...May we be allowed to remember just a few examples to better illustrate our argument: the operas by Rossini, *Semiramide* (1823); by Verdi, *Nabucco* (1841) and more recently, by Philip Glass, *Akhnaten* (1983/4), which present to the modern viewer, every time they are performed, a link to the strong figures and events of Egypt and Mesopotamia [67–69]. In the *Belles-Lettres*, we have the example of Thomas Mann's *Joseph der Ernahrer* (1943), an *opus* which the author considered his masterpiece [70]. In the nineteenth century, within the Pre-Raphaelites movement, Dante Gabriel Rossetti painted *Astarte Syriaca* (1877), evoking the Mesopotamian goddess Inanna/Ištar in its Syrian manifestation [71]. In the world of cinema, we can name *Metropolis* (1927), directed by Fritz Lang, who conjugated the science fiction imagery with the famous Tower of Babel, thus creating a masterpiece of German expressionism.

And, of course, the unforgettable *Cleopatra* (1963) directed by Joseph L. Mankiewicz and starred by the striking Elizabeth Taylor, which originated a ballet, *Cleopatra* (2011), directed by Claude-Michel with musical arrangements by David Nixon [72].

Curiously, despite all these artistic efforts, in so many fields, twenty-first century European countries, in general, still insist on a dated discourse, which portrays them as heirs of the Greco-Roman and the Jewish-Christian traditions only. This notion is systematically attested, when one watches the news, listens or reads political discourses, and enquiries students who are not enrolled in history courses. As a result of this flawed self-perception, a gap between

[19]We could, as well, mention others who were influence by this ancient past, like Paul Klee, Rauschenberg, or Basquiat. The choice on Kiefer is to stress a still alive and productive artist.

Europe and its African and Asian legacies came to exist, with deep consequences in our present-day political, social and most of all, humanitarian discourses and contexts.

The truth is that our Western world took its first steps in Egypt and Mesopotamia, with the first writing systems, the rise of the urbanization process and the consequent formation of the first political models and the establishment of long-trade interactions. History and all the other sciences had their genesis in the banks of the ancient Nile, Tigris and Euphrates. The inventive expressions of these ancient artists originated universal and paradigmatic symbols, like the pyramid or the ziggurat, and in these African and Asian lands arose the tensions between the divine multiple and the divine one, as mythological archetypes which were absorbed and appropriated by monotheism.

Time, that eternal and inexorable constructor, shifted the core of the ancient history to the banks of the Mediterranean, where these ancient pasts continued to be venerated. Through the commercial and cultural interactions accomplished by the Greek and the Phoenician colonization processes, through the conquers of Alexander the Great and, soon after, of the Roman power, the Mediterranean was extended and the transfer of knowledge deepened.

The Great Sea, thus, not only became a shared territory but also an arena of conflicts, wars and deaths. Above all, it was a wide and open space which challenged human mind, allowing adventures, dreams and utopias.

A deeper understanding of this ancient history, of these ancient roots, will allow the twenty-first century Western World to understand its cumulative identity, reconnecting modern society with their Middle Eastern and African backgrounds. And perhaps, the waters that once were the cradle of civilization, and are nowadays transformed in an immense graveyard of people, of souls, due to ignorance and prejudice against the alterity which *is* part of the Western identity, can once again shine, enlightening the paths of dreams and hopes for the modern world.

Acknowledgements

We would like to thank our colleague and friend Marcel Paiva do Monte, PhD, for giving us the permission and privilege of being the firsts to publish his map of the Assyrian empire. We would also like to thank the Oriental Institute of Chicago and the Utah State University for the licence images and maps disposed on their websites, which help academia to better illustrate the written arguments.

Author details

Helena Trindade Lopes* and Isabel Almeida

*Address all correspondence to: helenatrindadelopes@hotmail.com

CHAM, FCSH, NOVA University of Lisbon, University of Azores, Portugal

References

[1] UNESCO. World Heritage [Internet]. 2017. Available from: http://whc.unesco.org/en/list/404/ [Accessed: 3 March 2017]

[2] Glassner J-J. Écrire à Sumer: L'invention du cunéiforme. Paris: Éditions du Seuil; 2000

[3] Bard KA, editor. Encyclopedia of the Archaeology of Ancient Egypt. London/New York: Routledge; 1999. p. 26

[4] Vernus P. La naissance de l'écriture dans l'Égypte pharaonique: une problématique revisitée. ARCHÉO-NIL—Revue de la société pour l'étude des cultures prépharaoniques de la vallée du Nil. 2016;**26**:105-134

[5] Bottéro J. Religiosité et raison en Mésopotamie. In: Bottéro J, Herrenschmidt C, Vernant J-P, editors. L'Orient ancien et nous. Paris: Hachette Littératures; 2004. p. 36

[6] Brisch N, editor. Religion and Power—Divine Kingship in Ancient World and Beyond. Chicago: The Oriental Institute of the University of Chicago; 2008

[7] O'Connor DB, Silverman DP, editors. Ancient Egyptian Kingship. Leiden/New York/Koln: E. J. Brill; 1995

[8] Hill JA, Jones P, Morales AJ, editors. Experiencing Power, Generating Authority: Cosmos, Politics, and the Ideology of Kingship in Ancient Egypt and Mesopotamia. Philadelphia: University of Pennsylvania Press; 2013

[9] Lopes MHT. The Mediterranean Sea: The language of history. Cahiers de la Méditerranée—Dynamiques des Ports Méditerranées. 2010;**80**:11-16

[10] Fenollós JLM. Breve Historia de Babilonia. Madrid: Ediciones Nowtilus; 2012. p. 32

[11] Hudson M. Entrepreneurs: From the near East takeoff to the Roman collapse. In: Landes DS, Mokyr J, Baumol W J, editors. The Invention of Enterprise: Entrepreneurship from Ancient Mesopotamia to Modern Times. Princeton: Princeton University Press; 2012. pp. 8-36

[12] Sanmartín J, Serrano JM. Historia Antigua del Próximo Oriente: Mesopotamia y Egipto. Madrid: Ediciones Akal; 1998. pp. 173-178, 250-340

[13] Claus B. Nile. In: Orlin E, editor. The Routledge Encyclopedia of Ancient Mediterranean Religions. New York/London: Routledge; 2016. p. 655

[14] Jigoulov V. Phoenicia. In: Orlin E, editor. The Routledge Encyclopedia of Ancient Mediterranean Religions. New York/London: Routledge; 2016. pp. 721-722

[15] Hansen MH. Polis. Oxford: Oxford University Press; 2006

[16] Habicht C. The Hellenistic Monarchies: Selected Papers. Michigan: The University of Michigan Press; 2006

[17] Walbank FW, Astin AE, Frederiksen MW, Oglivie RM, editors. The Cambridge Ancient History. Part 1: The Hellenistic World. Vol. 7, 2nd ed. Cambridge: Cambridge University Press; 1984. pp. 87-95

[18] Nicolet C, editor Du pouvoir dans l'Antiquité: mots et réalités. Genève: Librairie Droz; 1990

[19] Pappa E. Early Iron Age Exchange in the West: Phoenicians in the Mediterranean and the Atlantic. Leuven: Peeters; 2013

[20] Niemeyer HG. The Phoenicians in the Mediterranean. Between expansion and colonization: A non-Greek model of overseas settlement and presence. In: Tsetskhladze G, editor. Greek Colonisation. An Account of Greek Colonies and Other Settlements Overseas. Vol. 1. Leiden/Boston: Brill; 2006. pp. 143-168

[21] Tsetskhladze G, editor. Greek Colonisation. An Account of Greek Colonies and Other Settlements Overseas. Vol. 2. Leiden/Boston: Brill; 2008

[22] Lopes MHT. The Mediterranean and the voices transported by time. In: Goffredo S, Dubinsky Z, editors. The Mediterranean Sea. Its History and Present Challenges. Berlin: Springer; 2014. p. 557

[23] Freeman P. Alexander the Great. New York: Simon and Schuster; 2011

[24] Erskine A. Alexandria. In: Orlin E, editor. The Routledge Encyclopedia of Ancient Mediterranean Religions. New York/London: Routledge; 2016. pp. 30-31

[25] Veiga P. Health and Medicine in Ancient Egypt—Magic and Science. Oxford: Hadrian Books; 2009

[26] Steele J. Mesopotamian calendars. In: Ruggles CLN, editor. Handbook of Archaeoastronomy and Ethnoastronomy. Vol. I. New York: Springer Science+Business Media LLC; 2015. DOI: 10.1007/978-1-4614-6141-8_189

[27] Wells RA. Astronomy. In: Redford DB, editor. The Oxford Encyclopedia of Ancient Egypt. Vol. 1. Oxford: Oxford University Press; 2001. pp. 145-151

[28] Steel J, editor. Calendars and Years: Astronomy and Time in the Ancient Near East. Vol. I. Oxford: Oxbow Books; 2007

[29] Richardson S. On seeing and believing: Liver divination and the era of the Warring States. In: Annus A, editor. Divination and Interpretation of Signs in the Ancient World. Chicago: Oriental Institute of the University of Chicago; 2010. pp. 225-229

[30] Rochberg F. The Heavenly Writing: Divination, Horoscopy, and Astronomy in Mesopotamian Culture. Cambridge: Cambridge University Press; 2004.

[31] Cryer FH. Divination in Ancient Israel and Its Near Eastern Environment. Sheffield: Sheffield Academic Press; 1994

[32] Bottéro J. La plus vieille religion en Mésopotamie. Paris: Éditions Gallimard; 1998. p. 344

[33] Vernant JP. Écriture et religion civique en Grèce. In: Bottéro J, Herrenschmidt C, Vernant J-P, editors. L'Orient ancien et nous. Paris: Hachette Littératures; 2004. pp. 191-192

[34] Caramelo F. A Linguagem Profética na Mesopotâmia (Mari e Assíria). Cascais: Patrimonia Historica; 2002. p. 247

[35] Robins G, Shute C. The Rhind Mathematical Papyrus: An Ancient Egyptian Text. London: British Museum Press; 1987

[36] Gillings RJ. Mathematics in the time of the Pharaohs. Cambridge, MA: MIT Press; 1972, and also Shute C. Mathematics. In: Redford DB, editor. The Oxford Encyclopedia of Ancient Egypt. Vol. 2. Oxford: Oxford University Press; 2001. pp. 348-351

[37] Brown D. The measure of times and distance in the heavens above Mesopotamia, with brief references made to other ancient astral sciences. In: Morley I, Renfrew C, editors. The Archaeology of Measurement—Comprehending Heaven, Earth, and Time in Ancient Societies. Cambridge: Cambridge University Press; 2010. pp. 183-194

[38] Potts DT. Mesopotamian Civilization—The Material Foundations. Ithaca, NY: Cornell University Press; 1997. pp. 236-253

[39] Verner M. The Pyramids. Their Architecture and History. London: Atlantic; 2002

[40] Josephson J. Imhotep. In: Redford DB, editor. The Oxford Encyclopedia of Ancient Egypt. Vol. 2. Oxford: Oxford University Press; 2001. pp. 151-152

[41] Lehner M, Hawass Z. Giza and the Pyramids. Chicago: University of Chicago Press; 2017

[42] Fenollós JLM. La ziggurat de Babylone: un monument à repenser. In: André-Salvini B, editor. La Tour de Babylone—études et recherches sur les monuments de Babylon. Roma: Istituto di studi sulle civiltà dell'Egeo e del vicino Oriente; 2013. pp. 127-146

[43] Nadali D, Polcaro A. The sky from the high terrace: Study on the orientation of the ziqqurat in ancient Mesopotamia. MAA. 2016;**16/4**:103-108. DOI: 10.5281/zenodo.220907

[44] Allen JP. Genesis in Egypt—The Philosophy of Ancient Egyptian Creation Accounts. New Haven, Connecticut: Yale Egyptological Seminar, Department of Near Eastern Languages and Civilizations, Graduate School, Yale University—Yale Egyptological Studies; 1988

[45] Bottéro J, Kramer SN. Lorsque les dieux faisant l'homme. Paris: Éditions Gallimard; 1989

[46] Hornung E. Conceptions of God in Ancient Egypt: The One and the Many. Ithaca/New York: Cornell University Press; 1996

[47] Porter BN, editor. One God or Many? Concepts of Divinity in the Ancient World: Essays on the Concept of Monotheism/Polytheism in Ancient Assyria, Egypt, Greece and Israel. Casco Bay: Casco Bay Assyriological Institute; 2000

[48] Lesko LH. Ancient Egyptian cosmogonies and cosmology. In: Shafer BE, editor. Religion in Ancient Egypt. Ithaca/London: Cornell University Press; 1991. pp. 90-96

[49] Rothöhler B. Neue Gedanken zum Denkmal memphitischer Theologie [thesis]. Heidelberg: Uni Heidelberg; 2006. DOI: 10.11588/heidok.00007030

[50] The Electronic Text Corpus of Sumerian Literature [Internet]. 2006. Available from: http://etcsl.orinst.ox.ac.uk/cgi-bin/etcsl.cgi?text=t.1.1.2# [Accessed: 8 March 2017]

[51] Dalley S, translator. Myths from Mesopotamia—Creation, the Flood, Gilgamesh and Others. Oxford: Oxford University Press; 2000. p. 233

[52] Laboury D. Akhenaton, el primer faraón monoteísta de la historia. Madrid: La Esfera de los Libros; 2012

[53] Goldwasser O. The Aten is the "Energy of Light": New evidence from the script. Journal of the American Research Center in Egypt. 2010;**46**:159-165

[54] Monte MP. Ideia e Presença: a imagem do rei na construção simbólica do espaço imperial Neo-Assírio (sécs. X-VII a.C.) [thesis]. Lisboa: Universidade NOVA de Lisboa; 2017

[55] Caramelo F. A religião mesopotâmica: entre o relativo e o absoluto. Revista da Faculdade de Ciências Sociais e Humanas. 2007;**19**:165-175. Available from: http://hdl.handle.net/10362/8154

[56] Almeida I. A construção da figura de Inanna/Ištar na Mesopotâmia: IV-II milénios a.C [thesis]. Lisboa: Universidade NOVA de Lisboa; 2015. Available from: http://hdl.handle.net/10362/16014

[57] Janssen JJ, Rosiland M. Growing up in Ancient Egypt. London: The Rubicon Press; 1990. pp. 67-89

[58] George A. In search of the é.dub.ba.a: The ancient Mesopotamian school in literature and reality. In: Sefati Y, editor. "An Experienced Scribe who Neglects Nothing". Ancient Near Eastern Studies in Honor of Jacob Klein. Bethesda, MD: CDL Press; 2005. pp. 127-137. Available from: http://eprints.soas.ac.uk/id/eprint/1618

[59] Ashurbanipal Library Project. 2013. Available from: http://oracc.museum.upenn.edu/asbp/about/history/index.html [Accessed: 8 March 2017]

[60] Frame G, George A. The royal libraries of Nineveh: New evidence for king Ashurbanipal's tablet collecting. Iraq. 2005;**67/1**:269

[61] El-Abbadi M. The Great Library and Mouseion: Intellectual center of the world. In: Steen GL, editor. Alexandria: The Site & the History. New York: New York University Press; 1993. pp. 10-104

[62] Picton J, Quirke S, Roberts PC, editors. Living Images, Egyptian Funerary Portraits in the Petrie Museum. New York/London: Routledge; 2007

[63] Arasse D. Anselm Kiefer. Paris: Éditions du Regard; 2001

[64] Kiefer A. L'art survivra à ses ruines [Art will Survive its Ruins]. Anselm Kiefer au Collège de France. Paris: Éditions du Regard; 2011. pp. 7-27

[65] Tate Museum. [Internet]. 2017. Available from: http://www.tate.org.uk/art/artworks/kiefer-man-under-a-pyramid-ar00037 [Accessed: 8 March 2017]

[66] Bouhours J-M, editor. Anselm Kiefer. Paris: Éditions du Centre Pompidou; 2016. p. 212

[67] Rossini G. Semiramide=Semiramis: A Tragico-Dramatic Opera in Two Acts. New York: Houel & Macoy; 1845

[68] Nabucodonosor. [Internet]. 2003. Available from: http://opera.stanford.edu/Verdi/Nabucco/libretto.html [Accessed: 9 March 2017]

[69] Akhnaten. Learning Resource [Internet]. 2016. Available from: https://d2ae1n566n-bglo.cloudfront.net/wp-content/uploads/2016/03/04152603/eno-akhnaten-learning-resource-160304.pdf [Accessed: 8 March 2017]

[70] Assmann J. Thomas Mann und Ägypten. Mythos und Monotheismus in den Josephsromanen: München: M C. H. Beck Verlag; 2006

[71] Astarte Syriaca. [Internet]. Available from: http://www.rossettiarchive.org/docs/s249.rap.html [Accessed: 5 March 2017]

[72] Dauth B. Joseph L. Mankiewicz: Interviews. Mississippi: University Press of Mississippi; 2008. pp. 87-88, 181-184, 198-200

Chapter 2

A Living Force of Continuity in a Declining Mediterranean: The Hospitaller Order of St John in Early Modern Times

Victor Mallia-Milanes

Additional information is available at the end of the chapter

http://dx.doi.org/10.5772/intechopen.69376

Abstract

The paper discusses two interrelated themes. From 1530 to 1798, the Order of St John, an international institution of the Church, played a significant role from its conventual base on central Mediterranean Malta—one that nourished a strong element of historical continuity within the wider context of a declining Mediterranean. This it did through its traditional twofold raison d'être—as a religious, charitable and Hospitaller Order with great expertise in medical knowledge and practice, and as a military institution whose traditional crusading zeal against Islam kept the 'clash of civilisations' alive. That is the first theme. The second theme concerns the concept of a declining Ottoman Empire and a declining Mediterranean. The paper argues against the idea that the collective impact of the great siege of Malta (1565) and the battle of Lepanto (1571) had marked the initial stage in the decline of the Ottoman Empire. It also claims that the decline of the sixteenth-century Mediterranean needs revisiting. It was only a partial change for the worse. The great geographical discoveries succeeded in robbing the Middle Sea of its primacy in international economy and exchange but in the long term failed to uproot most of its other characteristic features.

Keywords: Order of St. John, Hospitallers, Ottoman Empire, Mediterranean, North Africa, Christendom, Islam, hospitality, crusade, great siege of malta, battle of Lepanto, change, continuity, decline

1. Preamble

History is exploration, reconstruction and interpretation. It is a means 'to explain and understand', pointed out Fernand Braudel, the finest twentieth-century historian, in 1984[1], and

[1]Come possiamo proprio noi... erigerci a giudici del passato?'. See Ref. [1].

should not be turned into 'an instrument' to pass judgement on our ancestors, their ideas, motives and beliefs, their methods, their frailties, lifestyle, strengths and weaknesses, their failures and achievements. It is within this conceptual context that the present paper seeks to address the multiple function the Hospitaller Order of St John, an international, supranational and supra-diocesan organisation [2, 3], played as a dynamic force of continuity in the early modern Mediterranean. The chapter will also briefly revisit the conventional assumptions about the latter's decline.

2. The Hospitaller institution

From its modest inception in eleventh-century Jerusalem, the Order of the Hospital, as it came to be known, the only one of its kind in the Mediterranean,[2] began gradually to assume its four predominant roles, each evoking an alluring image of the Knight Hospitaller—the kneeling monk absorbed in silent prayer, the humble servant attending to his Lords the sick and the poor, the brave and experienced soldier on the battlefield fighting for the faith, and the efficient feudal seigneur, though at times admittedly arrogant and disdainful. It was precisely the integrated performance of all these various, intimately related roles that determined the character and function of the Hospitaller in society. He was 'a constructive factor in European civilisation' [4], one that turned his Order into a living force in the Mediterranean over the subsequent 700 years until the French Revolution evicted it from central Mediterranean Malta in 1798.

Elizabeth Schermerhorn, writing in 1940, asked: 'How did [the Order] manage to persist and endure among the ruins of feudalism down to the very eve of the [French] revolution?' [4; pp. 11–12]. In her endeavour to explain this 'endurance', she identified a number of determining 'elements' [4, p. 12]. Of these, the one that had perhaps lain at the root of its 'heritage' was its massive landownership. The extensive estates it had gradually acquired through donation, purchase, inheritance and absorption spread all over Europe and consisted of all sizes, shapes, forms, architectural styles and purposes, each betraying the geography and culture of its place of origin. The Order's intelligent management and efficient administration of these lands, their cultivation, in ways like 'farming and stock-raising' and various other forms, the meticulous statutory measures rigidly governing their regular maintenance and others as stringently observed against any form of alienation—exposed in a relentless way the Hospitaller institution's contribution to 'the development of the feudal manor' [3; p. 276, 4–6].

These lands, organised into commanderies, the Order's basic units of administration, gave the institution wealth and power. Each constituted a secure source of regular revenue, a centre to recruit and train new brethren [2; p. 340], and a means of social connectivity, of maintaining contact with the common peoples of Europe. They financed all its activities in the Holy Land, on Rhodes and on Malta, in Europe and throughout the Mediterranean. They sustained its crusading ideals and traditions. They were also important 'as retirement homes [and] residences for the Order's many priests' [2, p. 340]. By the end of the sixteenth century, the Order

[2]Only the [Order of the] Hospital,' writes Anthony Luttrell, 'could claim that it underwent no essential change between 1312 and 1798' [2, p. 361].

owned 564 commanderies. Through them the Order survived in its original privileged status until the enlightened thought of *égalité* was put permanently into rude practice. The *ancien régime* finally succumbed to the rise of the unprivileged, dragging with it the Hospitaller institution of pre-1789. The Order that eventually emerged with renewed vigour like a phoenix from the ashes was a reformed institution, an Order, religious and charitable, as it had originally been before being militarised, responding to the powerful ideals, spirit and mood of early nineteenth-century nationalism.[3] For the umptieth time, it showed convincingly its uniquely remarkable powers of resilience.

Resilience was the Order's strongest shield—its remarkable ability to recoil fairly quickly into its former shape after having experienced a severe crisis. This distinct quality had been callously tested on several occasions in its historical evolution—in 1187, the Hospitallers were evicted, along with the other military orders, from Jerusalem by Saladin; in 1291, they lost Acre to the Mamluks, never to return to the Holy Land; in 1522, they surrendered Rhodes to Suleyman the Magnificent and in 1798, French revolutionary forces under Napoleon drove them out of Malta. After each of these occasions, the Order succeeded in regaining its former state and strength.

Landownership was only one element which lay behind these powers of resilience. The other was the patronage the crowned heads of Europe were consistently willing to extend to the institution. Statute 2, on the rule governing it,[4] makes specific reference to the liberality with which the Holy See, the Catholic monarchies, principalities, duchies, republics and devout Christians dealt with the Order, enriching it with landed estates, together with a wide range of powers, jurisdictions, privileges and exemptions. It was vital for the Hospitallers to cultivate a perfectly healthy relationship with its powerful patrons. This was a reciprocal 'moral code', for in return the Order offered Europe 'direct and constant assistance against the enemies of the Christian faith'.

3. Hospitaller activity in the Mediterranean

The Order of the hospital was a religious-military institution. Its activities in the Mediterranean may be grouped into two large categories, two overwhelming obligations, which corresponded neatly with its character—charity and war—and which, in an early seventeenth-century document, were placed nearly on a par with the three monastic vows which every professed Hospitaller took on admission into the Order.[5] One was of a widely 'defensive' character, intended to extend forms of piety and protection, epitomised in its guiding principle of hospitality in its widest possible connotation. Inspired by the evangelical concept of 'love thy neighbour', it was offered to whoever needed any form of caring attention, with no restrictive qualifications of age, class, faith, gender, colour or country of origin. The second category was

---

[3]For the Order after the loss of Malta [7].

[4]See Volume che contiene gli Statuti della Sacra Religione Gerosolimitana; Le Ordinazioni dell'ultimo Capitolo Generale...; il Nuovo Cerimoniale prescritto dalla Santità di N. Sig. Papa Urbano VIII sopra l'Elezione de' Gran Maestri ....Borgo Nuovo MDCCXIX. Per Antonio Scionico, Stampatore Generale.

[5]This is discussed in my forthcoming book, *Lo Stato dell'Ordine di Malta, 1630*.

combative or 'offensive' in nature, always ready to launch an aggressive action. This comprised the militia principle or obligation, to wage an incessant war against the enemy of the Cross.

James Reston Jr.'s assertive definition of the state of the Knights Hospitallers after losing Rhodes as 'refugees without portfolio ... wondering aimlessly' [8] is wrong and misleading. In the eight long years from 1 January 1523, the Order's movement from one place to another was partly motivated by its strong desire to secure the permanent residence on Malta which Grandmaster L'Isle Adam had requested before leaving Crete[6] and partly dictated by persistent outbreaks of plague. Negotiating the difficult conditions which Charles V had originally decided to attach to his enfeoffment of the island was a protracted diplomatic process. These years constituted a perilous odyssey indeed, where the homeless institution was exposed to potential dissolution, but they were not wasted years. Rather, immediately following the humiliation the Knights suffered at Rhodes, these years were a true trial of Hospitaller strength, an audaciously bold public relations exercise, where their spirit of charity and hospitality, their medical expertise, and participation in corsairing were, as circumstances determined, openly manifested at every stage on their tortuous way from Rhodes to Malta via Crete, Sicily, Viterbo, Nice and Villefranche [11]. These activities underscored, and were perhaps meant to underscore now more than at any other time, the institution's relevance to the general interests of Christian Europe.

4. Health and hospitality

The Hospitallers' conventual fortress on Malta, as on Rhodes, militated against two implacable enemies. Its massive walls surrounding the new city of Valletta and the rest of the Grand Harbour, together with its numerous coastal towers, had been allowed to grow into their awesome architectural grandeur not only to defend the island and scare any potential enemy away. It was a standing symbol of another traditional scourge. The Mediterranean faced a worse threat than either the Crescent or the Cross could offer the opposing half of the great sea—north and west and east and south. The plague was more brutal and devastating than the characteristic violent confrontation between the two, worse than either piracy or slavery in its deadly consequences, worse than any crusading warfare in its destruction. The Hospitallers approached its potential outbreak with unwavering rigour. Their long medical tradition, their knowledge of hygiene, their expertise and direct involvement in health management and their inflexibility in dealing with quarantine and other related mechanisms endeavoured, though not always successfully[7], to keep the plague at a comfortably secure distance, to the extent that their aggressive severity at times provoked a storm of protest from Venice's own health authorities [13].

The Grand Hospital which the Knights built on migrating from Birgu to the new city of Valletta, and 'its numerous subsidiary institutions',[8] constituted their first line of defence

[6]Grandmaster Philippe Villiers de L'Isle Adam, before leaving Crete, had specifically requested Emperor Charles V to grant the Order either Malta or the port city of Brindisi in Southern Italy on the Adriatic coast, offering him to pay 100,000 ducats for the award [9], and [10].

[7]Hospitaller Malta experienced severe outbreaks of plague epidemics in 1592–1593, 1623, 1655, and 1675–1676. See [12].

[8]Ref. [14]. See also Riley-Smith's observations in Ref. [2], pp. 278–279.

against this scourge. The lazaretto, lying on an islet in Marsamxett Harbour, and the quarantine station on the Valletta quay on the Grand Harbour, offered a wide range of amenities—accommodation for crews and passengers under observation or in isolation, accommodation for animals, stores and warehouses for merchandise, facilities for disinfection and fumigation, and a chapel for spiritual needs dedicated to St Roche. These services marked an authentic continuity of those that had already been offered to the pilgrims, the sick, the poor and the homeless at the conventual hospital four centuries earlier in Jerusalem (see for example [15]). It was an unbroken tradition 'of extensive alms-giving to the poor, of doctors and surgeons, of nightly prayers in the hospital for benefactors and others led by the Order's priests, of provisions of orphans and lepers, of maternity wards and of financial support, medicine and diet, burial arrangements and the treatment of those wounded in battle' [15].

These services included, too, 'a mobile tented hospital that accompanied Christian field armies', singularly rare for the twelfth century [16, 17]. As his predecessors had done in Jerusalem, Acre, Limassol on Cyprus and Rhodes, and as he and his successors would do again on Malta, Grand Master l'Isle Adam continued, wherever the Order sojourned, to serve from his own hands 13 poor persons every morning in honour of Christ and his 12 apostles, offering bread and wine to the most wretched. The French Secretary to Grand Master Gregorio Carafa, Fra Gio. Batta Le Marinier de Cany, remarked in 1670 that these qualities formed the quintessence of the true grandeur of the Hospitaller ethos.[9] Over the very long perspective, the Order's hospices and hospitals were true centres of selfless commitment and dedication to those in need. The difference between the two conventual hospitals in Jerusalem and Malta lay solely in policies and practices that reflected the progress made in medical knowledge [15, p. 175].

Over 100 years after de Cany, the Order, now economically weakened and reduced to near impotence by the revolution in France and with hardly any powerful patron other than the papacy, was still loyal to its original principles. In 1930, on the strength of what he defines as 'a vast amount of evidence', including fruitful eye-witness accounts of travellers visiting Malta in the seventeenth and eighteenth centuries [19–21], Frederick Ryan claimed that the Order remained 'faithful to the duties of Hospitality' to the end of its days in Malta' [22], and that its hospital 'was well abreast of its time from the scientific point of view', and that from the religious standpoint this great hospital was ably fulfilling, with a multitude of other activities in Malta, the great function of a centre for corporal works of mercy' [22].

From the early origin of their hospice in Jerusalem 'shortly before 1071' [23] to their surrender of Malta in June 1798, over 700 years had passed. It is remarkable that, during this very long term, generation after generation of Hospitallers never failed to understand and positively respond as dispensers of charity and hospitality to the needs and condition of the marginalised in society and to the ever pressing demands of public health. Irrespective of where their convent stood, irrespective of the prevailing socio-economic context of the Mediterranean world and irrespective of the state of their priories throughout Christian Europe, the Hospitallers'

[9]National Library of Malta, Archives of the Order of Malta, Codex 1697: Riflessioni di un Cavaliere di Malta, Religioso dell'Ordine Militare degli Ospedalieri di S. Giovanni di Gerusalemme, sopra la grandezza e i doveri del suo stato, 1670. See also Ref. [18].

main concern was ultimately the preservation of humanity. The dynamic driving force behind this ancient and unbroken tradition was their unwavering concern for the common good. As pointed out elsewhere [13, p. 204], men occupied the centre stage of their charitable activities.

5. Hospitaller crusading warfare

Within decades of the fall of Acre to the Mamluks in 1291, the Templars ceased to exist and the Teutonic Knights shifted their focus to north-eastern Europe. The Knights Hospitaller was the only Order for whom crusading warfare in the eastern Mediterranean retained its original appeal and enthusiasm. They continued to operate from their conventual base first on Cyprus, then on Rhodes and from 1530 from nearer Western Europe on Malta. Their direct and persistent involvement in holy warfare prolonged the life of the crusade to the end of the eighteenth century. Organising and participating in holy warfare justified the Order's retention of its western priories and commanderies [2, p. 334]. This early-modern form of the crusade was distinct from the old, medieval one associated with Jerusalem. Its objectives were different: the Holy Land was gone and gone forever; no vows, temporal or spiritual privileges, no plenary indulgences or remission of sins and no papal sponsorship appear to have been attached to it; no age restrictions were imposed; no physical fitness or material wealth were demanded. This post-Acre role was in harmony with the normal expectations of a dynamic religious-military institution, whose *raison d'être* after all was to pray, help and fight. It was an uninterrupted continuation of the Order's old practices, methods and beliefs.

On Malta, more perhaps than on Rhodes, Europe was converted into a new 'holy land' for the Hospitallers to defend against Islam. On the central Mediterranean island, as on Dodecanese Rhodes, the Order's military role assumed four distinct forms [24]. It participated in practically all the holy leagues, like that formed by Paul III in 1538 or by Pius V in May 1571, placing all its naval forces at the disposal of the allied Christian fleets.

On other occasions, it contributed its entire galley squadron and other craft to the Habsburgs' punitive campaigns against the infidel. To instance one classic example: in preparation for Charles V's large-scale expedition to regain Tunis, which Khaireddin Barbarossa had seized in 1534, exposing the southern coasts of Spain, Italy and Sicily to Muslim attack [25], the Order first secured the necessary supplies of food from Sicily and then had its galley squadron strengthened, raising it from four to five by launching the *Santa Caterina* in mid-April 1535. The sixth was under construction. To the four Hospitaller galleys, 200 Knights were assigned [26]. On board the *caracca Sant'Anna*, which accompanied the Hospitaller squadron, there were another 70 Knights and a strong regiment of soldiers [26]. By the end of June 1535, the allied fleet, not without great difficulty [8; pp. 366–338], took the fortress of La Goletta. By 21st July, the fortress and city of Tunis were both in Christian hands. That gave the Habsburgs complete control over the entry into the western Mediterranean from the east. The only criticism of this great achievement was that Charles V failed to follow the logic dictated by geopolitics—to follow up that victory by an immediate assault on Algiers. That came 6 years later and ended up in disaster.

http://dx.doi.org/10.5772/intechopen.69376

The list of similar collective enterprises can be stretched indefinitely. From Tunis to Lepanto, the role played by the Hospitallers pursued a predictably regular pattern of participation in nearly all the activities of the allied Christian fleet. Its four-galley squadron was present at Corfu (1537) [25; iii: pp. 425–427, 26; iii: pp. 170–172], Prevesa (1538),[10] Otranto and Castelnuovo in Dalmatia (1539) [25; iii: p. 446, 26; iii: pp. 186–187], and the waters of Taranto (1540) [26; iii: pp. 186–187]. That same year, too, the Order, allied to the Sicilian forces, took part in the conquest of Monastir and Susa [26, p. 194] and in the unsuccessful siege of Sfax [26; pp. 194–195]. The next year Charles V decided to repeat at Algiers what he had achieved at Tunis six years before [26; pp. 199–200, 205–211]. The allied forces reached Algiers towards the end of October. When the fall of the fortress was about to be secured, the 'natural elements' intervened in favour of the besieged. 'The besiegers', observed Mori Ubaldini, 'were immobilised for three whole days by a violent storm with torrential rains and raging winds' [30]. Giacomo Bosio, the great historian of the Order, draws a very dramatically detailed portrait of the situation on the morning of 28 October [26, iii: p. 208]. In 1550, the Hospitallers found themselves once more engaged in the allied campaign against the port town of Mahdiya (today a Tunisian coastal city, south of Monastir and southeast of Sousse) [26; pp. 243, 257–258, 266–267, 27; p. 910, 31], and in 1564, they participated in the conquest of Peñón de Velez [32].

So much has been written on the Ottoman siege of Malta (1565) and on the battle of Lepanto (1571) that they hardly need any further remarks here, but their historical significance will be brought up towards the end of the present chapter. After the siege of Cyprus and Lepanto, the Order continued to respond equally positively to requests for help from the Republic of Venice. The long-drawn-out Venetian wars first at Crete (1645–1669) and then twice at the Morea, towards the end of seventeenth century and the second decade of the eighteenth century. The Order could hardly decline requests to contribute to either of these two types of formal anti-Muslim war. Invitations emanating from the papacy, the Order's ultimate authority, were synonymous with instructions. They could not be turned down unless for a very serious reason. Neither could it dismiss with any modicum of comfort requests from the powerful Spanish monarchy, its feudal lord.

The intervals between one formal war and another, between one Holy League or Habsburg campaign (which were hardly distinguishable) and another, gave the Order some respite to accommodate its statutory provisions, ambitions and aspirations, to live up to its naval and military tradition. This was the third form of Hospitaller crusading activity when it could design its own priorities and organise similar expeditions on a less spectacular scale in all parts of the Mediterranean. The objective of such ventures on seasonal crusading cruises was to harass Muslims on land and by sea, ravage their coastal towns and villages, harass Muslim merchants, raid their shipping, seize their merchandise and carry their men, women and children into slavery [33, 34].

The corso was the fourth genre of Hospitaller warfare. The initial involvement of individual member of the Order in formal anti-Muslim corsairing activity has been assigned to

[10]Refs. [25; iii: pp. 445–447 and n. 208, 26; iii: pp. 178–180], 'Prevesa gave Islam that control of the sea of which the Christian victory at Lepanto was to deprive it in 1571' [27–29].

the first years of Spain's *Siglo de Oro* [30; p. 108], when individual Hospitaller was authorised, like other ordinary privateers, to set out on their own on what may be termed a private crusade. On 25th November 1503, for example, the Knight Ynyogo Ayalla was licensed to arm the Order's *barcia Santa Maria* and sail out on a privateering expedition in Levantine waters on condition that two-thirds of his booty would go to the Order's common treasury. The event coincided with Spain's *Reconquista,* whose unintended repercussions delivered the birth of the Barbary regencies and dramatically widened the scope for the ruthless excesses of the pirates of Morocco, Algiers, Tunis, and, after 1551, Tripoli, spreading their activities over a vastly grander scale, and extended the Ottoman Empire's sphere of direct influence all along the Maghribi coast. The Barbary corsairs' activities in the Mediterranean, the sense of insecurity they created, and the general threat they offered to Christian trade proved beneficial to the Order as they highlighted the Hospitaller's political relevance in the region, providing tangible evidence in support of their claimed indispensability [32; pp. 388–397, 35–37].

These years indicated the first stages in the Mediterranean's gradual loss of the primacy it had long been enjoying in global economy [38, 39]. Hospitaller privateering, under the guise of crusading activity and covered by the magistral emblem or the Order's eight-pointed cross, eventually escalated into an endemic warfare, reaching spectacular peaks in the next century. These colourful exploits against Islam undertaken both by individual members of the Hospital and by a wide section of the native population were reconfirmed when the Order moved from Rhodes to the central Mediterranean and encouraged to develop into a major industry on Malta. Indeed, the island itself was transformed into a thriving corsair base with a flourishing international slave market. No wonder the Venetian Senate in the late 1580s dubbed the Hospitallers 'corsairs parading crosses' [40], who fostered a culture of piracy, ransom and plunder.

6. 'Coexistence and symbiosis'

In history, long-held assumptions often need to be revisited. In *Venetians in Constantinople,* Eric Dursteler argues how thin the distinction was between word and action, between rhetoric and reality, how 'porous' and 'pliable', the borderline or frontiers were between the two in matters of faith and geopolitics in the early modern Mediterranean world. Within the broad framework of the ideological chasm between Christianity and Islam not only were 'coexistence and symbiosis' possible; they were 'almost certainly the quotidian norm rather than the exception'. He questions the historical validity of the traditional reconstruction of these two powerfully determining forces as being in perpetual engagement 'in a life-and-death struggle' [41]. The relationship between Venice and Constantinople provides a perfect example. They cannot be defined as two friendly empires by whatever criterion is adopted. They were declared enemies. They fought several wars against each other. But when times 'were not distorted by hostility' [41; p. 20], their relationship was marked by 'interaction and coexistence' [41; p. 19]. This state of living in near harmony despite different ideologies and interests marked almost the entire history of the Republic.

It has now been claimed that the Hospitaller regime on Rhodes too adopted such sensible and realistic approach towards the Ottomans and the Mamluks, one based on practical rather than on theoretical consideration [42]. Unlike what we have hitherto believed, the inflexibility of the Order's stance towards Islam was more apparent than real. Geography, the weather, war and the overall political atmosphere prevailing throughout the eastern Mediterranean at the heart of which lay the island order-state[11] [32, 33] of Rhodes dictated an unavoidable form of 'interaction and coexistence' with its otherwise hostile neighbourly Muslim powers. The general perception entertained of the institution's unwavering militant opposition to Islam was of the Order's own creation. It was the intended outcome of its profession of its mission and propagated through the wide network of its experienced diplomatic representatives in Europe for the consumption of its Christian patrons in the West. There is no doubt, of course, that the Hospitallers were sworn enemies of Islam. But concealed underneath the art of Renaissance rhetoric stood another Order. The fifteenth-century picture that emerges of the hospital from its surviving chancery records does not in fact correspond very neatly to this traditional interpretation [42; pp. 109–119]. The reality of this other phase of the institution uttered a different dialect. The brutal truth, we are told, is that 'safe-conducts, licenses to arm ships and licenses to trade tell of the peaceful passage of Turkish merchants and diplomats through the Port of Rhodes. The Master of the Order licensed trade with Muslim ports in Syria and Egypt and enacted treaties with Muslim powers, with or without papal consent'[42; p. 110].

Rhodes ruling regime, it has been pointed out, 'could not afford to alienate the Ottomans completely because Constantinople was an important grain source for the region. Likewise the Hospitallers maintained diplomatic relations with the Mamluk Sultan of Egypt because Rhodes and Alexandria were trading partners' [42; p. 116].

The pragmatism of Hospitaller policies is still under-researched. It needs to be scrutinised further, more deeply, over a longer timeframe, and on a much wider spatial scale. Was the Order on Malta as pragmatic in its attitude towards the Ottoman Empire and the Muslim powers on North Africa as it had been on Rhodes? If so, such interaction must have been much more subtle and refined. Archival documentation does indicate a few isolated cases of apparently similar practices during its Maltese phase, but these were carried out through the mediation or intervention of French agencies, exchanges under cover of the French flag. In 1754, for example, severely strained relations between the Order and the Kingdom of Naples resulted in the suspension of Malta's trade links with that Kingdom. To counter the effects of the embargo,

[11]According to Luttrell's definition, the island order-state of Rhodes 'demanded the establishment of a naval tradition and the arrangement of the local economy and government in ways which would support defensive measures. The harbour brought shipping, pilgrims, pirates, trade, and taxes; the island was populated to produce foodstuffs and auxiliary forces; its forests furnished timber for shipbuilding; the inhabitants constructed and manned towers and castles or served as galley oarsmen' [2. pp. 334–335]. Most of these attributes applied equally well to Hospitaller Malta. Like the Greeks on Rhodes, the Maltese too were 'reasonably fed, protected… on the whole the population felt reasonably well treated and was prepared to collaborate' [2. pp. 334–335]. The central Mediterranean island-fortress offered all captains of vessels sailing the Mediterranean, all sailors, passengers and merchants of Christian States access to a strongly fortified, fully equipped base for all forms of operations, a safe neutral port of call, an arsenal, a lazaretto with other quarantine facilities, an efficient hospital, a flourishing market, spacious warehouses, courts of justice and consular assistance. They produced ashes and cultivated cumin and cotton for export. These attractive conditions explain why very few riots or protests occurred over 268 years of Hospitaller rule [43, 44]. See also the resident Venetian Minister's comments on the so-called 'uprising of the priests' in Ref. [45].

Grand Master Emanuel Pinto's *Uditore* sought the help of the French consul in Tunis to provide the order-state with some 5,000 heads of cattle and other livestock.[12] Another example concerns Hospitaller Malta's relations with Morocco in the 1760s [45; Letter CCVI, p. 596, n. 496].

The geographical proximity of central Mediterranean Malta to Europe, especially to the papacy, probably placed the Order under closer observation than it could have possibly been in the south-east Aegean. The inquisitor resided on the island to combat heresy; from 1575 he assumed too the function of a nuncio or apostolic delegate. His double surveillance role could not have been too enticing or comforting for the Hospitaller regime. His regular correspondence with the Secretariat of State at the Vatican shows a much wider interest than the eradication of heresy. And along with the nuncio, there were several other resident ministers and consuls representing the interests of various European royal courts on the island. France, for example, had her *Homme du Roi*; Spain, Sicily, Sardinia, Tuscany, the tiny principality of Monaco and others: each had their own minister or *chargé d'affaires*. From the mid-eighteenth century, Venice had her *Uomo della Repubblica* in the person of Massimiliano Buzzaccarini Gonzaga. Each corresponded regularly with his sending state to report, at times in minute detail [46], on what was going on in and around Malta. As late as 1770, for example, during the Russo-Turkish war of 1768–1774, the Venetian Minister received a note from Venice's Magistracy of Trade saying that it would be greatly appreciated if he continued to keep the Senate abreast of developments in the central Mediterranean.[13] Though himself a high-ranking member of the Hospital, like all the other similar representatives on the island except the nuncio [47], there were moments when the Venetian minister was quite critical both of Grand Master Emanuel Pinto's style and method of government [45; pp. 77–78] and of Benedict XIV [45; p. 79]. It is also interesting to observe, and quite revealing and enlightening, that no such foreign representation existed on Hospitaller Rhodes,[14] which goes some way to make the Order's stance there perhaps more understandable. There was a Venetian consulate for a while but was discontinued in about 1410 on the grounds that other nations wanted consuls as well.[15]

7. The siege of Malta and the Battle of Lepanto

Within the context of what Samuel P. Huntington has defined as the 'clash of civilisations' [48], the epic siege of Malta (1565) and the large naval Battle of Lepanto (1571) were the two most outstanding events in the history of the early-modern Mediterranean. In both, the Hospitallers were involved directly: in the first, they were the leading protagonists; in the second, they participated in Pius V's Holy League with three fully equipped galleys and

[12]National Library of Malta, Archives of the Order, codex 1511: The two letters, each dated 11 March 1754, addressed to the French consul Plowman in Tunis.

[13]The note, addressed too to various Venetian consuls, said: '[Il Senato] trova opportuno e necessario di aver da voi una regolare anticipata cognizione di tutte le cose che vanno succedendo a codesta parte' [45; Letter CXXXIX, p. 449, n. 355].

[14]I thank Dr Anthony Luttrell, Professor Juergen Sarnowsky and Professor Helen J. Nicholson for this observation. Emailed communications, 13, 16 March 2017.

[15]Dr Anthony Luttrell, in a private communication, 13 March 2017.

http://dx.doi.org/10.5772/intechopen.69376

manned with experienced Knights and soldiers along with the allied forces of Spain, Venice, the papacy and other Italian principalities [30; p. 270]. In 1565, the small, weakly fortified island, exposed and vulnerable, succeeded in withstanding, for four whole months, a furious assault unleashed by Suleyman's mighty armada. In 1571, the fragile Holy League destroyed nearly the entire Ottoman naval forces. News of both developments, at merely six years apart, spread all over Christian Europe like wild fire, followed by widespread rejoicing and celebrations on a grand scale and commemorated by contemporaries and later generations in various shapes and forms—in eyewitnesses' accounts, in literature, on canvas, in music, in archival records. But euphoria of victory and jubilation do not recast near-permanent structures.

Historians have tended to identify the outcome of both events as a determining stage in the gradual decline of the Ottoman Empire. Halil Inalcik claims that the withdrawal of Suleyman's forces from Malta in September 1565 'marked the beginning of a halt in the Ottoman advance into … the Mediterranean' [28; p. 41]. Colin Imber assigns 'the end of the Ottoman maritime expansion towards the west' to the conquest of Chios, the Genoese island in the Aegean, in 1566 [49]. Thomas Dandelet asserts that Lepanto 'signalled the end of Ottoman expansion beyond Crete' [50]. This historiographical discourse casts grey clouds over the reality, somehow obscuring the skyline. It should be discreetly relegated to the mythical spheres of the past where it comfortably belongs. The true significance of both events lies elsewhere.

The outcome of the siege greatly enhanced Philip II's 'imperial reputation … because it revealed him to be living up to the old … humanistic programme of reviving ancient Roman military power to crush the Ottoman threat' [50; p. 153]. It was to him that the Order, Malta and the central Mediterranean owed an enormous deal. His timely Spanish intervention had 'saved [them] from the Turks' [51]. On landing on the island, the relief force of some 10,000 men instantly scared the enemy away without even engaging it in battle.

Within the narrow sphere of Hospitaller and Maltese history, the siege helped the Hospitallers to survive and retain their Convent firmly in the central Mediterranean. It revamped their political relevance to Christian Europe much more than Tunis had done three decades earlier, and reconfirmed their institution's traditional role as a force of continuity. The siege also transformed their island order-state into an almost unrecognisable form, turned it into a formidable fortress, and ushered it into the forefront of Mediterranean politics and the modern world. It may well have set the Ottoman naval expansionism temporarily back, but failed to reverse the Turk's dominance in the Middle Sea. It simply defined his Empire's 'geographical limits' [52].

On the other hand, the Battle of Lepanto was not a collective effort to eliminate Christian Europe's common enemy as it has often been depicted. Venice depended so much on the empire for its trade. Judged from its short-term result, it was an implicitly professed reconfirmation of the major Latin protagonists that their own interests held pride of place in their scale of values. At Lepanto, the bitterly divided Christian West, allied in a tenuous holy alliance, confronted Islam through extreme, inbred religious fanaticism, a pervasive spirit of raw intolerance and a hysterical ideology of hatred of 'the other', so evident in the rejoicing and jubilation celebrating the horrors and atrocities of the battle. These were the ingredients which formed the texture of contemporary life in the sixteenth-century Mediterranean.

The Ottomans failed to realise their objective in 1565 and were defeated in 1571. Notwithstanding both disasters, they took the Venetian island of Cyprus, 'the greatest feat of Ottoman arms' [28; p. 41]. They defended their new acquisition with a newly built armada that within a few months had replaced the one destroyed at Lepanto [28, p. 41]. The new fleet, reflecting the innovative technological evolution [53], consisted of bigger vessels, including eight galeasses, larger than the Venetian ones and fully armed, apart from other galleys and smaller craft ([54] also in Ref. [55]). Its recovery and the perception of that extraordinary dynamism helped it retain the new acquisition and dictated Venice's 'betrayal' of the Holy League and the latter's dissolution.

Neither of the two events, nor their collective impact, succeeded in destroying the Ottomans' naval power. Ottoman 'presence in the Levant and North Africa' survived and remained as threateningly 'real and dreadful' as before [56]. 'The years after 1571 still gave enough evidence of the aggressive power of the Turks at sea and on land' [56; p. 203]. In 1573, they ravaged the coasts of Southern Italy and Sicily. They occupied La Goletta and permanently seized the fortress of Tunis in September 1574, with which they recovered their 'self-respect' [27; p. 1139]. In 1576, they raided Palermo and captured Fez in Morocco [57]. In 1588, they harassed 'Valencia and its surroundings'. 'The shores of Calabria and Sicily', points out Alexander De Groot, 'were raided in 1592, 1593, and 1594 by the corsair galleys from Bizerta … and the main Ottoman fleet … with 90–120 galleys'. In the last few years of the century and the beginning of the next, Naples and Sicily experienced similar aggressive pressure. 'The main fleet passed Malta and Gozo on its way to and from Algiers in 1598' [56; pp. 215–216]. In the seventeenth century, the Turks took the Venetian island of Crete and early in the eighteenth century they again confronted the Venetians in the Morea. 'The peace made at Passarowitz in 1718 confirmed the naval superiority of the Ottomans over the Venetians' [56; p. 230]. Drawing his data from Giuseppe Bonaffini's *La Sicilia e i Barbereschi* [58], Thomas James Dandelet claims that between 1570 and 1606, 'Sicily alone was attacked 136 times' [50; p. 152]. Not only were the Ottomans reconfirmed masters of the eastern Mediterranean, a dominance they had established since 1538 at Prevesa; they even strengthened 'their hold upon the Maghreb' [59]. That was Lepanto's legacy. Voltaire was not far from the truth when he claimed that 'the victory of Lepanto seemed rather to have been on the side of the Turks' [60]. Alexander De Groot argues that the term *Ottoman decline* 'is devoid of any real meaning' [56; p. 253].

My position on this issue does not diminish the importance of either event by one iota. Both the siege and the battle had indeed been stunning victories for the Christian West; both were extremely injurious to the dignity and pride of the Ottoman Empire. But humiliation is a psychological condition, an obvious discomfiture that was however fairly quickly overcome. The two events were strategically inconsequential in the long-term perspective of historical development, both of the Ottoman Empire and of the Mediterranean. The spirit of ideological confrontation which had inspired them both and, indeed, similar other occasions before and after them, did not disappear; nor did the prevailing mood grow fainter. The *status quo* in the Mediterranean was reconfirmed. 'The cultural shape of the lands around the Mediterranean' writes Gregory Melleuish, survived unimpaired, 'with a largely Islamic East and South staring across the waters at a Christian North and West' [53].

http://dx.doi.org/10.5772/intechopen.69376

8. Conclusion

The idea of the decline of the Mediterranean needs revisiting and redefining. Did the great discoveries, to which it is generally attributed, convert the sea into an isolated backwater, perhaps tranquil and undisturbed? Did the 'great sea' shrink into insignificance as a result? Was its unity, its internal coherence [27; p. 14], shattered? From the evidence to date, it would appear that the decline was only partial: the Middle Sea was deprived only of the primacy it had been for long enjoying in global economy and international exchange. But was economic pre-eminence, however central and determining, the only feature that had constituted the sea's true greatness? Today the validity of the claim of 'the economic decline of the eastern Mediterranean' has been questioned [61].

After 1492, the Mediterranean still retained the natural distinct elements of cultural unity that had been so intimately characteristic of it. The homogeneity of its climate [27; pp. 234–238] remained unaltered. Its winter temperatures did not grow colder with the 'massive invasion' of the Northerners from the 1580s. Nor did the latter turn its often fiercely hot summer months into a pleasantly mild season. The sea kept its 'blue transparency'. The cultivation of the vine, the olive tree and the palm tree does not appear to have been diminished or discontinued. The Mediterranean still held on to its 'relatively easy navigability' [62] and 'maritime interconnectedness' it enjoyed since ancient times [63]. The peoples inhabiting its ancient world still nourished their natural inclination and attitude towards the sea long after Christopher Columbus had crossed the Atlantic. Fishing, pastoralism and farming remained constant features of local economies through the entire region. Plague and other deadly diseases were also shared experiences. Inhabitants from each of the four corners of the Mediterranean who survived such characteristic catastrophes had frightening stories to narrate about these phenomena [64]. Piracy, corsairing and slavery, on either side of the ideological fence, remained vigorous activities that marked everyday life. And the open clash of religion between Christianity and Islam did not diminish in its intensity with the enormous structural changes that followed 1492—the great voyages to the New World, the discovery of the new route to the East round the Cape of Good Hope, the fall of Granada, the expulsion of the Jews from the Iberian peninsula and the massive migrations of exiled peoples that these changes set in motion [65], the sudden conversion of the brilliance of Renaissance Italy into a savage theatre of war between Habsburg Spain and Valois France, and the steady spread of Lutheranism.

One other relevant observation. Notwithstanding its loss of economic primacy in the world, the sixteenth-century Mediterranean retained its explosive civilising power and remained as inspiring as ever before. By the time the Order was about to settle on Malta, 'the curtain,' we are told [27; p. 828], had already fallen 'on the splendours of the Renaissance'. It soon became the cradle of the volcanic eruption of the 'sprawling and extravagant' Baroque [27; p.827]. From affluent Italy, it flourished rapidly for centuries, with the Spanish galleons carrying it to the newly discovered lands across the Atlantic. 'The Mediterranean,' asserts Braudel, 'was the donor, the transmitter and therefore a superior power, whose teachings, way of life and tastes were adopted in lands far from its shores' [27; p. 829]. The Baroque, an expression of the vitality of Italian society, demonstrated that Italy, at the core of the Mediterranean, had once more transformed itself into a 'competitive and predominant' civilising force, whose dynamic influence 'spread to cover almost all of Europe, the Catholic and indirectly the Protestant as well: curious evidence of the unity of a world divided' [66]. Whether this civilising overflow of exuberance, which ushered in a new age, was indeed 'a sign of ... economic failure' [27; p. 900] is debatable.

W. V. Harris draws up a list of other 'essential elements' which should be reconsidered [62; pp. 11–20]. The unchangeableness, or 'immobility' as he calls it [62; pp. 9–10], of certain Mediterranean attributes can be seen, by way of a tiny isolated illustration drawn from the island of Malta. Today, the colourful Maltese *luzzu* (or fishing boat) still sails elegantly in and out of Marsamxett harbour in the south of Malta, and several fishermen still cast their nets in the blue Mediterranean and have them spread out on the quay to dry, to mend, to have them cleared of weeds, much as their ancestors had done hundreds of years before them. The majestic Dingle Cliffs, over 250 meters above sea level, on the island's western coast, have hardly changed from prehistoric times, except perhaps as a result of natural erosion. These are classic symptoms of Braudelian timelessness: 'permanent', 'unchanging' and 'motionless' features, the 'constants' of the past. But Malta may perhaps be too small an island to demonstrate convincingly the vast recurring realities of the Mediterranean with an estimated total population ranging from 68 to 92 million between 1600 and 1800 [61; pp. 100–101]. An indefinite host of similar instances can be safely, and perhaps more convincingly, sought at Sicilian and southern Italian ports, in southern France and Spain, along the Adriatic, and on Greek islands.

Within this context of near-permanent structures, life did not stand still. The Mediterranean region would have stagnated, it would have been transformed into another 'dead sea', had there been no change at all. Life in the Mediterranean changed, for, after all, is not change the quintessence of history? But change and decline are not synonymous concepts. Only a persistent process of change for the worse over the long perspective can define the idea of decline accurately. In the midst of this historical process of change, there are always opposing forces of resistance, of continuity: they confront and delay the whole process of restructuring, conversion, or indeed decline and therefore slow down the whole course of development for better or for worse. The Hospitaller Order of St John had been one such living force. In its dynamism, and its traditional operations first from Rhodes and then from Malta, it helped preserve, as has been shown above, some of the basic features of early modern Mediterranean culture, lifestyle and society. When Buzzaccarini Gonzaga had first set foot on Malta as Venice's first resident minister to the Grand Master's court in the early 1750s, he observed that the island order-state was playing a vital role for the entire Christendom. Shortly before he passed away in 1776, he was still convinced that the Order was performing its mission in protecting Europe against the enemy of the Christian faith consistently and admirably [46, *passim*]. Despite this strong element of continuity, the Hospitaller institution, points out Anthony Luttrell, 'did not embody a decayed medieval ideal being lived out in a state of terminal anachronism' [2; p. 355].

Author details

Victor Mallia-Milanes

Address all correspondence to: victor.mallia-milanes@um.edu.mt

University of Malta, Msida, Malta

References

[1] Braudel F. Venezia (trans. Gemelli G.) Bologna: Il Mulino; 2013. pp. 57-58 (The French original, Venise. Paris: Flammarion, Arthaud; 1984)

[2] Luttrell A. Military orders 1312-1798. In: Riley-Smith J, editor. The Oxford History of the Crusades. Oxford: OUP; 1999. pp. 323-362: 334

[3] Riley-Smith J. Towards a history of military-religious orders. In: Borchardt K, Jaspert N, Nicholson HJ, editors. The Hospitallers, the Mediterranean and Europe: Festschrift for Anthony Luttrell. Aldershot: Ashgate; 2007. pp. 269-284: 276

[4] Schermerhorn EW. On the Trail of the Eight-Pointed Cross. New York: G.P. Putnam's Sons; 1940. pp. 10-11

[5] Hunyadi Z. Hospitaller estate management in the medieval Kingdom of Hungary (thirteenth to fourteenth centuries). In: Upton-Ward J, editor. The Military Orders, Volume 4: On Land and by Sea. Aldershot: Ashgate; 2008. pp. 145-153

[6] Couriers N. Hospitaller estates and agricultural production on fourteenth- and fifteenth-century Cyprus. In: Buttigieg E, Phillips S, editors. Islands and Military Orders, c.1291–c.1798. Surrey: Ashgate; 2013. pp. 215-224

[7] Sire HJA. The Knights of Malta: A Modern Resurrection. Profile Books. London: 2016

[8] Reston Jr J. Defenders of the Faith: Christianity and Islam Battle for the Soul of Europe 1520-1536. Harmondsworth: Penguin Books; 2009. p. 365

[9] Mallia-Milanes V. Introduction to Hospitaller Malta. In: Mallia-Milanes V, editor. Hospitaller Malta, 1530-1798: Studies on Early Modern Malta and the Order of St John of Jerusalem. Malta: Mireva Publications; 1993. p. 2

[10] Fulin R, et al. editors. I Diarii di Marino Sanuto. Visentini F, Vol. xxxiv. Venice: 1879-1903. p. 98

[11] Mallia-Milanes V. A trial of Hospitaller strength: From Rhodes to Malta via Nice and Villefranche, 1523-1530. In: La Navigation du Savoir: Réseau des arsenaux historiques de la Méditerranée. Actes du Colloque: Premières Journées internationales du patrimoine maritime méditerranéen, 27 au 30 mars 2003, Villefanche-sur-Mer. Nice; Malta International Foundation, 2006. pp. 73-81

[12] Cassar P. Medical History of Malta. London: Wellcome Historical Medical Library; 1964. pp. 164-175

[13] Mallia-Milanes V. Venice, Hospitaller Malta, and fear of the Plague: Culturally conflicting views. In: Shenk J, Carr M, editors. The Military Orders. Vol. 6.1: Culture and Conflict in the Mediterranean World. London: Routledge; 2017. pp. 197-198

[14] Luttrell A, The Hospitallers' medical tradition: 1291-1530. In: Barber M, editor. The Military Orders, Vol. I. Fighting for the Faith and Caring for the Sick. Aldershot: Variorum/Ashgate; 1994. pp. 64-81

[15] Mallia-Milanes V. The Knights Hospitaller's service of love. In: Contu M, et al., editors. Tra Fede e Storia: Studi in onore di Don Giovannino Pinna. Cagliari: AIPSA Edizioni; 2014. pp. 163-175

[16] Kedar BZ. A twelfth-century description of the Jerusalem hospital. In: Nicholson H, editor. The Military Orders. Vol. 2. Welfare and Warfare. Aldershot: Ashgate; 1998. pp. 3-26

[17] Riley-Smith J. 900 Years of the sovereign military order of Malta. In: *History Today*. November, 1999. pp. 2-3

[18] Dal Pozzo B. Historia della S. Religione Militare di S. Giovanni Gerosolimitano detta di Malta. Verona: Gerolamo Albrizzi, 1703-1715. pp. 582-583

[19] The Diary of Henry Teonge, Chaplain on Board His Majesty s Ships Assistance, Bristol, and Royal Oak, Anno 1675 to 1679. London: Charles Knight; 1927

[20] George Sandys. Travels. London: J. Williams Junior; 1670

[21] Campbell J. The travels and adventures of Edward Brown. London: Appleby; 1739

[22] Mallia-Milanes V. Decline and fall? The Order of the Hospital and its Surrender of Malta, 1798. Symposia Melitensia. 2016;**12**:129

[23] Luttrell A. The hospital's privilege of 1113: Texts and contexts. In: Schenk J, Carr M, editors. The Military Orders. Vol. 6.1. Culture and Conflict in the Mediterranean World. London: Routledge; 2017. p. 3

[24] Mallia-Milanes V. L'Ordine dell'Ospedale e le spedizioni antislamiche della Spagna nel Mediterraneo. Dal primo assedio di Rodi (1480) all'assedio di Malta (1565). In: Anatra B, Murgia G, editors. Sardegna, Spagna e Mediterraneo: Dai Re Cattolici al Secolo d'Oro. Rome: Cardocci; 2004. p. 112

[25] Setton KM. The Papacy and the Levant (1204-1571). Vol. iii. The Sixteenth Century to the Reign of Julius III. Philadelphia: The American Philosophical Society; 1984. pp. 393-398

[26] Bosio I. Dell'Istoria della Sacra Religione et Ill.ma Militia di San Giovanni Gierosolimitano. 3rd Impression. Venice: Girolamo Albrizzi; 1695. p. 141 (Volume iii)

[27] Braudel F. The Mediterranean and the Mediterranean World in the Age of Philip II. (trans. Reynolds S.) London: Collins; 1972-1973. p. 873

[28] Inalcik H. The Ottoman Empire: The Classical Age 1300-1600. London: Phoenix; 2000. p. 36

[29] Mercieca S. The Battle of Preveza 1538: The knights of Malta's perspective. In: Karabelas ND, Stork M, editors. Proceedings of the Second International Symposium for the History and Culture of Preveza (16-20 September 2009). University of Ioannina; 2010. pp. 107-120

[30] Mori Ubaldini U. La Marina del Sovrano Militare Ordine di San Giovanni di Gerusalemme di Rodi e di Malta. Rome: Regionale Editrice; 1971. p. 159

[31] Wolfe JB. The Barbary Coast: Algeria under the Turks 1500-1830. London: Norton; 1979. pp. 34-35

[32] Bono S. Naval exploits and privateering. In: Mallia-Milanes V, editor. Hospitaller Malta 1530-1798: Studies on Early Modern Malta and the Order of St John of Jerusalem. Malta: Mireva Publications; 1993. p. 357

[33] Wettinger G. Slavery in the Islands of Malta and Gozo ca. 1000-1812. Malta: Publishers Enterprises Group; 2002

[34] Bono S. Schiavi musulmani nell'Italia moderna: galeotti, vù cumprà, domestici. Naples: Edizioni scientifiche Italiane; 1999

[35] Earle P. Corsairs of Malta and Barbary. Sedgwick & Jackson, London: 1970

[36] Bono S. I Corsari Barbareschi. RAI Radiotelevisione Italiana, Turin: 1964

[37] Fontenay M. Corsaires de la foi ou rentiers du sol?: Les chevaliers de Malte dans le *corso* méditerranéen au XVIIe siècle. In: Revue d'Histoire Moderne et Contemporaine, xxxv (July–September 1988). pp. 361-384

[38] Valerian D. The Medieval Mediterranean. In: Horden P, Kinoshita S, editors. A Companion to Mediterranean History. Chichester: Wiley Blackwell; 2014. pp. 77-78

[39] Greene M. The Early Modern Mediterranean. In: In: Horden P, Kinoshita S, editors. A Companion to Mediterranean History. Chichester: Wiley Blackwell; 2014. p. 92

[40] Tenenti A. Piracy and the Decline of Venice 1580-1615. (trans. Pullan J, and Pullan B.) London: Longmans; 1967. p. 39

[41] Dursteler ER. Venetians in Constantinople: Nation, Identity and Coexistence in the Early Modern Mediterranean. Baltimore: The John Hopkins University Press; 2006. p. 11

[42] Vann TM. The role of rhetoric and diplomacy in the creation of Muslim identity in fifteenth-century Rhodes. In: Watkins J, Reyerson KL, editors. Mediterranean Identities in the Premodern Era: Entrepôts, Islands, Empires. Surrey: Ashgate; 2014. pp. 109-119

[43] Mallia-Milanes V. Introduction to Hospitaller Malta. In: Mallia-Milanes V. editor. Hospitaller Malta 1530-1798: Studies of Early Modern Malta and the Order of St John of Jerusalem. Malta: Mireva Publications; 1993. pp. 1-42, *passim.*

[44] Mallia-Milanes V. Society and the Economy on the Hospitaller Island of Malta: An Overview. In: Buttigieg S, Phillips S, editors. Islands and Military Orders, c.1291–c.1798. Surrey: Ashgate; 2013. pp. 238-256

[45] Mallia-Milanes V. Al servizio della Repubblica di Venezia: Le Lettere di Massimiliano Buzzaccarini Gonzaga, Commendatore di Malta, inviate alla Magistratura dei Cinque Savii alla Mercanzia 1754-1776. Vatican City: Libreria Editrice Vaticana; 2008. Letters CCXII–CCXIV, pp. 562-566

[46] Mallia-Milanes V. A man with a mission: A Venetian Hospitaller on Eighteenth-Century Malta. In: Upton-Ward J, editor. The Military Orders. Vol. 4. On Land and at Sea. Aldershot: Ashgate; 2008. p. 255

[47] Mallia-Milanes V. Venice and Hospitaller Malta 1530-1798: Aspects of a Relationship. Malta: Publishers Enterprises Group; 1992. pp. 212, 214-218

[48] Huntington SP. The clash of civilisations? Foreign Affairs. 1993;**72**:22-49

[49] Imber C. The Ottoman Empire, 1300-1650: The Structure of Power. Basingstoke: Palsgrave Macmillan; 2002. p. 60

[50] Dandelet TJ. The Renaissance of Empire in Early Modern Europe. New York: Cambridge University Press; 2014. p. 156

[51] Elliott JH. Europe Divided 1559-1598. London: Collins The Fontana History of Europe; 1968. p. 178

[52] Today in Mediterranean History: The Battle of Lepanto (1571) [Internet]. 2015. Available from: https//attwiw.com [Accessed: 9 September 2016]

[53] Lane FC. Venice: A Maritime Republic. Baltimore: John Hopkins University; 1973. pp. 370, 372

[54] Melleuish GC. The significance of Lepanto. Quadrant Online. 2008. https://quadrant.org.au/magazine/2008/04/the-significance-of-lepanto/ [Accessed: 20 March 2017]

[55] Mitchell JR, Mitchell HB, editors. Annual Editions World History: Volume 1-Prehistory to 1500. 10th ed. New York: McGraw-Hill; 2009

[56] De Groot AH. The Ottoman threat to Europe, 1571-1830: Historical fact or fancy? In: Mallia-Milanes V, editor. Hospitaller Malta 1530-1798: Studies on Early Modern Malta and the Order of St John of Jerusalem. Malta: Mireva Publications; 1993. p. 254

[57] Shaw S.J. History of the Ottoman Empire and Modern Turkey, i: Empire of the Gazis. The Rise and Decline of the Ottoman Empire, 1280-1808. Cambridge: CUP; 1976. pp. 178-179

[58] Bonaffini G. La Sicilia e i Barbereschi: Incursioni Corsare e Riscatto Degli Schiavi (1570-1606). Palermo: ila palma, 1983

[59] Setton KM. Venice, Austria, and the Turks in the Seventeenth Century. Philadelphia: The American Philosophical Society; 1991. p. 1

[60] The Works of M. de Voltaire. The Ancient and Modern History. Ch. CXXXII: Of the Battle of Lepanto. Newbery et al, p. 262

[61] Pamuk S. Braudel's Eastern Mediterranean revisited. In: Piterberg G, Ruiz TF, Symcox G, editors. Braudel Revisited: The Mediterranean World 1600-1800. Toronto: University of Toronto Press; 2010. pp. 101-108

[62] Harris WV. The Mediterranean and ancient history. In: Harris WV. editor. Rethinking the Mediterranean. Oxford: OUP; 2006. p. 4

[63] Purcell N. The Ancient Mediterranean: The view from the customs house. In: Harris WV, editor. Rethinking the Mediterranean. Oxford: OUP; 2006. p. 218

http://dx.doi.org/10.5772/intechopen.69376

[64] Varlik N. Plague and Empire in the Early Modern Mediterranean World: The Ottoman Experience, 1347-1600. New York: CUP; 2015

[65] Ray JS. After Expulsion: 1492 and the Making of Sephardic Jewry. New York/London: NY University Press; 2013

[66] Braudel F. Out of Italy: 1450-1650. (trans. Reynolds S.) Paris: Flammarion; 1989. p. 12

Chapter 3

Nutrient Cycling in the Mediterranean Sea: The Key to Understanding How the Unique Marine Ecosystem Functions and Responds to Anthropogenic Pressures[1]

Helen R. Powley, Philippe Van Cappellen and
Michael D. Krom

Additional information is available at the end of the chapter

http://dx.doi.org/10.5772/intechopen.70878

Abstract

The Mediterranean Sea is a marine desert: although it receives large nutrient inputs from a rapidly growing coastal population, its offshore waters exhibit extremely low biological productivity. Here, we use a mass balance modelling approach to analyse the sources and fate of the two main nutrients that support marine biomass production: phosphorus (P) and nitrogen (N). Surprisingly, the main source of P and N to the Mediterranean Sea is North Atlantic surface water entering through the Strait of Gibraltar, not emissions from surrounding land. The low biological productivity of the Mediterranean Sea is linked to the switch from less bioavailable nutrients entering the basin to highly bioavailable nutrients leaving it although similar amounts of total P and N enter and leave the Mediterranean Sea. This unique feature is a direct consequence of its unusual anti-estuarine circulation. An important environmental implication of the anti-estuarine circulation is that it efficiently removes excess anthropogenic nutrients entering the Mediterranean Sea, thus protecting offshore waters against eutrophication contrary to other semi-enclosed marine basins. In a similar vein, the "self-cleaning" nature of the Mediterranean Sea may prevent severe oxygen depletion of Mediterranean deep waters should ongoing climate warming lead to a weakening of the thermohaline circulation.

Keywords: Mediterranean Sea, nutrients, phosphorus, nitrogen, mass balance modelling, circulation, climate and environmental change, deep-water oxygenation

[1]Words or phrases marked for the first time in ***bold italics*** are defined in the glossary.

1. Introduction and background

The Mediterranean Sea (**Figure 1**) has been described as the cradle of western civilization. It is a sea of far greater importance to humanity than would be inferred from its size, as illustrated in the other chapters of this book and elsewhere. In modern times the Mediterranean Sea has become the receptacle of rising nutrient wastes from the surrounding countries. When discharged into coastal waters and inland seas, compounds containing phosphorus (P) and nitrogen (N), two chemical elements collectively known as plant macronutrients, enhance the growth of ***phytoplankton***. The synthesis of new biomass by phytoplankton is termed ***primary production***. Excessive phytoplankton growth leads to many and varied problems associated with ***eutrophication***, including toxic algal blooms, dead ***hypoxic*** areas and, sometimes, major fish kills. In the Mediterranean Sea, nutrient discharges are causing problems locally, such as in the North Adriatic Sea, Venice Lagoon or lagoons in the Nile Delta. However, unlike other semi-enclosed seas, there is little evidence for increasing primary production in the offshore waters of the Mediterranean Sea. In fact, the vivid blue waters of the Mediterranean Sea reflect the very low biomass of phytoplankton (the 'grass' of the world's oceans) making it as much a desert as the Sahara to the south (**Figure 2**). A key question, which we address in this chapter, is: why does primary production in the Mediterranean Sea remain one of the lowest of the world's ocean despite the high inputs of excess anthropogenic P and N? It is well known that the primary productivity of marine surface waters is controlled by the supply of macronutrients (i.e. P and N) plus the amount of light. The Mediterranean Sea has abundant sunlight, and the land-derived nutrient supply is similar to that of the Baltic Sea where problems of eutrophication, including nuisance algal blooms, are common. What makes the Mediterranean Sea special?

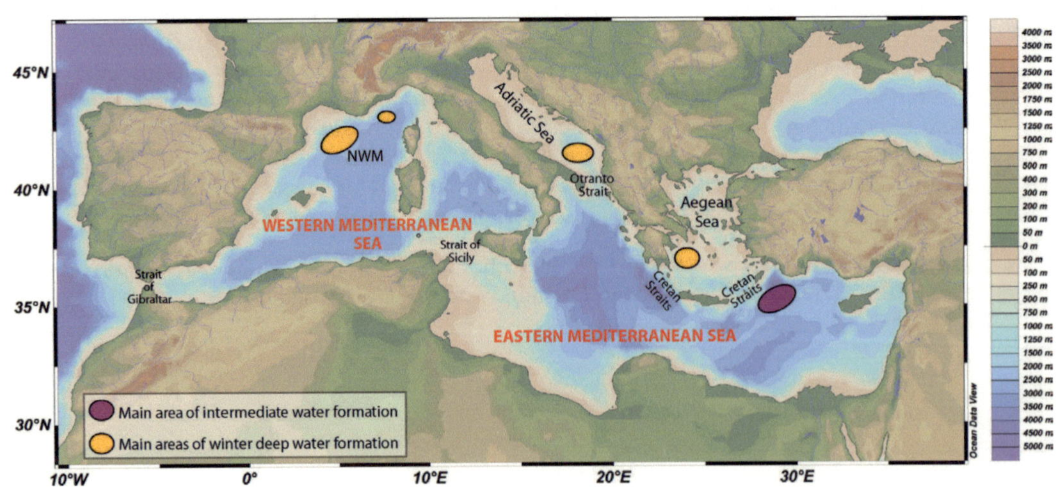

Figure 1. Map of the Mediterranean Sea, including the major areas of formation of intermediate and deep waters. After de Madron et al. [1]. NWM = north-west Mediterranean Sea. This figure is available in colour in the online version of this chapter.

http://dx.doi.org/10.5772/intechopen.70878

The work presented in this chapter is based on mass balance model calculations that quantitatively describe the sources, fate and transport of the nutrient elements P and N in the Mediterranean Sea. The calculations help explain not only why the Mediterranean Sea is a marine desert but also why the nutrient distributions differ between the Eastern and Western Mediterranean Seas. In addition, we explore how the Mediterranean Sea has responded to the large increases in anthropogenic nutrient inputs since the middle of the last century and whether changes in circulation due to climate change have the potential to lead to a depletion of the dissolved oxygen (O_2) levels in the deep waters of this semi-enclosed marine basin. As far as possible, the text explains briefly the major ocean processes occurring in ways that, hopefully, make the text accessible to readers with a non-expert knowledge of oceanography and marine science.

1.1. The Gaia hypothesis: how marine phytoplankton control their chemical environment

In 1979, James Lovelock in his landmark book *Gaia: A New Look at Life on Earth* advocated a new way of looking at the planetary ecosystem, which he called the Gaia hypothesis. The Gaia hypothesis states that living organisms and nonliving natural compounds are part of a highly coupled, dynamic system that shapes and regulates the earth's environment. It is this interactive system that maintains the Earth as a planet where life as we know it can exist. Lovelock developed his theory when comparing the chemical composition of the atmospheres on Earth and on Mars and asking the question: Why are they so different? In particular, why is there so much free O_2 on Earth and so little on Mars? His answer was that plants produce molecular O_2, which then accumulates and in turn is essential for all living animals to breathe. Overall, the production and consumption of O_2 closely track each other, hence allowing the atmospheric O_2 level to remain relatively stable over time. When this balance is severely perturbed, atmospheric O_2 may quickly collapse as probably happened on Mars. The Gaia theory has been central to much of the way we look at the mutual interactions between biological processes and their physical and chemical environment. Marine scientists have been looking at

Figure 2. The clear blue waters of the Mediterranean Sea reflect the very low biological productivity of its surface waters compared to most of the world's oceans. Thus, the Mediterranean Sea is as much of a 'desert' as the Sahara Desert to the south (photographs taken by G. Rilov and by N. Landsberg). This figure is available in colour in the online version of this chapter.

the ocean in a similar way for a long time: the so-called Redfield ratio is in effect Gaia as applied to the world's oceans.

What is the Redfield ratio? Look at the chemical equation in **Figure 3**: it represents the synthesis of new biomass by marine phytoplankton. The equation shows that for every 106 carbon (C) atoms that are incorporated in new biomass on average, 16 atoms of N and 1 atom of P are utilized. In 1936, Alfred Redfield, a marine scientist working at Harvard University, noticed that the dissolved nitrate-to-dissolved phosphate ratio in deep ocean waters in many parts of the globe is close to 16:1 (in units of mol/mol) that is the same value as the average N/P ratio of marine phytoplankton. He suggested that this surprising convergence reflects a biological control over the distribution of vital nutrient elements in seawater [2].

Marine phytoplankton produce biomass with an atomic (or molar) N/P ratio of 16:1. After death, the plankton biomass settles as biological debris towards the bottom of the oceans. The degradation of this debris, primarily by microorganisms, then releases the N and P as dissolved nitrate and phosphate to the water column in a 16:1 proportion. Upward mixing and ***upwelling*** return the dissolved inorganic nutrients to the surface waters where they are used again for primary production, and so the cycling of the N and P continues at the beat of the Redfield ratio. Additional biological processes further balance any departure of the supply of the inorganic nutrients from the required 16:1 ratio. When N is in short supply, ***nitrogen fixation*** transforms dinitrogen gas (N_2) from the atmosphere into chemical forms that can be used by phytoplankton, while excess dissolved nitrate is removed by the process of ***denitrification***. Given sufficient time, the balance between the various processes ultimately drives the nitrate to phosphate ratio of seawater towards the Redfield value of 16:1 [3].

1.2. On the uniqueness of the Mediterranean Sea

One of the unique features of the Mediterranean Sea is that the dissolved P and N concentrations in the water column do NOT obey the Redfield ratio. The dissolved nitrate-to-phosphate ratio of the deep waters is around 23:1 in the Western Mediterranean Sea and 28:1 in the Eastern Mediterranean Sea ([4]; **Figure 4**). Similarly, the ratios of dissolved organic N to dissolved organic P (DON/DOP) and particulate organic N to particulate organic P (PON/

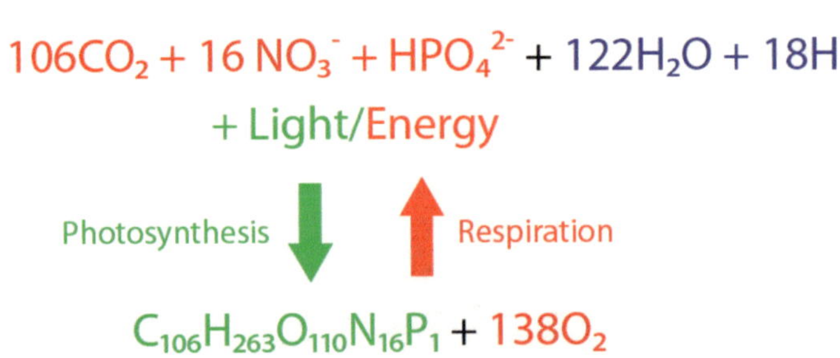

Figure 3. Net reaction showing the utilization of inorganic chemical ingredients by marine phytoplankton as they carry out photosynthesis and produce new biomass. The reverse reaction is respiration in which organisms use organic substrates as food to produce energy. Note that molecular oxygen (O_2) is generated during photosynthesis and consumed during respiration.

http://dx.doi.org/10.5772/intechopen.70878

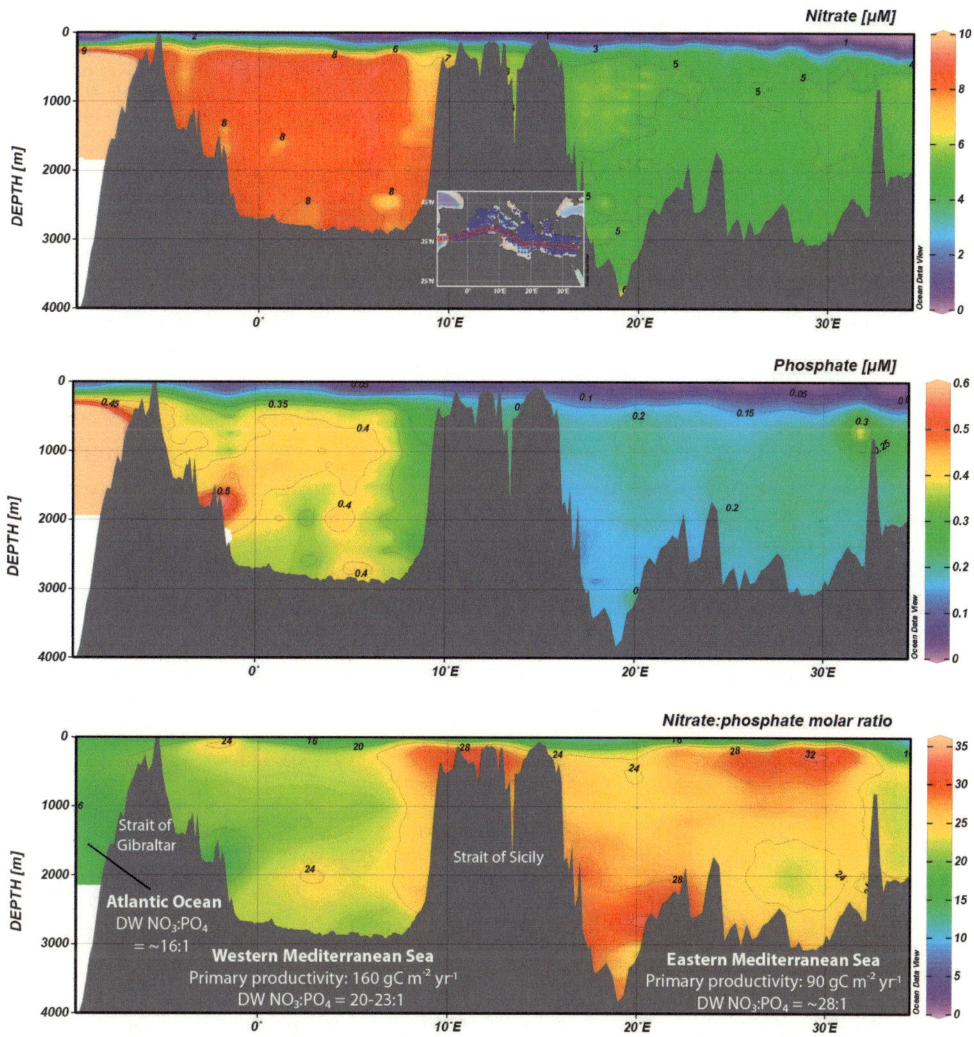

Figure 4. Distributions of dissolved nitrate and phosphate concentrations and the dissolved nitrate to dissolved phosphate molar ratio, in the water column of the North Atlantic Ocean, Western Mediterranean Sea and Eastern Mediterranean Sea. Note the differences in scales between the two uppermost panels. The figures are generated in Ocean Data View [11], using the data in Lavezza et al. [12]. This figure is available in colour in the online version of this chapter.

POP) typically exceed 16:1 in Mediterranean seawater [5–9]. In other words, compared to the world's ocean, the Mediterranean Sea is chronically short of P: it is in fact P-starved. One outcome of this highly unusual situation is that many bacteria and phytoplankton that live in the Mediterranean Sea have adapted to survive under severe P limitation [10].

Phytoplankton biomass in the Mediterranean Sea is also extremely low. A commonly used way to determine the phytoplankton abundance is to measure the amount of chlorophyll, the green pigment that plants use to capture light energy during photosynthesis. Compared with other semi-enclosed seas adjacent to the European continent, such as the Baltic Sea or North

Sea, the summer chlorophyll levels of the Mediterranean Sea are an order of magnitude lower (**Figure 5**). The chlorophyll concentrations in many European and other semi-enclosed seas often reach levels where eutrophication starts having harmful ecological impacts. With the exception of nearshore areas, this is not observed in the Mediterranean Sea.

The low phytoplankton biomass matches the very low primary production observed in the open waters of the Mediterranean Sea: in fact, offshore primary production is one of the lowest observed in the global ocean. Values reported for the Eastern Mediterranean Sea, 10–143 g C m^{-2} y^{-1} ([14]; **Table 1**), are even lower than those in the centre of large ***ocean gyres***, such as the Sargasso Sea, where values are around 150 g C m^{-2} y^{-1}. For comparison, the average primary productivity of the coastal ocean is 350 g C m^{-2} y^{-1}, while that of upwelling zones, some the most productive areas of the ocean typically exhibit values in excess of 500 g C m^{-2} y^{-1}. Bodies of water with low primary production are called ***oligotrophic***. Examples of oligotrophic systems are ocean gyres and the Western Mediterranean Sea. The Eastern Mediterranean Sea, with a primary productivity that is about three times lower than that of the Western Mediterranean Sea, is referred to as ultra-oligotrophic.

Ultimately, the low biological productivity of the Mediterranean Sea is due to the low availability of inorganic nutrients (in particular, dissolved phosphate, nitrate and ammonia) that are needed for phytoplankton to grow (**Figure 3**). The vertical distributions of dissolved inorganic nutrients in the oceans exhibit the lowest concentrations in the surface waters where

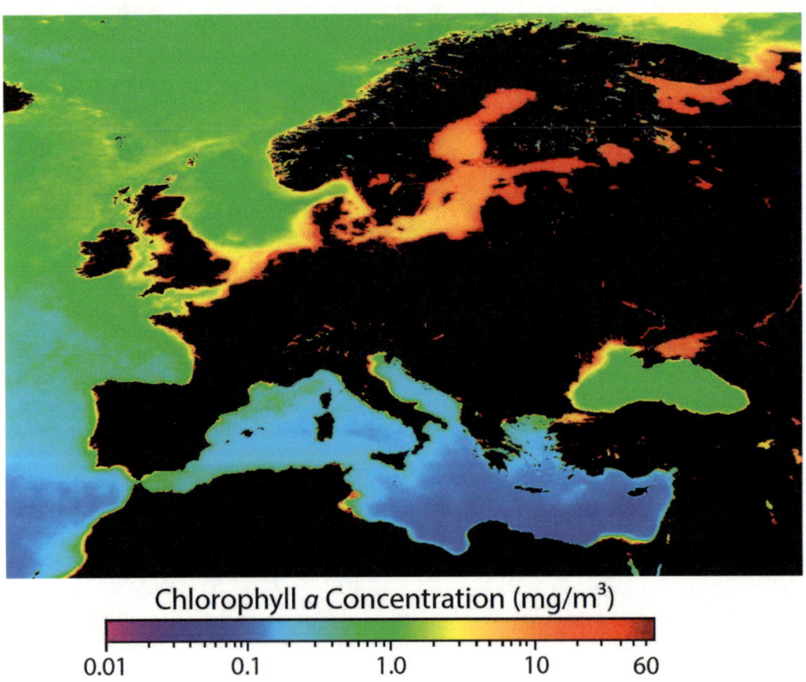

Figure 5. Summer surface water chlorophyll showing low concentrations in the Mediterranean Sea compared to other inland and coastal European waters. The concentrations displayed are average values recorded between July 2002 and December 2004. From NASA [13]. This figure is available in colour in the online version of this chapter.

Area of the ocean	Primary productivity (gC m^{-2} y^{-1})	Reference
Eastern Mediterranean Sea	10–143	[14]
Western Mediterranean Sea	37–475	[14]
Open ocean (average)	75	[15]
Continental shelves (average)	300	[15]
Upwelling regions (average)	500	[15]

Primary productivity is expressed in units of mass of carbon fixed in photosynthesis, per unit sea surface area per unit of time.

Table 1. Average primary productivity in the Eastern and Western Mediterranean Seas compared with average values for the global ocean.

the nutrients are actively taken up by phytoplankton. Below the depth of light penetration, the concentrations of dissolved inorganic P and N increase as they are regenerated by the breakdown of settling biological debris (i.e., ***respiration***). This pattern is observed throughout the world's oceans. However, the absolute concentrations of dissolved inorganic P and N in deep-water masses vary substantially from location to location. In the global ocean, the lowest deep-water concentrations of dissolved inorganic P and N are found in the North Atlantic Ocean, where deep water is newly formed by ***downwelling*** of nutrient-poor surface water. The deep waters subsequently accumulate dissolved inorganic nutrients as they travel through the South Atlantic Ocean and into the Indian Ocean and Southern Pacific Ocean. The highest dissolved inorganic P and N concentrations are therefore at the end of the so-called global conveyor belt in the Northern Pacific (**Table 2**). As can be seen, the concentrations of dissolved inorganic P and N in the waters of the major deep sea basins are systematically higher than those found in the Western Mediterranean Sea, which themselves are higher than those in the Eastern Mediterranean Sea (see also **Figure 4**).

Key to interpreting the unique features of the Mediterranean Sea discussed above is its ***anti-estuarine circulation*** (**Figure 6**). North Atlantic surface water flows into the Western Mediterranean Sea via the Strait of Gibraltar. This surface water flows east and gets progressively warmer and more saline because of the Mediterranean climate that, particularly in summer, is hot and dry. The modified Atlantic surface water enters the Eastern Mediterranean Sea through the Strait of Sicily and continues to flow eastwards. Eventually, it turns northwards until it reaches the southern coast of Turkey where, during the winter, it cools sufficiently to become denser than the surrounding water, which causes downwelling and formation of new intermediate water. This intermediate water then flows westwards and back out of the Eastern Mediterranean Sea through the Strait of Sicily, eventually reaching the Strait of Gibraltar and returning Mediterranean seawater to the North Atlantic Ocean. Thus, at both straits, the eastward surface inflow is balanced by a deeper, more saline and denser westward outflow. The only other major water body exhibiting a similar anti-estuarine circulation is the Red Sea, which is also very oligotrophic. Most semi-enclosed marine basins (e.g. the Baltic Sea) have an estuarine circulation with surface (less saline) water flowing out into the open ocean and deeper (more saline) seawater flowing land inwards. The implications of

Nutrient	Eastern Mediterranean Sea	Western Mediterranean Sea	North Atlantic Ocean	North-West Pacific Ocean
	μmol l^{-1}			
Nitrate	6	9	16	50
Phosphate	0.25	0.4	1	3
Silicic acid	6–12	10–32	20	160

Note that silicic acid, or dissolved silica, is essential to the growth of an important group of phytoplankton called diatoms. The concentration values are given in units of mircomoles per litre seawater where micro (μ) stands for 10^{-6}.

Table 2. Typical concentrations of dissolved inorganic nutrients in the deep waters of the Eastern and Western Mediterranean Seas compared with concentrations found in the deep waters of the global ocean.

the anti-estuarine circulation for the biogeochemical processes in the Mediterranean Sea are discussed below.

1.3. Mass balance modelling: nutrient cycling in the Mediterranean Sea

Mass balance models, also called box models, provide a simple but powerful tool to analyse complex environmental systems such as the Mediterranean Sea. These models require relatively little data and have low spatial and temporal resolution, yet they can yield some profound insights into the large-scale dynamics of coupled biogeochemical cycles. The starting point for the Mediterranean nutrient model is to determine how much P and N enter and leave the system. Once we then add our knowledge of the main in-system biogeochemical

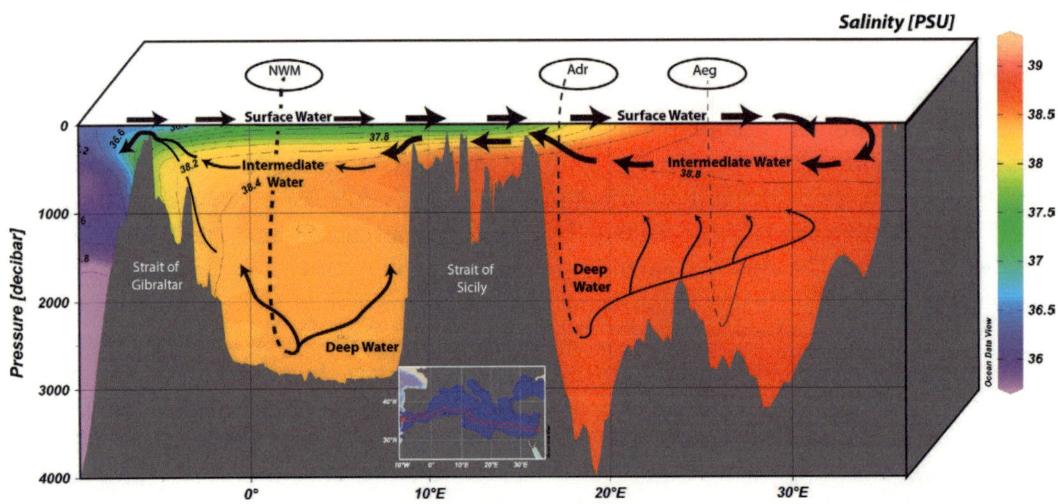

Figure 6. Cross section showing the distribution of salinity (colour) and the general circulation of the Mediterranean Sea (arrows). The thicknesses of the arrows indicate the relative flow rates. Major deep-water formation areas are identified (see also **Figure 1**): NWM = north-west Mediterranean Sea; Adr = Adriatic Sea; Aeg = Aegean Sea. Salinity data from MEDAR Group [16]. Figure generated in Ocean Data View [11]. This figure is available in colour in the online version of this chapter.

processes, it is possible to simulate the general trends of nutrient cycling in the Mediterranean Sea [17] and predict what changes may be expected under the influence of anthropogenic nutrient enrichment and global climate change [18, 19].

In the model, the Western and Eastern Mediterranean Seas are each divided into three layers: surface water, intermediate water and deep water. The water cycle describes the flows between the six water layers, plus the inflow and outflow exchanges with adjacent basins, including the Atlantic Ocean, the Adriatic Sea and the Aegean Sea. Once the water flows are assigned, the nutrient chemistry is added. Both inorganic and organic dissolved forms of P and N are included in the model, as well as particulate organic P and N to represent the nutrients associated with biological material, live and dead. Average concentrations of all the P and N chemical species in the six water layers are derived from measurements made during oceanographic cruises in the Mediterranean Sea. This gives a total of 42 individual chemical P plus N reservoirs or boxes. The transformations of the chemical species from one form to another are represented by dynamic rate expressions. Thus, for example, primary production transforms inorganic dissolved P (phosphate) and N (nitrate) into particulate organic P and N. In the model, the rates at which P and N are used in primary production are coupled to one another by the Redfield ratio.

The P and N model was initialized for conditions corresponding to the year 1950. We chose 1950 because the earliest oceanographic surveys of the Mediterranean Sea date back to the middle of the twentieth century, that is, nutrient time series data are available from that time on. In addition, anthropogenic nutrient inputs from the surrounding countries increased rapidly after 1950, in particular because of the systematic introduction of fertilizers in agriculture. To close the 1950 budgets of P and N, we made a crucial assumption, namely, that all the reservoirs were at steady state in 1950, that is, inputs and outputs to each reservoir were perfectly balanced. The model was then run forwards in time taking into account the temporal changes in P and N inputs due to human activities. The steady-state model simulations for 1950 and the time-dependent post-1950 simulations were used to answer a series of quantitative questions about how the Mediterranean Sea works and why it is such an unusual marine ecosystem.

2. Understanding the unusual properties of the Mediterranean Sea

2.1. Marine-derived P and N inputs exceed those from the surrounding land

A significant result of our analysis is that the main source of P and N to the Mediterranean Sea is the horizontal inflow of surface water from the North Atlantic Ocean through the Strait of Gibraltar (**Figures** 7 and **8**). This is unexpected given that the Mediterranean Sea is almost entirely surrounded by land with correspondingly large nutrient emissions from land-based human activities. The reason lies in the large volume of Atlantic water entering the Western Mediterranean Sea driven by the intense anti-estuarine circulation. According to our estimates, inflowing Atlantic surface water (also referred to as ASW) provides 58% of the total P input to the Western Mediterranean Sea (**Figure** 7) and accounts for 39% of its ***new production***, that is, the primary production supported by external sources of P to the ***photic zone***, rather than supported by recycled nutrients from the degradation of biological materials within the surface waters. In a similar vein, the inflow of Western Mediterranean surface water through the Strait of Sicily represents 77% of the total P input to the Eastern Mediterranean Sea (**Figure 7**) and supports 37% of its new production. In addition, much of the new production, 26 and 37% in the Western and Eastern Mediterranean Seas, respectively, can be attributed to inputs of DOP.

Interestingly, recent research has shown that horizontal (or lateral) nutrient inputs, and in particular those of DOP and DON, are a major part of the nutrient budgets and new production of subtropical oceanic gyres [20]. This, in some respects, means that the Mediterranean Sea behaves more like a subtropical ocean gyre than other semi-enclosed marine basins. While at first this may be surprising, it starts to make sense when we compare the circulation regimes of oceanic gyres and the Mediterranean Sea. In subtropical gyres the Coriolis force, the force created by the spinning of the Earth, forces surface water towards the centre of the gyre where it is subsequently forced downwards before moving outwards. That is, lateral inflow from outside the gyre leads to downwelling inside the gyre, followed by outflow at greater depth. This is similar to the general flow pattern in the Mediterranean Sea, which is also characterized by lateral surface inflow and deeper lateral outflow (**Figure 6**). Note, however, that the forcing mechanisms, the Coriolis force and the anti-estuarine circulation, are quite different.

2.2. Why is the Mediterranean Sea a biological desert?

The most biologically active forms of P (dissolved phosphate) and N (dissolved nitrate and ammonia) are easily assimilated by phytoplankton that form the base of the food web. Most phytoplankton do not die of 'old age', but are eaten by larger organisms, which in turn are eaten by even larger organisms. Eventually, biological material produced in the surface waters drops out of the photic zone either as the faeces of zooplankton or as part of what is called ***marine snow***. The organic matter settles into the intermediate water where much of it is converted back into phosphate and nitrate through ***solubilization*** and ***mineralization***. In addition, some of the dissolved and particulate nutrients present in the surface water are transported down when intermediate water forms in the winter. In contrast with much of the rest of the ocean, very little of the regenerated, and highly bioavailable, dissolved inorganic nutrients in the Mediterranean Sea make it back into the photic zone. There is almost no upwelling of intermediate water to the surface in the Eastern Mediterranean Sea and only some upwelling in the Western Mediterranean Sea. Instead, because of the anti-estuarine circulation, most intermediate water flows out of both the Eastern and Western Mediterranean Seas carrying with it the highly bioavailable nutrients, which are eventually lost to the Atlantic Ocean.

An important finding of the mass balance calculations is that the low productivity of the Mediterranean Sea is not so much caused by a net total loss of nutrients to the Atlantic Ocean, but rather by the differences in chemical forms of P and N entering and leaving the basin. In fact, the inflows of total P and total N to the Mediterranean Sea through the Strait of Gibraltar are of comparable magnitudes as the corresponding outflows of total P and total N with deeper Mediterranean Sea water (**Figures 7** and **8**). However, the P and N exiting the Mediterranean Sea are predominantly in their most biologically active forms, dissolved phosphate and nitrate,

http://dx.doi.org/10.5772/intechopen.70878

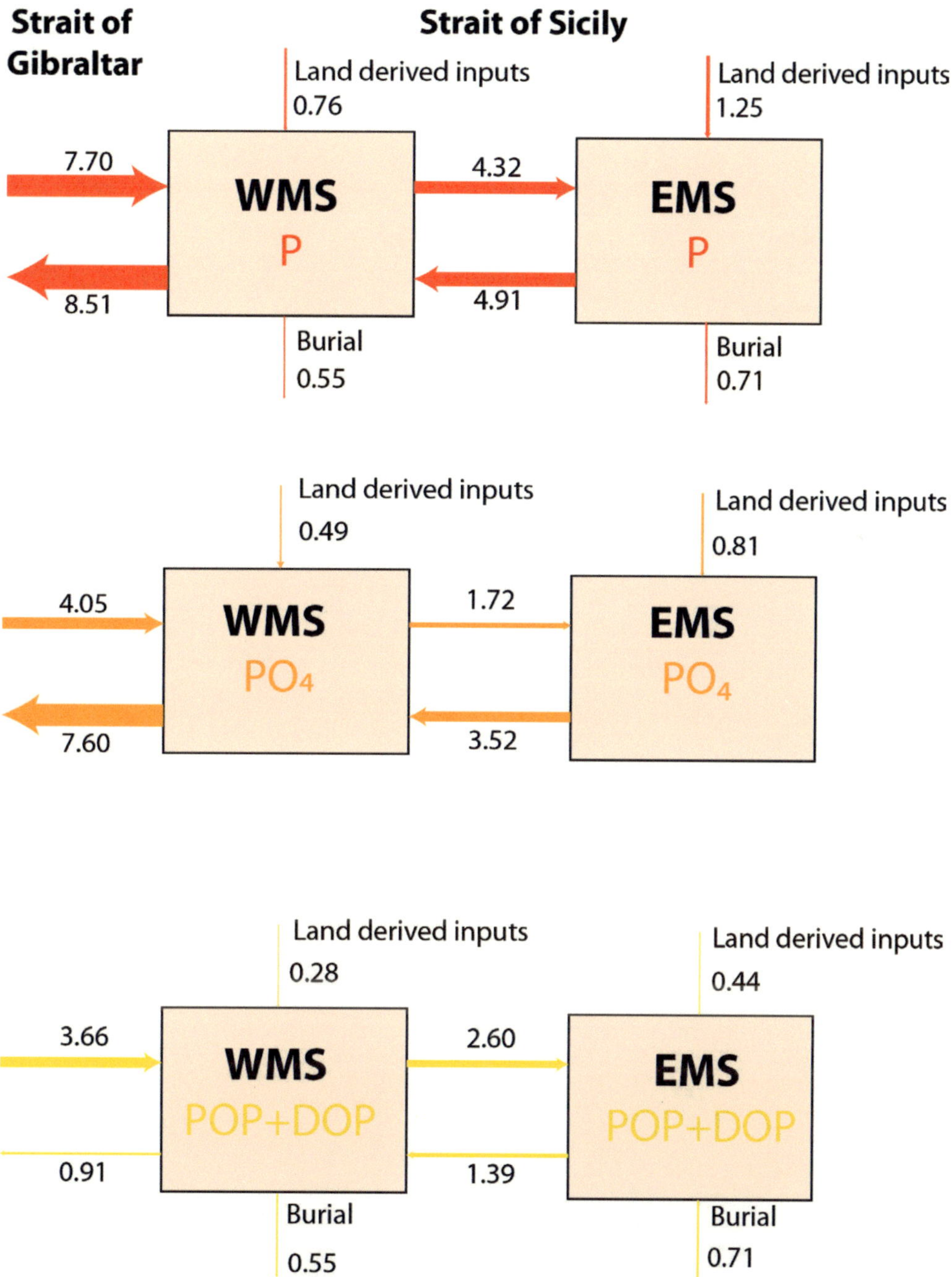

Figure 7. Input and output fluxes of reactive phosphorus (P), dissolved inorganic phosphorus (PO_4) and particulate and dissolved organic phosphorus (POP + DOP) to the Western Mediterranean Sea (WMS) and Eastern Mediterranean Sea (EMS). The thicknesses of the arrows scale to the corresponding fluxes. Numerical values are fluxes in units of 10^9 mol P y^{-1} [17]. Fluxes can be converted to units of mol m^{-2} y^{-1} by dividing them by the surface areas of the WMS and EMS, 815 × 10^9 m^2 and 1336 × 10^9 m^2, respectively.

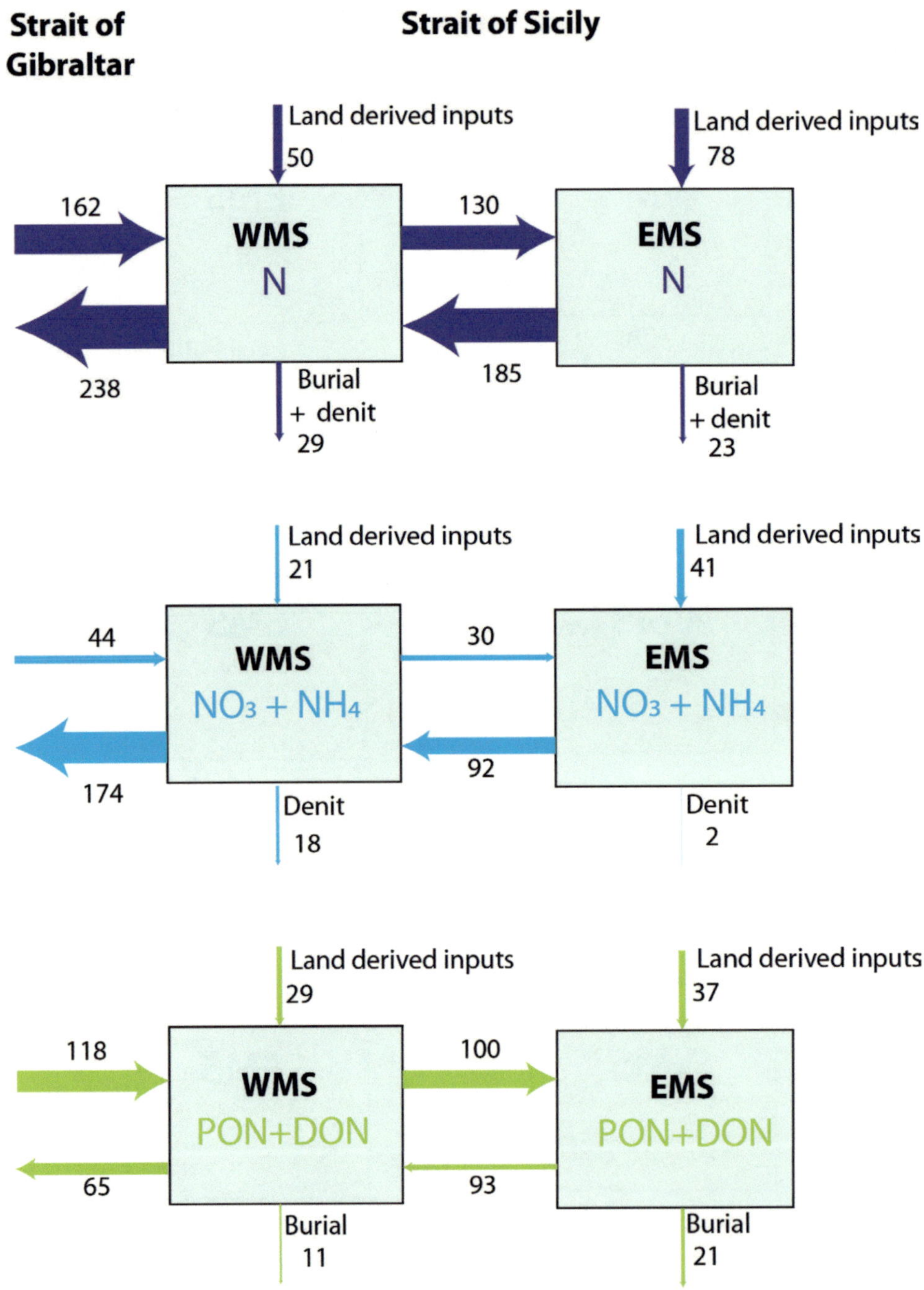

Figure 8. Input and output fluxes of reactive nitrogen (N), dissolved inorganic nitrogen (NO_3 + NH_4) and particulate and dissolved organic nitrogen (PON + DON) to the Western Mediterranean Sea (WMS) and Eastern Mediterranean Sea (EMS). The thicknesses of the arrows scale to the corresponding fluxes. Numerical values are fluxes in units of 10^9 mol N $yr.^{-1}$ [17]. Fluxes can be converted to units of mol m^{-2} y^{-1} by dividing them by the surface areas of the WMS and EMS, 815×10^9 m^2 and 1336×10^9 m^2, respectively.

whereas more than half of P and N entering the Mediterranean Sea are in their less bioavailable forms, DOP and DON. In total, 88% more dissolved phosphate leaves the Mediterranean Sea through the Strait of Gibraltar than flows in with the Atlantic surface water: the outflow of total P, however, is only 10% greater than the inflow of total P (**Figure 7**). A similar trend is observed for N, with a 47% difference in the total dissolved N flux leaving and entering through the Strait of Gibraltar but four times more dissolved nitrate leaving than entering (**Figure 8**). A similar switch in chemical speciation is also observed at the Strait of Sicily (**Figures 7** and **8**). This emphasizes the need to account for the chemical speciation of nutrient elements, in addition to their total concentrations and fluxes.

2.3. Primary productivity and nutrient cycling: west–east differences

The difference in magnitude of marine-derived sources of P and N to the Western and Eastern Mediterranean Seas is the primary reason behind the west to east gradients in primary productivity and phosphate and nitrate concentrations. Inputs of land-derived sources of P and N are rather similar per unit surface area between the Western and Eastern Mediterranean Seas, yet, per unit surface area, P and N inputs are 3.9 and 3.1 times greater, respectively, for the Western Mediterranean Sea (**Figures 7** and **8**). In the Western Mediterranean Sea, P and N inputs from land are added to a background of P and N that originates from the Atlantic Ocean and the deep layers of the Eastern Mediterranean Sea. In contrast, land-derived inputs into the Eastern Mediterranean Sea are added to a background which is depleted in P and N due to the relatively low concentrations of P and N within the Western Mediterranean surface water entering the Eastern Mediterranean Sea through the Strait of Sicily. Thus, the Western Mediterranean Sea exhibits properties that are intermediate between the 'normal' Atlantic waters and the ultra-oligotrophic Eastern Mediterranean Sea waters. In addition to the greater external inputs of P and N to the Western Mediterranean Sea, primary productivity is also supported by upwelling of nutrients from intermediate water into the photic zone, whereas this mechanism is not present in the Eastern Mediterranean Sea. In other words, upwelling provides an additional source of nutrients to the photic zone of the Western Mediterranean Sea compared to the Eastern Mediterranean Sea.

However, higher inputs into the Western Mediterranean Sea do not entirely explain the west to east trends in phosphate and nitrate concentrations observed across the Mediterranean Sea. Despite the much greater dissolved phosphate and nitrate concentrations in the Western Mediterranean Sea, the deep-water DOP and DON concentrations are quite similar in the Western and Eastern Mediterranean Seas. This can be explained by the faster recycling of organic P and organic N back into the dissolved inorganic pools in the Western Mediterranean Sea. Evidence for this faster recycling of organic matter can be seen in the higher population densities of bacteria observed in the Western Mediterranean Sea, which are four to six times greater than in the Eastern Mediterranean Sea, as well as the higher concentrations of alkaline phosphatase, which are about three times greater than in the Eastern Mediterranean Sea [21]. Alkaline phosphatases are a group of enzymes that are designed to break the chemical bonds that hold P locked into organic compounds, thereby releasing dissolved phosphate to seawater. Note that, based on the numbers just provided, there is about twice as much alkaline phosphatase per bacterial cell in the Eastern Mediterranean Sea. The reason is that the Eastern Mediterranean Sea is more P-starved than the Western Mediterranean. Therefore, the microbial community of the Eastern Mediterranean Sea must work harder to access this extremely limiting nutrient. It does so by producing more alkaline phosphatase per cell: this represents a specific adaptation of the microbial community in the Eastern Mediterranean Sea to the very low availability of P.

2.4. Heterotrophy versus autotrophy

In most areas of the ocean close to land, production of organic matter by phytoplankton exceeds its consumption by respiration, because of the abundant supply of land-derived nutrients. These areas are referred to as ***autotrophic***. The excess organic matter is buried in the sediments along the continental margins or exported further offshore into deeper waters. By contrast, many areas of the ocean far from the influence of land are net ***heterotrophic***, which means that, integrated over the entire area, more organic matter is respired than produced. Again, the Mediterranean Sea goes against the grain: although it is completely surrounded by land (with the exception of the Strait of Gibraltar), it is heterotrophic. This can be seen for P and N in **Figures 7** and **8**: more DOP and DON enter the Mediterranean Sea from marine and non-marine sources than are removed with the outflow through the Strait of Gibraltar or burial (**Figures 7** and **8**): more DOP and DON is thus mineralized than produced. It was originally suggested by Duarte et al. [22] that the excess dissolved organic matter sustaining heterotrophy in the Mediterranean Sea is supplied by river runoff. Our estimates, however, shows that most of the DOP and DON is delivered by water inflow from the North Atlantic Ocean.

A key factor controlling how much, and how fast, organic compounds are being consumed by bacteria and animals is their so-called lability. It is not just a matter of how much food there is but also what its quality is (i.e. is it prime rib or old shoe leather?). For example, fresh phytoplankton biomass is high-quality food for bacteria and animals. Higher inputs of relatively labile organic matter, both marine and land derived, explain the higher heterotrophy in the Western Mediterranean Sea than the Eastern Mediterranean Sea. More dissolved organic matter enters the Western Mediterranean Sea via the Strait of Gibraltar than enters the Eastern Mediterranean Sea via the Strait of Sicily. In addition, more phytoplankton is produced in the Western Mediterranean Sea, therefore providing a source of fresh organic matter. In comparison to the Eastern Mediterranean Sea, the Western Mediterranean Sea therefore exhibits higher DOP concentrations in its surface waters (more food), but lower DOP in its deep waters, because the higher lability of organic matter in the Western basin leads to its faster consumption (better food).

The net heterotrophy of the Mediterranean Sea is also modulated by the rather unusual timing of the annual phytoplankton bloom, that is, the period during which phytoplankton growth peaks. In many parts of the temperate ocean, the main phytoplankton bloom occurs in spring. Bioavailable nutrients are mixed into surface waters in winter during the cold and often stormy time of the year, but phytoplankton cannot grow rapidly until the depth of mixing decreases and enough sunlight becomes available to carry out photosynthesis. The latter usually happens in spring after the length of day and the angle of the sun have increased sufficiently. By contrast, in the Mediterranean Sea, the phytoplankton bloom occurs in winter (see **Figure 9**). This is because winter in the Mediterranean Sea is characterized by short periods of cold often windy weather, which causes the surface water to mix and bring up nutrients from below, followed by sunny periods during which mixing temporarily stops and phytoplankton can grow. However, winter is also the time when Mediterranean intermediate and deep waters tend to form. The downwelling surface waters entrain the phytoplankton biomass, hence supplying fresh, and thus labile, organic matter to the bacteria and animals living at greater water depths. Because downwelling occurs when the phytoplankton biomass is at its highest, heterotrophy is relatively more pronounced for the Mediterranean Sea than in areas of the oceans where the phytoplankton bloom takes place in spring.

2.5. Why does the nitrate-to-phosphate ratio of the Mediterranean Sea exceed 16:1?

One of the most remarkable properties of Mediterranean Sea is the significant departure of the deep-water molar nitrate-to-phosphate ratios from the average 16:1 value observed throughout most of the world's ocean ([4, 25]; **Figure 4**). For the Eastern Mediterranean Sea, Krom et al. [26] showed this was due to the combined N/P ratio of the total, non-marine inputs of P and N, which by far exceeds 16:1 (e.g., the ratio of inorganic nutrients supplied through atmospheric deposition is larger than 100:1), and the very limited amount of denitrification. Denitrification is the process in which anaerobic bacteria use nitrate to respire organic matter and produce N_2 gas as a by-product. In the ocean, denitrification is the primary mechanism by which deep-water nitrate-to-phosphate ratios that exceed 16:1 are brought closer to the Redfield ratio [3]. In the Eastern Mediterranean Sea, so little organic matter is produced that there are very few areas where dissolved O_2 is completely consumed. As a result, denitrifying bacteria cannot function, and the excess nitrate-to-phosphate ratio remains. Our mass balance model is able to reproduce the observed ratio of dissolved nitrate to phosphate in the Eastern Mediterranean deep water by accounting for the external supply and the transformations and transport of the two nutrient elements [17]. The results invalidate the previous explanation attributing the high nitrate-to-phosphate ratio to high rates of nitrogen fixation in the Eastern Mediterranean Sea. This is consistent with measured rates of nitrogen fixation that are amongst the lowest observed anywhere in the global ocean [27, 28].

The model calculations further demonstrate for the first time that the higher than Redfield nitrate-to-phosphate ratio of the Western Mediterranean Sea deep water is similarly caused by the high N/P ratio of the external nutrient supply combined with low levels of denitrification [17]. Although in our model some nitrogen fixation does occur in the Western Mediterranean Sea, compared to none in the Eastern Mediterranean Sea, it is more than offset by a 12 times greater rate of denitrification per m^2 sea surface area. The dissolved nitrate-to-phosphate ratio of Western Mediterranean deep water is nonetheless lower than that of Eastern Mediterranean deep water largely because of a lower total N/P ratio of the external nutrient inputs, in particular due to the inflow of low N/P North Atlantic surface water.

2.6. Mediterranean nutrient cycling: unravelling the importance of different processes

A useful in silico exercise that can be performed with the biogeochemical mass balance model is to remove one process at the time and observe the resulting effect on the cycling of

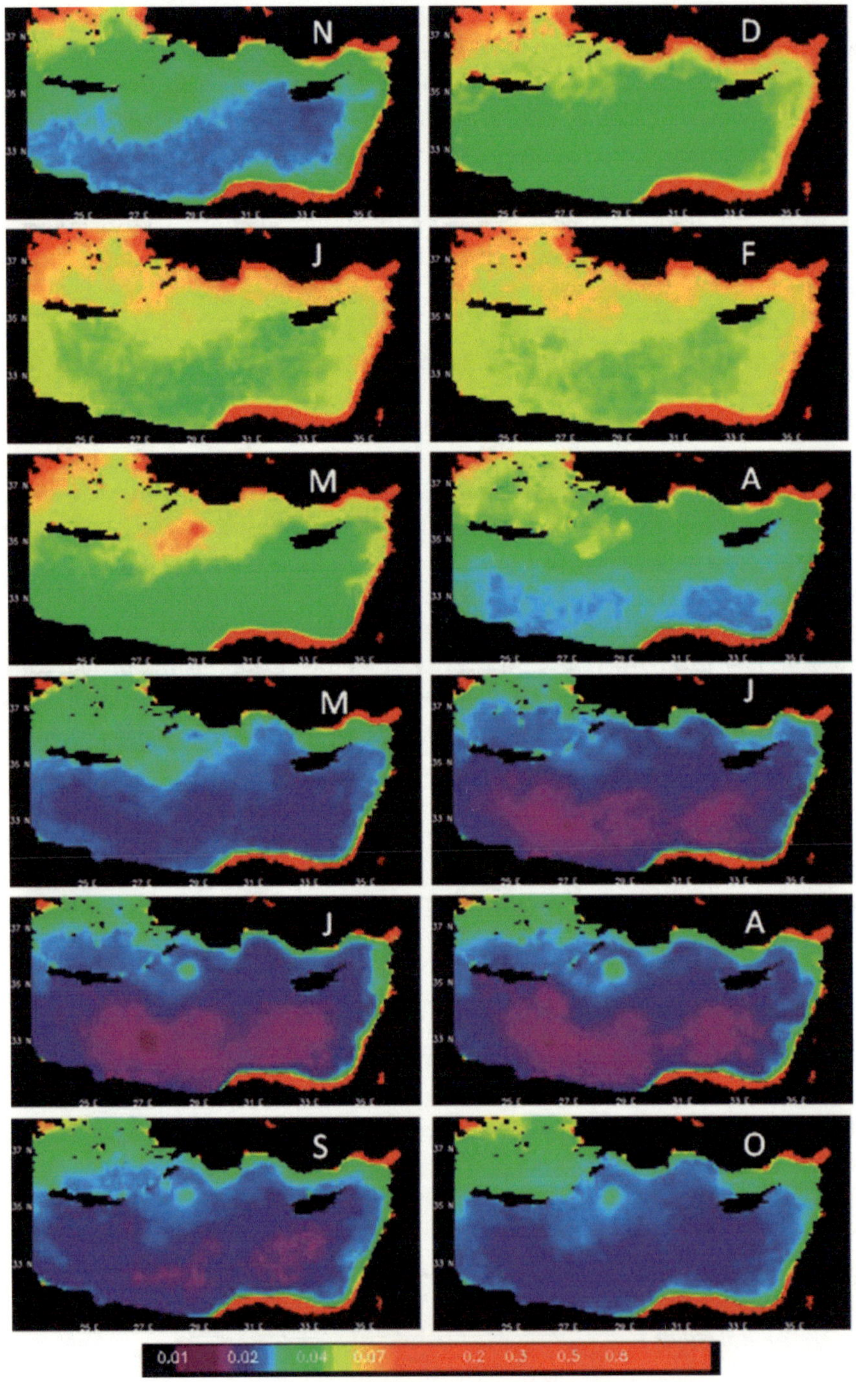

Figure 9. Monthly SeaWiFS 'climatological' images of chlorophyll from November to October. The images are based on input data from September 1997 to August 2004. Concentrations of chlorophyll-a were computed using the Bricaud et al. [23] algorithm for the Mediterranean Sea [24]. The images were provided by Steve Groom (NERC Earth Observation Data Acquisition and Analysis Service, Plymouth). This figure is available in colour in the online version of this chapter.

P and N [17]. For instance, we found that if the atmospheric input of bioavailable P and N is removed, the model yields deep water-dissolved nitrate-to-phosphate ratios of 15–16 for both the Western and Eastern Mediterranean Seas, despite the large differences in the starting ratios (21:1 and 28:1). By contrast the removal of P and N inputs associated with submarine ground water discharge, rivers or direct wastewater inputs has no significant effects on the deep-water nitrate-to-phosphate ratios. However, when the P and N inputs via the Strait of Gibraltar are removed, then the deep-water ratios increase to 53:1 (Western Mediterranean Sea) and 69:1 (Eastern Mediterranean Sea). This suggests that the higher than Redfield nitrate-to-phosphate ratios observed in the Mediterranean Sea are ultimately driven by the high N/P ratio of the external nutrient input via atmospheric deposition but also that without dilution by the low N/P nutrient supply from the North Atlantic Ocean, Mediterranean-dissolved nitrate-to-phosphate ratios could be even larger than actually observed.

Another model application consists in calculating by how much the denitrification rate would have to increase to bring the deep-water nitrate-to-phosphate ratios of the Mediterranean Sea down to the Redfield value (16:1). The results indicate that the denitrification rate in the Western Mediterranean Sea would have to be 2.2 times higher than in 1950 (increasing to 0.05 mol N m^{-2} y^{-1}), while in the Eastern Mediterranean Sea, it would have to be 7 times higher than in 1950 (increasing to 0.01 mol N m^{-2} y^{-1}). In the global ocean, denitrification rates mostly fall in the range of 0.04–0.10 mol N m^{-2} y^{-1} [29, 30]. This means that if the Mediterranean Sea supported rates of denitrification of the same order of magnitude as the rest of the oceans, then its deep waters would approach 'normal' Redfield ratios around 16:1. An important conclusion is that the unique nutrient distributions in the Mediterranean Sea are not due to some unknown process not encountered elsewhere in the oceans, but rather to 'normal' ocean processes—but then in an unusual combination.

3. Environmental implications

3.1. Anthropogenic nutrient enrichment

During the second half of the twentieth century, land-derived emissions of P and N to the Mediterranean Sea reached levels 2.6 and 2.3 times higher than in 1950, respectively ([19, 31]; **Figure 10**), as a result of the rapid growth of coastal populations and intensifying economic activities. The huge increases in nutrient supply by rivers and via atmospheric deposition, however, have had relatively little impact on the open waters of the Mediterranean Sea, with only a 10–20% increase in primary productivity (**Figure 10**). The reason is that the Mediterranean Sea has a very efficient natural buffer to counter increases in the supply of nutrients, especially in the Eastern Mediterranean Sea. Newly added nutrients to the surface water are rapidly transferred to the deeper-water layers, either by downwelling of surface water or settling of biological debris. Because of the anti-estuarine circulation, much of the additional dissolved P and N in the intermediate water of the Eastern Mediterranean Sea is removed by outflow through the Strait of Sicily, while most of the rest accumulates in the deep-water layer. A similar scenario plays out in the Western Mediterranean Sea with most of the additional P and N either exported to the North Atlantic Ocean or stored in the deep water, rather than contributing to primary production [19].

An additional result of the very low productivity is that the O_2 contents of the waters of the Mediterranean Sea remain elevated: no permanent or seasonal areas of hypoxia are observed in the open Mediterranean Sea. This contrasts with the Baltic Sea, which receives comparable influxes of external pollutant nutrients per unit area but with very different consequences. There are major eutrophication problems in the Baltic Sea including areas of hypoxia and toxic algal blooms. This does not mean there are no problems of excess nutrients in the Mediterranean Sea. There are, but they tend to be local and restricted to nearshore areas, such as the salt water lagoons of the Nile delta and Venice lagoon. In the past, coastal areas of the North Adriatic Sea were plagued by eutrophication problems, which have now been much reduced. However, offshore basin-wide, nutrient enrichment is not a problem, because the system is efficiently buffered. Model projections into the future show that this result is robust. Even taking into account the various possible changes in physical circulation and P and N emissions during the remainder of the twenty-first century, model simulations predict that the Mediterranean Sea will remain oligotrophic [19].

Similar considerations apply to the environmental impacts of mariculture, which is on the rise in the Mediterranean region as wild fish populations decrease. In most cases, concerns about excess nutrients being released by mariculture are likely unfounded given the very low natural productivity of the surrounding waters and the rapid assimilation of added nutrients into the regional food web. While other issues may arise, such as those related to the dispersion

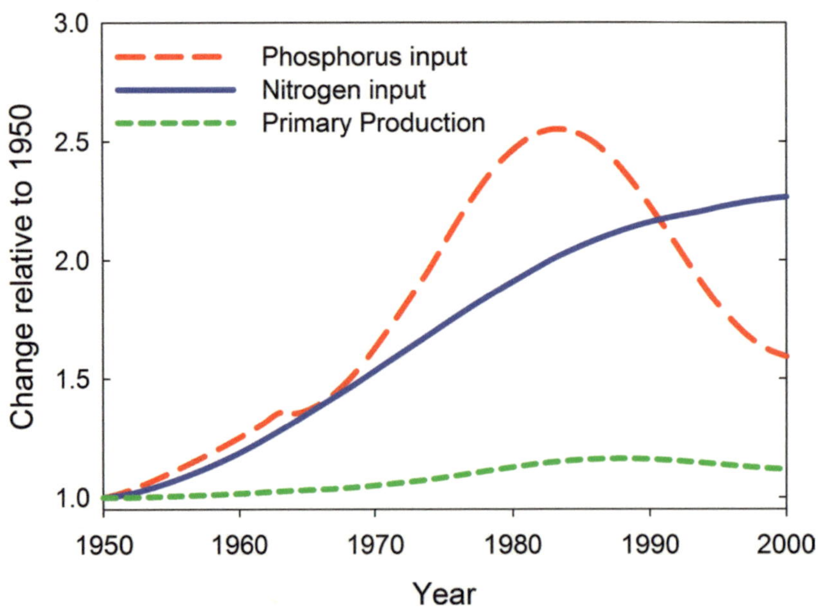

Figure 10. Reconstructed relative changes of land-derived reactive phosphorus and nitrogen inputs to the Mediterranean Sea and corresponding changes in primary productivity, from the year 1950 to 2000. Data from Powley [19].

http://dx.doi.org/10.5772/intechopen.70878

of pharmaceuticals or the possible introduction of invasive species, it seems clear that for any reasonable fish loading, mariculture will not lead to significant regional increases in nutrients and eutrophication [32].

3.2. Will ongoing global climate warming cause hypoxia in the Mediterranean Sea?

Changes in the ***thermohaline circulation*** can profoundly affect the biogeochemistry of the Mediterranean Sea. A good example from the past is sapropels, that is, organic-rich sediments that were deposited periodically between 6000 years and 13.5 million years ago, mostly in the Eastern Mediterranean Sea [33]. In deep sediment cores, sapropels occur as distinct dark layers separated from one another by intervals of lighter coloured, organic-poor sediments similar to those accumulating in the deep basins of the Mediterranean Sea at the present time. The sapropels are typically laminated and lack evidence of bottom dwelling animal life, implying they were deposited at times of ***anoxia*** when no (or very little) dissolved O_2 was present in the deep waters. The lack of O_2 in turn limits the degradation of organic matter at the seafloor, thereby leading to the accumulation of organic-rich sapropel. In contrast, today the deep waters of the Mediterranean Sea are well oxygenated (i.e. they are ***oxic***), and O_2-respiring bacteria decompose most of the organic matter that sinks to the bottom.

Extensive studies of sapropels, in particular those carried out on the most recently deposited sapropel (Sapropel S-1), conclude that the transitions in the deep-water ***redox*** chemistry of the Eastern Mediterranean Sea that accompanied the initiation and cessation of sapropel formation (i.e. the alternations between oxic and anoxic deep waters) were linked to changes in thermohaline circulation, which, in turn, were caused by natural climate fluctuations [34]. This raises concerns about the ongoing warming trend in the Mediterranean region, which could potentially trigger a reappearance of hypoxic or anoxic bottom waters in the Mediterranean Sea. Dwindling O_2 levels could have devastating consequences for marine life at the seafloor, as well as for water quality throughout the water column (under O_2-depleted conditions, microorganisms generate toxic by-products, such as ***hydrogen sulphide***).

Predicting the effects of anthropogenic climate change on the thermohaline circulation of the Mediterranean Sea, however, is far from straightforward. The difficulty resides in predicting how climate change will affect the ***density*** of Mediterranean surface waters. The rate at which deep water forms is directly related to the density contrast between the surface and deeper-water layers: the denser the surface water, the faster it sinks. Climate projections tend to agree with relatively high confidence that the Mediterranean region will experience higher temperatures and reduced rainfall in the coming decades [35]. The problem is that these changes have opposite effects on the density of seawater: higher temperatures decrease the density, while less rain and higher evaporation rates increase the density. Thus, depending on which effect wins out, thermohaline circulation under a warmer and drier climate could either accelerate or slow down. So far, there is no consensus in which direction the circulation of the Mediterranean Sea may be heading.

In order to explore the potential impacts of climate change on the oxygenation of the deeper waters of the Mediterranean Sea, we created a model of dissolved O_2 cycling similar to the nutrient model presented earlier in this chapter [18]. The model accounts for the consumption of O_2 by respiration in each water layer of the Western and Eastern Mediterranean Seas. The model-calculated O_2 concentration within each of the intermediate and deep-water layers then reflects the balance between the supply of O_2 by sinking (O_2-rich) surface water and consumption of O_2 by respiration. The dissolved O_2 concentration in the surface waters remains close to equilibrium with the O_2 concentration in the atmosphere, which in the model calculations is held constant.

The scenarios tested with the O_2 model correspond to the extreme effects of anthropogenic climate change on the physical circulation of the Mediterranean Sea, based on the work of Somot et al. [36] and Adloff et al. [37]. The extremes range from a slowing down to a speeding up of the Mediterranean thermohaline circulation, that is, from a weakening to a strengthening of the deep-water formation processes. The model calculations further account for the effects of increasing temperatures and salinity on the solubility of O_2 and respiration. Increasing temperatures reduce the amount of O_2 that dissolves in surface waters and thus reduce the O_2 supply to the interior of the Mediterranean Sea by deep-water formation. At the same time, higher temperatures stimulate microbial activity and thus increase respiratory O_2 consumption rate. The model scenarios were run until new steady states were reached. The new steady-state O_2 levels in the deep-water layer are shown in **Figure 11**, together with those obtained 100 years after the deep-water formation rates are switched to their new values.

Figure 11 also displays results obtained with two alternative representations of the O_2 respiration rate. The first formulation ('no feedback') assumes that the rate of O_2 consumption only depends on the availability of O_2 to the microorganisms. The chosen mathematical expression ensures that O_2 respiration slows down and ultimately stops when the dissolved O_2 concentration approaches zero. In the second formulation ('feedback'), the rate of O_2 respiration also depends on the availability of ***energy substrates*** that can be used during respiration. In deep ocean waters, the primary energy substrates accessible to microorganisms are dissolved organic carbon compounds (also referred to as DOC). The 'feedback' representation thus provides a more complete representation of O_2 respiration than the 'no-feedback' representation. As discussed below, comparison of results obtained with the two formulations illustrates the importance of ***negative feedbacks*** in marine environments.

The O_2 concentrations in the intermediate water layers exhibit little change (<10%) in response to the imposed changes in circulation. The largest changes in O_2 concentrations are observed for the deep-water layers in the scenario where climate warming causes a significant slowing down of the thermohaline circulation, as shown in **Figure 11**. Furthermore, the effect is most pronounced in the Eastern Mediterranean Sea where the deep-water O_2 concentration is predicted to drop below 50 μM (compared to present-day values around 200 μM). However, it takes at least 600 years after the change in circulation for hypoxic conditions to develop, that is, for the O_2 concentration to cross the threshold value of 2 mg/l (62 μM) below which the lack of O_2 may severely harm marine life. Interestingly, even when the deep waters of the Eastern Mediterranean Sea become hypoxic, those of the Western Mediterranean Sea remain oxic. Such a contrast in deep-water oxygenation between the two basins also characterized most periods of sapropel formation during which oxic conditions persisted in the deeper parts of the Western Mediterranean Sea. Furthermore, a slowing down of the thermohaline circulation

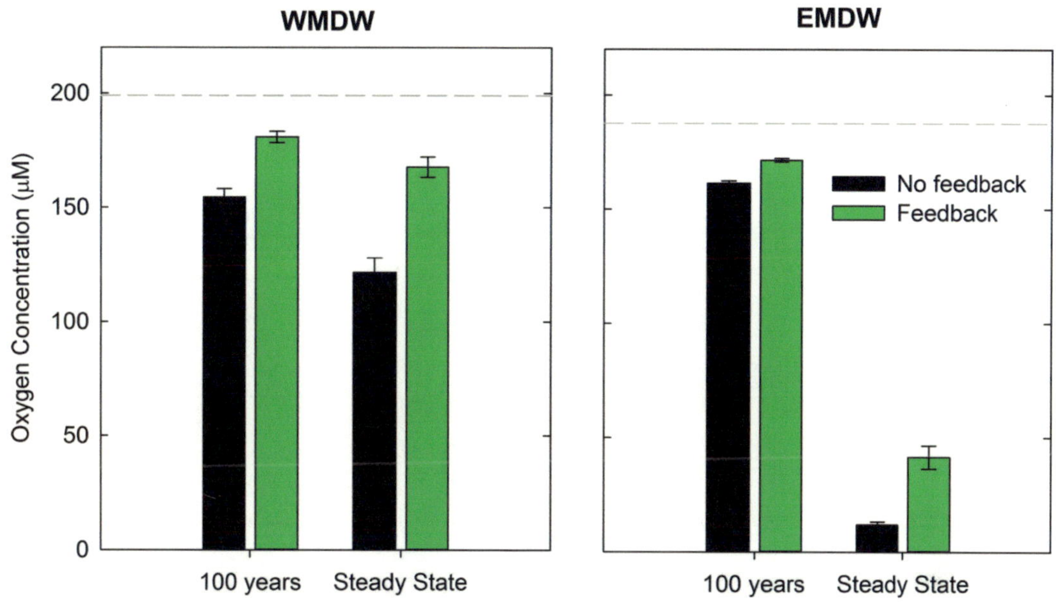

Figure 11. Model-predicted dissolved oxygen (O_2) concentrations in the Western Mediterranean deep water (WMDW) and Eastern Mediterranean deep water (EMDW), for the decreased thermohaline circulation climate change scenario. The dashed horizontal line represents the O_2 concentrations under the current thermohaline circulation regime. The two sets of bars in both panels are results obtained either 100 years after switching to the slower circulation regime or after the O_2 concentrations reach their new steady-state values. 'No feedback' corresponds to the model calculations ignoring the dissolved organic carbon (DOC) feedback on deep-water respiration, while 'feedback' corresponds to the results when the feedback is included. See text and **Figure 12** for more details. Error bars reflect uncertainty of temperatures and salinity in climate change projections and thus O_2 solubility and microbial rate kinetics. Data from Powley et al. [18].

similar to that simulated here has been proposed to have been a driver for sapropel formation [34, 38]. The analogy with the current warming trend, however, is only partially valid: all the evidence strongly suggests that the more recent sapropels (including S-1) were deposited under a wetter and colder climate during a period of sea level rise following the last ice age.

The model calculations also highlight a possible self-regulating negative feedback that minimizes the impacts of changes in the thermohaline circulation on the oxygenation of the deeper waters of the Mediterranean Sea. How the feedback works is illustrated in **Figure 12**. It results from the fact that deep-water formation not only supplies surface water enriched in dissolved O_2 but also in bioavailable DOC, that is, the two ingredients that sustain microbial respiration. This is not unlike the bloodstream in the human body, which supplies both O_2 and DOC in the form of glucose to our cells. If the thermohaline circulation weakens, and hence deep-water formation decreases, less O_2 is transferred to the deeper layers of the Mediterranean Sea but also less DOC. The outcome is less biological O_2 consumption in the model runs where the rate of respiration depends on both the availability of O_2 and DOC (i.e. the 'feedback' results in **Figure 11**). In other words, the decreased supply of O_2 by slower deep-water formation is in part compensated by the reduced supply of DOC and, hence, the reduced consumption of O_2. Similarly, a more vigorous thermohaline circulation supplies higher amounts of both O_2 and DOC to the deeper layers

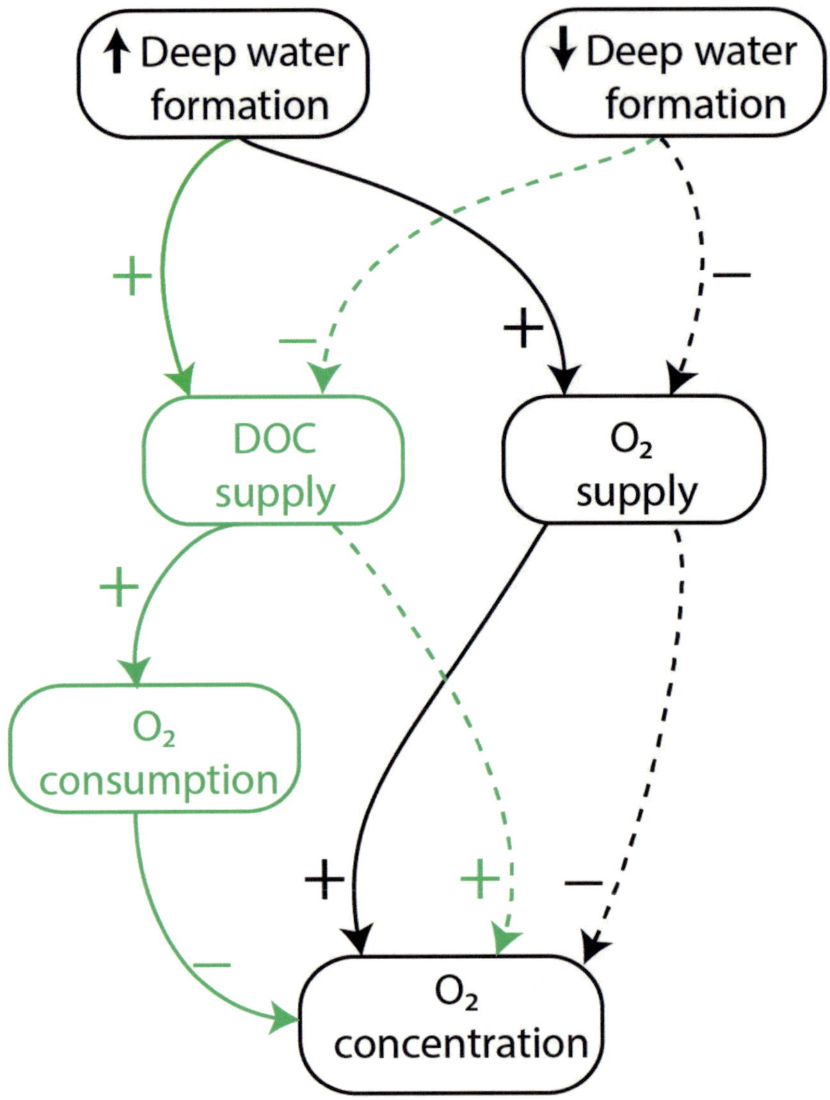

Figure 12. Schematic illustration of the effects of deep-water formation rates on deep-water oxygen (O_2) concentrations. Solid arrows show the effects of increased deep-water formation (stronger thermohaline circulation), and dashed arrows show the effects of decreased deep-water formation (weakened thermohaline circulation). Black arrows and boxes identify the effects of changes in the thermohaline circulation without the DOC feedback, while green (grey in print version) arrows and boxes show the additional effects of including the dissolved organic carbon feedback on O_2 consumption. This figure is available in colour in the online version of this chapter.

of the Mediterranean Sea. The higher deep-water O_2 concentration that may be expected from the enhanced supply of O_2, however, is offset by the increased supply of DOC, which increases the rate of O_2 consumption. Negative feedbacks such as the one illustrated here are commonly observed in biogeochemical systems. Nonetheless, the O_2-DOC feedback for deep-water oxygenation of the Mediterranean Sea remains, at present, speculative and requires further investigation.

http://dx.doi.org/10.5772/intechopen.70878

Biogeochemical mass balance (box) models.

Box models represent a given system as a series of interconnected reservoirs or boxes. The state variables of the model are the masses, M_i, in the reservoirs. For example, one of the state variables in the nutrient model is the mass of dissolved organic P (DOP) in the intermediate water layer of the Eastern Mediterranean Sea. Ordinary differential equations (ODEs) describe the fluxes of matter in and out of the reservoirs as a function of time, t. Thus, in **Figure B.1**, there is an input flux, F_{in}, into the box and output flux, F_{out}, away from the box.

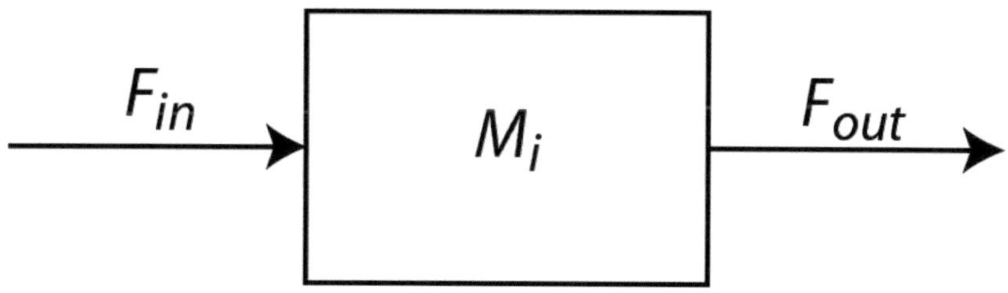

Figure B.1 Box model: the box or reservoir represents the mass of the variable of interest (M_i) and fluxes in (F_{in}) and out (F_{out}) of the reservoir.

Note that the fluxes may correspond to the physical transport of the state variable (e.g. DOP in Eastern Mediterranean intermediate water) from or to another physical location (e.g. DOP moving from the Eastern Mediterranean intermediate water layer into the Western Mediterranean Sea via the Strait of Sicily) or the transformation of the state variable at a given physical location (e.g. the mineralization within the Eastern Mediterranean intermediate water layer of DOP into dissolved inorganic phosphate).

The mass conservation equation describing the time evolution of the state variable, M_i, in **Figure B.1** is then given by

$$\frac{dM_i}{dt} = F_{in} - F_{out} \tag{1}$$

where the fluxes are given in units of mass per unit time. For each reservoir in the model, an ODE of the form of Eq. (1) is derived. For example, **Figure B.2** presents the dissolved phosphate reservoir in the deep-water layer of the Eastern Mediterranean Sea.

The corresponding ODE is

$$\frac{dPO_4^{EMDW}}{dt} = min + dwf - up \tag{2}$$

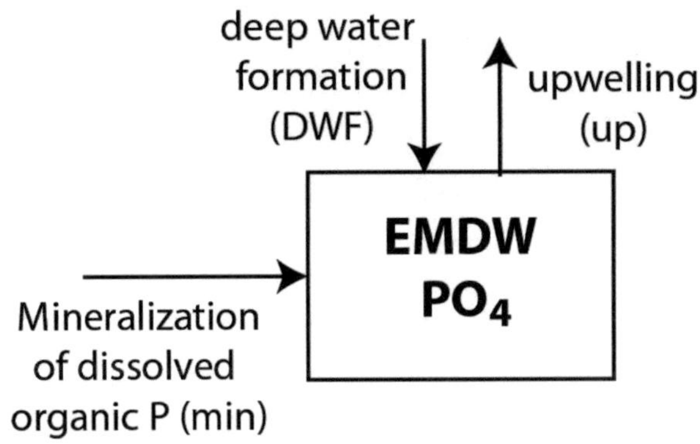

Figure B.2 Input and output fluxes to the Eastern Mediterranean Sea deep water (EMDW) dissolved phosphate (PO_4) reservoir in the nutrient model.

For each term on the right-hand side of Eq. (2), a mathematical expression must be assigned. When all the mass conservation equations are specified, one for each reservoir, the system of ODEs is solved as a function of time, *t*, using a numerical method. Mathematically, a system of ODEs can only be solved if initial conditions are specified, that is, the values of the state variables, M_i, must be known at some given initial time (i.e. at $t = 0$). In our mass balance calculations, we impose the estimated masses in the P and N reservoirs in 1950 as the initial conditions.

A special condition for any dynamic model (i.e. a model with time as an independent variable) is that of the steady state. For a box model, this means that, for each reservoir, the input fluxes exactly balance the output fluxes or

$$\frac{dM_i}{dt} = 0 \tag{3}$$

which is equivalent to stating that all state variables remain constant. This is the condition we assign to the 1950 P and N cycles in the Mediterranean Sea.

While the mass balance equations are written in terms of the chemical masses in the different reservoirs (M_i), the results of the calculations are compared to measured concentrations (Ci) in the different water layers of the Mediterranean. The factor that allows us to convert between M_i and C_i is the volume of the corresponding water layer.

Acknowledgements

This chapter is based on the PhD thesis research of HRP, with support from the Canada Excellence Research Chair (CERC) programme. HRP and PVC thank the generous input from numerous colleagues in the Ecohydrology Research Group. Leigh Kroeger suggested useful corrections to an earlier draft of this text. The senior author (MDK) was inspired while writing this text by some of the tunes in the Pirates of Penzance.

Glossary of terms

Anti-estuarine circulation This unusual form of circulation is typified by what happens in the Mediterranean Sea. Anti-estuarine circulation is where surface water flows into a basin and deeper-water flows out of the basin: estuarine circulation is the opposite.

Anoxia Occurs when there is no dissolved oxygen (O_2) left in a given water body.

Autotrophic When, in a given environment, more organic matter is produced as a result of photosynthesis than is consumed through respiration.

Denitrification A microbial process in which bacteria use nitrate (rather than oxygen) to respire organic matter to generate energy for their life processes. It only takes place in locations with very low (zero) oxygen and a supply of nitrate. The waste product of this reaction is dinitrogen gas.

Density Refers to how much mass is in a given volume (mathematically defined as mass/volume). In the oceans, temperature and salinity are the two main properties that influence the density of water. The higher the temperature, the lower the density of water as the water molecules have higher energy and are more spread out. Likewise the higher the salinity of water, the denser the water because more salt is present per unit volume of water.

Downwelling The deep water of the ocean is formed in specific locations almost always in the polar regions. These locations are where surface water becomes so cold and dense that it 'downwells' and fills the deep basins of the global ocean. The most important of these downwelling regions are in the North Atlantic Ocean between Greenland and Norway and in the waters around Antarctica.

Energy substrates Chemical substances that can be broken down during respiration to provide the energy required by organisms. The most common energy substrates are organic compounds containing carbon, hydrogen and oxygen and frequently phosphorus and nitrogen. Some major groups of organic compounds include carbohydrates, fats, proteins and amino acids.

Eutrophication A form of water pollution. Cultural or anthropogenic eutrophication occurs when excessive phosphorus and nitrogen from sources such as fertilizers and wastewater discharges run into a body of water. This encourages the growth of algae and other aquatic plants. As the plants die, bacterial degradation occurs (the plants decay) which consumes dissolved oxygen within the water. In extreme situations this can result in all the dissolved oxygen being consumed, leading to mass mortalities of fish and other marine organisms.

Heterotrophic When, in a given environment, more organic carbon is consumed by respiration than is produced by photosynthesis.

Hydrogen sulphide (H_2S) A colourless volatile compound that has a rotten egg smell and produces black-coloured sediments. It is produced from sulphate (SO_4) as a result of respiration under anoxic conditions when other higher energy-producing electron acceptors such as oxygen and nitrate have been entirely consumed. It is a toxic substance.

Hypoxic When the dissolved oxygen (O_2) content in a body of water is so low (<62 μM or 2 mg/l) that it is detrimental to the animals that live there.

Marine snow Is formed by aggregates of plankton and debris of other organic matter particles together with dust and other inorganic particles held together in a matrix of gel-like material. Together the aggregates are heavy enough to settle into deeper water rather like 'snow'.

Mineralization The transformation of dissolved organic phosphorus (DOP) and dissolved organic nitrogen (DON) back into dissolved inorganic P (phosphate) and N (nitrate). This process is primarily the result of enzymatic hydrolysis.

Negative feedback When the response of a system to a given perturbation (change) acts to counteract the initial perturbation and thus reduce its effect.

New production The primary production supported by nutrients added externally to the photic zone of the ocean, rather than by nutrients that are recycled by mineralization of organic matter produced within the photic zone.

Nitrogen fixation A bacterial process, which converts dinitrogen gas to organic nitrogen. It is a very ancient process having evolved in the early Precambrian and is the dominant natural source of fixed nitrogen for life processes.

Ocean gyres There are very large slowly spinning bodies of water in the major basins of the world's oceans. There are two major gyres in the Northern Hemisphere and three in the Southern Hemisphere. In the North Atlantic Ocean, the current that defines the western boundary of the gyre is called the Gulf Stream. In general, downwelling occurs in the centre of the ocean gyres, and these are the least productive parts of the global ocean.

Oligotrophic Refers to water bodies that have very low primary production, typically because they are severely limited by a lack of dissolved inorganic nutrients. Oligotrophic water bodies are defined as having chlorophyll concentrations less than 0.1 mg m^{-3}.

Oxic Bodies of water with plentiful dissolved oxygen (O_2): typically more than 62 μM or 2 mg O_2 per litre water.

Photic zone The area at the top of the ocean that receives sunlight and so is the area where phytoplankton grow.

Phytoplankton Microscopic marine plants which grow in the surface layers of the ocean. They are the main plant life in the ocean and form the base of the marine food chain.

Primary production Phytoplankton growth is called primary productivity. Phytoplankton carry out photosynthesis: sunlight is used to convert carbon dioxide and water plus other nutrient elements into organic matter. This forms the base of the food chain in the ocean. Organisms that photosynthesize are called autotrophs. Phytoplankton are autotrophs.

Redox This term is shorthand for reduction–oxidation reaction, that is, a chemical reaction in which the oxidation states of atoms are changed. Any such reaction involves both a reduction process and a complementary oxidation process. Photosynthesis and respiration are redox processes.

Respiration The chemical reactions by which cells produce energy. It involves the breakdown of energy substrates: often organic compounds: with the use of an oxidant. When molecular oxygen (O_2) is present, it is the preferred oxidant for respiration; when it is absent, microorganisms can use other oxidants, including nitrate.

Solubilization The transformation of particulate organic phosphorus (POP) and particulate organic nitrogen (PON) to dissolved organic phosphorus (DOP) and dissolved organic nitrogen (DON), respectively. A variety of processes are lumped together in this pathway including hydrolysis, passive diffusion or active exudation from phytoplankton, viral- and bacterial-induced cell lysis, sloppy feeding by zooplankton and bacteriavory.

Thermohaline circulation The large-scale circulation that is caused by density differences of oceanic water. The density is controlled by changes in water temperature (thermo) and salinity (haline). Thermohaline circulation of the global ocean is also called the global conveyor belt. The thermohaline circulation is driven by the downwelling of dense water, while water is returned to the surface by upwelling.

Upwelling The opposite of downwelling. There are regions of the ocean where deep and intermediate waters are 'upwelled' to the surface of the ocean. Since they carry with them high concentrations of dissolved macronutrients, these are areas of high primary productivity. Two important upwelling regions are off the coast of Peru and off the coast of Namibia.

Author details

Helen R. Powley[1], Philippe Van Cappellen[1] and Michael D. Krom[2,3*]

*Address all correspondence to: M.D.Krom@Leeds.ac.uk

1 Ecohydrology Research Group, Water Institute and Department of Earth and Environmental Sciences, University of Waterloo, Waterloo, Ontario, Canada

2 Department of Marine Biology, Charney School of Marine Sciences, University of Haifa, Mt Carmel, Haifa, Israel

3 School of Earth and Environment, University of Leeds, Leeds, United Kingdom

References

[1] de Madron XD, Guieu C, Sempéré R, Conan P, Cossa D, D'Ortenzio F, Estournel C, Gazeau F, Rabouille C, Stemmann L, Bonnet S, Diaz F, Koubbi P, Radakovitch O, Babin PM, Baklouti M, Bancon-Montigny C, Belviso S, Bensoussan N, Bonsang B, Bouloubassi I, Brunet C, Cadiou JF, Carlotti F, Chami M, Charmasson S, Charrière B, Dachs J, Doxaran

D, Dutay JC, Elbaz-Poulichet F, Eléaume M, Eyrolles F, Fernandez C, Fowler S, Francour P, Gaertner JC, Galzin R, Gasparini S, Ghiglione JF, Gonzalez JL, Goyet C, Guidi L, Guizien K, Heimbuerger LE, Jacquet SHM, Jeffrey WH, Joux F, Le Hir P, Leblanc K, Lefèvre D, Lejeusne C, Lemé R, Loÿe-Pilot MD, Mallet M, Méjanelle L, Mélin F, Mellon C, Mérigot B, Merle PL, Migon C, Miller WL, Mortier L, Mostajir B, Mousseau L, Moutin T, Para J, Pérez T, Petrenko A, Poggiale JC, Prieur L, Pujo-Pay M, Pulido V, Raimbault P, Rees AP, Ridame C, Rontani JF, Pino DR, Sicre MA, Taillandier V, Tamburini C, Tanaka T, Taupier-Letage I, Tedetti M, Testor P, Thébault H, Thouvenin B, Touratier F, Tronczynski J, Ulses C, Van Wambeke F, Vantrepotte V, Vaz S, Verney R. Marine ecosystems' responses to climatic and anthropogenic forcings in the Mediterranean. Progress in Oceanography. 2011;**91**(2):97-166. DOI: 10.1016/j.pocean.2011.02.003

[2] Redfield AC, Ketchum BH, Richards FA. The influence of organisms on the composition of seawater. In: Hill MN, editor. The Sea. New York: Interscience; 1963. p. 26-77

[3] Tyrrell T. The relative influences of nitrogen and phosphorus on oceanic primary production. Nature. 1999;**400**(6744):525-531. DOI: 10.1038/22941.

[4] Krom MD, Kress N, Brenner S, Gordon LI. Phosphorus limitation of primary productivity in the Eastern Mediterranean. Limnology and Oceanography. 1991;**36**(3):424-432

[5] Banoub MW, Williams PJ. Measurements of microbial activity and organic material in the Western Mediterranean Sea. Deep-Sea Research. 1972;**19**(6):433. DOI: 10.1016/0011-7471(72)90053-8

[6] Coste B, Minas HJ, Bonin M-C. Propriétés hydrologiques et chimiques des eaux du bassin occidental de la Méditerranée. In: Campagne Medripod IV - 15 Octobre - 17 Novembre 1981, pp. Centre national pour l'exploitation des océans: Brest CEDEX, France; 1984

[7] Moutin T, Raimbault P. Primary production, carbon export and nutrients availability in western and eastern Mediterranean Sea in early summer 1996 (MINOS cruise). Journal of Marine Systems. 2002;**33**:273-288. DOI: 10.1016/s0924-7963(02)00062-3

[8] Pujo-Pay M, Conan P, Oriol L, Cornet-Barthaux V, Falco C, Ghiglione JF, Goyet C, Moutin T, Prieur L. Integrated survey of elemental stoichiometry (C, N, P) from the western to eastern Mediterranean Sea. Biogeosciences. 2011;**8**(4):883-899. DOI: 10.5194/bg-8-883-2011

[9] Krom MD, Woodward EMS, Herut B, Kress N, Carbo P, Mantoura RFC, Spyres G, Thingstad TF, Wassmann P, Wexels-Riser C, Kitidis V, Law CS, Zodiatis G. Nutrient cycling in the south east Levantine basin of the eastern Mediterranean: Results from a phosphorus starved system. Deep-Sea Research Part II-Topical Studies in Oceanography. 2005;**52**(22-23):2879-2896. DOI: 10.1016/j.dsr.2005.08.009

[10] Dubinsky V, Haber M, Burgsdorf I, Saurav K, Lehahn Y, Malik A, Sher D, Aharonovich D, Steindler L. Metagenomic analysis reveals unusually high incidence of proteorhodopsin genes in the ultraoligotrophic Eastern Mediterranean Sea. Environmental Microbiology. 2017;**19**(3):1077-1090. DOI: 10.1111/1462-2920.13624

[11] Schlitzer R. 2015. Ocean Data View. <odv.awi.de>.

[12] Lavezza R, Dubroca L, Conversano F, Ludicone D, Kress N, Herut B, Civitarese G, Cruzado A, Lefevre D, Souvermezoglou K, Yilmaz A, Tugrul S, Ribera d'Alcala M. Compilation of quality controlled nutrient profiles from the Mediterranean Sea2011. DOI: http://doi.pangaea.de/10.1594/PANGAEA.771907 Supplement to: Lavezza, R; Dubroca, L Conversano, F; Iudicone, D; Kress, N; Herut, B; Civitarese, G; Cruzado, A; Lefèvre, D; Souvermezoglou, K; Yilmaz, A; Tugrul, S; Ribera d'Alcala, M. (2011): MED-Nut, a new Quality Controlled nutrient data base for the Mediterranean Sea. (PDI-828), in prep.

[13] NASA. 2007. The Mathematical Ocean: Deriving Planetary Health from Tiny Ocean Plants. <http://www.nasa.gov/vision/earth/lookingatearth/plankton.html> (5 August 2017).

[14] Berman-Frank I, Rahav E. Dinitrogen fixation as a souce for new production inthe Mediterranean Sea: A review. In: Stambler N, editor. Life in the Mediterranean Sea: A look at habitat changes. New York: Nova Science Publishers; 2012. p. 199-226

[15] Sverdrup KA, Armburst EV. An Introduction to the Worlds Oceans. 9th. 521pp ed. McGraw-Hill; 2007

[16] MEDAR Group. MEDATLAS/2002 database. Mediterranean and Black Sea database of temperature salinity and bio-chemical parameters. In: Climatological Atlas. 2002 IFREMER Edition (4 CD ROMS)

[17] Powley HR, Krom MD, Van Cappellen P. Understanding the unique biogeochemistry of the Mediterranean Sea: Insights from a coupled phosphorus and nitrogen model. Global Biogeochemical Cycles. 2017;**31**:1010-1031. DOI: 10.1002/2017GB005648

[18] Powley HR, Krom MD, Van Cappellen P. Circulation and oxygen cycling in the Mediterranean Sea: Sensitivity to future climate change. Journal of Geophysical Research: Oceans. 2016; **121**:8230-8247. DOI: 10.1002/2016JC012224.

[19] Powley HR. Phosphorus and Nitrogen Cycling in the Mediterranean Sea: Circulation, Biogeochemistry and Anthropogenic Forcing. PhD thesis, University of Waterloo. http://hdl.handle.net/10012/11780 291 pp. 2017

[20] Letscher RT, Primeau F, Moore JK. Nutrient budgets in the subtropical ocean gyres dominated by lateral transport. Nature Geoscience. 2016;**9**(11):815-819. DOI: 10.1038/ngeo2812

[21] Luna GM, Bianchelli S, Decembrini F, De Domenico E, Danovaro R, Dell'Anno A. The dark portion of the Mediterranean Sea is a bioreactor of organic matter cycling. Global Biogeochemical Cycles. 2012;**26** GB2017. DOI: 10.1029/2011gb004168

[22] Duarte CM, Regaudie-de-Gioux A, Arrieta JM, Delgado-Huertas A, Agusti S. The Oligotrophic Ocean Is Heterotrophic. Annual Review of Marine Science. 2013;**5**:551-569. DOI: 10.1146/annurev-marine-121211-172337

[23] Bricaud A, Bosc E, Antoine D. Algal biomass and sea surface temperature in the Mediterranean Basin—Intercomparison of data from various satellite sensors, and implications for primary production estimates. Remote Sensing of Environment. 2002;**81**(2-3): 163-178. DOI: 10.1016/s0034-4257(01)00335-2

[24] Groom S, Herut B, Brenner S, Zodiatis G, Psarra S, Kress N, Krom MD, Law CS, Drakopoulos P. Satellite-derived spatial and temporal biological variability in the Cyprus Eddy. Deep-Sea Research Part II-Topical Studies in Oceanography. 2005;**52**(22-23): 2990-3010. DOI: 10.1016/j.dsr.2005.08.019

[25] Bethoux JP, Copin-Montegut G. Biological fixation of atmospheric nitrogen in the Mediterranean Sea. Limnology and Oceanography. 1986;**31**(6):1353-1358

[26] Krom MD, Emeis KC, Van Cappellen P. Why is the Eastern Mediterranean phosphorus limited? Progress in Oceanography. 2010;**85**(3-4):236-244. DOI: 10.1016/j.pocean.2010.03.003

[27] Yogev T, Rahav E, Bar-Zeev E, Man-Aharonovich D, Stambler N, Kress N, Béjà O, Mulholland MR, Herut B, Berman-Frank I. Is dinitrogen fixation significant in the Levantine Basin, East Mediterranean Sea? Environmental Microbiology. 2011;**13**(4):854-871. DOI: 10.1111/j.1462-2920.2010.02402.x

[28] Bonnet S, Grosso O, Moutin T. Planktonic dinitrogen fixation along a longitudinal gradient across the Mediterranean Sea during the stratified period (BOUM cruise). Biogeosciences. 2011;**8**(8):2257-2267. DOI: 10.5194/bg-8-2257-2011

[29] Codispoti LA. An oceanic fixed nitrogen sink exceeding 400 Tg Na-1 vs the concept of homeostasis in the fixed-nitrogen inventory. Biogeosciences. 2007;**4**(2):233-253. DOI: 10.5194/bg-4-233-2007

[30] DeVries T, Deutsch C, Primeau F, Chang B, Devol A. Global rates of water-column denitrification derived from nitrogen gas measurements. Nature Geoscience. 2012;**5**(8):547-550. DOI: 10.1038/ngeo1515

[31] Powley HR, Krom MD, Emeis K-C, Van Cappellen P. A biogeochemical model for phosphorus and nitrogen cycling in the Eastern Mediterranean Sea (EMS) Part 2. Response of nutrient cycles and primary production to anthropogenic forcing: 1950-2000. Journal of Marine Systems. 2014;**139**:420-432. DOI: 10.1016/j.jmarsys.2014.08.017

[32] Karakassis I, Pitta P, Krom MD. Contribution of fish farming to the nutrient loading of the Mediterranean. Scientia Marina. 2005;**69**(2):313-321. DOI: 10.3989/scimar.2005.69n2313

[33] De Lange GJ, Thomson J, Reitz A, Slomp CP, Principato MS, Erba E, Corselli C. Synchronous basin-wide formation and redox-controlled preservation of a Mediterranean sapropel. Nature Geoscience. 2008;**1**(9):606-610. DOI: 10.1038/ngeo283

[34] Grimm R, Maier-Reimer E, Mikolajewicz U, Schmiedl G, Mueller-Navarra K, Adloff F, Grant KM, Ziegler M, Lourens LJ, Emeis K-C. Late glacial initiation of Holocene eastern Mediterranean sapropel formation. Nature Communications. 2015;**6**:7099. DOI: 10.1038/ncomms8099

[35] IPCC. Climate Change 2013. The Physical Science Basis. In: Stocker TF, Qin D, Plattner G-K, Tignor MMB, Allen ST, Boschung J, Nauels A, Xia Y, Bex V, Midgley PM, editors. Working Group I. Contribution to the Fifth Assessment Report of the Intergovernmental

Panel on Climate Change. Cambridge, United Kingdom and New York, NY, USA: Cambridge University Press; 2013. p. 1535

[36] Somot S, Sevault F, Déqué M. Transient climate change scenario simulation of the Mediterranean Sea for the twenty-first century using a high-resolution ocean circulation model. Climate Dynamics. 2006;**27**(7-8):851-879. DOI: 10.1007/s00382-006-0167-z

[37] Adloff F, Somot S, Sevault F, Jordà G, Aznar R, Déqué M, Herrmann M, Marcos M, Dubois C, Padorno E, Alvarez-Fanjul E, Gomis D. Mediterranean Sea response to climate change in an ensemble of twenty first century scenarios. Climate Dynamics. 2015:1-28. DOI: 10.1007/s00382-015-2507-3.

[38] Stratford K, Williams RG, Myers PG. Impact of the circulation on sapropel formation in the eastern Mediterranean. Global Biogeochemical Cycles. 2000;**14**(2):683-695. DOI: 10.1029/1999gb001157

Chapter 4

How Landscapes Make Science: Italian National Narrative, The Great Mediterranean, and Giuseppe Sergi's Biological Myth

Fedra A. Pizzato

Additional information is available at the end of the chapter

http://dx.doi.org/10.5772/intechopen.69186

Abstract

The perception and representation of landscape are not natural facts but are cultural constructions of human agents. In this chapter, I aim at deconstructing the role of pre-classical archaeology of Eastern Mediterranean in the process of Italian nation building between the nineteenth and the twentieth centuries. In particular, I focus on how a substantial group of Italian intellectuals deployed archeological discoveries to construct the Mediterranean as a representational space, which eventually served as a platform for their nationalistic political discourses. To this end, I discuss the spreading of these new conceptions in the Italian cultural debate at large. A prominent figure in this debate was Giuseppe Sergi. By reconstructing his views on the connections between national identity and biology, I demonstrate the considerable performative effect of the Mediterranean as a symbolic space and source of meaning on Italian culture. Furthermore, I argue that this new role of the Mediterranean resulted from a negotiation between the archaeological discovery of pre-classical past and the political aspirations of those scholars who opposed Italian foreign and interior politics of the period.

Keywords: nationalism, racism, Mediterranean studies, spatial turn, cultural geography, Giuseppe Sergi, Italian fascism

1. Introduction

One of the most significant historiographical developments of the last decade of the twentieth century is the so-called "spatial turn". This turn was anticipated by Henri Lefebvre in his 1974 essay *La production de l'espace*[1], in which he argued that space as a methodological variable is historically and culturally constructed. In particular, Lefebvre distinguishes three categories: (1) spatial practices, (2) representations of space, and (3) representational spaces. The latter is defined as an experienced space, that is, a space perceived and experimented through symbolic associations [1: p. 40]. In 1989, Edward Soja pushed Lefebvre's ideas a step further and argued that representational spaces are produced in a social context [2: pp. 10–11]. Subsequently, cultural historians deployed the concepts introduced by Lefebvre and Soja to the problem of defining cultural identities of communities and individuals—a problem in turn related to the symbolic action of producing national, social, and cultural boundaries [3]. In particular, many writers pointed out that the concepts of identity and cultural heritage are also essential to the elaboration and symbolic appropriation of experienced spaces [3: p. 2, 4: p. 6].

Furthermore, the studies on the psychological perception of space highlighted that representations of a territory derive from the cultural acknowledgement of the landscape [5: p. 69]. This acknowledgement is never completed but is always revised, reconfigured, and recalibrated by the fleeting historical context. In other words, landscape exists only as a historical object. Being socially, culturally, and politically constructed, space is interpreted differently at different times, and different "mental horizons" can be ascribed to each of these interpretations.

During the better part of the long nineteenth century and the beginning of the twentieth century, the Mediterranean served as a mental horizon and represented an effective source of meanings for the national aspirations of some Italian intellectuals, patriots, nationalists, politicians, and scientists. More specifically, the Mediterranean constituted the hinge of theories on the origin of nations aiming at countering the strong influence of Northern European narratives centered on Germans and Aryan peoples. It is important to realize the far-reaching political implications of these intellectual projects: ultimately, the "Mediterraneanists"[2] intended to challenge the continental hierarchies, to undermine Italy's alliance policy, and to revamp Italian nationalism.

The leading advocate of Mediterraneanism was the anthropologist, psychologist, and pedagogist Giuseppe Sergi (1841–1936). A dominant and unconventional figure in the cultural life from the Unification (1861) to the rise of Fascism (1922), Sergi was a fierce upholder of Darwinism, rebuilt the Italian Anthropological School in Rome (1893), and led it unchallenged for 50 years. Also as a fervent anti-German, he questioned the policy of alliances pursued by the Italian Government and the Royal House. Much later, his "left racism" (anti-German but not anti-Semite) was looked upon with embarrassment during the Fascist years. Thus, the deconstruction of Sergi's Mediterraneanism is key to understanding the role of the Mediterranean mental landscape in the Italian culture at the turn of the century and its connections with the larger European social, political, and cultural context.

[1]Here I refer to the 1991 English edition.

[2]Let me add a word of caution. Recently, this term was used to indicate a group of British anthropologists who elaborated a derogatory image of modern Mediterranean society [6]. However, in this essay I use the term in the original meaning, which refers to scholars who held the existence of cultural, social, ethnical, or biological features common to the peoples inhabiting the Mediterranean region.

http://dx.doi.org/10.5772/intechopen.69186

My main thesis is that this role changed importantly at this historical junction. In particular, within the archeological debate, the Mediterranean landscape became a representational space that could be used to fuel an alternative view of the Italian national identity *vis-à-vis* the pan-Germanism propounded by many Northern European scientists. This analysis is the first step of a broader research project on the social and cultural construction of the Mediterranean archeological landscape in the nineteenth and twentieth centuries.

I begin with a reconstruction of the social and cultural context in which Sergi elaborated his view of the "Great Mediterranean" (Section 2). I briefly discuss the evolution of the Pelasgian myth. According to this myth, often related to the Homeric epic, the Pelasgians were a people of the Eastern Mediterranean who constructed enormous buildings during the Bronze Age. Several Italian writers, for example, Monsignor Cesare Antonio De Cara (Section 3) and Giuseppe Sergi himself in his early works, believed that it was the Pelasgians who brought to Italy the civilization, which eventually paved the way to the greatness of Rome and Italy itself. More importantly, they saw a direct connection between Pelasgian "cyclopean buildings" and the walls of Tiryns, Mycenae, and Troy, thus tightly intertwining archeological landscape, ancient history, and national narrative. As evolutionary biology grew in influence, the Mediterranean landscape assumed new meanings preparing racial theories and colonial policies (Section 4). Within this context, the Mediterranean ceased to be perceived as a periphery of the emergent nations (i.e., France and Germany) and became the identity center of a new national project in the immediate wake of the Unification of the Kingdom of Italy (Section 5). It was the beginning of a long wave which eventually sustained, albeit not always in a consistent way, the Fascist myth of the Mediterranean (Mussolini's "Aryan and Mediterranean Rome"). However, the Fascist appropriation of the Mediterranean myth was possible only by turning upside down the Mediterranean narrative propounded by the intellectuals of the late nineteenth century (Section 5).

2. An ancient mental landscape

As early as the eighteenth century, the Mediterranean became the site in which archaeology—and thus antiquity—turned into the "image of reality itself" [7: p. 113]. The process began with the rediscovery of the ancient canon by architects and sculptors [7: pp. 113–115]. Subsequently, also under the spell of the social practice of the *Grand Tour*, which ascribed to the Mediterranean a natural aesthetic superiority [6], archaeology and landscape formed a binomial closely linked to the idea of Mediterranean identity. It was however during the Romanticism and the pan-European nation building process [8] that the central role of the Mediterranean was consolidated.

During the People's Spring (1848), archaeological heritage was integrated in the discourses on the identity ambitions of the people living by the Mediterranean and fighting for independence. For example, in his *Primato morale e civile degli italiani*—a sacred book for the Italian cause—the philosopher and patriot Vincenzo Gioberti (1801–1852) described ancient ruins as "fossils of nations", that is the most venerable and revered vestiges of the existence of a nation [9]. Relating landscape and culture, Gioberti highlighted that the geographical and naturalistic features of the Italian environment were ideal for the development of the ancient Pythagorean wisdom.

However, a crucial breakthrough happened in the second half of the nineteenth century. The emergence of Darwinism and the rediscovery of Nordic prehistory moved the origin of peoples considerably back in time. This process stimulated a competition among European nations, which prompted Italian intellectuals to reclaim pre-Roman traditions as a way to re-read archaeological sites. This process also facilitated the rise of new sciences, anthropology, and prehistoric archaeology (or "paleethnology" in Italian parlance), dedicated to account *scientifically* of Italian pre- and proto-history [10].

As a consequence of the concomitance between the institutionalization of Italian anthropology and paleethnology and the establishment of the state, the nation building process affected the features of prehistoric archaeology in multiple ways. First, prehistoric research focused particularly upon the issue of national origins, a tendency that Italian intellectuals shared with other European nations [11]. Much of the narratives produced by this research constituted, therefore, a pre-mythistory of the Italian nation [12]. Secondly, the institutionalization process of the prehistoric disciplines was greatly enhanced by their ability to fit several political goals of the new state [10]. Finally, anthropologists and archaeologists played a crucial pedagogical role at a national level, not only by reviving topics and ideas of Italy's mythical past but also by introducing the pre-Roman landscape in the Italian cultural imagery.

During the 1870s and 1880s, pre-Roman studies were dominated by the so-called Pigorinian paradigm, named after Luigi Pigorini (1842–1925). According to this theory, Italy, like Central European nations, was colonized by Aryan peoples who migrated from the North into the Italian Peninsula and were the ancestors of the Latin civilization. The Pigorinian paradigm was perfectly aligned with the alliances policy of the Government and the Royal House, which favored Austria and the German Confederation. Toward the end of the century, however, a new paradigm emerged, which aimed at reaffirming Italy's Mediterranean identity. As a consequence, the upholders of this new view questioned the Italian policy of appeasement to German Powers and advocated the alternative alliance with the Triple Entente [10, 13].

The resonances between the Mediterraneanist paradigm and the eighteenth-century tradition inspired by the classical artistic canon and the practice of the Grand Tour determined a particularly favorable reception of these new ideas by Italian architects, who saw themselves as the heirs of this tradition [7: pp. 111–115]. A vocal spokesman of the necessity of rediscovering the heritage of the pre-Roman civilization was Gaetano Koch, President of the Associazione Artistica fra i Cultori di Architettura [14: pp. 35–36; 45], who, in an open letter to the Minister of Education, invited the Government "to plough a field [the research in pre-Roman archaeology] in which other nations reaped and continue to reap lush harvest" [15: p. 65].

The support of the Associazione di Architettura to the archeological research should not come as a surprise. Born in a nationalistic climate to promote the study of architecture as the "first among the arts", the Associazione also organized appreciation excursions to the national monuments as a contribution to the patriotic education of the general public [16: p. 162]. Architects were not alone in their efforts. In general, while in the first part of the nineteenth century, pre-Roman archeology was exclusively a leisure activity for a handful of learned philologists and artists;[3] in the final decades of the century, it surged to become a national cause and a battlefield of numerous intellectuals.

[3] As an early example see Ref. [17].

3. The pillars of the Mediterranean: when the (archaeological) reality overcomes the myth

Koch's letter does not only reveal a growing interest for archaeology among Italian intellectuals: it is also an excellent window on the changing cultural geography of the last years of the nineteenth century. The excavations of Schliemann at Hissarlik and Mycenae, of Evans at Crete, and, above all, the first Italian archaeological missions in Eastern Mediterranean headed in 1884 by Federico Halbherr (1857–1930) contributed enormously to turn the cultural imagery toward east. Like many other Mediterraneanists of the time, Koch sees in the "neglected peoples of High Syria and Cappadocia" recently rediscovered the veritable ancestors of the Latin civilization and calls for a systematic comparison between Italian pre-Roman remnants and "the monuments of Greece and Asia Minor" [15: pp. 64–65].

Koch was therefore alluding to a profound shift in the Mediterranean landscape, which now consisted of some well-determined foci: Peloponnese, continental Greece, Crete, and finally, Asia Minor. This remarkable enlargement of the classical Mediterranean geography built on the archaeological discoveries in the East and in the Mesopotamian region included "High Syria", which was to be intended as the area including "all peoples from the shores of Eastern Mediterranean to Euphrates and along the shores of the Eastern, Western, and Northern Euxine Sea" [18: p. 447].

On the narrative side, this geography was justified by an ancient myth common to many tales on the origin of Italian cities: the Pelasgian myth. Until mid-nineteenth century, the term "Pelasgian" was used by Italian scholars to refer vaguely to alleged pre-Roman ruins. After the discovery of analogous constructions in the Eastern Mediterranean regions, the meaning of the term changed importantly and began to indicate a specific semantic and cultural connection between cyclopean buildings in both areas. An interesting "scientific" description of these cyclopean constructions can be found in a study dedicated to the archeology of Calabria and penned by a local lawyer named Luigi Grimaldi. Grimaldi distinguishes two chronological phases. In the first phase, walls are constructed by big irregular polygonal blocks without concrete, while in the second phase, the constructors use "more regular blocks, hammered or chiseled" [19]. This distinction, originally propounded by Louis-Charles-François Petit-Radel [20], was widely accepted and appropriated by professionals and amateurs alike as one can see in the works of the members of the Italian Alpine Club, one of the associations that contributed the most to the construction of the Italian national landscape [21].

Although often opposed by the Italian palethnological school [22: p. 263], the Pelasgian myth was held by some Catholic scientists and philosophers, mostly contributors to *La Civiltá Cattolica*, a Jesuit journal institutionalized by Pope Pius IX in 1866 and reconfirmed by Pope Leon XIII in 1890. The main upholder of the Pelasgian myth was certainly Monsignor Cesare Antonio De Cara (1835–1905). A Jesuit and an orientalist [23, 24], Monsignor De Cara authored a ponderous three-volume work on the origin of the Hethei-Pelasgians and their civilizing actions [18]. According to De Cara, the Pelasgians migrated westbound from Asia Minor through the Mediterranean and brought civilization not only to the people living by the sea but also to the people of Southern, Eastern, and Northern Europe [18: pp. 124–125]. This narrative dovetailed nicely with the Catholic view and had the merit to enlarge the eighteenth-century classical narrative both spatially and chronologically, without antagonizing the religious orthodoxy. The hypothesis of the Asian origin of civilization tallied with the Biblical tale according to which post-deluge civilization first appeared in Asia.[4] More importantly, in their attempts to integrate the new scientific discoveries and some aspects of modernity with the Catholic discourse [26] and in contributing to disseminate within the Catholic circles, a new geography of civilization derived from the new prehistoric archeology and anthropology, De Cara and his colleagues ultimately aimed at reassessing the historical mission of Catholic nations in Southern Europe.

As I have mentioned above, a crucial step in the process of re-imagining the Mediterranean identity and its fundamental role in the development of the European civilization was the discovery of Cretan remnants by Evans and Halbherr. As De Cara wrote:

> *At Crete it ends the migration of the Hethei-Pelasgians from the Asian continent to the Aegean Islands, a migration of central importance for the ethnography of Mediterranean peoples, for the introduction of the first civilization and for its role in the development of next civilizations in Greece and Italy [18: p. 211]*

As a beacon flashing in the night, Crete seemed to illuminate the path to civilization and Mediterranean identity, and the new discoveries unleashed in Italy (and elsewhere) a real Cretomania [27].

Thus, toward the end of the nineteenth century, the Pelasgian architecture became the central representation of the environment, a sort of "national fiction" of the landscape underlying the Mediterraneanist discourse.[5] The cyclopean walls represented the umbilical cord still joining the civilization of the Peninsula with Mycenae, Tiryns, and Troy. They were a monument, but also a narrative of the common origins of modern Italians and the great peoples of the pre-Greek antiquity.

At the beginning of the twentieth century, Koch's dream of an Italian archeological mission fully devoted to the study of the Pelasgian walls came finally true. In Norba, the most significant Pelasgian site of the Peninsula, archeologists Raniero Mengarelli and Luigi Savignoni tried hard to trace some decisive resemblance between the chalky blocks in Lazio and the walls of Mycenae, Tiryns, and Troy. Eventually, their hopes were disappointed. The excavations revealed that the site dated between VIII and IV BC, and this was a great predicament for the Pelasgian hypothesis [22]. Nevertheless, Norba remained firm in the imagery of the intellectuals and became part of the Roman epic toward the unification of Italy [30: p. 559].

4. Giuseppe Sergi and the Mediterraneanist anthropology

The archaeological discoveries in the Eastern Mediterranean and the renovation of the Pelasgian hypothesis encouraged a new account of the origin of the European civilization, an

---

[4]For example, De Cara considered the Hethei-Pelasgian civilization as Chaldeian-Assirian, thus establishing a clear analogy with the Biblical civilization [25: p. 105].

[5]In this way, the Italian case reminds closely the Irish [28] and Scottish [29] ones: national identity and origins were linked to a new perception of antiquity encapsulated in the contemporary space which now included more than just Roman ruins.

http://dx.doi.org/10.5772/intechopen.69186

account that countered head-on the results of the excavations in Scandinavia and Germany. Furthermore, in the closing years of the nineteenth century, these new discoveries determined a shift both in the ideal borders and in the perception and meanings of the Mediterranean geography. For even after the dismissal of the Pelasgian hypothesis, many Italian intellectuals maintained the view of the Mediterranean as the cradle of civilization. This position was part of an ongoing debate in Italian culture. Throughout the century, two cultural idioms[6] defining identity faced each other: the Aryan idiom, born in the 1850s and supported by linguists and orientalists, and the Mediterranean idiom, hinging on the heritage of classical antiquity and revived by the discovery of pre-Greek Aegean civilization [31: pp. 77–78].

The foremost upholder of the Mediterranean idiom was Giuseppe Sergi (1841–1936). Born in Messina (Sicily) in the Bourbon Kingdom, Sergi actively partook the Risorgimento and fought with Garibaldi in the battle of Milazzo (17–24 July 1860). Back to Messina, he studied philosophy and then ancient languages (Greek and Sanskrit), pedagogy, and psychology. Enlightened by Darwin's work, Sergi deepened his knowledge of physical anthropology and evolutionistic biology. He rose to the chair of anthropology at the University of Bologna and then moved to Rome where he founded (1893) and headed the Roman Society of Anthropology for 50 years. An atheist and an anticlerical, Sergi was first fascinated by socialism but then became a leftist and opposed the unbalanced, pro-North industrial development of Italy.

Sergi's fame was primarily due to his original classification of human species and races combined with a morphological method for the classification of skulls, which required only a minimal amount of anthropometric information and standardized the measurement of cephalic index.[7] In the 1890s, Sergi's new classification method led him to suggest the existence of a new race not included in previous classifications. More importantly, he was explicit in considering this race as "the morphologically most beautiful [one] ever appeared in Europe" [32: p. 45].

Sergi's "invented" race was included in the "Mediterranean" branch of the larger species called (by himself) Euro-African. The qualification of Mediterranean referred to the geographical context in a twofold way: it concerned the origin of the race as well as its basin of diffusion. Moreover, the qualification was studiously differentiated from the more customary term "Latin race", a term that was no longer associated with the greatness of Rome but rather, and in a derogatory sense, with the Latin people of Europe (Italians, Spaniards, Portuguese, and French) and America. As it was an anthropological commonplace to regard them as naturally "indolent" people as opposed to the entrepreneurial and active Nordic ones,[8] it comes as no surprise that Sergi wanted to distance himself from the expression "Latin race" [34]. By contrast, the term "Mediterranean" evoked a larger context both geographically—because it reached out the Eastern regions—and culturally—because it was linked to the eighteenth century aesthetic ideals of the *Grand Tour*. Moreover, one should stress the double ideological

[6]By cultural idiom it is here meant a set of ways to understand identity through the organization of the human interactions in terms of we/others, internal/external, inclusion/exclusion and so on.

[7]Roughly, the cephalic index is the ratio between the length and the width of the skull and it was customarily deployed to classify human races.

[8]A very similar dynamics was described by Edward Said in Ref. [33].

meaning of the term. On the one hand, it was connected with several themes of the positivist and Darwinist traditions, such as the emergence of a post-Christian, rationalist, and materialistic society [27: p. 4]. On the other, it encapsulated a formidable alternative to the Aryan myth not only championed by German scientists but also popular among other European intellectuals, poets, and artists [35]. This narrative about the North as a mythical homeland of a superior people can be found largely in German anthropology and archaeology—fields in which German and French scholars took an unchallenged leadership.

Building on the authority of Tacitus' *Germania* [36, 37], German scholars advocated the ideal of the Aryan people hardened by the rigidity of the climate and the harshness of the territory and thus suitable to dominate over the European continent [38: pp. 45–71]. These Herderian accounts stressed the strong link between blood and land, between race and *Urheimat* (original homeland) but also rely on the negative construction of the "other" [38] in this case the Mediterranean people. According to the pan-Germanic view, the Mediterranean climate favored a weak physique and an idle attitude, unfitting modernity and leadership.

In Europe, the discussions about the Aryan superiority permeated the national debate. As a consequence, non-German intellectuals tried hard to prove that their own nation belonged to the Aryan race (see, e.g., the English debate on Celts and Saxons, the French one on Celts and Francs, and so on [39]). Italy was no exception: Luigi Pigorini, the head of the Italian palethnological school, was actively engaged in the debate [10]. However, Italy also hosted a diffused discomfort for the downgrading of the Mediterranean mores implicit in the commitment to the Aryan paradigm.

Sergi quickly became the leader of the opposition. More than this, he did not confine himself to challenge the Aryan anthropological hegemony but brought forward a counter-paradigm, which won approval in Italy as well as abroad, especially in English-speaking countries.[9] Sergi's work on the Mediterranean myth was detailed in a number of books published by Fratelli Bocca, one of the most prestigious publishers of the time [42: pp. 115–142] and in numerous articles in *La Nuova Antologia*, a major cultural journal, whose programmatic goal was to promote Italy's cultural and scientific renaissance [43]. The gist of Sergi's theory concerns a new account of the origin of the Italian nation. As early as the 1870s, the common wisdom among Italian prehistoric scientists was that Italy was first populated by a series of migrations along the North-South direction. This view entailed that the original nucleus of the Italian people, from which Romans would eventually descend, was Aryan and Asian [44: pp. 25–30, 45]. By contrast, Sergi held that the original nucleus came from Central-Eastern Africa and migrated to Southern Italy through the Mediterranean. Thus, Sergi's theory turned upside down the cultural geography underlying Italian nationalistic discourses. He explicitly replaced an Aryan-centric geography, with a new account in which the superior race was biologically and culturally Mediterranean.

On the anthropological side, the theory of the Mediterranean race did not abandon the aesthetic preference for dolichocephaly but contested that it was an Aryan character. Instead,

[9] An emblematic case is the economist William Zebina Ripley (1867–1941), professor at Columbia, MIT, and Harvard, who endorsed and circulated Sergi's classification in Ref. [40]. Ripley's knowledge of Sergi's theory derived mainly from Ref. [41] and from several articles appeared in *The Monist*.

Sergi claimed that dolichocephaly was a distinctive feature of the Mediterranean people, which was later scattered throughout Northern Europe during the very first invasion of the Euro-African people. These people brought it to Europe culture and metals [34, 46: pp. 193–194] and dominated the continent until the arrival of the Aryans in the Neolithic Age. This invasion of "barbaric" people led to such deep reshuffling of the human species that no nation in Europe would be racially uniform ever since. It was nevertheless possible to distinguish between areas of racial domination of the Aryans (Central Europe) and the Euro-Africans, divided into a Northern (Scandinavia, British Isles) and Mediterranean branch [47].

The change of the geographical context implicit in the theory of the Mediterranean race had deep socio-political implications. The traditional paradigm according to which Rome had Aryan origins provided a firm background to the alliance policy with Austria and the German Confederation pursued by the Government and the King [13: p. 54]. By contrast, Sergi's theory heralded the central ideal of the Risorgimento further fuelled by, at the end of the nineteenth century, irredentism[10]: anti-Germanism [48: pp. 228–230]. According to Sergi, the natural allies of Italy were the Mediterranean France and the Nordic England, which shared the same racial background.[11]

5. The Great Mediterranean

In spite of his great admiration for Anglo-Saxon people [34], Sergi's cultural geography and mental horizon remained firmly anchored to the Mediterranean basin. This was probably due to two main factors. The first reason was due to Sergi's insular origin. In Sicily, his birthplace, the Mediterranean is the one and only border, but it is an open one: ever since antiquity, it connects the world, making Sicily one of those islands "actively involved in the dealings of the outside world" [51: p. 147]. For Sergi, thus, the Mediterranean was at the same time a separation from and a connection to his Italian homeland. Second, Sergi's cultural background was imbued with pre-Roman and Pelasgian discourses as one can see from his first work dedicated to the ancient wisdom of Pythagorean philosophy [52]. References to the Pelasgian myth can be found is Sergi's anthropological works at the beginning of the twentieth century and, above all, in *The Mediterranean Race* [53: pp. 169–170] to show, following De Cara, that the origin of the Greek civilization was not "in the primitive beliefs of the so-called Indo-Europeans" but "in the Mediterranean itself" and perhaps only partially in the valleys of Tigris and Euphrates [53: p. 172]. Furthermore, Sergi linked the Pelasgians not only to classical Greece but also to ancient Italian populations such as the Etruscans, considered by him "of a Mediterranean type" and "true and genuine Italici" belonging "to the Pelasgic branch" [53: p. 183].

A Pelasgian mental landscape hinging on cyclopean constructions persisted in the cultural imagery of the Mediterraneanists even after the abandonment of the Pelasgian hypothesis.

[10]For the connection between irredentism and the shift to the Mediterranean context see Ref. [13: p. 56].

[11]During the *fin-de-siècle* crisis [49], Sergi swerved again toward the idea of a decadence of the Latin nations [34], but he solved the problem of the Euro-African superiority by ascribing a guiding role to England. Later, he restated the superiority of the Mediterranean race on the occasion of France's and Italy's victory in WWI [50].

For example, Sergi claimed that masonry in general was a purely Mediterranean business. Shortly after the discovery of the Roman origin of the walls of Norba, he held that "until Roman age, Aryans did not know masonry for houses and temples […]. We know who was the masters of stonework and masonry in Northern Europe, that is the Romans" [46: p. 194]. Thus, cyclopean architecture and masonry are singled out as characteristic elements of the Mediterranean identity. Once again, the archeological landscape and the geographic context are merged in a discourse on the civilizing action of the Mediterranean people.

More than this, the geographical and ideological role of the sea becomes stronger as Sergi's thought evolves. While in his first works, the Mediterranean was primarily a background against which the epic of civilization took place, later he would begin to consider the Mediterranean as the sole geographical element that really contributed to the emergence of the European civilization. For example, in *Arii ed Italici*, he showed a conciliatory attitude by conceding that the use of metals—particularly bronze—was introduced by Aryans migrating from Asia [41]. However, a few years later in *The Mediterranean Race* and *Gli arii in Europa e in Asia*, Sergi excused himself for the previous concessions and assumed a more rigid stance. Building on Montelius' work [54], he affirmed that Tacitus' Germans were not particularly skillful with metals and, above all, triumphantly announced an alleged proof that metals came "from the South and especially from Mediterranean" [46: p. 195]. It is important to stress that Sergi always uses the comprehensive term "Mediterranean" to indicate the geographic source of the major elements of European culture (language, writing, metallurgy, scientific discovery, artistic, and architectonic expressions). In this way, he constructs an imagined space inseparably intertwined with the concepts of civilization and culture.

In this context, Italy assumed a relevant role in the dissemination of civilization in Europe, particularly in virtue of its geographical position—at the convergence of East and West—and its seas, such as the Adriatic Sea:

> *Eastern Mediterranean is larger [than Western] and can be divided into partly autonomous seas, which, like branches, serve as easier communications for the mainland. So is the Adriatic Sea, which is separated from Western Mediterranean by the diaphragm of the Italian Peninsula. In this way, it shares East and West, a true communication vehicle for Man's product of the land. [47: p. 17]*

The concept of Mediterranean is extended by Sergi in an unprecedented manner. While De Cara and other Mediterraneanists enlarged geographically the area to include the shores of Asia Minor and the Black Sea, Sergi overlaps the concepts of Mediterranean, culture, and civilization. This extension relied heavily on Sergi's own anthropological discoveries during a scientific journey through the Mediterranean and was fed by a number of direct observations of skulls sent to him from archeological missions in the Mediterranean[12] and in the Middle East[13] as well as an enormous collection of anthropological photographs. As a consequence, the geography of the Mediterranean, meant as the container of the millennial history "of a great race" [56: p. 310], was progressively enlarged:

[12]For example, the correspondence between Federico Halbherr and Sergi reveals that samples of skulls found in Crete were dispatched to Sergi to be classified. He did not fail to spot there the features of his Mediterranean race.

[13]Sergi was the only anthropologist to study (and to classify as Mediterranean) the skulls found in 1904 by Raphael Pumpelly (1837–1923) in Turkestan [55, 56: pp. 98-99]. His report was published entirely in the second volume of Pumpelly's *Exploration in Turkestan* [57: p. 445-446].

http://dx.doi.org/10.5772/intechopen.69186

> *One can affirm that the dominating race, which solely generated the great civilizations in a vast area of the Earth including the Mediterranean basin, with the European peninsulas, Northern Africa and particularly Egypt up to Nubia, Syria, Asia Minor and the Western territories, the valleys of Euphrates and Tigris up to Mount Zagros and Mount Elam, was the race I call Mediterranean, divided and fractioned in several branches, some of which, the three main ones, originated their own civilizations, regardless of their age or temporal priority, which did not have any effect. [56: p. 310]*

Such a big extension expressed Sergi's own mental image as one can see from a map designed by Sergi himself and published in *Gli arii in Europa e in Asia*. The map represents the spreading of the Euro-African species and served as an illustration of Sergi's idea of the Mediterranean basin as a crucial factor in determining the biological and anthropological features of the Mediterranean race (**Figure 1**).

Figure 1. Sergi's reconstruction of the spreading of the Euro-African species in the Middle East and Asia.

It is important to notice the hypothetical status of this map. At the time of its publication (1903), Sergi had not yet collected a complete set of anthropological data about Middle East and India. Therefore, by arranging the space according to new meanings and forms of appropriation, Sergi's map encapsulates a symbolical reconfiguration of a physical entity, that is, it makes it a cultural space. Moreover, a map communicates a geopolitical vision, a true project about the world, mirroring its author [58: p. 77]. Even before describing the territory, a map describes its author's gaze, its cultural geography [59]. As we shall see, by deconstructing the meaning of Sergi's map, one can find a clear anti-German message because maps—and this map in particular—are never value free for a positivist such as Giuseppe Sergi.[14] Moreover, the cartographic communication acts at the epistemological level as an instrument to reinforce and disseminate Sergi's message [60: p. 55]. At first sight, the functions of the map were (1) to show the extension of the Mediterranean symbolic imagery toward East and (2) to gather populations hitherto understood as different (e.g., Semitic and Hamitic people, Mediterranean and Indian people, etc.) in a single human group. Yet, if we look into the mental landscape of the author, a third function emerges.

To deconstruct the political and ideological project that grounds this representation, one needs to look at the evolution of Sergi's work. In the first decades of the twentieth century, Sergi attacked vigorously the last pillar of the Aryan hypothesis: the language. In a series of works, he stressed a connection between the Mediterranean and Euro-African race and the ancient populations of the Indian and Persian areas. His ultimate goal was to prove that the Indo-European *Ursprache* had been elaborated by the Indo-Iranian branch of the Euro-African race, that is, it was a creation of the Mediterranean race [61]. After masonry, arts, metallurgy, philosophy, and culture, Sergi aimed at dispossessing the Aryan people—and then Austria and Germany—of the last vestige of superiority.

6. Mediterranean *Urheimat*

In 1926, 4 years after the Fascism's rise to power, Sergi formulated the definitive version of his idea of Mediterranean. According to him, the ancient Mediterranean world was divided into three parts: (1) the Asian Mediterranean World (Syria, Asia Minor, Armenia, and even India); (2) the African Mediterranean World (Egypt[15]); and (3) the European Mediterranean World [56: p. 131–132]. These three parts surrounded the Mediterranean "like a divine lake" [56: p. 131]. On the shores of this lake, the longest-running civilization emerged at Crete, then in Greece, and finally in Italy. The Roman Empire made the Mediterranean civilization truly universal and cover all parts, Italy being "placed at the center between West and East" [56: pp. 316–317].

Thus, in the Fascist era, the myth of Rome, so dear to the patriotism of Risorgimento [62], was linked with the epic of the superior race. This connection, however, should not deceive.

[14]See Ref. [60: p. 53]: "Maps cease to be understood primary as inert records of morphological landscapes or passive reflections of the word of objects, but are regarded as refracted images contributing to dialogue in a socially constructed world (…) Maps are never value-free images; except in the narrowest Euclidean sense they are not in themselves either true or false."

[15]The reference to Egypt comes directly from the anticlerical tradition of the Italian masonic tradition in the Enlightenment [62], which was then resumed by positivism. This tradition evoked the ancient Egyptian wisdom as opposed to the Pope.

http://dx.doi.org/10.5772/intechopen.69186

Sergi's discourse remained strongly anti-German, overtly opposed to Mussolini's ideology. The Duce's goal was to fuse together the Aryan and the Mediterranean line.[16] Yet Sergi did not revise his position against the Aryan mythology but kept on working on his own myth of the Great Mediterranean. His distance from the Fascist regime can be clearly seen from the preface of his 1926 books dedicated to the greatness of the Mediterranean race. Instead of aligning himself to the regime, he thundered that the eagerness to follow a leader was "useful for social life, but it is also harmful because it leads to servility and any kind of turpitudes when it is developed through selfishness and personal advantage." He added that "sometimes it happens, and we cannot ignore it, that, especially in periods of great social and political upheavals, not the best, but the boldest men seize violently the leadership of a people." Eventually, these boldest men lead to "the decadence if not the ruin of the nations" because they are "inept and ill-prepared" to leadership [56: p. IX].

It is therefore clear that, superficial resemblances aside, Sergi's Mediterraneanist ideology was substantially different from Fascism's. His view was a positivist renewal of the patriotic interpretation of the Roman myth filtered through Vico, Gioberti, and the archeological debate on the Pelasgian hypothesis. Yet Sergi did not confine himself to combining the old topic with the archeological explorations of Eastern Mediterranean. The truly original feature of his discourse was its scientific character. Sergi was an active player in the debate on the soft-inheritance theory, that is, the attempt to combine Darwin and Lamarck. According to this theory, external factors such as nutrition and environment might affect the process of species and variety formation. "One needs to keep in mind that human events depend on biological as well as on environmental factors" wrote Sergi in 1926 [56: p. VI]. The soft-inheritance theory provided Sergi with the causal nexus between the Mediterranean geography and the emergence of a superior race. Furthermore, he could make a biologically informed use of Taine's philosophical reflections on the relation between milieu and people and Gioberti's thoughts about the favorable conditions for culture offered by the Italian climate and geology:

> *The Mediterranean provided the most favorable conditions for the development of civilization and a more cosmopolitan civilization than the one emerged in the valleys of the great rivers. (...) To these conditions, one may add those that made the Mediterranean one of the happiest regions of the world, that is the temperate climate, the fertility of the soil, the abundance of any production. As soon as one exits this fortunate basin, one finds deserts (...) or the inhospitable regions of the ancient world, those in Center Europe and Scythia. [32: p. 12]*

As his research went on, this theme became more prominent. He tried to show the existence of an original geographical Mediterranean overlapping the ancient Mediterranean culturally defined. According to Sergi, the sequence of geological eras actively affected great climatic changes and "the formation and retrocession of a large Asian Mediterranean, the Aral, the Caspian, and the Black Sea being its main remnants" [55: p. 305]. As "like in any human group, each type is self-contained and biologically in its natural place" [63: p. 145], it was obvious to conclude that the Mediterranean, "one of the happiest regions of the world", hosted the superior race.

But the Mediterranean myth intruded even more on Sergi's scientific work. Following the theories according to which the sea was the original locus of life, he held that man appeared

[16]See, for example, Mussolini's discourse at the Municipal Theater in Bologna on 3 April 1921 in which he stated that the birth of Fascism answered an intrinsic need of our "Aryan and Mediterranean stock."

along the neoteric fauna "in Africa, toward the center of the Mediterranean and his evolution took place or at least manifested itself in Western Europe; hence, the center of human origin was situated between Africa and Western Europe, i.e., a European-Mediterranean center" [56: pp. 8–9]. Ultimately, the Mediterranean was not only the cradle of civilization but also the site that determined the species generation process that detached the genus homo from primates. By intertwining evolutionary biology, archeological discourses, and geographical landscape, Sergi constructed a narrative that surpasses by far the national borders and became an epic of a race and a place, both essential factors of the civilizing action. In this way, the Mediterranean surged to the role of *Heimat* of civilization. This idea reminds closely Braudel's analysis that the Mediterranean history boiled down to a history of the relation between man and environment marked by long-term processes [64]. According to Braudel, ever since Lower Paleolithic, the Mediterranean civilization coalesced around processes that made it a "continual renaissance" [7: p. 107]. The idea of a cyclic Mediterranean renaissance was also part and parcel with Sergi's historical-biological vision, which led him, essentially a pacifist, to applaud Italy's renaissance after the end of the first World War [10].

7. Conclusion

There are geographical spaces that are also mental spaces and representational spaces possessing one or more symbolic meanings. As these meanings can then be adapted to social, cultural, and political agendas, interpretations of spaces are context- and community-dependent. Hence, cultural history creates interpretations of interpretations. In this essay, I have tried to deconstruct the Italian intellectuals' gaze on the Mediterranean from the end of the nineteenth century to the advent of Fascism. I have left aside the dynamics related to the creation of the Mediterranean colonial space, which will be dealt with in another essay. Instead, I have primarily focused on the cultural and scientific construction of the Mediterranean landscape as a way to define the Italian identity *vis-à-vis* other European powers.

For most of the Italian Mediterraneanists, the Mediterranean myth was mainly an answer to the Aryan narrative—and that entailed an opposition to the foreign politics of the Government and the King. Somewhat ironically, it was the success of the Aryan myth and the German archeology and anthropology that facilitated the acceptance of the Mediterranean paradigm in Italy for the major Mediterraneanists opposed the alliance with German powers either on religious grounds (De Cara) or for patriotic and ideological reasons. It was then at the turn of the century that the Mediterranean became an "amphibious concept", that is, a concept that lived in different social, political, and cultural contexts [6].

The new Aegean-Anatolian archeology and the dissemination of pre-classical archeological imagery contributed decisively to the construction of the Mediterranean representational space. First, they got rid of the old narratives based on ancient mythology (e.g., the Pelasgian myth). Second, they promoted the *mise en valeur* of the pre-Roman cultural heritage, which was, for the first time, appropriated by intellectual élites and nation builders.[17] Third, the intertwining

[17]It is important to notice that pre-Roman archaeology was later annihilated by the Fascist regime, which, by contrast, based its propaganda on the myth of Imperial Rome.

of the concepts of Mediterranean and civilization set the stage for a further extension of the geographical borders of the Mediterranean itself. In this national scenario, the originality of Sergi's theories lies in his integration of the new archaeological and anthropological discoveries in a biological structure. In this way, he created a cultural construction able to keep together a narrative of the nation and a narrative of the landscape. While this idea belonged to the tradition of Taine and Gioberti, Sergi added a causal nexus between human evolution and environment.

To the historian's ear, this role of the Mediterranean sounds very close to another classic of historiography, that is Braudel's Mediterranean. At the beginning, this comparison might appear far-fetched, but it can also be a useful analytical tool to shed light on the features of the Mediterranean representational space. For De Cara and other Catholic intellectuals, the Mediterranean was a site of passage of civilization from Asia to the Catholic nations and its importance lay in the re-assessment of the role of Catholic nations *vis-à-vis* Protestant ones. For the atheist Sergi, by contrast, the Mediterranean was a site of irradiation and played a dramatic role in the narration of the origins of the Italian civilization. While in Sergi's first works, the Mediterranean is something midway between a scenography and an actor, in his later books it acquires a more active presence in molding, biologically and environmentally, the only race able to produce and disseminate civilization. This view is somehow close to Braudel's according to whom the Mediterranean is a full-fledged actor facilitating encounters, clashes, and exchanges. Yet, contrary to Braudel, Sergi believed that the Mediterranean shapes not men's lives but men's bodies and minds.

However, the most conspicuous difference between Sergi and Braudel concerns the theme of plurality. Braudel's Mediterranean is a process: "What is the Mediterranean? It's a thousand things together. It is not a sea, but a succession of seas. It is not a civilization, but a series of civilizations piled up one upon the other (…) a primeval crossroads" [65: p. 8]. For Sergi, instead, the Mediterranean is a basic unit of his epistemology, a method of inquiry, a structure, and a framework. One can then conclude that, for some Italian intellectuals, the Mediterraneanism was an identity ideology that permeated the analysis of reality and conditioned the process of comprehension and apprehension of the physical geographical and the symbolic space.

Author details

Fedra A. Pizzato

Address all correspondence to: apemedea@yahoo.it

Ca' Foscari University of Venice, Venice, Italy

References

[1] Lefebvre H. The Production of Space. Chicago: Blackwell; 1991

[2] Soja E. Postmodern Geographies: The Reassertion of Space in Critical Social Theory. London: Verso; 1989

[3] Gellereau M. Nous et les autres: les représentations des identités culturelles au service de nouveaux territoires. Études de Communicacion. 2003;**26**:1-11

[4] Bailly A, Scariati R. Voyage en géographie. Paris: Anthropos; 1999

[5] Moles AA, Rohmer É. Psychologie de l'espace. Paris: Casterman; 1978

[6] Benigno F. Parole nel tempo. Un lessico per la storia. Rome: Viella; 2013

[7] Nencini D. Ipotesi di traduzioni architettoniche dopo il moderno mediterraneo. In: Carlotti P, Nencini D, Posocco P, editors. Mediterranei Traduzioni della Modernità. Milan: Franco Angeli; 2014. pp. 105-120

[8] Berger S, Lorenz C. Nationalizing the Past: Historians as Nation Builders in Modern Europe. London: Palgrave Macmillan; 2010

[9] Gioberti V. Del primato civile e morale degli italiani. 2nd ed. Lausanne: Bonamici; 1845

[10] Pizzato FA. Fossili delle nazioni. Paletnologia, antropologia e nazionalismo in Italia (1871-1915). Italy: Ca' Foscari University of Venice; 2016

[11] Diaz Andreu M. A World History of Nineteenth Century Archaeology. Nationalism, Colonialism and the Past. New York: Oxford University Press; 2007

[12] McNeill W. Mythistory, or Truth, Myth, History, and Historians. The American Historical Review. 1986;**91**(1):1-10

[13] Tarantini M. Tra teoria pigoriniana e mediterraneismo. Orientamenti della ricerca preistorica e protostorica in Italia (1886-1913). In: La nascita della Paleontologia in Liguria Atti del convegno. Bordighera; 2008. pp. 53-61

[14] Nizzo V. La "questione pelasgica" in Italia 1890-1910. In: Attenni L, Baldassarre D, editors. Quarto seminario internazionale di studi sulle mura poligonali. Alatri: Aracne; 2013. pp. 27-45

[15] Koch G. Le costruzioni poligonie dell'Italia centrale. Lettera a sua eminenza il comm. Guido Baccelli. Annuario dell'Associazione Artistica tra i cultori di Architettura. 1896;**VI**:63-65

[16] Barelli L. L'Associazione Artistica fra i Cultori di Architettura e il rilievo di S. Maria in Cosmedin. XY Dimensioni del Disegno—Il rilievo tra storia e scienza. 1991;**5**:162-165

[17] Candidi Dionigi M. Viaggi in alcune città del Lazio che diconsi fondate dal Re Saturno. Rome: Arti Grafiche Pasquarelli; 1809

[18] De Cara CA. Gli Hethei-Pelasgi. Ricerche di storia e di archeologia orientale, greca ed italica. II. Rome: Tipografia dell'Accademia dei Lincei; 1902

[19] Grimaldi L. Studi archeologici fatti sulla Calabria Ultra Seconda fatti dal segretario perpetuo della Società economica di quella provincia. Neaples: Stabilimento Librario tipografico di Borel e Bompard; 1845

[20] Cifarelli FM. Tra Rivoluzione e Restaurazione: Petit-Radel e la riscoperta delle città "pelasgiche". In: Cancellieri M, Cifarelli FM, Palombi D, Quilici Gigli S, editors. Tra memoria dell'antico e identità culturale Tempi e protagonisti della scoperta dei monti Lepini. Rome: Espera; 2012. pp. 33-47

[21] Fonteanive R. Sui munimenti ed altre costruzioni poligonie od epimonolite dette ciclopiche, saturnie o pelasgiche e sui resti di tali fabbriche esistenti nella provincia romana. Rome: Tipografia I. Sciolla; 1887

[22] Nizzo V. Il dibattito sull'origine degli Italici nell'età di Luigi Pigorini: dall'antiquaria all'archeologia. In: Guidi A, editor. 150 di Preistoria e protostoria in Italia, Istituto Italiano di Preistoria e Protostoria. Firenze: Osanna Edizioni; 2015. pp. 261-267

[23] De Cara CA. Le necropoli pelasgiche d'Italia e le origini italiche. La Civiltá Cattolica. 1894;**15**(9):287-297

[24] De Cara CA. Il primo scavo di una città pelasgica nel Lazio. La Civiltá Cattolica. 1901;**18**(3):296-305

[25] De Cara CA. Gli Hethei-Pelasgi. Ricerche di storia e di archeologia orientale, greca ed italica. III. Rome: Tipografia dell'Accademia dei Lincei; 1902

[26] Bertani S. La "Civiltà Cattolica" nel dibattito sull'evoluzionismo ottocentesco. Annali di storia moderna e contemporanea. 2006:89-122

[27] Momigliano N, Farnoux A. Cretomania. Modern Desire for the Minoan Past. New York: Routledge; 2017

[28] Partridge JA. The Making of the Irish Nation: And the First-Fruits of Federation. London: T. Fisher Unwin; 1886

[29] Wyle JA. History of the Scottish Nation. London: Hamilton, Adams & Co.; 1886

[30] Savignoni L, Mengarelli R. Ricerche topografiche nelle vicinanze di Norba. Notizie dagli scavi. 1901:514-560

[31] Raspanti M. Il mito ariano nella cultura italiana tra Otto e Novecento. In: Burgio A, editor. Nel nome della razza Il razzismo nella storia d'Italia 1870-1945. Bologna: Il Mulino; 1999

[32] Sergi G. Origine e diffusione della stirpe mediterranea. Rome: Società editrice Dante Alighieri; 1895

[33] Said E. Orientalism. New York: Vintage Books; 1979

[34] Sergi G. La decadenza delle nazioni latine. Turin: Fratelli Bocca; 1901

[35] Davidson P. The Idea of North. Edinburgh: Reaktion Books; 2015

[36] Canfora L. Tacito e la riscoperta degli antichi Germani dal II al III. Studi Urbinati di storia, filosofia e letteratura. 1979;**53**(1-2):219-254

[37] Mazza M. La Germania di Tacito: etnografia, storiografia e ideologia nella cultura tedesca dell'Ottocento. Studi Urbinati di storia, filosofia e letteratura. 1979;**53**(1-2):167-217

[38] Penka K. Origines Ariacae. Linguistisch-ethnologische Untersuchungen zur ältesten Geschichte der arischen Völker und Sprachen. Vienna: Prochaska; 1883

[39] Fehr H. Germanen und Romanen im Merowingerreich. Frühgeschichtliche Archäologie zwischen Wissenschaft und Zeitgeschehen. Reallexikon der Germanischen Altertumskunde. Berlin: de Gruyter; 2010

[40] Ripley WZ. The Races of Europe. A Sociological Study. London: Kegan Paul, Trench, Trübner & Co.; 1899

[41] Sergi G. Arii ed Italici. Intorno all'Italia preistorica. Turin: Fratelli Bocca; 1898

[42] Govoni P. Un pubblico per la scienza. La divulgazione scientifica nell'Italia in formazione. Rome: Carocci; 2011

[43] Spadolini G. La Nuova Antologia dal Risorgimento alla Repubblica (1866-1988). La Nuova Antologia. 1988;**2166**:5-96

[44] Guidi A. Storia della Paletnologia. Rome: Laterza; 1988

[45] Peroni R. Preistoria e protostoria. La vicenda degli studi in Italia. In: Angle M, editor. Le vie della preistoria. Rome: Manifestolibri; 1992. pp. 9-70

[46] Sergi G. Gli arii in Europa e in Asia. Turin: Fratelli Bocca; 1903

[47] Sergi G. L'origine dei popoli europei e le loro relazioni con i popoli d'Africa, d'Asia e d'Oceania. Turin: Fratelli Bocca; 1908

[48] Mannori L. Alla periferia dell'Impero. Egemonia austriaca e immagini dello spazio nazionale nell'Italia del primo risorgimento (1814-1835). In: Bellabarba M, Mazohl B, Stauber R, Marcello V, editors. Gli imperi dopo l'Impero nell'Europa del XIX secolo. Bononia: Il Mulino; 2008. pp. 309-346

[49] Mangoni L. Una crisi fine secolo. La cultura italiana e la Francia fra Otto e Novecento. Torino: Einaudi; 1985

[50] Sergi G. Le origini. Turin: Fratelli Bocca; 1919

[51] Braudel F. The Mediterranean and the Mediterranean World in the Age of Philip II. I. Berkeley: University of California Press; 1949

[52] Sergi G. Usiologia, ovvero scienza dell'essenza. Rinnovamento dell'antichissima filosofia italiana. Noto: Tipografia Morello; 1868

[53] Sergi G. The Mediterranean Race: A Study of the Origin of European Peoples. London: Walter Scott; 1901

[54] Montelius O. Die Chronologie der ältesten Bronzenzeit in Nord Deutschland und Skandinavien. Braunschweig: Vieweg; 1900

[55] Sergi G. Dalle esplorazioni del Turchestan. Atti della Società Romana di Antropologia. 1905;**XIII**:305-321

[56] Sergi G. Le prime e le più antiche civiltà. I Creatori. Turin: Fratelli Bocca; 1926

[57] Pumpelly R. Exploration in Turkestan. Expedition of 1904. Prehistoric Civilization of Anau voll. II. Washington: Carnegie Institution; 1908

[58] Farinelli F. I segni del mondo: immagine cartografica e discorso geografico in età moderna. Florence: La Nuova Italia; 1992

[59] Boria E. Cartografia e Potere. Rome: UTET; 2007

[60] Harley JB. The New Nature of Maps. Baltimore: The Johns Hopkins University Press; 2001

[61] Sergi G. Intorno alla monogenesi del linguaggio. Rivista italiana di Sociologia. 1909; **XI**(6):731-742

[62] Giardina A, Vauchez A. Il Mito di Roma. Da Carlo Magno a Mussolini. Milan: Laterza; 2008

[63] Sergi G. La vita animale e vegetale. Origine ed evoluzione. Milan: Sonzogno; 1922

[64] Braudel F. Les Mémoires de la Méditerranée. Paris: Edition de Fallois; 1998

[65] Braudel F. La Méditerranée: l'espace et l'histoire. Paris: Flammarion; 1985

Section 2

Biodiversity: Nature at the Crossroads

The Marine Biodiversity of the Mediterranean Sea in a Changing Climate: The Impact of Biological Invasions

Anna M. Mannino, Paolo Balistreri and Alan Deidun

Additional information is available at the end of the chapter

http://dx.doi.org/10.5772/intechopen.69214

Abstract

The Mediterranean Sea, one of the most complex marine ecosystems, is inhabited by a rich and diverse biota which is disproportionate to its dimensions. It is currently affected by different pressures, mainly driven by human activities such as climate change and bioinvasions. This Sea, also due to its geographic position (wedged between the temperate climate of central Europe and the arid climate of northern Africa), seems to be one of the regions most susceptible to global climate change. The increased rates of introduction and spread of marine alien species may represent a supplementary stress factor to Mediterranean marine native biota already challenged by climatic abnormalities. The Suez Canal is considered to be the main vector of introduction of non-indigenous marine species into the Mediterranean Sea. Due to the dramatically accelerating rate of such introductions and due to the sheer magnitude of shipping traffic, the Mediterranean Sea may be considered as a true hotspot of marine bioinvasions. The complexity of interactions between native and invasive species and the associated resulting impacts make environmental management of such an issue particularly difficult. A collaboration between researchers, resource management agencies and policy makers is called for to bolster the effectiveness of invasive species management procedures.

Keywords: climate change, invasive alien species (IAS), management measures, marine biodiversity, Mediterranean Sea

1. Introduction

The Mediterranean Sea, a 'sea in the middle of the land' (**Figure 1A**), a semi-enclosed sea at the crossroads between Europe, Africa and Asia, represents just 0.82% of the surface area of the world's oceans [1]. It is the deepest (average 1460 m) and largest (2969.000 km^2) enclosed sea on Earth [2] and from space looks like a narrow lake, stretching 3800 km from East to West and 800 km from North to South. Hot dry summers and low input from rivers make it a concentration basin. The Mediterranean Sea has a slight interchange of waters with the Atlantic Ocean through the Strait of Gibraltar in the west and with the Black Sea and the Sea of Marmara through the Dardanelles in the North-East [2]. The Strait of Gibraltar plays an important role in the circulation and productivity (the rate of generation of organic matter) of the Mediterranean, an extremely oligotrophic sea (largely due to a poor nutrient supply) [3]. Indeed, Atlantic surface waters, after having circulated within the Mediterranean in an anticlockwise direction, flow out denser and deeper below the entering waters in the form of the Mediterranean outflow water (MOW). As one progressively moves east into the Mediterranean Basin, oligotrophy increases whereas the productivity decreases. The Suez Canal, within the Levantine Basin, opened in 1869, provides an artificial navigable connection with the Red Sea and the Indo-Pacific Ocean. The Strait of Sicily, separating the island of Sicily from the coasts of Tunisia, is generally considered as the boundary zone between the two main sub-regions of the Mediterranean Sea, the Western and the Eastern, previously placed in the mid Ionian Sea [4]. The two sub-regions are divided into four main sub-basins (western Mediterranean, Adriatic Sea, Ionian Sea and Aegean-Levantine Sea), characterized by distinctive hydrodynamics and water circulation patterns.

The turbulent past geological events and a kaleidoscope of climatic and hydrologic conditions make this Sea one of the most complex marine ecosystems, at the crossroads of different biogeographic provinces [5], inhabited by a rich and diverse biota [2], currently affected by different pressures, mainly driven by human activities. The most important threats to Mediterranean biodiversity are habitat modification and loss, climate change, pollution, coastal urbanization, overexploitation and alien species, that is, organisms that arrive in a region beyond their native range due to direct or indirect human intervention [2, 6].

In this chapter, on the basis of a review of the available literature, we analysed (a) the origin of the Mediterranean Sea as well as the origin of its rich and diverse biota; (b) the main threats to Mediterranean biodiversity, focusing on climate change, one of the main drivers of the ongoing change in Mediterranean biodiversity; (c) the phenomenon of biological invasion (mediated through the entry of invasive alien species (IAS), which spread widely and cause environmental, economic, or human health impacts) in the Mediterranean Sea, particularly stressing impacts on ecosystems and pathways of introduction. Finally, we provide summaries for 10 invasive species known worldwide for their environmental and economic impact and we suggest possible actions to prevent and manage this phenomenon.

2. History of the Mediterranean Sea

The Mediterranean Sea is the result of a tormented geological history dating back around 230 million years before present (hereafter, myr BP), when a large single landmass Pangea and a large ocean Panthalassa co-existed (**Figure 1B**). When this landmass started breaking up, an eastward-open equatorial ocean, the Mediterranean Sea ancestor, inhabited by a highly diverse biota

http://dx.doi.org/10.5772/intechopen.69214

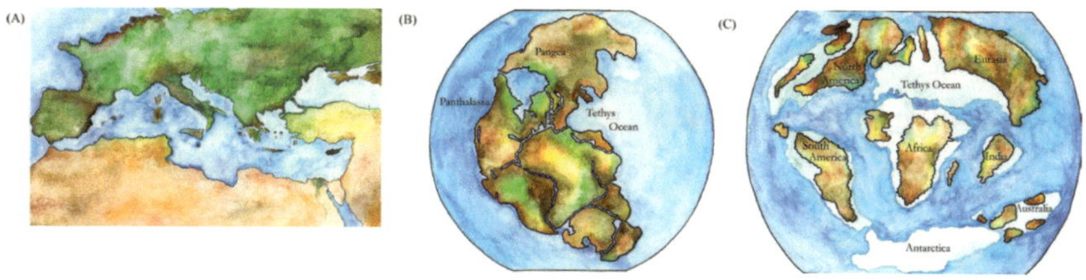

Figure 1. **(A)** The Mediterranean Sea, **(B)** Pangea and Panthalassa, **(C)** The Tethys Ocean (drawings Lucia Gastoldi).

of warm-water origin, was formed. The formation of this ocean—called Tethys—divided the Pangea into the Laurasia continent to the North and the Gondwana continent to the South and connected the Mediterranean connection with the Atlantic Ocean to the Indo-Pacific Ocean (**Figure 1C**). Around 130 myr BP (Cretaceous), the connection with the Indo-Pacific Ocean was interrupted. Subsequently, around 45 myr BP (Eocene), the original Tethys became smaller as a consequence of African and Eurasian plate collisions, responsible also for the formation of the Alps and Himalayas. Twenty-five myr BP (Miocene), the African plate made contact with the Eurasian plate, dividing the Tethys Sea into two parts: the ancestor of the Mediterranean Sea in the South and the so-called Paratethys in the North-East. Both seas underwent significant reductions and the Paratethys remnants formed the Black, the Caspian and the Aral Seas. When the narrow Isthmus of Suez was formed and the connection with the Indian Ocean was interrupted (around 13 myr BP, Miocene), this water body slowly lost its tropical biota. During the Miocene (6–7 myr BP), the connection (the Gibraltar Strait) with the Atlantic Ocean was closed as a consequence of a collision between Africa and the south-western segment Eurasian plate. This event, responsible for a significant negative water balance, led the Mediterranean Sea to dry up, ending up in the form of a series of hypersaline lakes separated by vast expanses of exposed sandy seabed (the so-called Messinian Salinity Crisis, around 5.6–5.5 myr BP). This spectacular event caused a severe reduction in the original rich biota and the disappearance of several paleotropical elements, even though a few elements were able to survive in refuges. Around 5 myr BP (Pliocene), the Strait of Gibraltar opened once again, allowing the waters of the Atlantic Ocean to flood the Mediterranean, repopulating it exclusively with species of Atlantic origin [5]. During the Quaternary, the alternation of glacial periods with warm interglacial periods allowed the influx into the Mediterranean of Atlantic species of boreal or subtropical origin [1].

3. The Mediterranean: a hotspot of biodiversity under threat

Currently, biodiversity is widely expressed as species richness (number of species occurring in a specific area), a term considered as a valuable indicator of environment health [7, 8]. Since marine ecosystems have probably received less attention than their terrestrial counterparts, the number of marine species is currently still considerably lower than that of terrestrial species [9].

The Mediterranean Sea, even though representing a small part of the world's oceans, is inhabited by an unusually rich and diverse biota. It hosts approximately 17,000 species, representing 4–18% of the world's marine biodiversity, and includes temperate, cosmopolitan, subtropical,

Atlantic and indo-pacific taxa [1, 2]. As a result, the Mediterranean Sea is considered as a true hotspot of biodiversity [1, 2, 10], even by virtue of the high rates of endemism it supports (an estimated 20–30% of marine species in the Mediterranean are considered endemic to the Basin [10]). The high marine biodiversity of the Mediterranean Sea may be due to different reasons: for example, it has been more intensively studied than the other seas; its tormented geological history and the numerous climatic and hydrologic events that led to the co-occurrence of temperate and subtropical species [1]. The low sea water temperatures during the glacial periods greatly contributed to the invasion of cold-water species from the northern Atlantic. The preponderance of species of Atlantic origin within the current Mediterranean, as a result of the reopening of the Strait of Gibraltar, makes the Sea, from a biogeographic point of view, an Atlantic Province. The opening of the Suez Canal coupled with a rise in sea temperature increase led to the settlement of thousands of tropical species, either in the form of 'Lessepsian migrants', named after Ferdinand de Lesseps, a French diplomat and designer of the Suez Canal [11], or as 'Erythrean aliens' [12]. According to [13], the current high Mediterranean settlement of tropical species coming from the Suez Canal is tantamount to a re-colonization by Tethyan descendants rather than an invasion by alien species. Sea warming is also responsible for the increase in the ingression of Atlantic thermophilic species into the Mediterranean through the Gibraltar Strait, the so-called 'range-expanding species'. The introduction of thousands of tropical species into the Mediterranean is without doubt the most remarkable biogeographic phenomenon within today's Basin [13]. With respect to marine biodiversity, an evident heterogeneity between the marine species composition of the western and eastern basins exists. In particular, we observe that in the western part of the Mediterranean Sea the biota composition is influenced by the Atlantic Ocean, whereas in the Levant Sea, this is strongly linked to the Red Sea. The western Mediterranean shows the highest species richness followed by the central basin, the Adriatic and Aegean Seas, and the Levantine basin, which is more influenced by species introduction through the Suez Canal and which displays the lowest values [10]. Endemic marine species in the Mediterranean either consist of rare paleo-endemisms of Tethyan origin (i.e. they precede the Messinian Salinity Crisis) and of more frequent neo-endemisms of Pliocenic origin. The western basin shows a higher rate of endemism than the eastern basin, appearing to be an active centre of endemism [10]. The neo-endemism *Cystoseira*, considered a key-stone genus, presents 21 species endemic to the Mediterranean [14]. In particular, the paleo-endemic *Caulerpa sedoides* is confined to the coasts of Algeria, Tunisia, and the Island of Pantelleria. Another paleo-endemism, the seagrass *Posidonia oceanica*, could be considered a Messinian species, that is, a species which persisted, probably in refugia, when the Mediterranean dried up.

Nowadays, changes in the Mediterranean biodiversity are essentially driven by human activities. The Mediterranean is amongst the most impacted regional sea areas, as a consequence of different anthropogenic pressures on different coastal and marine ecosystems within the Basin: habitat modification and loss, climate change (e.g. warming, acidification and sea level rise), pollution, coastal urbanization, overexploitation and the intentional or indirect introduction of alien species (synonyms: exotic, non-native, non-indigenous, and allochthonous) [2, 6, 15, 16]. Habitat modification and loss, pollution and overexploitation are considered to be the main threats to marine biodiversity and natural ecosystem functioning in the Basin, but other threats such as climate change and invasive species (the most insidious of alien species) introductions are similarly expected to have significant impacts [5, 17, 18]. In the Mediterranean Sea, several valuable and unique habitats,

http://dx.doi.org/10.5772/intechopen.69214

supporting an extensive repository of biodiversity, are also under threat [19]. A recent study showed that coastal areas and shelves are the most threatened habitats, and the areas of highest concern are concentrated in the northern region [18]. The Mediterranean Sea represents the highest proportion of threatened marine habitats in Europe (32%, 15 habitats) with 21% being listed as vulnerable and 11% as endangered. Among the endangered habitats, we find, for example, the photophilic communities consisting of canopy-forming algae in the infralittoral and upper circalittoral rock, whereas the following habitats are considered vulnerable: biogenic habitats of Mediterranean mediolittoral rock, photophilic communities dominated by calcareous, habitat-forming algae, and *Posidonia* beds in the Mediterranean infralittoral zone.

4. Climate change and its effects on Mediterranean biodiversity

Global climate change and its associated impacts on the marine domain, for example, sea warming, ocean acidification, and sea level rise (at the rate of about 1 mm per year), is an ongoing phenomenon which is certainly affecting biodiversity, human activities and health, but we do not know exactly how profound the consequent changes in marine ecosystems will be [6, 20]. Climate change, may act at different biological levels: individual, population, and ecosystem. In particular, species with a low dispersion ability are highly affected by climate change, which may also lead to local extinctions, greatly contributing to biodiversity loss. The Mediterranean Sea, also due to its geographic position between the temperate climate of central Europe and the arid climate of northern Africa, seems to be one of the most vulnerable regions to global climate change but we don't know if all sub-basins will be equally affected by this phenomenon [6, 20]. The Mediterranean climate is expected to become warmer and drier with an increase in inter-annual variability due to extreme heat and drought events [20–22]. Even though the Mediterranean Sea is probably entirely affected by this warming trend, data on this phenomenon are mainly reported for the north-western Mediterranean [6]. Mediterranean temperature anomalies observed during summer 1999 and 2003 led to catastrophic mass mortality events, in particular of benthic invertebrates (e.g. sponges, gorgonians, bryozoan and molluscs) [23–25]. The 2003 event resulted in mortality of *P. oceanica* but also in large-scale flowering episodes for this seagrass [26, 27].

Sea warming may also have effects on the virulence of pathogens, as observed for *Vibrio shiloi*, responsible for the whitening of the coral *Oculina patagonica* in the eastern Mediterranean [28]. Moreover, it is responsible for the expansion of toxic dinobionts such as *Ostreopsis ovata*, which produces palytoxins, a serious public health hazard [6]. Temperature anomalies seem also to negatively affect the chemical defences of marine organisms [29], allowing pathogens to act undisturbed. The warming trend experienced by the Mediterranean is also influencing the distribution of its marine native species, with some (e.g. *Thalassoma pavo*) embarking on a westward range expansion within the same basin, the so-called 'meridionalisation' phenomenon as opposed to the 'tropicalisation' of Mediterranean marine biota fuelled by the invasion of Lessepsian migrants [6]. Tropical species, that is, Lessepsian migrants or those of Atlantic origin, have long been confined to the extremities of the eastern Levantine or western basin, but as a consequence of the warming trend of the Mediterranean as well as the ever-expanding shipping networks and volumes, they are rapidly spreading throughout the entire basin [2].

Warming may also lead to the decrease or extinction of cold-water boreal species which cannot migrate further northwards and seek deeper waters, as is the situation in the Adriatic.

The increased introduction and spread of alien species may represent a supplementary stress factor for native species already weakened by climate variations.

The rise ofdissolved carbon dioxide, lowering the pH of the ocean waters, is altering the seawater carbonate system, as part of the so-called 'ocean acidification' phenomenon, which may affect organisms and ecosystem functioning [30]. Calcifying organisms (planktonic and benthic) such as corals, foraminifera, coccolithophores and coralline red algae, important depositors of calcium carbonate, may be hampered under increasing levels of water acidity in their attempts to synthesize their calcium carbonate shell or skeletons [31]. Coralline red algae, contributing to the formation of coralligenous habitats and coastal reefs, are particularly sensitive to ocean acidification, partly because they are composed of high magnesium-content calcite, the most soluble form of $CaCO_3$ [32]. The decline in coralline red algae is particularly significant for the ecosystems they form and for the carbonate cycle in general [33].

Moreover, we have to consider that climatic factors may act in combination and/or interact with other factors such as pollution and overexploitation. Acidification and warming, for example, may exert synergistic effects on the calcification rates of corals and crustose coralline algae [33].

5. Alien species in the Mediterranean Sea: pathways and impacts

Alien taxa are species, subspecies or lower taxa introduced outside of their natural, past or present, range and outside of their natural dispersal potential. Their presence in the given region is due to direct or indirect (e.g. passive dispersal through a man-made shipping canal) introduction by humans [34]. A common denominator amongst the new arrivals within the Mediterranean is their thermophilic affinity, being native of warmer sea regions, such as the Red Sea or the Indo-Pacific region for the so-called 'Lessepsian' migrants or the sub-tropical equatorial Atlantic for the 'range-expanders' coming in through the Straits of Gibraltar [35]. According to the convention on biological diversity (CBD), the compilation and dissemination of information on alien species that threaten ecosystems, habitats or species is a priority [36]. The estimate for new introductions in the Mediterranean is constantly being revised and may be as high as 1000 species. The majority of alien species in the Mediterranean are of Indo-Pacific origin (41%), followed by those of Indian Ocean (16%) and Red Sea (12%) origin; some species have a pantropical or circumtropical distribution (19%) [2]. The number of alien species differs among the Mediterranean sub-regions (the Eastern, the Western, the Central and the Adriatic Sea) described under the Marine Strategy Framework Directive [37]. The majority of alien species occurs in the Eastern sub-region, with lower numbers being registered in the Western and Central sub-regions and in the Adriatic Sea [15].

Alien species are entering the Mediterranean Sea naturally or as directly mediated through human activities. Three main mechanisms of introduction for non-indigenous species (NIS) are currently recognized (Convention on Biological Diversity: [38]): wilful importation as a commodity, arrival via a transport vector or unassisted spread from a neighbouring region through man-made channels and canals, for example. Six pathways are associated to these mechanisms: commodities intentionally released or escaped (e.g. aquaculture, botanic gardens, agriculture and aquarium trade) or contaminants of commodities (seeds and fisheries) unintentionally transported, transport-stowaway (shipping), corridors (man-made marine or inland canals) and secondary natural spread of species introduced through the other pathways [38].

The Suez Canal is considered to be the main vector for marine introductions into the Mediterranean Sea. The Suez Canal in fact has been responsible for 53% of all exotic marine species entering the Mediterranean, and some species (11%) initially entered through the Suez Canal but were later dispersed within the Basin by vessels [2]. Prior to the opening of the Suez Canal, the entry of 'Lessepsian migrants' was insignificant due to the existence of a high salinity barrier, constituted by the Bitter Lakes which stymied the passive entry of species through this route, but later, with the construction of the Aswan Dam, this barrier ceased to exist, allowing the migration of hundreds of Red Sea organisms. The widening of the existing canal at Suez in 2015, so as to cater for increased volumes of shipping, is expected to have a strong impact on the Mediterranean marine ecosystems [39].

The rates of introduction of new marine species through the Bosphorus are negligible, especially when compared to those through the Strait of Gibraltar and the Suez Canal.

Transportation via shipping-related infrastructure, for example, ballast waters, ballast tanks, anchors and fouling, is expected to increase as a result of our increasing reliance on maritime trade, and is the second vector of introduction in terms of importance. Even though the majority of the vessels originate in North Atlantic ports, North Atlantic species represent only 14% of the alien biota entering the Basin, whereas around 50% are of Indian and Indo-West Pacific origin. That is probably due to the differences or similarities in salinity and temperature between these two areas of origin and the Mediterranean Sea. Research vessels and platforms for offshore oil and gas exploration may also contribute to the spread of alien species [40]. Due to the important role played by shipping, ports and adjacent areas may be considered as 'hotspots' of biotic invasion. Shipping is also important as a vector for secondary introduction; for instance, *Sargassum muticum* and *Caulerpa taxifolia* have been spread in the Mediterranean through fishing and recreational vessels [41, 42]. The Mediterranean biodiversity is particularly vulnerable to shipping impacts such as pollution, anchor damage and introduction of exotic species, due to the high volume of shipping within the Basin, which greatly expanded over the past half century and which is expected to grow in the next years. The current ongoing intensive movement of people and goods contributed to greatly increase the rate of spread of invasive species [43].

The aquaculture industry represents the third vector of introduction but probably it could be more important than previously thought as a consequence of the unintentionally introduced species, for example, attached to the intentionally introduced species. Introductions through the aquarium trade, involving more freshwater species than marine ones, represent only a small fraction of the total number of invasive species, but these are on the increase due to the popularity of household marine aquaria. Here again, the number of alien species entering as a consequence of aquarium trade is probably higher than that officially reported

[44], also because many alien species previously assigned to other introduction pathways could in fact have arrived via aquarium trade.

However, it is not always possible to identify with certainty the pathway for a species introduction, except for intentional introductions, also because an introduction may have occurred through multiple pathways or stages. Even worse, assigning a native or non-indigenous status to a species might not even be possible, as is the case for cryptogenic species, whose origins are unknown.

Whatever is the pathway of introduction, when an alien species become established, that is, capable of reproducing and maintaining self-perpetuating populations in the wild, it is really difficult to extirpate it completely [45]; hence, prevention, for example, removal of pathways, is the only effective action for reducing the introduction and spread of alien species.

However, the rise of establishment conditions for an alien species does not necessarily translate into invasion success. The success of alien species may depend on the biological characteristics and the dispersion vectors of the species, and the susceptibility of the receiving habitat, for example, availability of niches, low species richness, physical and chemical conditions, release from competitors and parasites [46, 47]. For example, rich and highly structured ecosystems are more resistant to invasion than poor and disturbed ones [48].

It has been ascertained that alien species may have significant environmental, socio-economic and human health impacts [49–52]; consequently, globally, they are recognized as one of the major threats to biodiversity. However, it is not always true that 'native is good, alien is bad'. Some alien species may have positive effects on native biodiversity and can enhance or provide new ecosystem services [53]. Alien species may affect native biota directly or indirectly and may act at different levels, that is, at the gene, individual, population and community level [54]. Moreover, alien species impacts can vary spatially and temporally, that is, there is a lag between alien population growth and native species response [55, 56]. The impacts of alien species generally increase when they are established and they spread in the new environment, but they may also have impacts at the initial stages, as soon as they are introduced [57, 58]. The level of the impact of an alien species may vary considerably, ranging from a negligible to a profound impact. Indeed, many alien species are known for having caused significant changes in native ecosystems, whereas others apparently have had little or no visible effects on receiving environments [58, 59]. Moreover, very often, the impacts become visible long after the onset of invasion or they are not well known, even though some effects are visible. Conversely, the effect of the introduction of a herbivore, which overgrazes algal communities, for example, is easy to detect. To plan effective management actions against biological invasions, a system for evaluating, comparing and possibly predicting the level of impact of different alien species is essential. Recently, a dynamic method for classifying alien species on the base of the level of their environmental impacts (ranging from 0 to 5), considering their impact mechanisms, for example, rates of herbivory, competition and impact on ecosystems, was proposed [51] and adopted by IUCN (International Union for Conservation of Nature). Since it is not always easy to assess and/or predict the impact of alien species, readily accessible information on spread ability and ecological impacts is also necessary.

http://dx.doi.org/10.5772/intechopen.69214

6. Invasive alien species (IAS) in the Mediterranean Sea

Invasive alien species (IAS) are those alien taxa that have established large populations, significantly expanding their range, and/or exert substantial negative impact on native biota, for example, substitution of native species, biodiversity loss, habitat modifications and alterations in community structure, socio-economic amenities, or human health [53, 60–64]. According to the 'tens rule' proposed by Williamson [65], approximately 10% of all introduced species in a given area will become established, and approximately 10% of established species will become invasive. Currently, invasive species are considered as one of the main causes of biodiversity loss after habitat destruction [2, 62]. Even though the impact of most invasive species remains unknown [47], they are regarded worldwide as a threat. Due to the dramatically accelerating rate of species introductions, the Mediterranean Sea may be considered a true hotspot of marine bioinvasions [66], and marine protected areas (MPAs) within the Basin have not been spared this environmental hazard. To date, almost 1000 marine non-indigenous species (NIS) have been introduced in the Mediterranean, of which more than half are considered to be both establishing and spreading, criteria that categorize them as invasive species [67].

The success of invasive species in the new areas is linked to some peculiar characteristics such as the capacity to thrive in different environments, to tolerate a wide range of environmental conditions, to exploit a variety of food sources, together with their high growth and reproduction rates and the lack of natural predators [68]. However, possession of these characteristics does not automatically translate into a successful invasion. Indeed, the success of invaders within a new area to which they are introduced is not predictable [69].

Many inventories of alien/invasive species have been compiled to date. In order to guarantee the scientific quality of the information constituted by these inventories, a guide for identifying the causes of uncertainty that can negatively affect data quality, as well as a protocol to select eligible records for such inventories, based on standardized criteria, was recently proposed [70].

We report subsequently on the impacts of a selected group of marine species featuring in the list of 100 'worst' Invasive Alien Species of the Mediterranean [71], mainly belonging to zoobenthos (particularly Mollusca, fish and Crustacea) and phytobenthos.

Among the macroalgae, *Caulerpa* species are well-known invaders in the Mediterranean. They have been reported to cause serious problems to native marine ecosystems [72]. They are able to compete with native key species [73], becoming the dominant species themselves, but they are also known to have negative impacts on commercial and recreational fishing.

Turf algae such as *Womersleyella setacea* and *Acrothamnion preissii*, reported to seriously change native communities and to clog up fishing nets and gear [74], may exert a considerable socio-economic impact. Several invasive bivalves, imported for mariculture purposes, are well known for their negative impacts on native marine ecosystems, as is the case of the Pacific oyster *Crassostrea gigas*, considered to be the culprit behind the most spectacular invasion in the Venice lagoon [75]. These bivalves, forming over time consolidated biogenic reefs, may act as 'ecosystem engineers' [71]. The jellyfish *Rhopilema nomadica* has been reported to negatively affect coastal installations, fisheries, human health and tourism [76].

The Atlantic species *Percnon gibbesi* has spread rapidly in the Mediterranean ever since its first record from the island of Linosa in the central Mediterranean in 1999 [77] and is probably the most invasive decapod species in the region. It consumes primarily algae such that it may seriously affect the structure of algal assemblages within a given area, and it may also compete with native species for food and shelter [78].

Invasive dinoflagellates such as *Alexandrium catenella, O. ovata* and *Coolia monotis* exert serious impacts on human health but also on tourist activities in the areas where they occur [79]. Invasive fish species also represent a serious public health hazard, for example, *Lagocephalus sceleratus,* is a potential risk to humans, since it contains tetrodotoxin (TTX) that may be a source of food poisoning [80].

7. Ten of the worst marine invasive species

We present subsequently the descriptive summaries for 10 invasive species known worldwide for their environmental and economic impact. Phytobenthos: *C. cylindracea, S. muticum, Undaria pinnatifida, W. setacea.* Zooplankton/Cnidaria: *R. nomadica.* Zoobenthos/Mollusca: *Brachidontes pharaonis.* Zoobenthos/Crustacea: *P. gibbesii.* Zoobenthos/Fish: *Fistularia commersonii, L. sceleratus, Pterois volitans.*

The species are grouped into phytobenthos, zooplankton and zoobenthos. For each species, we report a photograph, a description of the main morphological and ecological characteristics, the origin, the first Mediterranean record, the distribution in the Mediterranean (as species may spread quickly, this information may soon become out of date), the likely pathway of introduction and spread, and the potential or documented impacts [68, 81–86].

7.1. Phytobenthos

Scientific name: *Caulerpa cylindracea* Sonder, 1845 (**Figure 2A**).

Classification: *Chlorophyta* (*Ulvophyceae, Caulerpaceae*).

Common name: grape algae.

Description: a green algae with a creeping axis (stolon), attached to the substratum by numerous short rhizoids, and bearing erect axes (fronds) slightly inflated above their attachment to the stolon. Fronds, up to 20 cm high, bears rounded, vesicular branchlets (also called ramuli).

Ecology: Subtidal communities, on all kinds of substrata, from 0 to 70 m depth.

Origin: Australia and New Caledonia.

First Mediterranean record: in 1990 off the coasts of Libya [87].

Distribution: throughout the Mediterranean.

Mode of introduction: remains unknown; however, maritime traffic and the aquarium trade are the most likely vectors for its introduction.

Impacts: Invasive species that competes with native species and may change native benthic communities. There are also reports of fishing nets being clogged and broken by this algae.

Scientific name: *Sargassum muticum* (Yendo) Fensholt, 1955 (**Figure 2B**).

Classification: *Ochrophyta* (*Phaeophyceae, Sargassaceae*).

Common name: Japanese wireweed.

Description: Algae up to 4 m high, attached to the substratum by a discoid holdfast; axis erect and spirally branched with leaf-like simple branchlets; subspherical air vesicles; cylindrical and smooth receptacles.

Ecology: Shallow subtidal communities. It can form floating mats on the sea surface.

Origin: North-western Pacific.

First Mediterranean record: in 1980 from France, Etang de Thau [88].

Distribution: Balearic Islands, Corsica, France, Italy, Spain.

Mode of introduction: unintentionally, with imported oysters likely from Japan.

Figure 2. **(A)** *Caulerpa cylindracea* (photo Fabio Russo), **(B)** *Sargassum muticum* (photo Conxi Rodríguez-Prieto), **(C)** *Undaria pinnatifida* (photo: Ester Cecere and Antonella Petrocelli), **(D)** *Womersleyella setacea* (photo: Enric Ballesteros).

Impacts: Invasive species. It is a fouling organism. There are reports of fishing nets being clogged and interferences with propellers and intakes.

Scientific name: *Undaria pinnatifida* (Harvey) Suringar, 1873 (**Figure 2C**).

Classification: *Ochrophyta* (*Phaeophyceae, Alariaceae*).

Common name: sea mustard or wakame.

Description: Sporophytes up to 1 m high, with a simple stipe continuing as a midrib through the lanceolate blade. Sporophylls along the stipe. Gametophyte microscopic.

Ecology: shallow subtidal communities. Annual.

Origin: North-western Pacific.

First Mediterranean record: in 1971 from France, Etang de Thau [89].

Distribution: France, Italy.

Mode of introduction: unintentionally, with imported oysters from Japan.

Impacts: Invasive species that cause problems to aquaculture and shipping.

Scientific name: *Womersleyella setacea* (Hollenberg) R.E. Norris, 1992 (**Figure 2D**).

Classification: *Rhodophyta* (*Florideophyceae, Rhodomelaceae*).

Common name: not assigned.

Description: Small filamentous algae (up to 1 cm high) with cylindrical axes. It forms mats of prostrate and erect branches. Prostrate filaments anchor the thallus to the substrate by means of rhizoids that terminally form a multicellular disc. Erect branches are poorly ramified.

Ecology: Subtidal communities. Usually epiphytic and forming perennial, dense tufts on native communities by modifying benthic assemblages and outcompeting key native species (i.e. *Paramuricea clavata, Cystoseira* and sponge species). Forms dense cover on coralligenous habitat.

Origin: Pacific Islands.

First Mediterranean record: in 1986 from Italy, Livorno [90].

Distribution: Alboran Island, Balearic Islands, Corsica, Croatia, Cyprus, France, Greece, Italy, Maltese Islands, Monaco, Rhodes Island, Spain.

Mode of introduction: shipping or accidental escape from an aquarium.

Impacts: invasive species that competes with native species. Its mats lead to an impoverishment of subtidal communities. There are also reports of fishing nets being clogged.

7.2. Zooplankton/Cnidaria

Scientific name: *Rhopilema nomadica* Galil, Spanier & Ferguson, 1990 (**Figure 3A**).

Classification: *Cnidaria* (*Scyphozoa, Rhizostomatidae*).

http://dx.doi.org/10.5772/intechopen.69214

Common name: nomad jellyfish.

Description: A neritic epipelagic, swarming jellyfish. The bell is up to 90 cm in diameter, usually 40–60 cm. The body is light blue with blunt tuberculation of the exumbrella. The mouth arms end in vermicular filaments.

Ecology: can form dense aggregations in coastal areas during summer months, although it can also appear all year round. It has a two-stage life cycle consisting of a swimming medusa stage and a benthic polyp stage (the scyphistoma). Depending upon food availability and other environmental variables, the scyphistomas form large numbers of pelagic medusas.

Origin: Originally from East Africa and the Red Sea.

First Mediterranean record: in 1976 from the coasts of Israel [91].

Distribution: Egypt, Greece, Israel, Lebanon, Turkey, Malta, Tunisia.

Mode of introduction: entered the Mediterranean through the Suez Canal.

Impacts: It is a voracious predator that consumes vast amounts of shrimp, mollusc and fish larvae and can cause major trophic cascades in the marine food web, with a resulting impact on biodiversity. It can inflict painful injuries to bathers. Furthermore, large swarms can clog fishing nets, consequently reducing catches, and block cooling water intakes of coastal industrial facilities and desalination plants.

7.3. Zoobenthos/Mollusca

Scientific name: *Brachidontes pharaonis* (P. Fischer, 1870) (**Figure 3B**).

Classification: *Mollusca* (*Bivalvia, Mytilidae*).

Common name: rayed erythrean mussel.

Description: A small gregarious intertidal bivalve with a 40-mm shell, externally dark brown-black and internally violet-black. Shell equivalve, inequilateral, attached to substrate by stout byssus. Sculpture of numerous fine radial bifurcating ribs, which become coarser posteriorly and margin crenulate. The hinge has dysodont teeth.

Ecology: Suspension feeders. Lives in shallow water (at sea level or just below) attached by

Figure 3. (**A**) *Rhopilema nomadica* (photo: Maria Ghelia), (**B**) *Brachidontes pharaonis* (photo: Bella Galil), (**C**) *Percnon gibbesi* (photo: Fabio Russo).

its byssus to rocks and stones, mostly in clusters. It may reach very high densities and cover completely a rocky shore.

Origin: Indian Ocean, Red Sea.

First Mediterranean record: in 1876 from Egypt, Port Said [92].

Distribution: Egypt, Lebanon, Israel, Italy, Greece, Syria, Turkey, Rhodes, Cyprus, Croatia.

Mode of introduction: propagules entered the Mediterranean through the Suez Canal. Ship transport as a major vector.

Impacts: Locally invasive. Can deplete the phytoplankton concentration in the water column, constraining the growth of other filter-feeding animals such as *Mytilaster minimus*. Dense mats of these bivalve populations in industrial facilities and salt works might result in high-energy consumption and economic losses.

7.4. Zoobenthos/Crustacea

Scientific name: *Percnon gibbesi* (H. Milne Edwards, 1853) (**Figure 3C**).

Classification: *Arthropoda* (*Malacostraca, Percnidae*).

Common name: sally lightfoot crab.

Description: This mimetic, relatively small crab is up to 3 cm across. Its body is flat and square-shaped with a smooth surface. The carapace is brownish green and the long, flattened legs are banded with golden yellow rings. The ventral surface is pale. The walking legs have a row of spines along the leading edge. The eyestalks and claws are orange.

Ecology: It occurs along shallow infralittoral rocky shores, under boulders, or in narrow crevices. It is an opportunistic herbivorous feeder, consuming filamentous and calcareous algae, and the fine film of algal growth. But it may also feed on animal food such as pagurids, polychaetes, gastropods, crustaceans and jellyfish.

Origin: Atlantic Ocean.

First Mediterranean record: in 1999 from Italy, Linosa Island and the Balearic Islands [77, 93, 94].

Distribution: Albanian, Montenegro, Italy, Balearic Islands, Malta, Crete, Greece, Turkey, Libya, Tunisia, Algeria, Cyprus, Israel, Lebanon.

Mode of introduction: It may spread in the ballast waters of ships and on fishing nets as well as being transported in its larval stage by water currents.

Impacts: Its widespread and expanding distribution along the Mediterranean coasts and the high rates of increase in abundance tell us that it is a highly invasive species. Since its habitat overlaps with those of native *Pachygrapsus marmoratus* and *Eriphia verrucosa,* exclusion of native crabs may occur in some areas.

http://dx.doi.org/10.5772/intechopen.69214

7.5. *Zoobenthos*/Fish

Scientific name: *Fistularia commersonii* Rüppell, 1838 (**Figure 4A**).

Classification: *Chordata* (*Actinopterygii, Fistulariidae*).

Common name: bluespotted cornetfish.

Description: a fish with a grey to olivegreen body, commonly 20–100 cm long (max. 150 cm). The body is extremely elongated, becoming depressed after head, which is very long. Snout tubular and long, ending in a small mouth. Teeth are very small. Dorsal and anal fins are posterior in position, opposite to each other. The skin is smooth, without bony plates along the midline of the back.

Ecology: Lives solitary or in small groups, near reefs. Feeds mainly on fish as well as on squids and shrimps.

Origin: Indo-Pacific and eastern central Pacific.

First Mediterranean record: in 2000 from Israel [95].

Distribution: Turkey, Rhodes Island, Crete, Greece, Italy, Algeria, Spain, Libya, Cyprus, Israel, Tunisia.

Mode of introduction: Entered the Mediterranean through the Suez Canal.

Impacts: It is an extremely voracious predator and is aggressive when in schools. It preys on native commercially important fish. Its rapid increase in abundance may potentially have adverse effects competing with native fish communities by exploiting local resources faster.

Scientific name: *Lagocephalus sceleratus* (Gmelin, 1789) (**Figure 4B**).

Classification: *Chordata* (*Actinopterygii, Tetraodontidae*).

Common name: silverstripe blaasop.

Description: The body is elongated (20–85 cm), somewhat compressed laterally and inflatable. The body is brownish in colour with black, regularly distributed spots of equal size dorsally; a conspicuous wide silver band is present on the lower parts of the flanks, from the mouth to the caudal fin. The head is large with a blunt snout.

Figure 4. (**A**) *Lagocephalus sceleratus* (photo: Alan Deidun), (**B**) *Fistularia commersonii* (photo: Puccio Distefano), (**C**) *Pterois volitans* (photo: Mariolina Corsini Foka and Gerasimos Kondylatos).

Ecology: Capable of inflating when threatened. Feeds on benthic invertebrates. It is a benthic species living above sandy substrate in the vicinity of the coral reef but also in deep waters up to 250 m.

Origin: Indian and Pacific Oceans.

First Mediterranean record: In 2003 from Turkey, in the Gökova Bay [96].

Distribution: Turkey, Israel, Rhodes, Greece, Tunisia, Crete.

Mode of introduction: Through the Suez Canal.

Impacts: It contains tetrodotoxin (TTX) that may cause death. It attacks fishes caught in nets and on lines and can cause serious damage to both fishing gear and catch.

Scientific name: *Pterois volitans* (Linnaeus, 1758) (**Figure 4C**).

Classification: *Chordata* (*Actinopterygii, Scorpaenidae*).

Common name: Red lionfish.

Description: It has greatly elongated dorsal-fin spines and a long pectoral fin. Head and body cream coloured to red to reddish-brown vertical stripes. Eight small white spots are present on the lateral line. It has often large tentacles above eyes.

Ecology: It inhabits natural and artificial substrates. It is a voracious predator feeding on small fish, shrimps and crabs. Lionfishes are well known for their venomous spines to ward off predators.

Origin: Indo-Pacific and Atlantic Ocean.

First Mediterranean record: In 2016 from Turkey, Iskenderun Bay [97].

Distribution: Turkey.

Mode of introduction: Through the Suez Canal.

Impacts: Is an invasive mesopredator that voraciously consumes native coral-reef fishes.

8. Managing and prevention strategies

Globalization of trade and travel, together with climate change, habitat modification, aquaculture and mariculture, the aquarium trade and biological invasions themselves have facilitated the introduction and spread of marine invasive species, and without an effort to manage such introductions, the number of invasive species is expected to become much greater in the future. Since invasive species are globally recognized as a threat, policies to reduce the introduction and spread of alien species, and to manage those species whose populations are already established, are a priority for governments within Europe, and for the European Union [98, 99]. Currently, several policies and legal instruments concern the introduction and spread of potentially invasive species: for example, the convention on biological diversity (CBD), the Habitats Directive (Directive 92/43/EEC), the European Union marine strategy framework directive (MSFD), the ecosystem approach (EcAp) within UNEP-MAP's Barcelona Convention, the Biodiversity Strategy and the Regulation on the Prevention and Management of the Introduction and Spread of invasive alien species [100].

For a better understanding of the invasive potential and spread dynamics of invasive species, the establishment of regular monitoring programs is necessary. Moreover, public awareness campaigns, citizen science initiatives and conventional scientific surveys by scientists are fundamental to monitor invasive species and to manage continuous spill-over effects, especially in the areas most vulnerable to marine biological invasions (e.g. MPAs).

Invasive species can be managed through different strategies: prevention, early detection, eradication and control (The European Commission, [101]). Because biological invasions are generally irreversible, prevention is an important task in many management strategies [102]. It is the cheapest and most effective option to reduce the future ecological and economic costs of invasive species [103]. To prevent marine bioinvasions, one should focus first on the invasion pathway/vector in order to minimize the risks of further introductions [104]. Identification of the most common pathways for introduction can help mould international policies aimed at preventing this phenomenon. Since commercial shipping seems to be the most significant vector for exotic species introduction (through ballast water, hull fouling and passive dispersal through canals), ballast water and fouling management practices (e.g. installation onto all carrier ships of ballast treatment technology) have a pivotal role to play, but these will invariably increase the cost of shipped goods. The recent expansion of the Suez Canal has spurred the biological community to raise the alarm on the ecological implications of such an expansion [105], with this alarm enjoying unprecedented media attention within high-profile portals, including *The Guardian* and *The Economist*. Mitigation measures proposed by the biological community in an attempt to check the unremitting introduction of allochthonous species through the Suez Canal into the Mediterranean include the installation of salinity barriers and of air bubble screens.

Controlling the spread of invasive species requires international cooperation, and this is often difficult to achieve [106]. An important step towards prescribing the management of ballast water was achieved through the adoption by the International Maritime Organisation (IMO) of the International Convention for the Control and Management of Ships Ballast Water and Sediments (BWM Convention) in 2004, which will finally come on stream in September 2017 after enough signatory parties ratified it. With respect to hull-fouling management, anti-fouling practices are largely based on the application of biocides, toxic to biofouling organisms [107], but also with a broader ecological impact on the aquatic environment. As conventional and more aggressive anti-fouling alternatives, for example, those based on TBT, tributyltin, are phased out, there is a real concern that the rate of introductions of marine alien species via hull fouling could increase in future, until other effective anti-fouling systems are developed.

Aquaculture and mariculture practices can be more effectively controlled than any other pathway [108]. The FAO Code [109] discourages the use of invasive alien species in aquaculture and requests consultation with neighbouring states before introducing alien species. In terms of aquaculture imports, quarantine measures have been recommended [110, 111]. The international aquarium trade represents a multi-billion dollar industry that has expanded rapidly since the 1970s [112]. Many species are incidentally imported as contaminants, along with intentionally introduced species [44], whilst intentional releases into the wild by aquarium hobbyists in the Mediterranean are not known (e.g. *Lutjanus sebae*—[113]). The aquarium trade could be regulated following similar procedures as to aquaculture and by raising awareness of the public about the dangers of releasing aquarium species to the sea.

To mitigate the ecological impacts of IAS in the Mediterranean, different actions can be undertaken: eradication (manual, biological and chemical methods), containment and control. Invasive species are successfully eliminated when actions start as soon as possible after their introduction [114]. Once an alien species become established in a non-native area, the implementation of effective control and mitigation strategies are next to impossible. Early detection of alien invasive species, together with the capacity to take rapid action, is the key to successful and cost-effective eradications.

An obstacle to a rapid, early-stage eradication by authorities is a lack of biogeographical information about the species in question. Therefore, a collaboration between researchers, resource management agencies, and policy makers is needed to make management procedures more effective. Since the evaluation of management measures is essential for appropriate actions [103], a set of criteria (effectiveness, ecological impact and impact on human health) have been recently proposed to evaluate the effectiveness of the same measures [115].

The complexity of interactions between native and invasive species and the associated impacts makes environmental management quite difficult [53]. Therefore, continuous research is fundamental to predict future invasions and their effects. Such research efforts, however, should also be directed towards informing stakeholders and assessing the effectiveness of different management measures to reduce invasions and their impacts [66]. A dire need for environmental auditing methodologies (e.g. CBA, DPSIR) applied to biological invasions and potential mitigation measures exists, as well as for high-quality information on the introduced species, for example, life histories, invasive strategies and impacts on ecosystem services and biodiversity. Moreover, the quantification and mapping of impacts, for example, through the ecosystem services approach, might greatly assist managers and policy makers in their decisions on prevention or mitigation actions to be taken [53].

Acknowledgements

The authors thank Lucia Gastoldi for providing the drawings, and Enric Ballesteros, Ester Cecere, Mariolina Corsini Foka, Puccio Distefano, Bella Galil, Maria Ghelia, Gerasimos Kondylatos, Antonella Petrocelli, Conxi Rodríguez-Prieto and Fabio Russo for providing the photos.

http://dx.doi.org/10.5772/intechopen.69214

Author details

Anna M. Mannino[1]*, Paolo Balistreri[2] and Alan Deidun[3]

*Address all correspondence to: annamaria.mannino@unipa.it

1 Department of Biological, Chemical and Pharmaceutical Sciences and Technologies, Section of Botany and Plant Ecology, University of Palermo, Palermo, Italy

2 Vicolo Giotto 6, Favignana, Italy

3 Department of Geosciences, University of Malta, Msida, Malta

References

[1] Bianchi CN, Morri C. Marine biodiversity of the Mediterranean Sea: Situation, problems and prospects for future research. Marine Pollution Bulletin. 2000;**40**:367-376

[2] Coll M, Piroddi C, Steenbeek J, Kaschner K, Ben Rais Lasram F, Aguzzi J, Ballesteros E, Bianchi CN, Corbera J, Dailianis T, Danovaro R, Estrada M, Froglia C, Galil BS, Gasol JM, Gertwagen R, Gil J, Guilhaumon F, Kesner-Reyes K, Kitsos M-S, Koukouras A, Lampadariou N, Laxamana E, López-Fé de la Cuadra CM, Lotze HK, Martin D, Mouillot D, Oro D, Raicevich S, Rius-Barile J, Saiz-Salinas JI, San Vicente C, Somot S, Templado J, Turon X, Vafidis D, Villanueva R, Voultsiadou E. The biodiversity of the Mediterranean Sea: Estimates, patterns and threats. PLoS One. 2010;**5**:e11842. DOI: 10.1371/journal. pone. 0011842

[3] Dugdale RC, Wilkerson FP. Nutrient sources and primary production in the Eastern Mediterranean. Oceanologica Acta. 1988;**9**:179-184

[4] Pérès JM, Picard J. Nouveau manuel de bionomie benthique de la mer Méditerranée. Recueil des Travaux de la Station Marine d'Endoume. 1964;**31**(47):1-137

[5] Bianchi CN. Biodiversity issues for the forthcoming tropical Mediterranean Sea. Hydrobiologia. 2007;**580**:7-21. DOI: 10.1007/s10750-006-0469-5

[6] Lejeusne C, Chevaldonne P, Pergent-Martini C, Boudouresque CF, Perez T. Climate change effects on a miniature ocean: The highly diverse, highly impacted Mediterranean Sea. Trends in Ecology and Evolution. 2010;**25**:250-260. DOI: 10.1016/j.tree.2009.10.009

[7] Baltanás A. On the use of some methods for the estimation of species richness. Oikos. 1992;**65**:484-492

[8] Grime JP. Biodiversity and ecosystem function: The debate deepens. Science. 1997; **277**:1260-1261

[9] Cognetti G, Curini-Galletti M. Biodiversity conservation problems in the marine environment. Marine Pollution Bulletin. 1993;**26**(4):179-183

[10] Boudouresque CF. Marine biodiversity in the Mediterranean: Status of species, populations and communities. Scientific Reports of Port-Cros National Park. 2004;**20**:97-146

[11] Por FD. Lessepsian Migrations. The Influx of Red Sea Biota into the Mediterranean by Way of the Suez Canal. Ecological Studies V. 23. Berlin: Springer; 1978. p. 228

[12] Galil BS. The marine caravan - the Suez Canal and the Erythrean invasion. In: Gollasch S, editor. Bridging Divides: Maritime Canals as Invasion Corridors. Monographiae Biologicae 83. Heidelberg: Springer; 2006. pp. 207-300

[13] Por FD. Tethys returns to the Mediterranean: Success and limits of tropical re-colonization. BioRisk. 2009;**3**:5-19. DOI: 10.3897/biorisk.3.30

[14] Giaccone G. Biogeografia di alghe brune e tettonica a zolle. Bollettino dell'Accademia Gioenia di Scienze Naturali. 1991;**24**(337):65-86

[15] Zenetos A, Gofas S, Morri C, Rosso A, Violanti D, García Raso JE, Çinar Me, Almogi-Labin A, Ates As, Azzurro E, Ballesteros E, Bianchi Cn, Bilecenoglu M, Gambi Mc, Giangrande A, Gravili C, Hyams-Kaphzan O, Karachle Pk, Katsanevakis S, Lipej L, Mastrototaro F, Mineur F, Pancucci-Papadopoulou Ma, Ramos Esplá A, Salas C, San Martín G, Sfriso A, Streftaris N, Verlaque M. Alien species in the Mediterranean Sea by 2012. A contribution to the application of European Union's Marine Strategy Framework Directive (MSFD). Part 2. Introduction trends and pathways. Mediterranean Marine Science. 2012;**13**:328-352. DOI: 10.12681/mms.327

[16] Katsanevakis S, Zenetos A, Belchior C, Cardoso AC. Invading European Seas: Assessing pathways of introduction of marine aliens. Ocean and Coastal Management. 2013;**76**:64-74. DOI: 10.1016/j.ocecoaman.2013.02.024

[17] Lasram FBR, Guilhaumon F, Albouy C, Somot S, Thuiller W, Mouillot D. The Mediterranean Sea as a 'cul-desac' for endemic fishes facing climate change. Global Change Biology. 2010;**16:**3233-3245. DOI: 10.1111/j.1365-2486.2010.02224.x

[18] Coll M, Piroddi C, Albouy C, Ben Rais Lasram F, Cheung WWL, Christensen V, Karpouzi VS, Guilhaumon F, Mouillot D, Paleczny M, Palomares ML, Steenbeek J, Trujillo P, Watson R, Pauly D. The Mediterranean under siege: Spatial overlap between marine biodiversity, cumulative threats and marine reserves. Global Ecology and Biogeography. 2012;**21**:465-481. DOI: 10.1111/j.1466-8238.2011.00697.x

[19] Gubbay S, Sanders N, Haynes T, Janssen JAM, Rodwell JR, Nieto A, García Criado M, Beal S, Borg J, Kennedy M, Micu D, Otero M, Saunders G, Calix M. European Red List of Habitats. Part 1. Marine Habitats. Luxembourg: European Commission; 2016. p. 46

[20] Giorgi F, Lionello P. Climate change projections for the Mediterranean region. Global and Planetary Change. 2008;**63**:90-104. DOI: 10.1016/j.gloplacha.2007.09.005

[21] IPCC. Climate Change 2007: The Physical Science Basis. Contribution of Working Group I to the Fourth Assessment Report of the Intergovernmental Panel on Climate Change. Cambridge: Cambridge University Press; 2007. p. 996

[22] IPCC. Climate Change 2013: The Physical Science Basis. Working Group I Contribution to the Fifth Assessment Report of the Intergovernmental Panel on Climate Change. Cambridge: Cambridge University Press; 2013. p. 1535

[23] Cerrano C, Totti C, Sponga F, Bavestrello G. Summer disease in *Parazoanthus axinellae* (Schmidt, 1862) (Cnidaria, Zoanthidea). Italian Journal of Zoology. 2006;**73**(4):355-361

[24] Rodolfo-Metalpa R, Bianchi CN, Peirano A, Morri C. Tissue necrosis and mortality of the temperate coral *Cladocora caespitosa*. Italian Journal of Zoology. 2005;**72**:271-276

[25] Bensoussan N, Romano JC, Harmelin JG, Garrabou J. High resolution characterization of northwest Mediterranean coastal waters thermal regimes: To better understand responses of benthic communities to climate change. Estuarine, Coastal and Shelf Science. 2010;**87**:431-441. DOI: 10.1016/j.ecss.2010.01.008

[26] Marbà N, Duarte CM. Mediterranean warming triggers seagrass (*Posidonia oceanica*) shoot mortality. Global Change Biology. 2010;**16**:2366-2375. DOI: 10.1111/j.1365-2486.2009.02130.x

[27] Díaz-Almela E, Marbà N, Duarte CM. Consequences of Mediterranean warming events in seagrass (*Posidonia oceanica*) flowering records. Global Change Biology. 2007;**13**:224-235. DOI: 10.1111/j.1365-2486.2006.01260.x

[28] Kushmaro A, Rosenberg E, Fine M, Haim YB, Loya Y. Effect of temperature on bleaching of the coral *Oculina patagonica* by Vibrio AK-1. Marine Ecology Progress Series. 1998;**171**:131-137

[29] Thomas OP, Sarrazin V, Ivanisevic J, Amade P, Pérez T. Sponge chemical defenses in stress conditions: The case study of the last disease outbreak observed in the NW Mediterranean. In: Proceedings of the 5th European Conference on Marine Natural Products; 16-21 September 2007; Ischia, Italy; 2007. p. 61

[30] Hall-Spencer JM, Rodolfo-Metalpa R, Martin S, Ransome E, Fine M, Turner SM, Rowley SJ, Tedesco D, Buia M-C. Volcanic carbon dioxide vents show ecosystem effects of ocean acidification. Nature. 2008;**454**:96-99. DOI: 10.1038/nature07051

[31] Fabry V. Marine calcifiers in a high-CO_2 ocean. Science. 2008;**320**:1020-1022. DOI: 10.1126/science.1157130

[32] Milliman JD. Marine carbonates. Recent Sedimentary Carbonates Part 1. Berlin, New York: Springer; 1974. p. 375

[33] Martin S, Gattuso J-P. Response of Mediterranean coralline algae to ocean acidification and elevated temperature. Global Change Biology. 2009;**15**:2089-2100. DOI: 10.1111/j.1365-2486.2009.01874.x

[34] Occhipinti-Ambrogi A, Galil B. Marine alien species as an aspect of global change. Advances in Oceanography and Limnology. 2010;**1**(1):199-218. DOI: 10.1080/19475721003743876

[35] Azzurro E, Ben Souissi J, Boughedir W, Castriota L, Deidun A, Falautano M, Ghanem R, Zammit-Mangion M, Andaloro F. The Sicily Strait: A transnational observatory for

monitoring the advance of non-indigenous species. Biologia Marina Mediterranea. 2014;**21**:105-106

[36] CBD (Convention on Biological Diversity). Interim Guiding Principles. Conference of the Parties Decision V/8 Alien Species that Threaten Ecosystems, Habitats or Species. 2000. Available from: http://www.cbd.int/decision/cop/default.shtml?id=7150 [Accessed: 31 October 2016]

[37] EU. Directive 2008/56/EC of the European Parliament and of the Council of 17 June 2008 establishing a framework for community action in the field of marine environmental policy (Marine Strategy Framework Directive). Official Journal of the European Union. 2008a;**L164**:19-40

[38] CBD (Convention on Biological Diversity) Pathways of Introduction of Invasive Species, their Prioritization, and Management. 2014. Available from: www.cbd.int/doc/meetings/sbstta/sbstta-18/official/sbstta-18-09-add1-en.pdf) [15-September-2016]

[39] Galil B, Boero F, Fraschetti S, Piraino S, Campbell M, Hewitt C, Carlton J, Cook E, Jelmert A, Macpherson E, Marchini A, Occhipinti-Ambrogi A, Mckenzie C, Minchin D, Ojaveer H, Olenin S, Ruiz G. The enlargement of the Suez Canal and introduction of non-indigenous species to the Mediterranean Sea. Limnology and Oceanography Bulletin. 2015;**24**;41-43. DOI: 10.1002/lob.10036

[40] Relini Orsi L, Mori M. Due reperti Mediterranei di *Thalamita gloriensis* Crosnier, 1962 (*Crustacea Decapoda*: Portunidae). Oebalia. 1979;**1**:7-13

[41] Knoepffler-Peguy M, Belsher T, Boudouresque CF, Lauret M. *Sargassum muticum* begins to invade the Mediterranean. Aquatic Botany. 1985;**23**:291-295

[42] Meinesz A. Modes de dissémination de l'algue *Caulerpa taxifolia* introduite en Méditerranée. Rapports et P.V. des Reunions de la Commission internationale pour l'Exploration scientifique de la Méditerranée, Monaco. 1992;**33**(B):44

[43] Essl F, Dullinger S, Rabitsch W, Hulme PE, Hülber K, Jarošík V, Kleinbauer I, Krausmann F, Kühn I, Nentwig W, Vilà M, Genovesi P, Gherardi F, Desprez-Loustau ML, Roques A, Pyšek P. Socioeconomic legacy yields an invasion debt. Proceedings of the National Academy of Sciences. 2011;**108**:203-207. DOI: 10.1073/pnas.1011728108

[44] Duggan IC. The freshwater aquarium trade as a vector for incidental invertebrate fauna. Biological Invasions. 2010;**12**:3757-3770. DOI: 10.1007/s10530-010-9768-x

[45] Thresher RE, Kuris AM. Options for managing invasive marine species. Biological Invasions. 2004;**6**:295-300

[46] Ribera MA, Boudouresque CF. Introduced marine plants, with special reference to macroalgae: Mechanisms and impact. Progress in Phycological Research. 1995;**11**:187-268

[47] Ruiz GM, Carlton JT, Grosholtz ED, Hines AH. Global invasions of marine and estuarine habitats by non-indigenous species: Mechanisms, extent and consequences. American Zoologist. 1997;**37**:621-632

[48] Elton CS. The Ecology of Invasions by Animals and Plants. New York, NY: John Wiley & Sons; 1958. p. 181

[49] Vilà M, Espinar JL, Hejda M, Hulme PE, Jarošík V, Maron JL, Pergl J, Schaffner U, Sun Y, Pyšek P. Ecological impacts of invasive alien plants: A meta-analysis of their effects on species, communities and ecosystems. Ecology Letters. 2011;**14**:702-708. DOI: 10.1111/j.1461-0248.2011.01628.x

[50] Simberloff D, Martin JL, Genovesi P, Maris V, Wardle DA, Aronson J, Courchamp F, Galil B, García-Berthou E, Pascal M, Pyšek P, Sousa R, Tabacchi E, Vilà M. Impacts of biological invasions: What's what and the way forward. Trends in Ecology & Evolution. 2013;**28**:58-66, DOI: 10.1016/j.tree.2012.07.013

[51] Blackburn TM, Essl F, Evans T, Hulme PE, Jeschke JM, Kühn I, Kumschick S, Marková Z, Mrugała A, Nentwig W, Pergl J, Pyšek P, Rabitsch W, Ricciardi A, Richardson DM, Sendek A, Vilà M, Wilson JRU, Winter M, Genovesi P, Bacher S. A Unified classification of alien species based on the magnitude of their environmental impacts. PLoS Biology. 2014;**12**(5):e1001850. DOI: 10.1371/journal.pbio.1001850

[52] Hulme PE. Resolving whether botanic gardens are on the road to conservation or a pathway for plant invasion. Conservation Biology. 2014;**9**:816-824. DOI: 10.1111/cobi.12426

[53] Katsanevakis S, Coll M, Piroddi C, Steenbeek J, Ben Rais Lasram F, Zenetos A, Cardoso AC. Invading the Mediterranean Sea: Biodiversity patterns shaped by human activities. Frontiers in Marine Science. 2014;**1**:32. DOI: 10.3389/fmars.2014.00032

[54] Goodenough AE. Are the ecological impacts of alien species misrepresented? A review of the "native good, alien bad" philosophy. Community Ecology. 2010;**11**(1):13-21. DOI: 10.1556/ComEc.11.2010.1.3

[55] Crooks JA. Lag times and exotic species: The ecology and management of biological invasions in slow-motion. Ecoscience. 2005;**12**:316-329. DOI: 10.2980/i1195-6860-12-3-316.1

[56] Schutzenhofer MR, Valone TJ. Positive and negative effects of exotic *Erodium cicutarium* on an arid ecosystem. Biological Conservation. 2006;**132**:376-381. DOI: 10.1016/j.biocon.2006.04.031

[57] Jeschke JM, Keesing F, Ostfeld RS. Novel organisms: Comparing invasive species, GMOs, and emerging pathogens. AMBIO. 2013;**42:**541-548. DOI: 10.1007/s13280-013-0387-5

[58] Ricciardi A, Hoopes MF, Marchetti MP, Lockwood JL. Progress toward understanding the ecological impacts of non-native species. Ecological Monographs. 2013;**83**:263-282. DOI: 10.1890/13-0183.1

[59] Strayer DL Eight questions about invasions and ecosystem functioning. Ecology Letters. 2012;**15**:1199-1210. DOI: 10.1111/j.1461-0248.2012.01817.x

[60] Boudouresque CF, Verlaque M. Biological pollution in the Mediterranean Sea: Invasive versus introduced macrophytes. Marine Pollution Bulletin. 2002;**44**:32-38

[61] Lodge DM, Williams S, MacIsaac HJ, Hayes KR, Leung B, Reichard S, Mack RN, Moyle PB, Smith M, Andow DA, Carlton JT. Biological invasions: Recommendations for US policy and management. Ecological Applications. 2006;**16**(6):2035-2054

[62] Galil BS. Loss or gain? Invasive aliens and biodiversity in the Mediterranean Sea. Marine Pollution Bulletin. 2007;**55**:314-322. DOI: 10.1016/j.marpolbul.2006.11.008

[63] Wallentinus I, Nyberg CD. Introduced marine organisms as habitat modifiers. Marine Pollution Bulletin. 2007;**55**:323-332. DOI: 10.1016/j.marpolbul.2006.11.010

[64] Vilà M, Basnou C, Pyšek P, Josefsson M, Genovesi P, Gollasch S, Nentwig W, Olenin S, Roques A, Roy D, Hulme PE. How well do we understand the impacts of alien species on ecosystem services? A pan-European, crosstaxa assessment. Frontiers in Ecology and the Environment. 2010;**8**(3):135-144. DOI: 10.1890/080083

[65] Williamson M. Biological Invasions. London: Chapman & Hall; 1996

[66] Rilov G, Galil B. Marine bioinvasions in the Mediterranean Sea e history, distribution and ecology. In: Rilov G, Crooks JA, editors. Biological Invasions in Marine Ecosystems. Berlin: Springer; 2009. pp. 549-576

[67] Ardura A, Planes S. Rapid assessment of non-indigenous species in the era of the eDNA barcoding: A Mediterranean case study. Estuarine, Coastal and Shelf Science. 2017;**188**:81-87

[68] Otero M, Cebrian E, Francour P, Galil B, Savini D. Monitoring marine invasive species in Mediterranean Marine Protected Areas (MPAs): A strategy and practical guide for managers. Medpan North Project. Malaga, Spain: IUCN; 2013. p. 136

[69] Williamson M, Fitter A. The varying success of invaders. Ecology. 1996;**77**:1661-1666

[70] Marchini A, Galil BS, Occhipinti-Ambrogi A. Recommendations on standardizing lists of marine alien species: Lessons from the Mediterranean Sea. Marine Pollution Bulletin. 2015;**101**:267-273. DOI: 10.1016/j.marpolbul.2015.09.054

[71] Streftaris N, Zenetos A. Alien marine species in the Mediterranean—The 100 'worst invasives' and their impact. Mediterranean Marine Science. 2006;**7**(1):87-118. DOI: 10.12681/mms.180

[72] Klein J, Verlaque M. The *Caulerpa racemosa* invasion: A critical review. Marine Pollution Bulletin. 2008;**56**:205-225. DOI: 10.1016/j.marpolbul.2007.09.043

[73] Francour P, Pellissier V, Mangialajo L, Buisson E, Stadelmann B, Veillard N, Meinesz A, Thibaut T, De Vaugelas J. Changes in invertebrate assemblages of *Posidonia oceanica* beds following *Caulerpa taxifolia* invasion. Vie et Milieu - Life and Environment. 2009;**59**:31-38

[74] Verlaque M. Contribution à la flore des algues marines de Méditerranée: Espèces nouvelles pour la Méditerranée occidentale. Botanica Marina. 1989;**32**:101-113

[75] Occhipinti Ambrogi A. Biotic invasions in a Mediterranean Lagoon. Biological Invasions. 2000;**2**:165-176. DOI: 10.1023/A:1010004926405

http://dx.doi.org/10.5772/intechopen.69214

[76] Deidun A, Arrigo S, Piraino S. The westernmost record of *Rhopilema nomadica* (Galil, 1990) in the Mediterranean – off the Maltese Islands. Aquatic Invasions. 2011;**6**(Suppl. 1):S99–S103. DOI: 10.3391/ai.2011.6.S1.023

[77] Relini M, Orsi L, Puccio V, Azzurro E. The exotic crab *Percnon gibbesi* (H. Milne Edwards, 1853) (Decapoda, Grapsidae) in the Central Mediterranean. Scientia Marina. 2000;**64**(3):337-340

[78] Katsanevakis S, Poursanidis D, Yokes MB, Mačić V, Beqiraj S, Kashta L, Sghaier YR, Zakhama-Sraieb R, Benamer I, Bitar G, Bouzaza Z, Magni P, Bianchi CN, Tsiakkiros L, Zenetos A. Twelve years after the first report of the crab *Percnon gibbesi* (H. Milne Edwards, 1853) in the Mediterranean: Current distribution and invasion rates. Journal of Biological Research-Thessaloniki. 2011;**16**:224-236

[79] Simoni F, Di Paolo C, Gori L, Lepri L. Further investigation on blooms of *Ostreopsis ovata, Coolia monotis, Prorocentrum lima,* on the macroalgae of artificial and natural reefs in the Northern Tyrrhenian Sea. Harmful Algae News. 2004;**26**:5-7

[80] Bilecenoglu M, Kaya M, Akalin S. Range expansion of silverstripe blaasop, *Lagocephalus sceleratus* (Gmelin, 1789), to the northern Aegean Sea. Aquatic Invasions. 2006;**1**(4):289-291. DOI: 10.3391/ai.2006.1.4.14

[81] Golani D, Orsi-Relini L, Massuti E, Quingnard JP. CIESM Atlas of Exotic Fishes in the Mediterranean. 2006. Available from: http://www.ciesm.org/atlas/appendix1.html [Accessed: 20-February 2017]

[82] Galil BS, Froglia C, Noel PY. CIESM Atlas of Exotic Crustaceans in the Mediterranean. 2006. Available from: http://www.ciesm.org/atlas/appendix2.html [Accessed: 22-February-2017]

[83] Zenetos A, Gofas S, Russo G, Templado J. CIESM Atlas of Exotic Species in the Mediterranean. Volume 3: Molluscs. Briand F, editor. Monaco: CIESM Publishers; 2003. p. 376

[84] Verlaque M, Ruitton S, Mineur F, Boudouresque CF. CIESM Atlas of Exotic Species in the Mediterranean,Volume 4: Macrophytes. F. Briand (Ed.). Monaco: CIESM Publishers; 2015. 364 p.

[85] Daisie. Delivering Alien Invasive Species Inventories for Europe [Internet]. Anno. 2005. Available from: www.europe-aliens.org/ [Accessed: 21 February 2017]

[86] Sinanet. Information System for Marine Alien Species Observation [Internet]. 2001. Available from: www.marinealien.sinanet.isprambiente.it/uploads/ [Accessed: 24 February 2017]

[87] Nizamuddin M. The Green Marine Algae of Libya. Bern: Elga Publishers; 1991. p. 230

[88] Pérez R, Kaas R, Babaroux O. Culture expérimentale de l'algue *Undaria pinnatifida* sur les côtes de France. Science & Pêche, Bulletin de l'Institut des Pêches Maritimes. 1984;**343**:3-15

[89] Anonymous. Activité de l'Institut des Pêches en 1979. Principales actions en matière de cultures marines. Science & Pêche, Bulletin de l'Institut des Pêches Maritimes. 1980;**306**:1-39

[90] Benedetti-Cecchi L, Cinelli F. Note on a *Polysiphonia* sp. (Rhodophyta, Ceramiales) collected at Rosignano Solvay (Western Mediterranean). Giornale Botanico Italiano. 1989;**123**:49-54

[91] Galil B, Spanier E, Ferguson W. The Scyphomedusae of the Israeli Mediterranean coast, including two Lessepsian migrants to the Mediterranean. Zoologische Mededlingen. 1990;**64**(7):95-105

[92] Fuchs Th. Die geologische Beschaffenheit der Landenge von Suez. Denkschriften der Kaiserkichen Akademie der Wissenschaften, Mathematisch-Naturwissenschaftliche Classe. 1878;**38**:25

[93] Garcia L, Reviriego B. Presència del rank subtropical *Percnon gibbesi* (H. Milne Edwards, 1853), (Crustacea, Decapoda, Grapsidae) a les Illes Balears. Primera cita a la Mediterrània occidental. Bolleti de la Societat d'Història natural de les Balears. 2000;**43**:81-89

[94] Mori M, Vacchi M. On a new occurrence of the alien flat crab, *Percnon gibbesi* (H. Milne Edwards), in the southern Sicily (Central Mediterranean Sea) (Crustacea, Brachyura, Grapsidae). Annali del Museo Civico di Storia Naturale "Giacomo Doria". 2002;**114**:295-302

[95] Golani D. First record of the bluespotted cornetfish from the Mediterranean Sea. Journal of Fish Biology. 2000;**56**:1545-1547. DOI: 10.1111/j.1095-8649.2000.tb02163.x

[96] Akyol O, Ünal V, Ceyhan T, Bilecenoglu M. First confirmed record of *Lagocephalus sceleratus* (Gmelin, 1789) in the Mediterranean Sea. Journal of Fish Biology. 2005;**66**:1183-1186. DOI: 10.1111/j.0022-1112.2005.00667.x

[97] Gurlek M, Ergüden D, Uyan A, Doğdu SA, Yağlıoğlu D, Öztürk B, Turan C. First record of red lionfish *Pterois volitans* (Linnaeus, 1785) in the Mediterranean Sea. Natural and Engineering Sciences. 2016;**1**(3):27-32

[98] Commission of the European Communities: Towards a strategy on invasive species. 2008. Available from: http://ec.europa.eu/environment/nature/invasivealien/docs/1_EN_ACT_part1_v6.pdf. Communication from the Commission to the Council, the European Parliament, the European Economic and Social Committee and the Committee of Regions. COM (2008) 789 final

[99] Council of the European Union: A Mid-Term Assessment of Implementing the EU Biodiversity Action Plan and Towards an EU Strategy on Invasive Alien Species - Council Conclusions. 2009. Available from: http://register.consilium.europa. eu/pdf/en/09/st11/st11412.en09.pdf. Information Note 11412/09.

[100] EU. Regulation on the Prevention and Management of the Introduction and Spread of Invasive Alien Species. PE-CONS 70/14. Brussels: European Commission; 2014. p. 11

[101] COM (Commission of the European Communities). Towards an EU Strategy on Invasive Species. COM (2008) 789 final. SEC (2008) 2887 et SEC (2008) 2886

[102] Simberloff D. How much information on population biology is needed to manage introduced species?. Conservation Biology. 2003;**17**:83-92. DOI: 10.1046/j.1523-1739.2003.02028.x

[103] Hulme PE. Beyond control: Wider implications for the management of biological invasions. Journal of Applied Ecology. 2006;**43**:835-847. DOI: 10.1111/j.1365-2664.2006.01227.x

[104] EC. Proposal for a Regulation of the European Parliament and of the Council on the Prevention and Management of the Introduction and Spread of Invasive Alien Species. 2013. COM(2013) 620 final. 2013/0307 (COD)

[105] Galil BS, Boero F, Campbell ML, Carlton JT, Cook E, Fraschetti S, Gollasch S, Hewitt CL, Jelmert A, Macpherson E, Marchini A. 'Double trouble': The expansion of the Suez Canal and marine bioinvasions in the Mediterranean Sea. Biological Invasions. 2015;**17**(4):973-976

[106] Perrings C, Burgiel S, Lonsdale WM, Mooney H, Williamson M. International cooperation in the solution to trade-related invasive species risks. Annals of the New York Academy of Sciences. 2010;**1195**:198-212. DOI: 10.1111/j.1749-6632.2010.05453.x

[107] GISP. Global Invasive Species Programme. Marine Biofouling: An Assessment of Risks and Management Initiatives. GISP and UNEP Regional Seas. 2008

[108] Savini D, Occhipinti-Ambrogi A, Marchini A, Tricarico E, Gherardi F, Olenin S, Gollasch S. The top 27 animal alien species introduced into Europe for aquaculture and related activities. Journal of Applied Ichthyology. 2010;**26**(Suppl. 2):1-7. DOI: 10.1111/j.1439-0426.2010.01503.x

[109] FAO. Code of Conduct for Responsible Fisheries. Rome: FAO; 1995

[110] EU. Council Regulation Concerning Use of Alien and Locally Absent Species in Aquaculture. Regulation 708/2007, OJ L 168; 2007

[111] EU. Commission Regulation Laying Down Detailed Rules for the Implementation of Council Regulation (EC) No 708/2007 Concerning Use of Alien and Locally Absent Species in Aquaculture. Regulation 535/2008, OJ L 156; 2008

[112] Padilla DK, Williams SL. Beyond ballast water: Aquarium and ornamental trades as sources of invasive species in aquatic systems. Frontiers in Ecology and the Environment. 2004;**2**(3):131-138. DOI: 10.1890/1540-9295(2004)002[0131:BBWAAO]2.0.CO;2

[113] Zenetos A, Apostolopoulos G, Crocetta F. Aquaria kept marine fish species possibly released in the Mediterranean Sea: First confirmation of intentional release in the wild. Acta Ichthyologica et Piscatoria. 2016;**46**(3):255-262. DOI: 10.3750/AIP2016.46.3.10

[114] Wilson JR, Ivey P, Manyama P, Nänni I. A new national unit for invasive species detection, assessment and eradication planning. South African Journal of Science. 2013;**109**(5/6):1-13. DOI: 10.1590/sajs.2013/20120111

[115] Schmiedel D, Wilhelm E-G, Roth M, Scheibner C, Nehring S, Winter S. Evaluation system for management measures of invasive alien species. Biodiversity and Conservation. 2016;**25**:357-374. DOI: 10.1007/s10531-016-1054-5

Chapter 6

Biodiversity in Central Mediterranean Sea

Nunziacarla Spanò and Emilio De Domenico

Additional information is available at the end of the chapter

http://dx.doi.org/10.5772/intechopen.68942

Abstract

The Central Mediterranean Sea is an area that connects the western and the eastern Mediterranean Sea where migratory fluxes of marine organisms are very peculiar. The high biodiversity of these areas is owed to particular hydrological and geomorphological characteristics (Messina Strait and Sicily Channel). The morphology of the Strait of Messina resembles a funnel with the narrow end to the North and the largest one to the South, and its underwater profile can be compared to a mountain whose opposite sides have markedly different slopes. The great biodiversity that characterizes this ecosystem is linked to the particular hydrology of the area. The Sicily Channel (or Strait of Sicily) is a wide water body located between southern Sicily and northern African coasts and represents the transition between the Western (WMED) and the Eastern (EMED) basins of the Mediterranean Sea. Morphologically, the Strait of Sicily belongs to the continental shelf with some other sub-units, as basins, seamounts and 'banks'. The bottoms are generally irregular and canyons are present. Mediterranean Sea has been divided into different biogeographic districts that present great variability in water parameters and biocenosis too. This chapter resumes the main physical, chemical and biological properties of the Central Mediterranean Sea.

Keywords: biodiversity, Central Mediterranean Sea, Strait of Messina, Sicily Channel, biogeography

1. Introduction

This chapter will show the main physical, chemical and biological properties of the Central Mediterranean Sea. These characteristics play a key role in this area that is located in a particular position by collecting all the species that migrate from the Atlantic Ocean to the East and those from the Red Sea moving westwards.

In the recent years, the Mediterranean Sea has been divided into biogeographic districts that have very different characteristics of the waters and of the organisms that inhabit also; for this reason the Mediterranean Sea is divided into eastern, central, western, Adriatic Sea and Alboran Sea. In this chapter, we have presented all the research carried out over the years by several research groups, including the University of Messina, on marine organisms in this area, and the specific conditions that affect their distribution in the western Mediterranean Sea, and as well as physical and chemical mechanisms and climate regulating the very high biodiversity. The faunal composition of the Mediterranean Sea is very complex, because the population is as follows: 20% endemic species; 62% species in the Atlantic-boreal affinity; 13% species in the Mediterranean-Atlantic-Pacific affinity; 5% to the Indo-Pacific species affinity. Each biogeographical district shows faunal elements characterizing it.

As described by Bergamasco and Malanotte-Rizzoli [1], the Mediterranean Sea is an enclosed basin composed of two biggest basins (western and eastern) and different sub-basins (**Figure 1**, **Table 1**).

It is a concentration basin, where evaporation exceeds precipitation. In the surface layer, there is an inflow of Atlantic water which is modified along its path to the eastern basin that was transformed through surface heat loss and evaporation particularly in the Levantine basin (**Figure 2**).

Figure 1. Mediterranean Sea.

http://dx.doi.org/10.5772/intechopen.68942

Size	
Surface	2.516×10^6 (km^2)
Volume	3.5×10^6 (km^3)
Maximum depth [3]	5267 (m)
Middle depth	1498 (m)
Length (*W-E*)	3860 (km)
Breadth (*N-S*)	600 (km)
Hydrography	
Water volume gibraltar (*IN*)	35,000 (km^3/year)
Rain input	850 (km^3/year)
River input	350 (km^3/year)
Water volume gibraltar (*OUT*)	36,400 (km^3/year)
Physical-chemical properties	
Salinity	35.6–39.1
Water temperature at bottom	13°C
Source: According to Della Croce et al. [2].	

Table 1. Basic facts about the Mediterranean Sea.

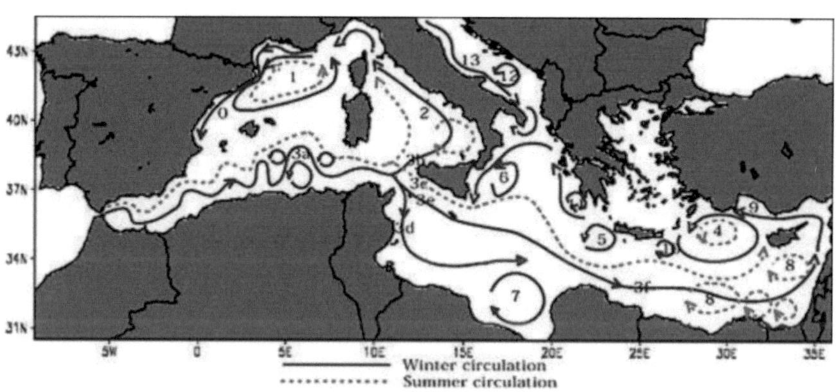

Figure 2. Circulation of Mediterranean Sea (Pinardi e masetti, 2000).

2. Mediterranean Sea

The Mediterranean is the site of water mass formation processes which can be studied experimentally because of its easy research and data collecting. There are two main reasons why the Mediterranean is important. The first one is the impact of the Mediterranean on the global thermohaline circulation; second reason is that the Mediterranean basin can be considered as a laboratory for investigating processes occurring on the global scale of the world ocean (**Figure 2**).

Bergamasco and Malanotte-Rizzoli [1] provided a short historical review of the evolving knowledge of the Mediterranean Sea circulation that has emerged from experimental investigations over the last decades. They start to describe the old picture of the basin circulation which had stationary and smooth large scale patterns and then they show the major experiments that led to the discovery of the sub-basin scale circulation and its mesoscale features. In conclusion, they described the dynamical discovery of eastern Mediterranean Transient (EMT) in 1990s.

From a biogeographical point of view, the Central Mediterranean Sea (Messina Strait and Sicily Channel) is a geographic, well defined, area, located between the western and the eastern Mediterranean Sea where migratory fluxes of marine organisms are very peculiar, interesting migrations from W to E and vice versa. The highest biodiversity of this area is owed to geomorphology and to oceanographic and climatic properties of the studied areas (Messina Strait and Sicily Channel).

This chapter will show the main physical, chemical and biological properties of the Central Mediterranean Sea. These characteristics play a key role in this area that is located in a particular position by collecting all the species that migrate from the Atlantic Ocean to the East and those from the Red Sea moving westwards.

In the recent years, the Mediterranean Sea has been divided into biogeographic districts that have very different characteristics of the waters and of the organisms that inhabit also; for this reason, the Mediterranean Sea is divided into eastern, central, western, Adriatic Sea and Alboran Sea.

The trophic, genetic, lessepsian and accidental migrations of organisms within the Mediterranean are a major 'cause' of the area's biodiversity. The faunal composition of the Mediterranean Sea is very complex, because the population is as follows: 20% endemic species; 62% species in the Atlantic-boreal affinity; 13% species in the Mediterranean-Atlantic-Pacific affinity; 5% to the Indo-Pacific species affinity. Of course, each biogeographical district shows faunal elements characterizing it.

3. Strait of Messina

The underwater profile of the Messina's Strait can be compared to a mountain whose opposite sides have distinctly different slopes. In the Tyrrhenian Sea the sea floor slopes gently up to 1000 m in the area of Milazzo Gulf reaching 2000 m depth near to the Stromboli Island. Differently, in the southern part (Ionian Sea) the slope is very steep and the depth between the cities of Messina and Reggio Calabria already exceeds 500 m, and reaches 2000 m in opposite to Capo dell'Armi, Cape Taormina connecting line.

The smaller width (3.150 m) of the Strait is along the connecting line between Ganzirri (Sicily) and Punta Pezzo (Calabria). This zone corresponds to a submarine 'relieve' (sill) with depths comprised between 80 and 115 m. In this stretch, the bottom has an irregular line, with the maximum depth of 115 m, which separates a western irregular zone (Ganzirri Bench) from the eastern of Punta Pezzo that is deeper and smother.

http://dx.doi.org/10.5772/intechopen.68942

The Strait of Messina is the geographical point of union between the Ionian and Tyrrhenian basins. These, although in continuity, are distinct in their different physical and chemical parameters and physiographic characteristics. The large difference in heights of surface water masses generates very strong currents at each change of tide [4].

The hydrodynamic characteristics and the particular ecological conditions of the Straits are closely related with its geological structure and the intense tectonic activity that characterizes this area. Even the distribution of sediments is affected by seismic activity as well as the intense hydrodynamic regime.

The northern sector of the Strait is characterized by a wide valley (named Valley of Scilla), with a portion deeper and steep beginning at about 200 m depth. The valley became flatten out and then to be less steep towards the Tyrrhenian Sea where it takes the name of Palmi basin. The side walls of the valley, deep and steep, rise abruptly giving a U-shape to the cross section of the valley. In the South sector there is a wide and irregular depression, less engraved (Messina Valley), also having a U-shaped section. Deeper than 500 m, the Messina Valley becomes deeper and develops to a canyon steep (Canyon of Messina) that reaches up to the Ionian Bathyal Plain. It represents the main route of sediment transport, and is spatially extended from the Sicilian to the Calabrian coasts [5]. **Figure 3** shows the Messina Straits and the underwater profile.

3.1. Tides

The Strait of Messina is the junction point (or perhaps better separation), as already said, between two basins, the Ionian and Tyrrhenian, contiguous but dissimilar, having different physical, chemical and oscillatory water characteristics. For this reason, steady and tidal currents are present, also thanks to the particular geomorphology of the whole area which determine several peculiar hydrodynamic phenomena.

Near to the submarine relieve or sill, steady currents flowing southwards from the surface to 30 m and in the opposite direction from this depth to the bottom, with speed that can reach up to 50 cm/s with particular weather and sea conditions. The co-oscillation of Messina's Straits water masses with the tides of the adjacent seas originates tidal currents with almost

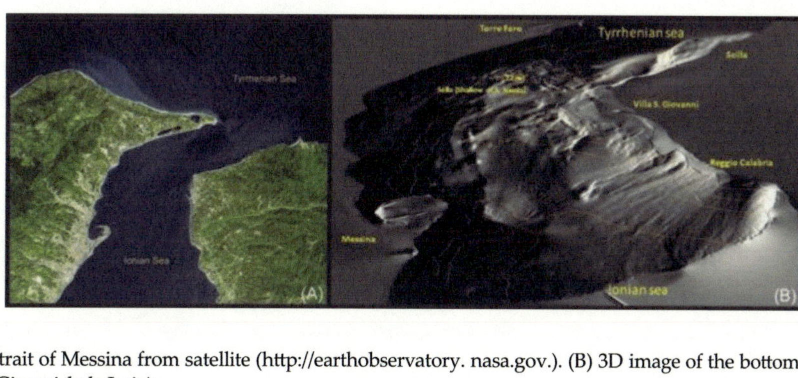

Figure 3. A) Strait of Messina from satellite (http://earthobservatory. nasa.gov.). (B) 3D image of the bottom of the Strait of Messina (Ph. Gianmichele Iaria).

opposite phase and equal amplitude. The relative speeds reach, along the section of the sill Ganzirri, Punta Pezzo, maximum values of more than 200 cm/s in both the flow to North (current 'montante'), and in the southwards (current 'scendente'). Mosetti [6] performed a research based on a long time series of data collected by a network of 27 currentmeters placed at three depths in nine sites in the central part of the Strait, indicate that the speed of the water flow in the Strait can reach, in particular moments and thanks to the coincidence of several components, up to a maximum of 20 km/h [7] (**Figure 4**).

These phenomena are created by the different levels of Tyrrhenian and Ionian Sea; in fact, when in the Tyrrhenian Sea there is low tide in the northern part of the Straits the adjacent Ionian Sea is in the development of high tide and the opposite happens during the next tide change. The difference in height (up to 27 cm) that is created determines that waters may flow into the contiguous basin.

More precisely, in the period of 'scendente' tide the Tyrrhenian waters, that are lighter than the Ionian ones, slide on these. Finally, the entire central part of the Strait is filled by Tyrrhenian water flowing towards South. In contrast, during the 'montante' tide, heavier Ionian waters, rising the central zone of the Strait flowing under the sea surface previously occupied by Tyrrhenian waters. Once over the sill these waters sink towards the northern part of the Strait.

The meeting (or better the clash) of these two water masses (Ionian and Tyrrhenian) determines the onset of numerous phenomena which are attributable to dynamic instability that is created and dispersed in the well-known spectacular manifestations of turbulence; these

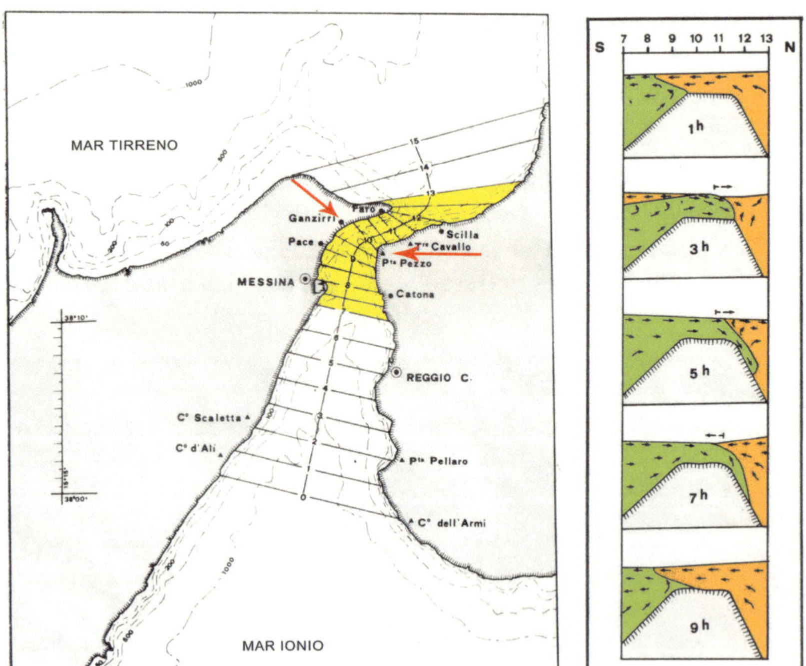

Figure 4. Current trends in the Strait of Messina (Defant, 1941, with data of Vercelli, 1925).

'disturbances' of the current may occur with development in the horizontal sense (in the case of 'cuts' and 'sea stairs') or vertical (in the case of 'garofali', 'bastards' and 'oil stains'). The phenomena of the first group are characterized by real waves, similar to those occurring in the change of tide in estuaries, which develop when, in the case of the 'montante', the heavier waters of the Ionian Sea rush against lighter Tyrrhenian waters in recession or when, in the case of 'scendente', the Tyrrhenian waters glide swiftly on those heavier from Ionian, already present in the basin. These discontinuity waves develop in particular zones (Ganzirri, Torre Faro and Punta Pezzo) reaching also the central part of the Strait. Sometimes these phenomena are more intense, thanks to the action of winds that mix the different waters.

The vertical phenomena are eddies deriving from the meeting of opposite tidal currents, also favourite by the irregularity of the bottom. The main eddies form in well-known areas. During the 'montante' flow, the mythological 'Scylla' and 'Charybdis' occur such as first is formed on the Calabrian coast, and the other to the South of Capo Peloro. A big 'Garofali' can be observed during the 'scendente' phase at Punta S. Raineri, in front of Messina port. The 'garofali' have a cyclonic rotation, and in them the heavier waters sink above the lighter ones that emerge with turbulent movements. In the case of 'oil spots' the movement is instead anti-cyclonic and waters emerge at the centre of the vortex showing an oily appearance calm surface. The 'bastards' are currents that develop along the coasts, with intensity and direction opposite to that of the main flow; these vary from zone to zone in relation to the bottom morphology [4] (**Figure 5**).

Figure 5. Current trends in surface. 'Cuts and sea stairs' (A–B). 'Oil stains and bastards' (C–D). (Ph. Andrea Potoschi).

3.2. Benthic ecosystem

The benthic ecosystem of the Strait is very composite and it is the result of a long and complex evolution. Coastal bottom populations show ancient origins and come from different geographical sites. The peculiar topography of the bottom, hydrodynamics as well as low water temperature, are the main causes for the presence in the Straits of many exclusive communities of Atlantic origin. For this reason, the Strait of Messina is considered the 'Sanctuary of Atlantic species in the Mediterranean'. It has been largely demonstrated that the high biodiversity can be related to hydrodynamic characteristics of the area. The intense hydrological regime allows the upwelling of deep water rich in organic matter and nutrients which is directly available for coastal benthic populations.

The Strait of Messina, owing to its particular geographic position, represents an exceptional observation point for the migration of species that cross the two basins (Ionian and Tyrrhenian Seas).

The physical characteristics of the Strait obviously influence the organisms that live in it, and therefore the entire biological environment. For this reason, in the Messina's Strait there is a unique ecosystem for biodiversity, biocoenosis and abundance of species. The intense water flow, the low temperature and the abundance of nitrogen and phosphorus salts (derived from upwelling) make available a large amount of organic matter used by both pelagic organisms and, above all, coastal benthic populations.

These features with the associated phenomena, determine an incredible 'ecological rearrangement' that species with predominantly western distribution tends to mimic an Atlantic type condition. In fact, many species, strictly Atlantic, such as 'Kelp forest' (*Laminaria sp*), live in the Straits of Messina, and their number is largely underestimated as evidenced by the continuous reports in this area by the scientific and local community [8].

As already said, the Strait of Messina, due to its geographical position, is an exceptional point for observation of migrating species that cross the two basins. Many planktonic species live and transit in this zone, even from distant areas, as confirmed by the ancient and recent reports of Atlantic organisms, e.g. *Corolla spectabilis* [9].

About the benthic species, the presence in the Strait of Messina of *Pilumnus inermis* is of particular importance [10]. Until recent years, this species was considered to be only Atlantic; furthermore *P. inermis* shows an association with the hydrozoan *Errina aspera* (the only Stylasteridae in the Mediterranean), well known endemism of the Messina Strait [11]. This particular environment is present in the sill zone between 80 and 115 m depths. This area also represents the perfect environment to harbours and numerous other particular species including the brittle star *Ophiactis balli*, the crabs *Parthenope expansa* and *Portunus pelagicus* [12].

Of course, of great importance are the organism's associations with *Laminariales* of the Strait. In fact, despite their sporadic presence in other areas of the Mediterranean, only in the Strait of Messina, *Laminariales* forms well-structured communities.

Another particular case is represented by *Albunea carabus* that, despite its warm waters origin (Senegal), has been repeatedly reported in the Strait. Its presence has not been reported out the Messina Strait [13].

In term of faunal assemblage, the Strait of Messina has been always considered the 'Paradise of Zoologists', thanks to its enormous biodiversity.

The most interesting species are the benthic invertebrates. The seabed is characterized by a wide variety of colours and shapes due to the abundance of coelenterates (sea anemones and corals), such as the forests of yellow and red gorgonian (*Paramuricea clavata*) from Scylla coastal seabed.

Fish biodiversity is allowed by a great number of species, as groupers, bream, snapper, damselfish, yellowtail and the beautiful and seasonal *Zeus faber* (John Dory or Peter's Fish), visible in these waters, between January and April, when the temperature is maintained at about 14°C [14] (**Figure 6**).

Figure 6. Characteristic species of Strait of Messina. (A) *Antipathella subpinnata*. (B) *Errina aspera*. (C) *Portunus pelagicus* (www.ciesm.org Ph. Bella Galil). (D) *Laminaria ochroleuca*. (E) *Hexanchus griseus*. (F) *Savalia savaglia*. (G) *Zeus faber*. (H) *Paramuricea clavata* (Photos Gianmichele Iaria).

3.3. Beach rock

In the Sicilian and Calabrian coasts of the Strait there is a coastal biotope of considerable interest (in the Sicilian shore, this zone is part of the Oriented Natural Reserve of 'Capo Peloro Lagoon'). It is an extensive stretch of coastline between Capo Peloro and S. Agata, affected by the presence of a rocky bench, from the beachfront, it takes up to several meters deep.

This zone, named **beach rock** (**Figure 7**), is located in a position between the intertidal zone and the upper fringe of the infralittoral. This structure is the only natural hard substrate for the benthic communities inside this bathymetric range, along the Sicilian side of the Strait [15].

Furthermore, due to its particular morphology, the topographic distribution, and on the basis of certain special influences from hydrodynamic regime of the Straits, the structure houses benthic communities completely original compared to what is known to the generality of similar Mediterranean habitats. In addition to its significant interest in terms of geological (proof of age Tyrrhenian) and anthropological documentation (formerly used as a quarry for millstones), the structure is of great importance as it hosts extended Vermetus formations, that is a protected biotope by the European community laws. These formations are also a unique case in the Mediterranean Sea, since they are located on the surface of the substrate, rather than placed in the typical *trottoir* formation [16]. It has been recently assessed that this zone represents a nursery zone for many fish species [17].

Figure 7. Beach rock (Ph. Nunziacarla Spanò).

http://dx.doi.org/10.5772/intechopen.68942

3.4. Migration of large pelagic fishes

The Strait of Messina is a crucial point for the migration of many species, located along one of the main Mediterranean routes. Among these, certainly the most important, from an economic and environmental point of view are the large pelagic fishes, i.e. tuna (*Thunnus thynnus*), the albacore (*Thunnus alalunga*), the Atlantic bonito (*Sarda sarda*), the Mediterranean spearfish (*Tetrapturus belone*) and the swordfish (*Xiphias gladius*). Just the 'wealth', high trophic resources of the Strait, determines that here are passing fishes in shallow waters and can be caught with special boats called 'passerelle' or 'feluche' active only in this part of the world. It is important to remember that in the Strait, using with different tools, the tuna can be caught throughout the year, from juvenile to adult. This is possible due to the presence of a local tuna population that moves regularly between the Tyrrhenian and Ionian Seas.

In addition to pelagic fishes, also the cetaceans cross the Messina's Strait; this latter probably represents the most important route in the Mediterranean sea in terms of species diversity. All dolphin species present in the Mediterranean have been reported in the Strait, besides Whales and Sperm Whales; the latter probably uses this way to Strait to reach Aeolian Islands area that represents a good environment, possibly, for their reproduction also.

Finally, the migration of large sharks, as *Carcharodon carcharias* (white shark) and *Hexanchus griseus*, through the Strait of Messina has been largely documented. Specimens of *Hexanchus griseus*, can be found in the moonlight of dark nights at depths between 15 and 30 m in some areas of the Strait (Paradiso and S. Agata).

3.5. Bathypelagic fishes

A further distinctive element of the Strait of Messina is the presence of bathypelagic fauna (commonly called abyssal fauna), which, transported to the surface by the upwelling, can be easily found in the turbulence points, or stranded along the shore in particular weather conditions. These organisms habitually live at bathypelagic depths in the Mediterranean Sea (usually between 300 and 1000 m) [18]. Their massive presence (with all ages represented) in surface water of the Strait is of considerable biological and ecological importance. The easy availability of these 'monstrous' fishes, most of which presenting special light organs (photophores) [19], called in Messina, between the second half of 1800 and early 1900, the attention of scientists from all over Europe. They defined the Strait of Messina as 'the zoologists' paradise'.

Some of these species are not transported to the surface by currents in a totally passive way but perform well-defined vertical movements, especially during the night named 'nictemerali' migrations (diel vertical migration) [20]. These are mainly myctophids whose role in the pelagic trophic network, given the considerable quantities involved, is certainly of primary importance. In the same manner it is possible to find a lot of *Chauliodus sloani* and *Argyropelecus hemigymnus* specimens (**Figure 8**).

Figure 8. Bathypelagic fishes. (A) *Argyropelecus hemigymnus*. (B) *Chauliodus sloani*. (Ph. Fernand Baguet). (C) *Myctophum punctatum* (Ph. Francesco Costa).

4. Strait of Sicily

The Strait of Sicily is a relatively shallow area with a complex bottom orography [21]. Maximum depths of this area are reached in three different points: Pantelleria basin (1317 m), Malta basin (1721 m) and Linosa basin (1529 m). The valley communicates with the western and eastern Mediterranean Sea by a narrow sill, NW of Pantelleria Island (400–500 m deep), and a wider channel, SE of Malta (500–600 m deep), respectively. By the way, a 'sill' is a critical bathymetric restriction preventing exchange of waters of the same depths inside and outside a basin [22]. Morphologically, the Strait of Sicily belongs to the continental shelf and slope but other morphologic sub-units of the slope are usually identified: basins, seamounts (guyots) and 'banks' [21]. The slope shape is extremely irregular, incised by many canyons, trenches and steep declivity. It is furrowed transversally by the deep, above mentioned, basins (>800 m) and interrupted by 'guyots' and 'banks'. Sediments tend to pile up and fill these basins; the most important are those of Pantelleria, Gela, Linosa and Malta. Volcanism has been and is still active in the area [21]. The raise and successive wash-out of the Ferdinandea Island (now the Graham Bank), in front of Sciacca, occurred in the eighteenth century is an example of the volcanic activity in this area. 'Banks' are typical morphologic units in the Strait of Sicily. Due to its pre-eminent role as a bridge for the western and eastern basin (the two basic hydrological units recognized for the Mediterranean Sea; [23]), the Strait of Sicily has been widely investigated [24–26] (**Figure 9**).

Figure 9. Sicily Channel. (http://earthobservatory.nasa.gov).

A recent evaluation revue of the seasonal circulation of the upper 800 m of the Mediterranean Sea has evidenced a complex dynamic situation in the water circulation.

Flowing mainly in West-South direction, its saline content increases gradually as a consequence of the strong evaporation (38.2–38.9% in the eastern basin). An analogous increase in temperature occurs in summer; in this season a thermocline can be found at about 50 m. For the sill in Strait of Sicily, Morel [21] has estimated that the Atlantic Water (AW) outflowing changes from 37.0 to 37.3% in transit whereas the incoming subsurface water changes from 38.8 to 38.7%. AW arrive to have 39% in the Levantine basin where, in Winter, it cools at 15–16°C and sinks to form the Levantine Water (LW) [27]. Recent further elaborations [28] have suggested that the path of AW in the Ionian basin and in the Strait of Sicily forms a meander in all seasons except in winter; a more complex situation has been also described (for example, Ref. [29]). A schematic representation of water flows can be observed in **Figure 10**.

Especially in the latter area the flow of AW could have not been well resolved but a deeper influence (down to 250–300 m) has been evidenced [28]. In the Strait of Sicily, the LW influences mainly the depth range between 150–200 and 1000–1200 m or less (200–600 m), according to other authors, e.g. Bombace and Sarà [30]. LW is more saline (38.5–38.8%) than AW (38.0%); both masses of water flow overlap. The bulk of these bodies of water follows the same path but in opposite directions, diverging only at the two extremities [22, 24].The boundary layer has been estimated to coincide with the isohaline of 38%. According to the inverse model results [28], the deep water circulation seems to become well discriminated from that of the surface at about 500–600 m, resulting in a level of discrimination (no motion level) at about 400 m. The most striking features resulted by the application of the inverse model were that the LW, after leaving the Strait of Sicily, flows mainly northwards crossing the waters in front of Sardinia, France, Spain and the Balearic Islands [28]. In the Strait of Sicily, the AW extends its influence down to 150–200 m (and more, up to 250–300 m according to other authors [30]) where it meets and partially mixes with the LW. Its surface temperature varies according the

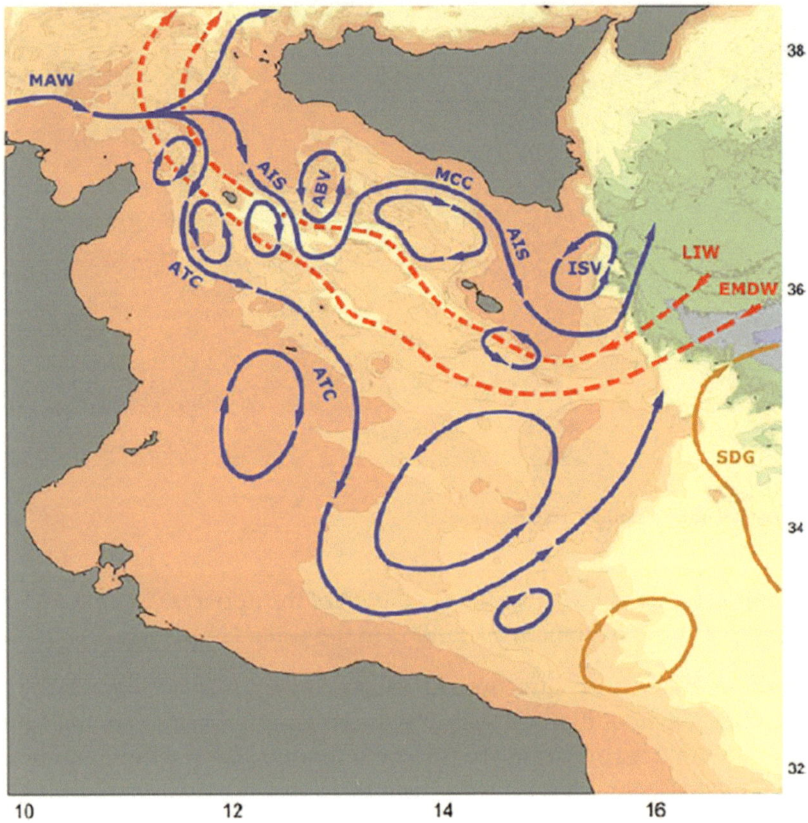

Figure 10. Scheme of the flows crossing the Sicily Channel. The paths have been drawn according to the SSH map from myOcean average data for 2010–2013 period. MAW, modified Atlantic Water; AIS, Atlantic Ionian Stream; ATC, Atlantic Tunisian Current; ABV, Adventure Bank Vortex; MCC, Maltese Channel Crest); ISV, Ionian Chef Break Vortex; SDG, Sidra Gyre; LIW, Levantine Intermediate Water; EMDW, Eastern Mediterranean Deep Water. (UNEP/MAP RAC/SPA, Tunis, 2015).

seasons (about 15°C and more than 22°C in winter and summer, respectively) and its saline content is about 37.5%; flowing towards the eastern basin waters become cooler and more saline. The LW has about 14.2°C and 38.7% when reaching the Strait of Sicily, where its influence extends down to 600–800 m.

4.1. Bionomy

Different studies on the benthic communities in the Strait of Sicily [31, 32] highlighted the dominance of rheophilic species, and with an Atlantic affinity. Large prairies of the seagrass *Posidonia oceanica,* which represents the most widespread and typical ecosystem in the Mediterranean phytal zone [33], characterize wide extension of the infralittoral bottoms down to 40–50 m [32, 34]. The hard substrates in the infralittoral and the deeper bottoms of the circalittoral are colonized by wide populations of large brown algae such as *Cystoseira, Sargassum,* and especially *Laminaria rodriguezi* [35]. The sandy bottoms swept by unidirectional and oscillating currents between 40–120 m (but also at lower depth, between the sea-grasses), especially in the centre of the channel (Lampedusa Island), host populations of calcareous red algae, the so-called 'Maerl' [33]. Bottoms between 30–50 and 150–200 m are characterized, using the Péres and Picard's classification [33], as the 'circalittoral' layer. The deeper level, generally associated with the continental slope and the deep steeps around the banks, is the 'bathyal' layer. In the hard substrates of the circalittoral layer, there is present a peculiar biocoenotic series named, thanks to the presence of red coral (*Corallium rubrum*), 'coralligenous bottoms'. They are characterized by: calcareous red algae, sponges, cnidaria (*Eunicella cavolini, Astroides calycularis*), polychaetes (*Serpula vermicularis*), brachiopods (*Argyrotheca cuneata*), bryozoa, crustaceans (*Lissa chiragra*), echinoderms (*Ophidia sterophidianus*), sea squirts (*Rhodosoma callense*) and bivalves (*Manupecten pesfelis, Lima vulgaris*). Péres and Picard [33] described another circalittoral biocoenosis of the hard bottoms, the 'off-shore rocks'; the characteristic group is represented by many sponges species, together with the cnidarians already mentioned. Some species, e.g. the Sea urchin *Cidaris cidaris*, are considered as outsiders coming from the bathyal zone. In the area of the Sicilian Straits the circalittoral bottoms between 65 and 105 m are often sandy, with grains from coarse to very fine, nearly muddy, interspersed with abundant larger detritus of organic origin, such as shell fragments, calcareous plants and, less frequently, bryozoan remnants. Soft seaweeds are abundant too: at low depth the green seaweed *Cladophora fracta* is predominant, elsewhere other algae are common (e.g. *Dictyota dichotoma, Laminaria rodriguezii*). The presence of sponges is massive, with *Crambe crambe* and others. A non-peculiar distribution is shown by the cnidaria *Pennatula phosphorea*. This species is typical of a particular kind of terrigenous mud-shelf assemblage, characterized by a viscous mud, and it is often associated with *Veretillum cynomorium* [36]; bivalves are mainly occurring with the species, *Acanthocardia echinata,, Hiatella arctica, Pecten jacobaeus* (assumed to be characteristic of whole western Mediterranean [36]; cephalopods presences are patchy, with *Eledone moschata, E. cirrhosa, Octopus vulgaris, Sepia officinalis* and *Loligo vulgaris*. Around Lampedusa, the crustacean galatheid *Munida curvimanna* is observed on sandy-muddy grounds, while *Galathea strigosa* lives on the 'Maerl' bottoms [37]. Many fishes of commercial interest live in the meso littoral layer, among which the most important red mullets, both the rocky one (*Mullus surmuletus*) and the muddy one (*M. barbatus*), and the pandora bream (*Pagellus erythrinus*); common are other demersal species such as the Sea robins *Lepidotrigla cavillone*, the flatfishes *Solea kleini*, the scorpion fish *Scorpaena scrofa*. The Sicilian Channel Strait presents some species of sub-tropical attitude, such as the Portuguese sole *Synaptura lusitanica*, the corb *Umbrina ronchus, Cynoponticus ferox, Facciolella oxyrhyncha, Epigonus constanciae* [38]. In the Sicilian Strait, the hard bottoms of deeper bathymetric zones, those of the bathyal layer, are characterized [39] by many huge 'buildings' produced by madrepores (*Madrepora oculata, Lophelia prolifera*), generally forming scattered clumps, that give origin to the 'white coral assemblages' biocoenosis [33], locally known as 'cannelleri' (Ragonese, S. personal communication): those presents strong and stony corms that make extended surfaces of the grounds between 300 and 450 m dangerous for the fishing activity, or even untrawlable altogether. Another white coral, *Dendrophyllia cornigera*, is less hard but still an obstacle for fishing, lives at higher depths; in fact, it colonizes rocky substrates exposed to hydrodynamics, while the former corals prefer finer sediments. The most typical biological indicator species is the Sea pen *Funiculina quadrangularis*, even if it is now quite rare especially on the trawlable grounds; the occurrence of such a species is closely related to the abundance of the food supply. In a few zones, the brachiopod *Terebratula vitrea* is dominant, but always associated with *F. quadrangolaris*. Among the crustaceans of this layer, the pink shrimp *Parapenaeus longirostris* (often associated with the presence of *Funiculina*) is the most interesting; the Norway lobster, *Nephrops norvegicus*, is caught here too; cartilaginous fishes are well and constantly represented in the layer with the dogfishes (*Etmopterus spinax, Scyliorhinus canicula*) and the skates (*Raja oxyrinchus, R. miraletus*). The northern part of the channel, i.e. the western Sicily, does not show significant differences from the other areas [40] while in the southern most part (in front of Capo Passero) are present the wreck fish *Polyprion americanus*, the grouper *Epinephelus guaza* and various Sea robins *Trigla* spp. [41]. The characteristic facies of this bionomic stratum is composed of *Isidella elongata* (Gorgonacea) that is associated with the red shrimps *Aristaeomorpha foliacea* and, depending on the zones, *Aristeus antennatus*. Parts of the same facies are the brachiopod *Tenebratula vitrea* and the echinoderm *Brisingella coronata* [37]. In western Sicily (i.e. the northern part of the Sicilian Strait Channel), the mesobathyal layer shows some typical and exclusive species, such as the sponges *Pheronema grayi* and *Poecillastra compressa*, the cnidaria *Lophogorgia sarmentosa, Madrepora oculata, Lophelia prolifera, Dendrophyllia cornigera, Calliactis parasitica* and *Amphianthus dohrni*, the crab *Anamanthia rissoana*, the gastropods *Natica millepunctata, Lunatia fusca, Buccinum humphreysianum* and *Fusinus rostratus*, the cephalopods *Bathypolypus sponsalis, Heteroteuthis dispar, Alloteuthis media, Abralia veranyi, Histioteuthis reversa, Chiroteuthis veranyi* and *Octopoteuthis sicula* [40]; in the northernmost area, the fishes *Chlorophthalmus agassizi* and *Aulopus filamentosus* are present too [41].

5. Conclusion

The Strait of Messina and the Sicilian Channel, part of the Central Mediterranean Sea, are considered as two unique exclusive ecosystems. The Strait of Messina is the point of contact between the western and eastern parts of the Mediterranean Basin. It presents unique biological communities and animal organisms and extremely rare seaweeds and is a great reservoir of biodiversity. For its peculiarities and for the seaweeds and animals that live in, it is considered as a sort of 'Atlantic Island' in the middle of the Mediterranean Sea. The two areas are to be considered as migratory routes for many species of fish (many of commercial interests) and headed whales in the Southern Tyrrhenian Sea, some species for genetic reasons also. The Central Mediterranean is a hot spot of biodiversity that sustains important endemic species and communities that interacts with two adjacent basins, eastern and western, concentrating and redistributing an important species budget. Finally, the reasons for which the Central Mediterranean is to be considered as a hot spot of biodiversity are to be found in its particular hydrology and geomorphology of the seabed.

Acknowledgements

A sincere thanks to Sergio Ragonese (CNR IAMC Mazara del Vallo) for the contribution in the critical review of some parts of the manuscript on the Sicilian Channel. Particular thanks to Gianmichele Iaria for providing photographic documentation and for the friendly patience in long discussions on the Strait of Messina.

http://dx.doi.org/10.5772/intechopen.68942

Similarly, we wish to thank Francesco Costa and Andrea Potoschi to have kindly provided significant photos shown in this note. A proper and sincere thanks to Gioele Capillo, Giuseppe Panarello and Marco Graziano for the precious help in the preparation of the text and shapes.

Author details

Nunziacarla Spanò* and Emilio De Domenico

*Address all correspondence to: spano@unime.it

Department of Chemical, Biological, Pharmaceutical and Environmental Sciences, University of Messina, Messina, Italy

References

[1] Bergamasco A, Malanotte-Rizzoli P. The circulation of the Mediterranean Sea: A historical review of experimental investigations. Advances in Oceanography & Limnology. 2010;**1**(1):11-28

[2] Della Croce N, Cattaneo Vietti R, Danovaro R. Ecologia E Protezione dell'Ambiente Marino Costiero. UTET Libraria SRL, editor. Editel-Torino; 1997. pp. 15-196

[3] Barale V. The European marginal and enclosed seas: An overview. In: Barale V, Gade M, editors. Remote Sensing of the European Seas [Internet]. Springer Netherlands: Dordrecht; 2008. pp. 3-22. Available from: http://dx.doi.org/10.1007/978-1-4020-6772-3_1

[4] De Domenico E. Caratteristiche fisiche e chimiche delle acque nello Stretto di Messina. In: Geronimo D, Barrier, Mantenat, editors. Le Detroit de Messine, Evolution Tectono-Sedimentaire Recente (Pliocene et Quaternaire) et Environment Actuel. Paris; 1987. pp. 225-235

[5] Colantoni P. Marine geology of the Strait of Messina. In: Di Geronimo, Barrier, Mantenat, editors. Le Detroit de Messine, Evolution Tectono-Sedimentaire Recente (Pliocene et Quaternaire) et Environment Actuel. IGAL Paris; 1987. pp. 191-209

[6] Mosetti F. Some news on the currents in the straits of messina. Boll di Oceanol Teor ed Appl. 1988;**6**(3):119-201

[7] Tomasino M. The exploitation of energy in the straits of Messina. In: Guglielmo L, Manganaro A, De Domenico E, editors. The Straits of Messina Ecosystem; Università degli Studi di Messina, Messina; 1995. pp. 49-60

[8] Fredij G, Giaccone G. Bionomie des fonds a Laminaires du Detroit de Messine. In: Geronimo D, Barrier, Mantenat, editors. Le Detroit de Messine, Evolution Tectono-Sedimentaire Recente (Pliocene et Quaternaire) et Environment Actuel. Paris; 1987. pp. 237-238

[9] Berdar A, Giacobbe S, Leonardi M. Prima segnalazione per il Mediterraneo di Corolla spectabilis Dall, 1871 (Thecosomata). Boll Malacol. 1982;**18**(1-4):35-40

[10] Giacobbe S, Spanò N. Pilumnus inermis (Decapoda, Brachyura) in the straits of Messina and the Southern Tyrrhenian Sea (Mediterranean Sea): Distribution and some aspects of its ecology. Crustaceana [Internet]. 2001;**74**(7):659-672. Available from: http://booksandjournals.brillonline.com/content/journals/10.1163/156854001750377939

[11] Rinelli P, Spanò N, Giacobbe S. Alcune osservazioni su crostacei decapodi ed echinodermi dei fondi a Errina aspera dello Stretto di Messina. Biol Mar Medit. 1999;**6**(1):430-432

[12] Smriglio C, Mariottini P, Giacobbe S. Jujubinus errinae n. sp. (Gastropoda Trochidae) from the Strait of Messina, Mediterranean Sea. Biodiversity Journal. 2016;**7**(1):59-66

[13] Giacobbe S, Spanò N. New Records of Albunea Carabus (L., 1758) (Decapoda, Anomura) in the Mediterranean Sea. Crustaceana [Internet]. 1996;**69**(6):719-726. Available from: http://booksandjournals.brillonline.com/content/journals/10.1163/156854096x00736

[14] Costa F. Atlante dei pesci dei mari italiani. Mursia Editore, editor. Milano; 1991. pp. 1-430

[15] Bottari A, Bottari C, Carveni P, Giacobbe S, Spanò N. Genesis and geomorphologic and ecological evolution of the Ganzirri salt marsh (Messina, Italy). Quaternary International [Internet]. 2005;**140-141**:150-158. Available from: http://www.sciencedirect.com/science/article/pii/S1040618205001102

[16] Scuderi D, Terlizz A, Faimali M. Osservazioni su alcuni tratti della biologia riproduttiva di vermeti biocostruttori e loro ruolo nella edificazione dei "trottoir." Biol Mar Medit. 1998;**5**(1):284-289

[17] Capillo G, Sanfilippo M, Panarello G, Spinelli A, Manganaro A, Spano N. Preliminary investigation of an intertidal zone's faunal assemblage (Beach Rock, R.N.O. Capo Peloro, NE Sicily). Journal of Biological Research. 2016;**89**(s1): 1-20

[18] Scotto Di Carlo B, Costanzo G, Fresi E, Guglielmo L, Ianora A. Feeding ecology and stranding mechanisms in two lanternfishes (Hygophum benoiti and Myctophum punctatum). Marine Ecology Progress Series. 1982;**9**:13-24

[19] Baguet F. Bioluminescence of deep-sea fishes in the straits of messina. In: Guglielmo L, Manganaro A, De Domenico E, editors. The Straits of Messina Ecosystem. Università degli Studi di Messina, Messina; 1995. pp. 203-212

[20] Andersen V, Sardou J, Nival P. The diel migrations and vertical distributions of zooplankton and micronekton in the Northwestern Mediterranean Sea. 2. Siphonophores, hydromedusae and pyrosomids. Journal of Plankton Research [Internet]. 1992;**14**(8):1155. Available from: +

[21] Morel, A. Charactères hydrologiques des eaux échangées entre le bassin oriental et le bassin occidental de le Méditerranée. Cah. Ocean. 1971:**23**(4):329-342

[22] Hopkins TS. Physics of the sea. In: Margalef R, editor. Western Mediterranean. New York: Pergamon; 1985. pp. 100-125

[23] Sverdrup HU, Johnson MW, Fleming RH. The Oceans, their Physics, Chemistry, and General Biology. New York: Prentice-Hall; 1942

[24] De Maio A, Moretti M, Sansone E, Bregant D. Idrologia del Canale di Sicilia, Nave Dectra, ottobre 1969. Volume di Studi in onore di G. Aliverti. Napoli; 1972

[25] Grancini GF, Michelato A. Current structure and variability in the Strait of Sicily and adjacent areas. Annals of Geophysics. 1982;**5**:75-88

[26] Ragonese S, Bianchini ML, Di Stefano L, Bertolino F, Campagnuolo S. Study of the selec- tivity and assessment of the coefficient of retention of the trawl nets used for red shrimp fishing (Aristaeomorpha foliacea Risso, 1827 and Aristeus antennatus Risso, 1816; Crustacea-Aristeidae) in the Sicilian Channel (Central Mediterr. Sea), Project MED, 92/010, final report to the European Community, D.G. XIV, Bruxelles. 1994, 190 pp

[27] Gorgy S, Shaheen AH. Exploration des lieux de Pêche de la République Arabe Unie. — Résultats hydrographiques de l'expédidition du Shuyo-Maru dans la Mer Méditerranée et la Mer Rouge. In: Proc gen Fish Coun Médit. 1963. pp. 139-70

[28] Tziperman E, Malanotte-Rizzoli P. The climatological seasonal circulation of the Mediterranean Sea. Journal of Marine Research. 1991;**49**(3):411-434

[29] Ovtchinnikov IM. Circulation des eaux de la mer Méditerranée. In: Burkov VA, editor. Hydrobiologie de la Méditerranée. Gidrometeoizdat, Leningrad; 1976. pp. 240-313

[30] Bombace G, Sarà R. La pesca a strascico sui fondali da -500 a -700 metri nel settore a Sud-Est di Pantelleria. Mem Minist Mar Merc. 1972;**33**:77

[31] Giaccone G. Struttura, ecologia e corologia dei popolamenti a Laminarie dello Stretto di Messina e del Mare di Alboran. Mem di Biol Mar e Oceanogr. 1972;**2**:37-59

[32] Cinelli F, Fresi E, Mazzella L, Ponticelli MP. Deep algal vegetation of the western Mediterranean. G Bot Ital [Internet]. 1979;**113**(3):173-188. Available from: http://dx.doi.org/10.1080/11263507909426641

[33] Pères JM, Picard J. Noveau manuel de bionomie bentique de la Mer Mediterraneè. Requeil des Trav la Stn Mar d'Endoume Bull. 1964;**31**(47):5-137

[34] Cinelli F. Le Fanerogame marine: Problemi di trapianto e di riforestazione. Mem di Biol Mar e Oceanogr. 1980;**10**:17-27

[35] Suriano C, Mazzola S, Levi D, Giusto GB. La biocenosi dei substrati duri circalitorali a grandi Phaeophyceae (Laminaria rodriguezii Bornet, 1888) nel Canale di Sicilia e nel Canale Maltese. Oebalia. 1992;**17**(Suppl):429-432

[36] Pères JM. History of the Mediterranean biota and the colonization of the depths. In: Margalef R, editor. Western Mediterranean Key environments. Oxford, UK: Pergamon Press; 1985. pp. 198-232

[37] GFCM. Living deep-water resources of the Western Mediterranean and their exploitation. Stud Rev Gen Fish Coun Mediterr. 1970;**44**:38

[38] Fredj G, Maurin C. Les poissons dans la banque de données Medifaune. Application à l'étude des caractéristiques de la faune ichtyologique méditerranéenne. Cybium, Journal of Ichthyology. 1987;**11**(3):215-342

[39] Arena P. Studio sulla possibilità di razionalizzare e rendere più produttiva la pesca a strascico nel Canale di Sicilia e nel Mediterraneo centro-meridionale: Mimeo, Palermo; 1985

[40] Arena P, Li Greci F. Indagini sulle condizioni faunistiche e sui rendimenti di pesca dei fondali batiali della Sicilia occidentale e della bordura settentrionale dei banchi della soglia Siculo-Tunisina. Quad Lab Teen Pesca. 1973;**1**(5):157-201

[41] Scaccini C, Piccinetti R. Stato attuale della pesca in acque profonde nei mari italiani. Boll Pesca Piscic Idrobiol. 1970;**25**:5-35

Chapter 7

Deep-Sea Biodiversity in the Aegean Sea

Onur Gönülal and Cem Dalyan

Additional information is available at the end of the chapter

http://dx.doi.org/10.5772/intechopen.70492

Abstract

The present chapter aims to describe macrofauna in the Aegean deep waters. The review is based mainly on the studies of deep waters below 200 m. A total of 386 species are included on the present checklist belonging to 9 phyla. Among these species, Porifera has 9 species, Cnidaria 4, Brachiopoda 3, Bryozoa 1, Polychaeta 34, Mollusca 92, Arthropoda 86, Echinodermata 30 and Pisces 127 species.

Keywords: Aegean Sea, deep sea, macrofauna, biodiversity, geology

1. Introduction

The Aegean Sea is an arm of the Mediterranean Sea located between the mainland of Greece and Turkey. Therefore, it has strategic, economic and political importance for these two neighboring countries. It is connected to the Sea of Marmara via the Çanakkale Strait (max depth 105 m) in the northeast, while several deeper gateways provide communications to the rest of the Eastern Mediterranean Sea in the south [1]. These gateways are between the Argolid Peninsula situated in Greece in the Peloponnesus and Elafonisos island (42 m, 0.3 nm), Elafonisou strait located between Elafonisos island and Kythira island (291 m, 4.5 nm), Kythira strait located between Kythira island and Antikythira island (263 m, 17 nm), Antikythira strait located between Antikythira island and Crete island (725 m, 16.5 nm), Kasou strait located between Crete island and Kasos island (1100 m, 26.5 nm), between Kasos island and Karpathos island (55 m, 3.5 nm), Karpathos strait located between Karpathos island and Rhodes island (896 m, 23.5 nm), between Rhodes island and Bozburun Peninsula situated in Turkey (439 m, 10 nm).

The Aegean Sea has a very irregular coastline with a number of small and large bays, peninsulas and islands or islets. This irregular coastline forms small basins and passages. It covers about 191,000 km^2 in area and measures about 610 km longitudinally and 300 km latitudinally.

The total water area is 193,950.33 km^2. While there are only 96 islands and rocks belonging to Turkish waters, there are 460 of them belonging to Greek waters [2]. The sea was traditionally known as "Archipelago" because of many islands, while in some south Slavic languages it is often called White Sea [3]. Deep sea is often described to begin at the edge of the continental shelf, which varies around the world, but in average it is around 200 m of depth [4]. It also defined as a portion of the ocean that stands below 200 m of depth, both in the water column and in the benthos [5]. Deep sea is almost totally devoid in light which means that it is also a dark sea. Thus, photosynthesis may not act there with enough efficiency to sustain life. This level, called compensation depth, is found at about 150–200 m of depth in the clearest ocean waters and, it is shallower in more turbid waters. Deep sea remains largely unexplored. Most scientific research so far focused on the sunlight zone of the oceans and seas.

Deep zones are considered as lifeless domains, while deep-sea habitats are considered as exceptional ecosystems [6]. The first data on the bathyal species was recorded by Risso in France at depth between 600 and 1000 m [7].

2. Geographic setting and morphology of the Aegean deep sea

The Aegean Sea is surrounded by the Anatolia from the east, the Thrace and Eastern Macedonia from the north, the Thessaly and Peloponnesus peninsula from the west and the islands including Crete and Rhodes from the south [2]. Approximately 33.6% of the Aegean Sea is shallower than 200 m, while the mean depth is 362 m [8]. The continental shelf is represented by two types, narrow (1–10 km) and broad (25–95 km). While narrow shelf occurs along the western coast, broad one dominates the northeastern Aegean Sea [8].

The Aegean Sea is a volcanically active region, characterized by above average continental heat-flow values. It has deeper Moho depths compared to the Eastern Mediterranean and Black Sea [9]. The Aegean Volcanic Arc is situated on the north of the Hellenic Arc, and swinging to northeast toward mainland Turkey [10].

The present form of the Mediterranean Sea is the result of continuous interaction of complex geodynamic processes during the last 50–70 myr. The eastern Mediterranean Basin (Ionian and Levantine) is the only true remnant of the older Tethys Ocean, which is being actually consumed along the active Hellenic, Cypriotic and Calabrian arcs. The East Mediterranean Ridge represents the accretionary prism formed above the shallow, north-northeastward dipping subduction of the oceanic Mediterranean crust below the Aegean microplate. The rest of the Mediterranean basins, like the Aegean, resulted from back-arc extension behind the southward migrating Hellenic Arc and the east-southeastward migrating Calabrian Arc [11].

After the disintegration of the Pangae, the Eastern Mediterranean Sea made up the southern boundary of the Tethys Ocean for the last 200 myr. The history of the Aegean Sea began during Oligocene about 35 myr ago. The Island of the Aegean actually started being formed in the middle Miocene about 12–11 myr after the sea began to penetrate to single land mass of Aegeis. The Island was formed at the end of the Miocene 6—5.3 myr. Main causes of the change in the Aegean region's geography were eustatism and tectonism during the Pleistocene, from 1.8

to 0.9 myr. As the sea level rises, the Aegean region gradually acquires its current geography during the Holocene. Eastern Aegean islands were cut off from Asia Minor and the Cyclades islands were permanently isolated from one another [9].

The formation and distribution of the deep south Aegean basins are the result of the geotectonic regime, which was active over the region during the last 5 myr. Nevertheless, the tectonic and seismic activity in the southern Aegean is presently much lower in respect to the northern Aegean. Tectonic activity has migrated southward and affected the Island Arc with faulting, which is responsible for uplift or subsidence of successive regions along the arc. The tectonic fragmentation of the Island Arc has resulted in the formation of relatively shallow straits, west and east of Crete, which enable water exchange between the Aegean Sea and the eastern Mediterranean.

The Aegean Sea has a complex nature of the shoreline and sea-mountain tops emerging from the seabed. Among a number of basins, five major basins are clearly identified (**Figure 1**).

1. The Karpathos basin is the deepest basin located on the southernmost part of the Aegean Sea, with a maximum depth of 2500 m. The basin is bordered by a steep faulted slope toward Karpathos Island.
2. The Kamilonisi basin is located between the northern coast of eastern Crete and the Kamilonisi Islet. Its depth reached 2200 m.
3. The Chios basin, in the central Aegean Sea.
4. The Skyros basin, situated in the central part of the Aegean Sea.
5. The North Aegean basin, with depths reaching 1500 m.

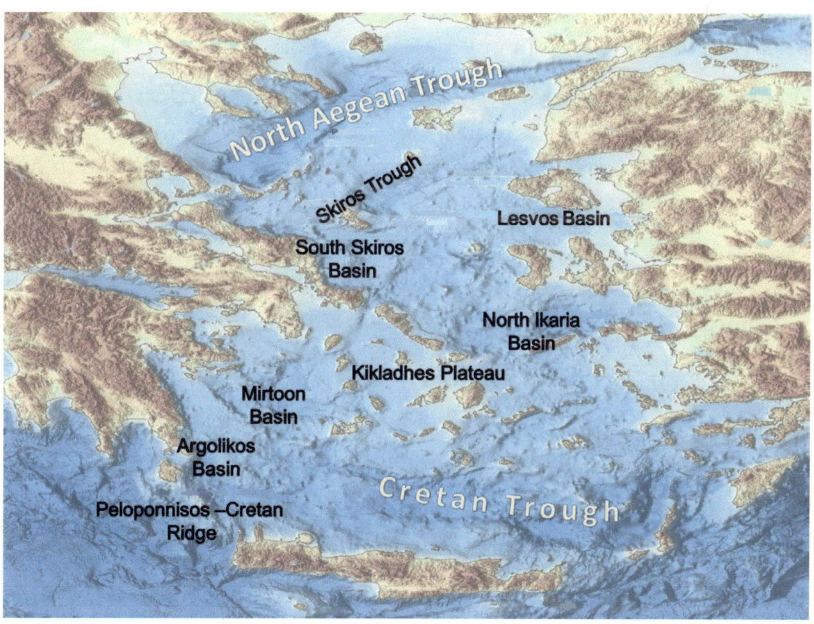

Figure 1. Morphology of the Aegean deep sea.

The Aegean Sea is topographically divided in two basins approximately along the 38° parallel: the north and south Aegean Sea. The hydrological and topographical conditions differ strongly between the two basins. The Aegean Sea can be divided into three distinct regions with different morphological characteristics as well as geotectonic regimes: the northern, the central and the southern part. The current morphological shape of the Aegean Sea was processed by means of three main parameters: the tectonism, the volcanic activity and the eustatism [9].

Eventhough the northern part of the Aegean Sea is characterized by an extreme continental shelf, the north Aegean Trough is the dominant morphological character in this part. The trough is elongated along the trace of the Northern Anatolian Fault and includes a series of three main depressions [12]. The eastern depression is a narrow but long feature which extends from the Limnos Island to the Gulf of Saros and reaches 1500 m of depth. Another depression named The Athos Basin is located on the south of the Chalcidice Peninsula with an average depth of 1200 m. The Sporades Basin is the western depression reaching depths of 1468 m. All of the depressions in the North Aegean Trough are separated from each other by morphological heights, a sill of 350 m from the 800 m deep North Skyros Basin, on their south. The southern slopes of the trough contact the 100–300 m shallow platform extending between the Sporades islands to the west and the Limnos and Gokceada islands to the east. In addition, there are many basins between the trough and the Central Aegean Plateau [13]: the 1000 m deep Skopelos Basin (2), the 800 m deep Kymi Basin (3), the 800 m deep S. Skyros Basin (4), the 1000 m deep N. Skyros Basin (5), the 800 m deep Psara Basin (6). The Ikaria Basin is the southernmost deep basin of the northern Aegean region, toward the shallow Central Aegean Plateau.

All the above-mentioned basins are surrounded by steep slopes and isolated morphological depressions separated from each other by 200–400 m shallow platforms.

The Central Aegean Plateau shows a curved shape and extends throughout from Evvoia Island in Greece to the Menderes region in Asia Minor. The southern limit of the Plateau bordering the Volcanic Arc consists of the Nisyros, Santorini, Milos, Poros and Aigina Islands along with smaller islets and submarine volcanos [13, 14].

The South Aegean Sea consists of a series of deep elongated basins which are distributed between the Aegean Volcanic Arc to the north and the Hellenic Arc to the south. The Hellenic Arc extends from the Argolikos Gulf, off the eastern Peloponnesus, over the Cretan Sea, between Crete and Santorini Islands, and continues to the Sea of Karpathos, west of Karpathos Island. This concave row of islands characterizes the morphology of the southern Aegean Sea. The Aegean Arc extends from mainland Greece toward the broadly arcuate Cyclades islands and thereafter swings northeast toward mainland Turkey [10]. The Aegean Sea is the deepest in the Karpathos Basin (2500 m) which is bordered by a steep faulted slope toward Karpathos Island. A 1300 m deep shallow ridge separates the Karpathos Basin from the 2200 m deep Kamilonisi Basin. The latter is located between the northern coast of eastern Crete and the Kamilonisi Islet. Next to this and north of central Crete, the 1800 m deep Irakleio Basin occupies the central part of the southern Aegean Sea. The Cretan Trough is situated between the Hellenic Arc and the Aegean Volcanic Arc, composed of a series of smaller sub-basins. The

http://dx.doi.org/10.5772/intechopen.70492

depression depth in the trough generally exceeds 1000 m. Further to the west, a shallower but long and narrow basin follows the shallow ridge, which connects western Crete with the Antikythira and Kythira Islands and the eastern Peloponnesus [11].

3. Deep-sea biodiversity

The Mediterranean Sea is the area of the first deep-sea explorations. In 1843, Forbes conducted numerous dredge operations in the sea floor of the Aegean Sea. The abyssal was defined as azoic since no living organism in the area could be observed. At the beginning of the 1800s, Antoine Risso, the first naturalist to describe deep Mediterranean species, recorded deep-sea fishes and crustaceans from the Gulf of Genova at depths between 600 and 1000 m [15].

The eastern basin was later investigated by the Russian Vessel Vitiaz in 1889, and the Austrian Vessel Pola between 1890 and 1893. Danish expeditions conducted with the Vessel Thor in 1908–1910 entering the Black Sea through the Mediterranean and completed in 1921 and 1922 with the Dana Cruises [15].

Tselepides represented that the macrofaunal species composition and mean benthic biomass, in the Cretan deep water to the depth of 1600 m was similar to the western Mediterranean and the neighboring Atlantic. Ben-Eliahu and Fiege found significant correlations between macrofauna diversity and food availability while the positive correlations is found to be the principle regulating factor in the system [16]. Although Mediterranean ecosystems are among the most studied areas of the world, deep-sea fauna research lags behind those of other areas [17].

It is not possible to understand the fauna of the Mediterranean Sea deep water, without considering its geological history. The present Mediterranean Sea is the only remnant of the Tethys Ocean. The Ocean existed till the Miocene. Approximately 6 myr ago (the end of the Miocene) the connection between the Mediterranean and Atlantic Ocean closed. The Mediterranean Sea dried up until there was almost no water left. Abiotic conditions affected all marine life, especially brachi. Finally, 5.3 myr ago the connection between the Mediterranean and Atlantic Ocean reopened. The Mediterranean basin was refilled with Atlantic waters and it started to take its current state.

Thus, the Mediterranean deep-sea is very young compared to other oceans. The youngest basin in the Mediterranean is the Aegean Sea, which mostly developed during the Pliocene and Quaternary, from 5 myr to the present [14].

The Mediterranean Sea is an oligotrophic marine system, while the eastern Mediterranean basin being its most oligotrophic part [18]. A similar trend of decreasing primary production values is present along the N–S transect of the Aegean Sea [19]. In the northern Aegean Sea, nutrients are supplied by freshwater runoff of rivers and by inflow of nutrient-rich Black Sea surface waters [20, 21]. Most of the deep sea is heterotrophic except the hydrothermal vents, thus the life of the deep-sea benthos depends on the food supply derived from surface production [22]. As a consequence, even though the Mediterranean Sea is oligotrophic, the northern Aegean Sea is more productive. Higher faunal densities are expected in this area. The biodiversity of deep benthic communities is also related to depth and sediment characteristics.

The quantity and quality of the descending organic matter that reaches the deep-sea floor are a function of a variety of factors, including primary production, phytoplankton sinking rate, zooplankton grazing rate, water column depth, mixed layer depth and proximity to land and input sources [23, 24].

The richness of the Mediterranean deep benthic fauna is decreasing gradually from west to east [15]. Food availability seems to be the major factor influencing the structure and function of macrofauna communities for deep and oligotrophic areas, such as the eastern Mediterranean [25].

The current Mediterranean deep-water fauna is less closely related to the Atlantic bathyal fauna than it was in the Pleistocene [23], due to the lack of Atlantic deep water fauna entering the Mediterranean Sea [26].

Bouchet and Taviani claim that much of the Mediterranean deep-sea fauna are made up of non-reproducing pseudopopulations that have entered the Mediterranean as meroplankton with the Atlantic inflow at Gibraltar [27]. The populations of the most common benthic molluscs at depths greater than 1000 m off the Israeli coast are composed of both adult and juvenile specimens. One particular species, *Yoldia micrometrica* Seguenza G., 1877, the most common and abundant species in the eastern Mediterranean, is not recorded in the westernmost part of the Sea [26]. Therefore, benthic decapod crustaceans and egg carrying fishes were collected from the depths of the Levantine Sea [25, 28].

The onset of current hydrological conditions in the Holocene led to the almost complete extermination of the richer Mediterranean deep-sea fauna, which was more similar to the present Atlantic fauna [27].

The close affinity between Mediterranean and Atlantic congeneric deep-water species suggests that the ancestors of the Mediterranean bathyal endemic species moved from the Atlantic when conditions were favorable [29].

Despite the presence of bathyal and abyssal species in the Mediterranean, typical deep-sea groups are absent [15]. According to Coll, the Mediterranean deep water has only 20–30 genuine abyssal species. In the western basin, where the depth does not exceed 3000 m, the abyssal fauna is less abundant than in the deeper eastern basin, where abyssal species are dominant in the Matapan trench, reaching to 5050 m depth [29].

The fauna of the Aegean Sea represents a distinct and separate subsystem in relation to the Mediterranean fauna [4]. The native fauna of the Aegean Sea is enriched both with Indo-Pacific immigrants and with Atlantic origin species [30]. The Aegean Sea can be considered as a separate subsystem of the eastern Mediterranean Sea and is isolated from it by extremely deep waters, high evaporative loss and its climatic, hydrographic and geomorphological diversity and faunal make-up [31]. Although the Aegean Sea is distant from the Strait of Gibraltar (the main pathway of enrichment for the Mediterranean Sea), it is inhabited by a greater number of species. According to Koukouras, main reasons for this may be its communication with the western basin and the high temperature and salinity variations [30]. In addition, The Aegean Sea (especially its northern sector) also shows high invertebrate species richness, which is otherwise low in most of the remaining central and eastern basin [29].

The Aegean Sea is isolated from the deep Atlantic waters by the shallow Gibraltar Strait, the African-Sicilian threshold and the topographical/hydrological barrier of the Hellenic trench. Atlanto-Mediterranean species dominate the deep waters of the Aegean Sea. Drivers of biodiversity in the deep Mediterranean are within three main categories: (1) bathymetric gradients, which are associated with increasing pressure and decreasing food availability in deeper sediments; (2) geographical and physico-chemical features, which are responsible for the north-northwest—south-southeast gradient in trophic conditions; and (3) environmental heterogeneity (e.g., grain size distribution, habitat complexity, distribution of food inputs) [29]. Yet, our knowledge about deep-sea biodiversity patterns is still limited [32, 33].

Food availability in deep sea decreases with depth and depends on the supply of energy from the pelagic area. This may explain most of the variability between the observed spatial patterns of the benthic biodiversity in the deep Mediterranean Sea [29, 34].

3.1. Deep-sea sponge

Sponges living in deep waters are an important component of the seas' bathyal benthos and prefer to inhabit soft bottom, whereas sponges are usually dependent for attachment on hard substrate [35]. Records of deep-sea sponges in the Mediterranean Sea are hard fragments within soft bottoms and also hard substrate such as vertical walls [36, 37].

Littoral zones and bathyal zones are clearly distinguishable for their sponge fauna in the Aegean Sea. A total number of sponge species are decreasing with depth [37]. According to Witte, the reason lies in scarcity of water-borne particles in deep-sea environments [38].

Although many endemic sponges recorded from deeper water increased the number of the known sponges in the western Mediterranean Sea [39], the number of endemic sponges recorded in the Aegean Sea is limited. One possible reason for this is lack of research in deep waters.

While sponges with sponging skeletons are widely distributed in warm waters, those with siliceous skeletons ones are found at high latitudes or deeper waters [40]. Order Poecilosclerida that showed a distinct diversity center in cold and deep waters is the most diverse in species number in the Aegean Sea [41].

3.2. Deep-sea coral

Cold-water reef-forming corals have been known for nearly 250 years in the deep ocean. However, there were some recent discoveries of white coral communities in the Mediterranean Sea. At the present, the Mediterranean Sea is known for its richness of fossil white coral communities exposed in land outcrops, and extant coral communities [42]. The Mediterranean basin represents an excellent biological archive of the past and recent deep-coral growth and evolutionary patterns of present deep-coral bioconstructions [43].

So far, only one isolated spot containing live *Lophelia pertusa* (Linnaeus, 1758) and *Madrepora oculata* (Linnaeus, 1758) was reported from a trough off Thassos, in the northern Aegean Sea [44]. Also, fossil or subfossil white corals have been sampled from the Cretan Arc [45]. The Cretan Arc relevant to the depth intervals suitable for white corals is not far from the isolated unusual occurrence of the live *L. pertusa* and *M. oculata* in the Aegean Sea. *Desmophyllum dianthus* (Esper, 1794) is relatively common and widespread in the whole Mediterranean and it has been reported alive from the Aegean Sea [45, 46].

One of the potential regions for white coral communities in the Aegean Sea is the northern Aegean trough (especially between Gökçeada Island and Samothrace) that has steep bathymetric gradient in terms and narrow shelves.

3.3. Bryozoan diversity

Information on bryozoan diversity from the Aegean Sea have been reported from a variety of habitats, including soft sediments, sea-grass leaves and rhizomes, macroalgae and coralligenous [47].

The richest bryozoan diversity is found in the coralligenous and in the dark and semi-dark cave biocoenoses [48]. Bryozoan diversity is much lower in bathyal habitats [48]. This habitat is characterized by complete darkness, long-lasting hard substrata provided by exposed coral skeletons, coral rubble and occasionally by firm- and hardgrounds. Most information on bryozoan diversity in this habitat come from communities associated to cold water corals [49]. Although Mediterranean deep-water habitats are known as relatively depleted in macrofaunal species, a number of new bryozoans species have been described in these habitats including seamounts and ridges [49].

The only bryozoan species *Herentia hyndmanni* that has been recorded so far by Johnston (1847) is found in the Rhodes Channel at 270 m, in the deep Aegen Sea. Specific regions such as the deep of the Aegean Sea can be potentially suitable for bryozoan diversification.

3.4. Deep-sea Arthropoda

Eighty-six Arthropoda species have been reported deeper than 200 m in the Aegean Sea. Fifteen of 86 deep-water species are found below 1000 m depth (**Table 1**). Arthropoda are the dominant invertebrate taxon in deep Aegean waters. All of these species are Atlanto-Mediterranean originated. *Parapenaeus longirostris, Pandalid shrimps* (*Plesionika* spp.) and *Polycheles typhlops* are among dominating species. It is considered that, due to their low levels of food consumption, crustacean decapods are more important in the deep Mediterranean (an oligotrophic region) [6].

Since deep Aegean Sea waters are still not yet well known, it is quite difficult to give comparative remarks.

Phylum	Reference	Depth
Porifera		
Thenea muricata Bowerbank, 1858	[59]	I
Stylocordyla pellita (Topsent, 1904)	[43]	I
Discodermia polydiscus du Bocage, 1869	[43]	I
Leiodermatium lynceus Schmidt, 1870	[43]	I
Desmacella annexa (Schmidt, 1870)	[43]	I
Hamacantha falcula (Bowerbank, 1866)	[43]	I
Mycale syrinx (Schmidt, 1862)	[59]	I
Eurypon coronula (Bowerbank, 1874)	[43]	I
Scalarispongia scalaris (Schmidt, 1862)	[43]	I
Cnidaria		
Desmophyllum dianthus(Esper, 1794)	[48]	I
Lophelia pertusa (Linnaeus, 1758)	[46]	I
Madrepora oculata Linnaeus, 1758	[46]	I
Funiculina quadrangularis (Pallas, 1766)	[57]	II
Brachiopoda		
Argyrotheca cuneata (Risso, 1826)	[58]	I
Megerlia truncata (Linnaeus, 1767)	[58]	I
Gryphus vitreus (Born, 1778)	[57]	II
Bryozao		
Herentia hyndmanni (Johnston, 1847)	[59]	I
Polychaeta		
Augeneria profundicola sp. nov	[60]	I
Adercodon pleijeli Mackie, 1994	[61–63]	I
Amage gallasi Marion, 1875	[65]	I
Amphitritides gracilis (Grube, 1860)	[66]	I
Aponuphis bilineata (Baird, 1870)	[67]	I
Euratella salmacidis (Claparède, 1869)	[67]	I
Exogone campoyi San Martin, Ceberio & Aguirrezabalaga, 1996	[64]	I
Exogone lopezi San Martin, Ceberio & Aguirrezabalaga, 1996	[64]	I
Exogone sorbei San Martin, Ceberio & Aguirrezabalaga, 1996	[64]	I
Heterospio mediterranea Laubier, Picard & Ramos, 1973	[68]	I
Inermonephtys foretmontardoi Ravara, Cunha & Pleijel, 2010	[69]	I
Langerhansia caeca Katzmann, 1973	[70]	I

Phylum	Reference	Depth
Levinsenia demiri Çinar, Dagli & Açik, 2011	[71]	I
Melinna palmata Grube, 1870	[72]	I
Metavermilia multicristata (Philippi, 1844)	[73]	I
Naiades cantrainii Delle Chiaje, 1828	[74]	I
Neomediomastus glabrus (Hartman, 1960)	[64]	I
Nephtys hombergii Savigny, 1818	[66]	I
Nepthys pulchra Rainer, 1991	[70]	I
Notophyllum foliosum (M. Sars, 1835)	[75]	I
Ophelina cylindricaudata (Hansen, 1878)	[76]	I
Otopsis chardyi (Katzmann, Laubier and Ramos 1974)	[70]	I
Paradoneis lyra (Southern, 1914)	[72]	I
Pherusa plumosa (O. F. Müller, 1776)	[72]	I
Pista cristata (O. F. Müller, 1776)	[67]	I
Pista lornensis (Pearson, 1969)	[70]	I
Poecilochaetus fauchaldi Pilato and Cantone, 1976	[70]	I
Prionospio dubia Day, 1961	[76]	I
Salmacina incrustans Claparède, 1870	[67]	I
Scoloplos armiger (O. F. Müller, 1776)	[67]	I
Serpula vermicularis Linnaeus, 1767	[69]	I
Sternaspis scutata (Renier in Ranzani, 1817)	[77]	I
Tachytrypane jeffreysii McIntosh, 1879	[65]	I
Terebellides stroemi M. Sars, 1835	[67]	I
Mollusca		
Abra alba (W. Wood, 1802)	[75]	II
Abra longicattus (Scacchı 1834)	[76]	II
Abralia veranyi (Rüppell, 1844)	[77]	I
Acar clathrata (Defrance, 1816)	[78]	I
Acteon monterosatoi Dautzenberg, 1889	[79]	I
Alloteuthis media (Linnaeus, 1758)	[77]	I
Alvania cimicoides (Forbes 1844)	[80]	II
Anatoma crispata Flemmıng 1828	[81]	II
Ancistrocheirus lesueurii (Orbigny, 1842)	[77]	I
Aporrhats serresianus (Mıchaud 1828)	[81]	I
Barbatia scabra (Poli 1795)	[80]	II

Phylum	Reference	Depth
Bathyarca grenophia (Rlsso 1826)	[81]	II
Bathypolypus sponsalis (Fischer, P. & Fischer, H. 1892)	[77]	I
Benthomangelia macra (Watson 1881)	[81]	II
Brachioteuthis riisei (Steenstrup, 1882)	[77]	I
Cavolinia gibbosa (d'Orbigny, 1834)	[75]	I
Chiroteuthis veranii (Férussac, 1835)	[77]	I
Chrysallida doliolum (Phılıppı 1844)	[82]	II
Chrysallida interstincta (Adams, J., 1797)	[82]	I
Clio pyramidata Linnaeus, 1767	[82]	I
Crenilabium exile (Jeffreys 1870 Ex Forbes ms)	[83]	I
Cuspidaria rostrata (Spengler 1793)	[80]	I
Cyclopecten hoskynsi (Forbes 1844)	[80]	II
Danilia otaviana (Cantkaıne 1835)	[80]	II
Delectopecten vitreus (Gmelın 1791)	[81]	II
Dentalium agile Sars M. in Sars O.G. 1872	[81]	II
Dentalium panormum Chenu 1842	[84]	II
Diacria trispinosa (Blainville, 1821)	[75]	I
Drilliola emendata (Monterosato 1872)	[80]	I
Eledone cirrhosa (Lamarck, 1798)	[77]	I
Ennucula aegeensis (Forbes, 1844)	[78]	I
Entalina tetragona (Brocchi, 1814)	[85]	I
Epitonium celesti (Aradas 1854)	[86]	I
Eulimella neoattenuata Gaglini, 1992	[87]	I
Euspira fusca (Blaınvılle 1825)	[81]	I
Euspira nitida (Donovan, 1804)	[80]	I
Falcidens gutturosus (Kowalevsky 1901)	[88]	I
Fusinus rostratus (Olivi, 1792)	[89]	I
Galeodea echinophora (Linnaeus, 1758)	[78]	I
Heteroteuthis dispar (Rüppell, 1844)	[77]	I
Histioteuthis bonnellii (Férrussac, 1835)	[77]	I
Illex coindetii (Vérany, 1839)	[77]	I
Japonactaeon pusillus (MacGilivray, 1843)	[90, 91]	I
Kelliella abyssicola (Forbes 1844)	[81]	II
Kurtiella bidentata (Montagu, 1803)	[82]	I

Phylum	Reference	Depth
Limatula subauriculata (Montagu 1808)	[80]	I
Lissopecten hyalinus (Polı-i 1795)	[80]	I
Loligo forbesi Steenstrup, 1856	[77]	I
Mangelia nuperrima (Tıberı 1855)	[81]	II
Neorossia caroli (Joubin, 1902)	[77]	I
Notolimea crassa (Forbes 1844)	[80]	II
Nuculoma aegeensis Forbes 1844	[80]	I
Octopus salutii Vérany, 1837	[77]	I
Octopus vulgaris Cuvıer 1798	[80]	I
Odostomella bicincta (Tiberi, 1868)	[79]	I
Odostomia silesui Nofroni, 1988	[92]	I
Parthenina flexuosa (Monterosato, 1874)	[85]	I
Parvicardium scabrum (Phılıppı 1844)	[80]	I
Phdine catena (Montagu 1803)	[80]	I
Philine scabra (Müller, O.F., 1784)	[78]	I
Pleurotomella eurybrocha (Dautzenberg & Fıscher 1896)	[93]	II
Prochaetoderma raduliferum (Owalevsky 1901)	[89]	II
Propeamussium fenectratum (Fobes I844)	[80]	I
Propilidium pertenue Jeffreys 1883	[76]	II
Pulsellum lofotense (Sars M. 1865)	[76]	I
Punctiscala cerigottana (Sturany, 1896)	[94]	I
Putzeysia wiseri (Calcara 1842)	[95]	II
Pyramidella octavtana (Dı Geronımo 1973)	[96]	II
Pyroteuthis margaritifera (Rüppell, 1844)	[77]	I
Ranella olearium (Linnaeus, 1758)	[97]	I
Rondeletiola minor (Naef 1912)	[98]	I
Rossia macrosoma (Delle Ch1aje 1830)	[99]	I
Scaeurgus unicirrhus (Delle Chiaje in de Férussac & d'Orbigny, 1841)	[77]	I
Scaphander lignarius (Linnaeus, 1758)	[79]	I
Scaphander punctostriatus (Mighels & Adams, 1842)	[75]	II
Sepia elegans Blainville 1827	[100]	I
Sepia orbignyana Férrussac, 1826	[101]	I
Sepietta neglecta Naef, 1916	[77]	I

Phylum	Reference	Depth
Sepietta oweniana Naef, 1916	[77]	I
Sepiola rondeletii Leach, 1817	[77]	I
Syrnola minuta Adams, H., 1869	[87]	I
Tellina pulchella Lamarck, 1818	[81]	I
Teretia teres (Reeve 1844)	[80]	I
Thracia convexa (Wood W. 1815)	[101]	II
Thysanoteuthis rhombus Troschel, 1857	[77]	I
Todarodes sagittatus (Lamarck, 1798)	[77]	I
Todaropsis eblanae (Ball, 1841)	[77]	I
Tonna galea (Linnaeus, 1758)	[57]	I
Trophon echinatus (Kiener 1840)	[80]	I
Turbonilla gradata Bdd 1883	[96]	I
Turbonilla micans (Monterosato, 1875)	[87]	I
Turbonilla paucistriata (Jeffreys, 1844)	[87]	I
Arthropoda		
Acanthephyra eximia Smith, 1884	[102]	I
Acanthephyra pelagica (Risso, 1816)	[103]	I
Aegaeon lacazei (Gourret, 1887)	[29]	I
Argissa hamatipes (Norman, 1869)	[104]	I
Aristaeomorpha foliacea (Risso, 1827)	[104]	I
Aristeus antennatus (Risso, 1816)	[104]	I
Bathynectes maravigna (Prestandrea, 1839)	[104]	I
Calappa granulata (Linnaeus, 1758)	[105]	I
Calocaris macandreae Bell, 1853	[106]	I
Chaceon mediterraneus Manning & Holthuis, 1989	[107]	I
Chlorotocus crassicornis (A. Costa, 1871)	[103]	I
Deflexilodes subnudus (Norman, 1889)	[103]	I
Diastylis cornuta (Boeck, 1864)	[108]	I
Dorhynchus thomsoni Thomson, 1873	[103]	I
Ebalia nux A. Milne Edwards, 1883	[109]	I
Ekleptostylis walkeri (Calman, 1907)	[108]	I
Epimeria cornigera (Fabricius, 1779)	[78]	II
Eryoneicus puritanii Lo Bianco, 1903	[110]	I

Phylum	Reference	Depth
Euphausia krohnii (Brandt, 1851)	[111]	II
Eurydice spinigera Hansen, 1890	[78]	II
Eusergestes arcticus (Krøyer, 1855)	[103]	I
Galathea dispersa Bate, 1859	[112]	I
Gennadas elegans (Smith, 1882)	[106]	II
Geryon longipes A. Milne-Edwards, 1882	[107]	I
Goneplax rhomboides (Linnaeus, 1758)	[113]	I
Macropipus tuberculatus (Roux, 1830)	[114]	I
Macropodia longipes (A. Milne-Edwards and Bouvier, 1899)	[114]	I
Macropodia rostrata (Linnaeus, 1761)	[115]	I
Maja goltziana d'Oliveira, 1888	[116]	I
Medorippe lanata (Linnaeus, 1767)	[114]	I
Meganyctiphanes norvegica (M. Sars, 1857)	[111]	II
Monodaeus couchii (Couch, 1851)	[113]	I
Monodaeus guinotae Forest, 1976	[110]	I
Munida curvimana A. Milne-Edwards & Bouvier, 1894	[117]	I
Munida intermedia A. Milne Edwards and Bouvier, 1899	[112]	I
Munida rugosa (Fabricius, 1775)	[118]	I
Munida rutllanti Zariquiey Alvarez, 1952	[112]	I
Munida tenuimana G.O. Sars, 1872	[110]	II
Nebalia abyssicola Fage, 1929	[119]	II
Nematoscelis atlantica Hansen, 1910	[115]	II
Nematoscelis megalops G. O. Sars, 1883	[115]	II
Nephrops norvegicus (Linnaeus, 1758)	[113]	I
Odontozona minoica Dounas and Koukouras, 1989	[110]	II
Pagurus alatus Fabricius, 1775	[116]	I
Pagurus prideaux Leach, 1815	[116]	I
Pandalina profunda Holthuis, 1949	[110]	II
Parapandalus narval (Fabricius, 1787)	[118]	I
Parapenaeus longirostris (Lucas, 1846)	[120]	I
Parthenopoides massena (Roux, 1830)	[118]	I
Pasiphaea multidentata Esmark, 1866	[107]	I
Pasiphaea sivado (Risso, 1816)	[107]	I
Pilumnus hirtellus (Linnaeus, 1761)	[118]	I

Phylum	Reference	Depth
Platyscelus serratulus Stebbing, 1888	[78]	I
Plesionika acanthonotus (Smith, 1882)	[113]	I
Plesionika antigai Zariquiey Alvarez, 1995	[121]	I
Plesionika edwardsii (Brandt, 1851)	[107]	I
Plesionika gigliolii (Senna, 1902)	[121]	I
Plesionika heterocarpus (A. Costa, 1871)	[121]	I
Plesionika martia (A. Milne-Edwards, 1883)	[121]	I
Plesionika narval (Fabricius, 1787)	[121]	I
Polycheles typhlops Heller, 1862	[121]	I
Polycheles typhlops typhlops Heller, 1862	[110]	II
Processa canaliculata Leach, 1815	[121]	I
Processa macrophthalma Nouvel & Holthuis, 1957	[114]	I
Processa nouveli Al-Adhub & Williamson, 1975	[114]	I
Richardina fredericii Lo Bianco, 1903	[121]	I
Rissoides pallidus (Giesbrecht, 1910)	[121]	I
Scalpellum scalpellum (Linnaeus 1767)	[120]	I
Scopelocheirus hopei (A. Costa, 1851)	[78]	I
Sergestes arachnipodus (Cocco, 1832)	[106]	I
Sergestes arcticus Krøyer, 1855	[122]	I
Sergestes atlanticus H. Milne Edwards, 1830	[122]	I
Sergestes sargassi Ortmann, 1893	[122]	I
Sergestes vigilax Stimpson, 1860	[106]	I
Sergia robusta (Smith, 1882)	[122]	I
Sergia tenuiremis (Krøyer, 1855)	[122]	I
Solenocera membranacea (Risso, 1816)	[118]	I
Spinolambrus macrochelos (Herbst, 1790)	[118]	I
Squilla mantis (Linnaeus, 1758)	[122]	I
Stylocheiron abbreviatum G. O. Sars, 1883	[115]	II
Stylocheiron longicorne G. O. Sars, 1883	[115]	II
Stylocheiron maximum Hansen, 1908	[115]	II
Stylocheiron suhmi G. O. Sars, 1883	[115]	I
Urothoe corsica Bellan-Santini, 1965	[108]	I
Westwoodilla caecula (Bate, 1857)	[108]	I
Xantho pilipes A. Milne-Edwards, 1867	[118]	I

Phylum	Reference	Depth
Echinodermata		
Amphilepis norvegica (Ljungman, 1865)	[123]	II
Amphipholis squamata (Delle Chiaje, 1828)	[123]	I
Amphiura chiajei Forbes, 1843	[83]	I
Anseropoda placenta (Pennant, 1777)	[123]	I
Antedon mediterranea (Lamarck, 1816)	[123]	I
Brissopsis lyrifera (Forbes, 1841)	[124]	I
Ceramaster grenadensis (Perrier, 1881)	[123]	II
Cidaris cidaris (Linnaeus, 1758)	[83]	I
Echinus acutus Lamarck, 1816	[123]	I
Echinus melo Olivi, 1792	[123]	I
Hemiaster expergitus Lovén, 1874	[123]	II
Hymenodiscus coronata (Sars G.O., 1872)	[60]	I
Leptometra phalangium (Müller, 1841)	[124]	I
Luidia ciliaris (Philippi, 1837)	[123]	I
Luidia sarsi sarsi Düben & Koren, in Düben, 1845	[123]	II
Marginaster capreensis (Gasco, 1876)	[123]	II
Ocnus koellikeri (Semper, 1868)	[123]	I
Oestergrenia digitata (Montagu, 1815)	[123]	I
Ophiacantha setosa (Retzius, 1805)	[123]	I
Ophiomyxa pentagona (Lamarck, 1816)	[123]	I
Ophiothrix fragilis (Abildgaard, in O. F. Müller, 1789)	[123]	I
Ophiura ophiura (Linnaeus, 1758)	[123]	I
Parastichopus regalis (Cuvier, 1817)	[123]	I
Peltaster placenta (Müller & Troschel, 1842)	[123]	I
Psammechinus microtuberculatus (Blainville, 1825)	[123]	I
Pseudostichopus occultatus Marenzeller, 1893	[123]	I
Sclerasterias neglecta (Perrier, 1891)	[123]	I
Sclerasterias richardi (Perrier, 1882)	[123]	I
Stylocidaris affinis (Philippi, 1845)	[123]	I
Tethyaster subinermis (Philippi, 1837)	[123]	II
Pisces		
Acantholabrus palloni (Risso, 1810)	[54, 55]	I
Arctozenus risso (Bonaparte, 1840)	[54, 55]	I

Phylum	Reference	Depth
Argentina sphyraena Linnaeus, 1758	[54, 55]	I
Argyropelecus hemigymnus Cocco, 1829	[54, 55]	I
Ariosoma balearicum (Delaroche, 1809)	[54, 55]	I
Arnoglossus rueppelii (Cocco, 1844)	[54, 55]	I
Aulopus filamentosus (Bloch, 1792)	[54, 55]	I
Bathophilus nigerrimus Giglioli, 1882	[54]	I
Bathypterois dubius Vaillant, 1888	[54]	I
Bellottia apoda Giglioli, 1883	[54, 55]	I
Benthocometes robustus	[54, 55]	I
Benthosema glaciale (Reinhardt, 1837)	[54, 55]	I
Brama brama (Bonnaterre, 1788)	[54, 55]	I
Capros aper (Linnaeus, 1758)	[54, 55]	I
Centracanthus cirrus Rafinesque, 1810	[54, 55]	I
Centrolophus niger (Gmelin, 1789)	[54, 55]	I
Centrophorus granulosus (Bloch & Schneider, 1801)	[54, 55]	I
Centroscymnus coelolepis Barbosa du Bocage & de Brito Capello, 1864	[55]	I
Ceratoscopelus maderensis (Lowe, 1839)	[54, 55]	I
Chalinura mediterranea Giglioli, 1893	[107]	II
Champsodon capensis Regan, 1908	[56]	I
Chauliodus sloani Bloch & Schneider, 1801	[54, 55]	I
Chimaera monstrosa Linnaeus, 1758	[54, 55]	I
Chlopsis bicolor Rafinesque, 1810	[55]	I
Chlorophthalmus agassizi Bonaparte, 1840	[54, 55]	I
Coelorinchus caelorhincus (Risso, 1810)	[54, 55]	I
Conger conger (Linnaeus, 1758)	[54, 55]	I
Cubiceps gracilis (Lowe, 1843)	[54]	I
Cyclothone braueri Jespersen & Tåning, 1926	[54, 55]	I
Dalatias licha (Bonnaterre, 1788)	[54, 55]	I
Diaphus holti Tåning, 1918	[54, 55]	I
Diaphus metopoclampus (Cocco, 1829)	[54, 55]	I
Diaphus rafinesquii (Cocco, 1838)	[54, 55]	I
Dipturus batis (Linnaeus, 1758)	[54, 55]	I
Dipturus oxyrinchus (Linnaeus, 1758)	[54, 55]	I
Dysomma brevirostre (Facciolà, 1887)	[54]	I

Phylum	**Reference**	**Depth**
Echinorhinus brucus (Bonnaterre, 1788)	[54]	I
Echiodon dentatus (Cuvier, 1829)	[54, 55]	I
Electrona risso (Cocco, 1829)	[54]	I
Epigonus constanciae (Giglioli, 1880)	[54]	I
Epigonus denticulatus Dieuzeide, 1950	[54, 55]	I
Epigonus telescopus (Risso, 1810)	[54, 55]	I
Etmopterus spinax (Linnaeus, 1758)	[107]	II
Evermannella balbo (Risso, 1820) 38	[54, 55]	I
Facciolella oxyrhyncha (Bellotti, 1883)	[54, 55]	I
Gadella maraldi (Risso, 1810)	[54, 55]	I
Gadiculus argenteus Guichenot, 1850	[54, 55]	I
Gaidropsarus biscayensis (Collett, 1890)	[54, 55]	I
Galeus melastomus Rafinesque, 1810	[107]	II
Glossanodon leioglossus (Valenciennes, 1848)	[54, 55]	I
Gnathophis mystax (Delaroche, 1809)	[54]	I
Gonichthys cocco (Cocco, 1829)	[54, 55]	I
Gonostoma denudatum Rafinesque, 1810	[54, 55]	I
Helicolenus dactylopterus (Delaroche, 1809)	[54, 55]	I
Heptranchias perlo (Bonnaterre, 1788)	[54, 55]	I
Hexanchus griseus (Bonnaterre, 1788)	[54, 55]	I
Hoplostethus mediterraneus Cuvier, 1829	[54, 55]	I
Hygophum benoiti (Cocco, 1838)	[54, 55]	I
Hygophum hygomii (Lütken, 1892)	[55]	I
Hymenocephalus italicus Giglioli, 1884	[54, 55]	I
Ichthyococcus ovatus (Cocco, 1838)	[55]	I
Lampanyctus crocodilus (Risso, 1810)	[54, 55]	I
Lampanyctus pusillus (Johnson, 1890)	[55]	I
Lampris guttatus (Brünnich, 1788)	[55]	I
Lepidion lepidion Risso, 1810	[107]	II
Lepidopus caudatus (Euphrasen, 1788)	[54, 55]	I
Lepidorhombus boscii (Risso, 1810)	[54, 55]	I
Lepidorhombus whiffiagonis (Walbaum, 1792)	[54, 55]	I
Lestidiops jayakari (Boulenger, 1889)	[54, 55]	I
Lestidiops sphyrenoides (Risso, 1820)	[54, 55]	I

Phylum	Reference	Depth
Leucoraja circularis (Couch, 1838)	[54, 55]	I
Leucoraja fullonica (Linnaeus, 1758)	[54, 55]	I
Leucoraja melitensis (Clark, 1926)	[55]	I
Lobianchia dofleini (Zugmayer, 1911)	[54, 55]	I
Lobianchia gemellarii (Cocco, 1838)	[54, 55]	I
Lophius budegassa Spinola, 1807	[54, 55]	I
Lophius piscatorius Linnaeus, 1758	[54, 55]	I
Macroramphosus scolopax (Linnaeus, 1758)	[54, 55]	I
Maurolicus muelleri (Gmelin, 1789)	[54, 55]	I
Merluccius merluccius (Linnaeus, 1758)	[54, 55]	I
Microichthys coccoi Rüppell, 1852	[54]	I
Micromesistius poutassou (Risso, 1827)	[54, 55]	I
Microstoma microstoma (Risso, 1810)	[54, 55]	I
Molva macrophthalma (Rafinesque, 1810)	[54, 55]	I
Mora moro (Risso, 1810)	[107]	II
Myctophum punctatum Rafinesque, 1810	[54, 55]	I
Nansenia oblita (Facciolà, 1887)	[54, 55]	I
Nemichthys scolopaceus Richardson, 1848	[54, 55]	I
Nettastoma melanurum Rafinesque, 1810	[54, 55]	I
Nezumia aequalis (Günther, 1878)	[54]	I
Nezumia sclerorhynchus Valenciennes, 1838	[54, 55]	I
Notacanthus bonaparte Risso, 1840	[54, 55]	I
Notoscopelus elongatus (Costa, 1844)	[54, 55]	I
Notoscopelus kroyeri (Malm, 1861)	[54]	I
Odontaspis ferox (Risso, 1810)	[54, 55]	I
Ophidion barbatum Linnaeus, 1758	[54, 55]	I
Oxynotus centrina (Linnaeus, 1758)	[54, 55]	I
Pagellus bogaraveo (Brünnich, 1768)	[54, 55]	I
Peristedion cataphractum (Linnaeus, 1758)	[54, 55]	I
Petromyzon marinus Linnaeus, 1758	[54, 55]	I
Phycis blennoides (Brünnich, 1768)	[54, 55]	I
Physiculus dalwigki Kaup, 1858	[55]	I
Polyprion americanus (Bloch & Schneider, 1801)	[54, 55]	I
Rachycentron canadum (Linnaeus, 1766) 109	[54, 55]	I

Phylum	Reference	Depth
Regalecus glesne Ascanius, 1772	[54, 55]	I
Rostroraja alba (Lacepède, 1803)	[54, 55]	I
Ruvettus pretiosus Cocco, 1833 20,168	[54, 55]	I
Saurenchelys cancrivora Peters, 1864	[55]	I
Schedophilus ovalis (Cuvier, 1833)	[54, 55]	I
Scorpaena elongata Cadenat, 1943	[54, 55]	I
Scyliorhinus canicula (Linnaeus, 1758)	[54, 55]	I
Scyliorhinus stellaris (Linnaeus, 1758)	[54, 55]	I
Somniosus rostratus (Risso, 1827)	[55]	I
Squalus acanthias Linnaeus, 1758	[54, 55]	I
Squalus blainville (Risso, 1827)	[54, 55]	I
Stomias boa (Risso, 1810)	[54, 55]	I
Sudis hyalina Rafinesque, 1810	[54, 55]	I
Symbolophorus veranyi (Moreau, 1888)	[55]	I
Symphurus ligulatus (Cocco, 1844)	[55]	I
Symphurus nigrescens Rafinesque, 1810	[54, 55]	I
Synchiropus phaeton (Günther, 1861)	[54, 55]	I
Trachipterus trachypterus (Gmelin, 1789)	[54, 55]	I
Trachyrincus scabrus (Rafinesque, 1810)	[54, 55]	I
Trigla lyra Linnaeus, 1758	[54, 55]	I
Vinciguerria attenuata (Cocco, 1838)	[54, 55]	I
Vinciguerria poweriae (Cocco, 1838)	[55, 56]	I
Zu cristatus (Bonelli, 1819)	[55, 56]	I

Table 1. Aegean deep-sea fauna that is recorded at a greater depth than 200 m.

3.5. Deep-sea mollusca

The mollusca species list is based on a review of published findings and 92 species found in deep-sea waters of depth over 200 m. But, there are no Indo-Pacific origin molluscan species in the deep of the Aegean Sea. In general, mollusca species showed a decline with depth. This has also been reported for other macrofauna species in the Mediterranean Sea [14].

3.6. Other benthic invertebrates

Suspension feeders (such as; hexactinellid sponges, pennatulids) are dominant in terms of invertebrate biomass on the upper and middle slope (to 1400 m) in the Atlantic, while echinoderms are important at all depths and dominant on the middle and lower slope [50]. Suspension feeders are of relatively little importance in the Mediterranean, due to its oligotrophic waters, and they are important only locally. Probably due to their low levels of food consumption, crustacean decapods are more important in the deep Mediterranean (an oligotrophic region) than in the deep Atlantic, where echinoderms dominate [50].

There are 4 Cnidaria species and 3 Brachiopoda species found in the depth of the Aegean Sea. Brachiopoda species are distributed in shallow water and since cnidarian species are in a symbiotic relationship with endosymbiotic algae for food and calcification, they prefer shallow waters too [51].

Nevertheless, given the low number of studies focusing on deep water, further research is needed to increase our knowledge on the brachiopod and cnidaria fauna of the Aegean Sea.

3.7. Deep-sea fishes

The depth ranges of the Aegean deep-sea fishes were specified by Froose and Pauly [52]. B Turkish and Greek waters have almost the same number of marine fish species (512 and 510, respectively) [53, 54]. Also, there are little differences between the lists of deep-sea fishes. This checklist is composed of fishes that are usually observed at depths deeper than 200 m in the Aegean Sea. Although the lower limit of their depth range is below 200 m, some of the fish species are not included to the list due to their rareness at those depths. Fishes represented by 126 fish species included: Agnatha by 1 species, Elasmobranchii by 21 species, Holocephali by 1 species and Osteichthyes by 103 species.

The deep-sea fish fauna of the Aegean Sea has three main origins: Atlanto-Mediterranean, Cosmopolite and Mediterranean Endemic species. *C. capensis* is the only Red Sea immigrant species [55].

Their ecological characters are mostly restricted and are generalized as mesopelagic, bathypelagic, benthopelagic and bathydemersal species. The order Gadiformes formed 16% of the total number of species with 20 species. Furthermore, most of the deep-sea fishes that have economic value belong to this order. Myctophidae is the largest family represented by 17 small mesopelagic fish species in the deep waters of the Aegean Sea. It is expected that further studies will reveal a number of new Myctophid species.

L. lepidion and *M. moro* are the only species living in waters deeper than 500 m, while *A. palloni, C. cirrus, C. capensis, C. bicolor, C. caelorhincus, L. guttatus, L. fullonica, N. oblita* and *S. stellaris* are found at depths shallower than 500 m.

4. Conclusion

There is a lack of detailed studies on the fauna in the deep Aegean Sea. This chapter is the first attempt to describe composition of macro faunal assemblages occurring in the deep waters of the deep Aegean Sea including Turkey and Greek coast.

We present an updated checklist including the most recent status of the relevant deep biodiversity. The aims of the present chapter were to give recent taxonomic and biogeographic knowledge of deep-sea species reported from Aegean Sea (deeper than 200 m) to date.

Author details

Onur Gönülal[1]* and Cem Dalyan[2]

*Address all correspondence to: ogonulal@istanbul.edu.tr

1 Gökçeada Marine Research Department, Istanbul University, Canakkale, Turkey

2 Department of Biology, Faculty of Science, Istanbul University, Istanbul, Turkey

References

[1] Hopkins TS. Process in the Mediterranean Basins, Estuarine Transport. California: Uni. South Cali. Press; 1978 310 p

[2] Göksel Ç, Ipbuker C, Bildirici Ö, Ulugtekin N. In: Katağan T, Tokaç A, Beşiktepe Ş, Öztürk B, editors. The Aegean Sea Marine Biodiversity, Fisheries, Conservation and Governance. Istanbul: TUDAV; 2015. p. 16-26. DOI: 978-975-8825-33-2

[3] Chisholm H. The Encyclopædia Britannica: A Dictionary of Arts, Sciences, Literature and General Information. Encyclopaedia Britannica ed.; Cambridge: England. 1911

[4] Pérès JM. History of the Mediterranean biota and the colonization of the depths. In: Margalef R, editor. Key Environments: Western Mediterranean. New York: Pergamon; 1985. p. 198-232

[5] Gage JD, Tyler PA. Deep-Sea Biology: A Natural History of Organisms at the Deep-Sea Floor. Cambridge: Cambridge University Press; 1985 232 p

[6] Cartes JE, Sardà F. Abundance and diversity of decapod crustaceans in the deep-Catalan Sea (Western Mediterranean). Journal of Natural History. 1992;**26**(6):1305-1323

[7] Risso A. Histoire naturelle des Crustacés des environs de Nice. Paris: Librairie Grecque-Latine-Allemande; 1816 175 p

[8] Aksu AE, Piper DJW, Konuk T. Late quaternary tectonic and sedimentation history of Outer Izmir Bay and Çandarlı Basin, Western Turkey. Marine Geology. 1987;**76**:89-104

[9] Ergün M. General tectonics of the Aegean Sea area and geophysical signatures. In: Katağan T, Tokaç A, Beşiktepe Ş, Öztürk B, editors. The Aegean Sea Marine Biodiversity, Fisheries, Conservation and Governance. Istanbul: TUDAV; 2015. p. 31-38

[10] Yaşar D. Marine geology of the Aegean Sea. In: Katağan T, Tokaç A, Beşiktepe Ş, Öztürk B, editors. The Aegean Sea Marine Biodiversity, Fisheries, Conservation and Governance. Istanbul: TUDAV; 2015. p. 55-61

[11] Sakellariou D, Papoulia I. Geotectonic setting and seismicity. In: Papathanassiou E, Zenetos A, editors. State of the Hellenic Marine Environment. Atina: HCMR; 2005. p. 26-42

[12] Lyberis N. Tectonic evolution of the north Aegean trough. In: Dixon JE, Robertson AHF, editors. The Geological Evolution of the Eastern Mediterranean. Oxford: Blackwell Scientific Publ. Geol. Soc; 1984. p. 709-725

[13] Mascle J, Martin L. Shallow structure and recent evolution of the Aegean Sea: A synthesis based on continuous reflection profiles. Marine Geology. 1990;**94**:271-299

[14] Le Pichon X, Angelier J, Osmaston MF, Stegena L. The Aegean Sea. Philosophical Transactions of the Royal Society of London A: Mathematical Physical and Engineering Sciences. 1981;**300**(1454):357-372

[15] Fredj G, Laubier L. The deep Mediterranean benthos. In: Moraitou-Apostopulou M, Kiortsis V, editors. Mediterranean Marine Ecosystems. London: Plenum Press; 1985. p. 169-188

[16] Ben-Eliahu MN, Fiege D. Erpulid tube-worms (Annelida Polychaeta) of the Central and Eastern Mediterranean with particular attention to the levant basin. Senckenbergiana Maritima. 1996;**28**(1-3):1-5

[17] Danovaro R, Corinaldesi C, D'Onghia G, Galil B, Gambi C, Gooday AJ, Polymenakou P. Deep-sea biodiversity in the Mediterranean Sea: The known, the unknown, and the unknowable. PloS One. 2010;**5**(8):e11832

[18] Psarra S, Tselepides A, Ignatiades L. Primary productivity in the oligotrophic Cretan Sea (NE Mediterranean): Seasonal and interannual variability. Progress in Oceanography. 2000;**46**(2):187-204

[19] Antoine D, Morel A, André JM. Algal pigment distribution and primary production in the eastern Mediterranean as derived from coastal zone color scanner observations. Journal of Geophysical Research. 1995;**100**(C-8):16193-16209

[20] Lykousis V, Chronis G, Tselepides A, Price B, Theocharis A, Siokou-Frangou I, Van Wambeke F, Danovaro R, Stavrakakis S, Duineveld G, Georgopoulos D, Souvermezoglou A. Major outputs of the recent multidisciplinary biogeochemical researches in the Aegean Sea. Journal of Marine Systems 2002;**33-34**:313-334.

[21] Poulos SE, Drakopoulos PG, Collins MB. Seasonal variability in sea surface oceanographic conditions in the Aegean Sea (eastern Mediterranean): An overview. Journal of Marine Systems. 1997;**13**:225-244

[22] Sevastou K, Lampadariou N, Polymenakou PN, Tselepides A. Benthic communities in the deep Mediterranean Sea: Exploring microbial and meiofaunal patterns in slope and basin ecosystems. Biogeosciences. 2013;**10**(7):4861-4878

[23] Barrier P, Di Geronimo I, Montenat C, Roux M, Zibrowius H. Présence de faunes bathyales atlantiques dans le Pliocène et le Pleistocène de Méditerranée (détroit du Messine, Italie). Bulletin de la Société Géologique de France. 1989;**85**(4):787-796

[24] Sara M. Ecological factors and their consequences in the Mediterranean ecosysterns. 1985;

[25] Galil B, Goren M. The bathyal fauna of the Levant-new records and rare occurrences. Senckenberg. 1994;**25**(1-3):41-52

[26] Salas C. Marine Bivalves from off the Southern Iberian Peninsula collected by the Balgim and Fauna 1 expeditions. Haliotis. 1996;**25**:33-100

[27] Bouchet P, Taviani M. The Mediterranean deep-sea fauna: Pseudopopulations of Atlantic species? Deep Sea Research Part A. Oceanographic Research Papers. 1992;**39**(2):169-184

[28] Fishelson L, Galil BS. Gonad structure and reproductive cycle in the deep-sea hermaphrodite tripodfish, Bathypterois Mediterraneus (Chlorophthalmidae, Teleostei). Copeia. 2001;**2**:556-560

[29] Coll M, Piroddi C, Steenbeek J, Kaschner K, Lasram FBR, Aguzzi J, Danovaro R. The biodiversity of the Mediterranean Sea: Estimates, patterns, and threats. PloS One. 2010;**5**(8):e11842

[30] Koukouras A, Dounas C, Türkay M, Voultsiadou-Koukoura E. Decapod crustacean fauna of the Aegean Sea: New information, check list, affinities. Senckenbergiana Maritime. 1992;**22**(3-6):217-244

[31] Tselepides A, Papadopoulou KN, Podaras D, Plaiti W, Koutsoubas D. Macrobenthic community structure over the continental margin of Crete (South Aegean Sea, NE Mediterranean). Progress in Oceanography. 2000;**46**(2):401-428

[32] Lampadariou N, Tselepides A. Spatial variability of meiofaunal communities at areas of contrasting depth and productivity in the Aegean Sea (NE Mediterranean). Progress in Oceanography. 2006;**69**:19-36

[33] Danovaro R, Dinet A, Duineveld G, Tselepides A. Benthic response to particulate fluxes in different trophic environments: A comparison between the Gulf of Lions–Catalan Sea (western-Mediterranean) and the Cretan Sea (eastern-Mediterranean). Progress in Oceanography. 1999;**44**:287-312

[34] Galil BS. The limit of the sea: The bathyal fauna of the Levantine Sea. Scientia Marina. 2004;**68**(3):63-72

[35] Ilan M, Gugel J, Galil BS, Janussen D. Small bathyal sponge species from East Mediterranean revealed by a nonregular soft bottom sampling technique. Ophelia. 2003; **57**:145-160

[36] Boury-Esnault N, Pansini M, Uriz MJ. Spongiaires bathyaux de la mer d' Alboran et du golfe ibero-marocain. Memorias del Museo de Historia Natural. 1994;**160**:1-174

[37] Maldonado M, Young CM. Effects of physical factors on larval behavior, settlement and recruitment of four tropical demosponges. Marine Ecology Progress Series. 1996;**138**:169-180

[38] Witte U, Brattegard T, Graf G, Springer B. Particle capture and deposition by deep-sea sponges from the Norwegian-Greenland Sea. Marine Ecology Progress Series. 1997;**154**:241-252

[39] Pansini M. Considérations biogéographiques et systématiques pour une mise à jour des données sur le peuplement de spongiaires méditerranéens. Bulletin de l'Institut océanographique. 1992;**9**:43-51

[40] Vacelet J. Indications de profondeur donne'es par les Spongiaires dans les milieux benthiques actuels. Géologie Mediterranéenne. 1988;**15**:13-26

[41] Voultsiadou E. Demosponge distribution in the eastern Mediterranean: A NW–SE gradient. Helgoland Marine Research. 2005;**59**(3):237-241

[42] FreiwAld A, Taviani M. White coral community. Oceanography. 2009;**22**(1):58-65

[43] Taviani M, Freiwald A, Zibrowius H. Deep coral growth in the Mediterranean Sea: An overview. In: Freiwald A, Roberts JM, editors. Cold-Water Corals and Ecosystems. Berlin: Springer; 2005. p. 137-156

[44] Vafidis D, Koukouras A, Voultsiadou-Koukoura E. Actiniaria, Corallimorpharja, and Scleractinia (Hexacorallia, Anthozoa) of the Aegean Sea, with a checklist of the eastern Mediterranean and Black Sea species. Israel Journal of Zoology. 1997;**43**(1):55-70

[45] Zibrowius H. Les Scléractiniaires de la Méditerranée et de l'Atlantique nord-oriental. Mémoires de l'Institut océanographique: Monaco; 1980 284 p

[46] Zibrowius H. Les Scléractiniaires de la Méditerranée et de l'Atlantique nord-oriental. Rapports et Proces-Verbaux des Reunions-Commission Internationale pour l'Exploration Scientifique de la Mer Mediterranee: Monaco; 1979

[47] Gerovasileiou V, Rosso A. Marine Bryozoa of Greece: An annotated checklist. Biodiversity Data Journal. 2016;**4**:e10672

[48] Rosso A, Di Martino E. Bryozoan diversity in the Mediterranean Sea: An update. Mediterranean Marine Science. 2016;**17**(2):567-607

[49] Mastrototaro F, D'Onghia G, Corriero G, Matarrese A, Maiorano P, Panetta P, Gherardi M, Longo C, Rosso A, Sciuto F, Sanfilippo R, Gravili C, Boero F, Taviani M, Tursi A. Biodiversity of the white coral bank off Cape Santa Maria di Leuca (Mediterranean Sea): An update. Deep-Sea Research II. 2010;**57**(5-6):412-430

[50] Lampitt RS, Billett DSM, Rice AL. Biomass of the invertebrate megabenthos from 500 to 4100 m in the northeast Atlantic Ocean. Marine Biology. 1986;**93**(1):69-81

[51] Muscatine L, Porter JW. Reef corals: Mutualistic symbioses adapted to nutrient-poor environments. Bioscience. 1977;**27**(7):454-460

[52] Froese R, Pauly D. www.fishbase.org [Internet]. 2017

[53] Bilecenoğlu M, Kaya M, Cihangir B, Çiçek E. An updated checklist of the marine fishes of Turkey. Turkish Journal of Zoology. 2014;**38**:901-929

[54] Papaconstantinou C. Fauna Graeciae. An Updated Checklist of the Fishes in the Hellenic Seas. Athens: Monographs on Marine Sciences; 2014 340 p

[55] Kebapcioglu T, Dereli H. First record of gaper (Champsodon Capensis Regan, 1908) in the Aegean Sea. Mediterranean Marine Science. 2016;**17**(3):794-821

[56] Voultsiadou E. Sponge diversity in the Aegean Sea: Check list and new information. Italian Journal of Zoology. 2005;**72**(1):53-64

[57] Gönülal O, Güreşen SO. A list of macrofauna on the continental shelf of Gökçeada Island (northern Aegean Sea) with a new record (Gryphus vitreus Born, 1778)(Brachiopoda, Rhynchonellata) for the Turkish seas. Journal of Black Sea/Mediterranean Environment. 2014;**20**(3):228-252

[58] Logan A, Bianchi CN, Morri C, Zibrowius H, Bitar G. New records of recent brachiopods from the eastern Mediterranean Sea. Annali del Museo civico di Storia naturale. 2002;**94**:407-418

[59] Harmelin JG. Bryozoaires récoltés au cours de la campagne du Jean Charcot en Méditerranée Orientale (Août–Septembre 1967). Bulletin du Muséum National d'Histoire Naturelle. 1969;**41**:295-311

[60] Kurt-Sahin G, Çinar M, Gonulal O. A new species of Augeneria (Polychaeta: Lumbrineridae) from deep waters of the Aegean Sea (eastern Mediterranean). Mediterranean Marine Science. 2016;**17**(3):708-713

[61] Simboura N, Zenetos A. Increasing Polychaete diversity as a consequence of increasing research effort in Greek waters: New records and exotic species. Mediterranean Marine Science. 2005;**6**(1):74-88

[62] Çinar ME, Ergen Z, Dagli E. New records of polychaetes from the Turkish Aegean coast. Rapp Comm Int Mer Médit. 2004;**37**:508

[63] Ergen Z. Investigations on the taxonomy and ecology of Polychaeta from Izmir Bay and its adjacent areas. Scientific Reports of the Faculty of Science Ege University. 1976;**14**:1-66

[64] Kiseleva MI. Repartition qualitative et quantitative du benthos dans la region des Dardanelles en Mer Egee. Trudy Sevastopol'skoy Biologicheskoy Stantsii, Akademiya Nauk SSSR. 1961;**14**:135-146

[65] Dağlı E, Ergen Z, Çınar ME. The taxonomic and ecological characteristics of Longosomatidae and Spionidae (Annelida: Polychaete) distributed in Saros Bay-Turkey. Journal of Fisheries Sciences. 2008;**2**(3):198-209

[66] Ergen Z, Çinar ME, Dagli E, Kurt G. Seasonal dynamics of soft-bottom polychaetes in Izmir Bay (Aegean Sea, eastern Mediterranean). Scientia Marina. 2006;**70**(S3):197-207

[67] Arvanitidis C. Polychaete fauna of the Aegean Sea: Inventory and new information. Bulletin of Marine Science. 2000;**66**(1):73-96

[68] Çinar ME, Dagli E, Açik S. Annelids (Polychaeta and Oligochaeta) from the sea of Marmara, with descriptions of five new species. Journal of Natural History. 2011;**45**: 2105-2143

[69] Önen M. The qualitative and quantitative investigation of the macrobenthic fauna found in the soft substratum of the Urla harbour. Ege University, Journal of the Faculty of Science. 1983;**6**:29-39

[70] Koçak F, Ergen Z, Çınar ME. Fouling organisms and their developments in a polluted and an unpolluted marina in the Aegean Sea (Turkey). Ophelia. 1999;**50**:1-20

[71] Wesenberg-Lund E. Pølseorme (Gephyrea): Sipunculider, Priapulider, Echiurider. GEC Gad. 1939;**45**:

[72] Cinar ME, Ergen Z. Polychaetes associated with the sponge Sarcotragus Muscarum Schmidt, 1864 from the Turkish Aegean coast. Ophelia. 1998;**48**(3):167-183

[73] Doğan A, Çinar ME, Önen M, Ergen Z, Katağan T. Seasonal dynamics of soft bottom zoobenthic communities in polluted and unpolluted areas of Izmir Bay (Aegean Sea). Marine Biodiversity. 2005;**35**(1):133-145

[74] Ergen Z. The distribution of the Polychaeta in the soft substratum of Gulbahce area in the Bay of Izmir. Rapp. Comm. Int. Mer Medit. 1985;**29**(5):229-230

[75] Gönülal O. Bathymetric distribution of non-decapod crustacean invertebrata species in the deep Northern Aegean Sea. In: CIESM; 4-11 November; Marsilya. Marsilya: 2013. p. 7.

[76] Jeffrevs JG. Mediterranean Mollusca, no 3, and other Invertebrata. Annals and Magazine of Natural History. 1883;**5**(11):393-401

[77] Salman A, Katagan T, Benli HA. Vertical distribution and abundance of juvenile cephalopods in the Aegean Sea. Sciente Maritime. 2003;**67**:167-176

[78] Demir M. Shells of Mollusca collected from the seas of Turkey. Turkish Journal of Biology. 2003;**27**:101-140

[79] Öztürk B. Shelled Molluscs of the Turkish Coasts: Informal Group. İzmir: Ege University Press; 2014. 131 p. DOI: 978-605-338-058-0

[80] Forbes E. Report on the Mollusca and Radiata of the Aegean Sea and their distribution considered as bearing on geology. In: 13th Meeting British Assn. Adv. Sci; London. 1844. p. 21-23.

[81] Sturany R, Zoologische Ergebnisse VII, Mollusken I. (Prosobranchier und Opisthobranchier, Scaphopoden Lamellibranchier) gesammelt von S.M. Schiff "Pola 1890-1894". Denkschr. k. k. Akad. Wiss., math naturwiss. 1896;**63**:1-36

[82] Van-Aartsen JJ, Kinzelbach R. Marine molluscs from the İztuzu beach near Dalyan (Mediterranean coast of Turkey). Zoology in the Middle East. 1990;**4**(1):103-112

[83] Janssen R. Benthic molluscs from the deepwater of the eastern Mediterranaean Sea, collected during "METEOR"-cruise 5. Senckenbergiana Maritime. 1989;**20**(5-6):265-276

[84] Peres JM, Picard J. Recherches sur les peuplements benthiques de la Mediterranee Nord-Orientale. Annales de l'Institut Oceanographique. 1958;**34**:213-281

[85] Öztürk B, Bitlis-Bakır B, Er-Filiz M. The genus Chrysallida Carpenter, 1856 on the Turkish coasts. Zool Middle East. 2011;**54**:53-78

[86] Zenetos A, Papathanassiou E, Aartsen JJ. Analysis of the benthic communities in the Cyclades Plateau (Aegean Sea) using ecological and paleoecological. Marine Ecology. 1991;**12**(2):123-137

[87] Öztürk B, Bakır BB. Heterostropha species of the Turkish coasts: Anisocycla, Eulimella, Puposyrnola, Syrnola and Turbonilla (Gastropoda, Heterobranchia). Turkish Journal of Fisheries and Aquatic Sciences. 2013;**13**(3):423-440

[88] Salvini-Plawen LV. Caudofoveata des Forschungsprojektes. Bull. Mus. nat. Hist. natur. 1977;**310**:413-421

[89] Öztürk B, Ergen Z, Önen M. Muricidae (Neogastropoda) fauna of Izmir Bay (Aegean Sea). Triton. 2001;**3**:1-4

[90] Öztürk B, Doğan A, Önen M, Çevik C. Lepidopleurus Cimicoides (Monterosato, 1879) and Lepidochitona Furtiva (Monterosato, 1879): Two new reports for the Polyplacophora (Mollusca) fauna of the Aegean se. Boll Malacologico. 2007;**43**:33-38

[91] Öztürk B, Bitlis-Bakır B, Micali P. Heterostropha species of the Turkish coasts: Odostomiinae Pelseneer, 1928 (Gastropoda, Heterobranchia, Pyramidellidae). Turkish Journal of Fisheries and Aquatic Sciences. 2013;**13**:139-157

[92] Tenekidis NS. A Collection of Shells from the Greek Seas. Athens; 1989 188 p

[93] Öztürk B, Doğan A, Bitlis B, Önen M. Ptenoglossa species (Mollusca: Gasropoda) distributed along the Turkish coast of the Aegean Sea. Turk J Zool. 2008;**32**:201-211

[94] Koutsoubas D, Koukouras A, Voultsiadou-Koukoura E. Prosobranch mollusc fauna of the Aegean Sea: New information, check list, distribution. Israel Journal Zoology. 1997;**43**:19-54

[95] Koutsoubas D, Koukouras A, Karakassis I, Dounas C. Contribution to the knowledge of Gastropoda and Bivalvia (Mollusca) of Crete island (S. Aegean Sea). Bollettino Malacologico. 1992;**28**(1-4):69-82

[96] Kabasakal H, Kabasakal EA. New record for the mollusca fauna of the seas of Turkey: Oil-vessel triton, Ranella olearum (Linnaeus, 1758), (Gastropoda: Ranellidae). C Conch Inform. 2002;**33**:17-19

[97] Mangold K. Boletzky Sv. Cephalopodes (Seiches, Calmars et Poulpes/Pieuvres). In: Fischer W, Bauchot M, Schneider M, editors. Fishes FAO d'identification des especes pour les besoins de la peche Mediterranee et Mer Noire. Zone de Peche 37. Rome: FAO; 1987. p. 635-717

[98] D'onghia G, Capezzuto F, Cardone F, Carlucci R, Carluccio A, Chimienti G, Panetta P. Macro-and megafauna recorded in the submarine Bari canyon (southern Adriatic, Mediterranean Sea) using different tools. Mediterranean Marine Science. 2015;**16**(1): 180-196

[99] Sumner A. Some marine Molluscs from Crete. Conchol. Newsl. 1983;**84**:61-62

[100] Salman A, Katagan T, Benli HA. Bottom trawl teuthofauna of the Aegean Sea. Archive of Fishery and Marine Research. 1997;**45**:183-196

[101] Zenetos A. Fauna Graeciae. Vil. the Marine Bivalvia (Mollusca) of Greece. Hellenic Zoological Society: Athens; 1996 319 p

[102] Koukouras A. The pelagic shrimps (Decapoda, Natantia) of the Aegean Sea, with an account of the Mediterranean species. Crustaceana. 2000;**73**(7):801-814

[103] Kocataş A, Katağan T. The decapod crustacean fauna of the Turkish seas. Zoology in the Middle East. 2003;**29**(1):63-74

[104] Kocataş A, Katağan T. Littoral Benthic Amphipods of Turkish Seas and their Distributions. TBAG: Ankara; 1978 223 p

[105] Özcan T, Irmak E, Ateş AS, Katağan T. The occurrence of Bathynectes maravigna (Decapoda: Brachyura: Portunidae) in the Turkish part of the Levantine Sea. Marine Biodiversity Records. 2009;**2**:e101

[106] Gönülal O, Sezgin M, Öztürk B. Diversity and bathymetric distribution of decapod crustaceans attracted to baited traps from the middle slope of the northern Aegean Sea. Crustaceana. 2014;**87**(1):19-34

[107] Jones EG, Tselepides A, Bagley PM, Collins MA, Priede IG. Bathymetric distribution of some benthic and benthopelagic species attracted to baited cameras and traps in the deep eastern Mediterranean. Marine Ecology Progress Series. 2003;**251**:75-86

[108] Katağan T, Kocataş A, Benli HA. Note préliminaire sur les Décapodes bathyaux de la côte Turque de la mer Egée. Rapport de la Commission Internationale pour l'Exploration Scientifique de la Mer Méditerrannée. 1988;**31**:1-23

[109] Adensamer T. Decapoden: Gesammelt auf SM Schiff Pola in den Jahren 1890-1894. Germany: KK Hof-und Staatsdruckerei; 1898

[110] Koukouras A, Dounas C. Decapod crustaceans new to the fauna of the Aegean Sea. Crustaceana. 2000;**73**(4):497-502

[111] Mavidis M, Aplikioti M, Kirmitzoglou I, Koukouras A. The Euphausiacean fauna (malacostraca) of the Aegean sea, and comparison with those of the neighbouring seas. Crustaceana. 2005;**78**(1):19-27

[112] Geldiay R, Kocataş A. Anomura (Crustacea, Decapoda) species are determined in Turkish Aegean Sea coast. EÜ Fen Fak. İlmi Rap. Ser. 1970;**98**(66):1-35

[113] Geldiay R, Kocataş A. A preliminary research on the benthos of İzmir Bay. EÜ Fac Sci Monogr Ser. 1972;**12**:1-34

[114] Özcan T, Katağan T. Deep-water decapod crustacean fauna of the Sıgacık Bay, Aegean Sea coast of Turkey. EU Journal of Fisheries and Aquatic Sciences. 2009;**26**:149-151

[115] Mater S, Kocataş A. İzmir Körfezi Brachyura'sı hakkında bir ön çalışma. EÜ Fen Fak. İlmi Rap. Ser. 1967;**38**(23):1-16

[116] Kocataş A. Two new stomatopod species for the Turkish fauna: Nannosquilloides occultus (Giesbrecht) and Meiosquilla Desmaresti (Risso). Crustaceana. 1981;**40**(2):213-215

[117] Koukouras A, Kallianiotis A, Vafidis D. The decapod crustacean genera Plesionika Bate (Natantia) and Munida leach (Anomura) in the Aegean Sea. Crustaceana. 1998;**71**(6):714-720

[118] Ateş AS, İşmen A, Özekinci U, Yiğin C. A new record of Munida Rugosa (JC Fabricius, 1775) (Decapoda, Anomura, Galatheidae) from the eastern Aegean Sea, Turkey. Crustaceana. 2005;**78**(10):1265-1267

[119] Moreira J, Sezgin M, Katağan T, Gönülal O, Topaloğlu B. First record of a bathyal leptostracan, Nebalia Abyssicola Fage, 1929 (Crustacea: Malacostraca: Phyllocarida), in the Aegean Sea, eastern Mediterranean. Turkish Journal of Zoology. 2012;**36**(3):351-360

[120] Topaloğlu B, Gönülal O. A new record for the Turkish Aegean Sea Fauna: Scalpellum Scalpellum Linnaeus, 1767 (Crustacea: Cirripedia: Thoracica). Journal of Black Sea/Mediterranean Environment. 2012;**18**(1):90-96

[121] Koukouras A, Doulgeraki S, Kitsos MS. Notes on the vertical distribution of pelagic shrimps (Decapoda, Natantia) in the Aegean Sea. Crustaceana. 2000;**73**(8):979-993

[122] Sezgin M, Katağan T, Kırkım F, Aydemir E. Soft-bottom crustaceans from the Saros bay (NE Aegean Sea). Rapp Comm int Mer Médit. 2007;**38**:589

[123] Koukouras A, Sinis AI, Bobori D, Kazantzidis S, Kitsos MS. The echinoderm (Deuterostomia) fauna of the Aegean Sea, and comparison with those of the neighbouring seas. Journal of Biological Research. 2007;**7**:67-92

[124] Özaydın O, Katağan T, Ünsal S. The echinoderms of the Turkish Seas. Israel Journal of Zoology. 1995;**41**:57-68

Chapter 8

Exotic Plant Species in the Mediterranean Biome: A Reflection of Cultural and Historical Relationships

Irene Martín-Forés

Additional information is available at the end of the chapter

http://dx.doi.org/10.5772/intechopen.69185

Abstract

The Mediterranean basin was the world's cradle of agriculture and the first human civilisation. In the Neolithic age, the agrarian culture expanded throughout the Mediterranean basin from the East to the West. Later, an expansion of agrarian culture and trade occurred, associated with the European colonialism, giving rise to a great plant exchange among Mediterranean-type regions. Despite being a biodiversity hotspot, the Mediterranean biome has been subjected to several anthropic impacts, such as alteration of land-use and cross-introductions of exotic species. The millenary anthropic modification of the landscape occurred in the Mediterranean basin gave rise to the formation of seminatural systems in which plants co-evolved with anthropogenic activities over a long time. Thus, species that originated in the Mediterranean basin might have developed a key role in other agro-silvo-pastoral systems along the whole Mediterranean biome. Research is biased towards highlighting the negative impact of exotic species on the ecosystems. To defy the traditional belief, outstanding recent literature focused on the positive effects of exotics on native communities was reviewed. Exotic species seem to have a key role in Mediterranean-type seminatural systems, as evidences of tolerance and facilitation processes were found. Exotic species that have co-evolved with human practices over millennia seem to enhance biodiversity in the Mediterranean biome.

Keywords: biodiversity, coexistence, exotic species, facilitation, grasslands, Mediterranean biome, native species, naturalisation

1. Introduction

Human history and activity have given rise to a wide range of new planetary-scale forces [1] that exert an increasing impact on the ecosphere creating cascades of repercussions on natural and socioeconomic systems [2]. The magnitude of that anthropic influence has increased so much that a new geological era has been recently recognised, the Anthropocene [3]. The impact and ecosystem alteration exerted by humans on the Earth is known as global change and is of major concern within an ecological framework. Basically, it is a question of an unbalanced growth of human population in relation to the energy consumption by different societies [4] and the fact that human activities occurred at local or regional scale affect the global functioning of the planet [5]. Global change comprehends, among others, changes in carbon cycle [6], climate [7, 8], land-use [9], and areas of species distributions [10]. The areas of species distribution in the context of the Mediterranean biome will be the main focus of the current study. Thus, to fully understand this ecological issue, it is necessary to adopt a holistic approach including the socio-cultural and historical contexts which will allow perceiving the interconnection of all components involved [11].

The Mediterranean basin was, around 10,000 years ago, the world's cradle of agriculture and first human civilisations. During the Neolithic era, the first forms of agriculture and human settlements came up in the territory known as fertile crescent, located in the Eastern Mediterranean, and comprised within the ancient territories of Mesopotamia and Near East. Back to that time, eight founder crops, including four cereals (wheat, einkorn, emmer wheat, and barley) and four pulses (lentil, pea, bitter vetch, and chickpea), were domesticated [12]; moreover, farming activities also started taking place with the associated domestication of several livestock species, including mainly sheep, goats, cattle, and pigs [13]. As a consequence of the human displacements which occurred in Southern Europe from the East to the West, archeophytes (i.e. the exotic species that were introduced before 1500) were established in Western European territories such as the Iberian peninsula. It took about three millennia until the agrarian and farming cultures expanded throughout the whole Mediterranean basin, reaching the Iberian peninsula (**Figure 1**) [14, 15]. Once agriculture and farming were integrated in humans' lives, food supply was under control and populations grew drastically, involving an increase in the dependence on and the intensification of agriculture [16].

Later on, changes in the area of species distribution have occurred in three other key phases. The first key phase took place at end of the Middle Ages (1500 AD), with the European rediscovery and exploration of the Americas [17]. The birth of colonialism at the end of the fifteenth century had consequences in human demography, agriculture expansion, and trade and industrial intensification. The expansion of the European colonial powers (remarkably Spain) radically increased the transport of living material. One major aim associated with colonialism was to exploit new economic crops for the empires. Particularly, the discovery of the New World by Spaniards coming from the Mediterranean basin gave rise to a great surge of plant exchange ([18] and references therein) among different Mediterranean-type regions. This phenomenon was specially marked when first female settlers established in the colonised lands [19]. The second key phase took place during the Industrial Revolution (1800 AD) [20], when traditional forms of rural economy were substituted by urban industrialised and mechanised economy. The third key phase has occurred over the last three decades, related to the rise in ease and efficiency of long-distance transport, income growth [21], and tourism [22], which prompt the globalisation era [20]. As a result, species movement and worldwide interconnectedness have become more intensive, occurring across wider space and in a shorter time than before [23]. Thus, humans' activities have radically modified species' distance dispersal and areas of distribution [18, 24–27].

http://dx.doi.org/10.5772/intechopen.69185

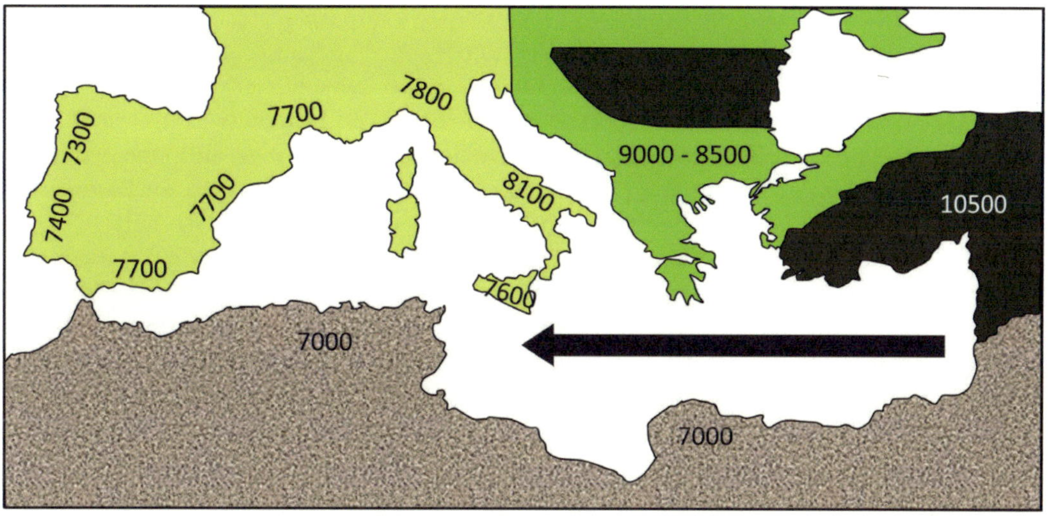

Figure 1. An integrated model of the Neolithic expansion in the Mediterranean basin. The location and approximate dates of colonist farming enclaves are shown by numbers (calibrated years before present). The darkest areas represent the place settled by colonist farmers; the lightest area indicates where indigenous foragers adopted elements of the Neolithic package and the areas with intermediate tonality indicate areas of the proposed integration of colonist farmers with indigenous foraging groups (modified from Ref. [13]).

As a result of the human-favoured transit of organisms [28–31] and habitat alteration, a reshuffling of species on the Earth has taken place. This gives rise to economic and ecological damage [10, 32–37] and a loss of cultural diversity. In this way, the subsequent biological invasions and biodiversity loss have important consequences at a variety of levels, affecting ecosystem structure [32], function, services, and human wellbeing [38–43], being therefore considered of major environmental and social concern and hence a focus of ecological research [44–47].

In the case of plants, species that have been transported from one region to another are defined as weeds, non-native, exotic, or alien to that new occupied region [48]. The introduction of plant species in an exotic area took place as a result of the movement of specimens around the world with ornamental, gardening, agricultural, and forestry purposes [18]. Since agricultural practices were introduced in human cultures, whenever people moved, plants also did, either deliberately (domesticated crops) or accidentally (associated spread of weeds and ruderal species) [18]. Due to their fast way of reproduction, their ability to withstand difficult environmental conditions through dormancy period in seed form, and the variety of ways of seed dispersal, exotic plants have been traditionally considered as a threat in the biological invasion context, with a greater number of invasive species than animals [32].

Notwithstanding the factors enabling establishment, one of the consequences associated with the colonisation process that has been highlighted in previous reports is that exotic species naturalisation involves landscape and global floristic (taxonomic and phylogenetic) homogenisation of regional flora at a biogeographical scale [49, 50]. Moreover, naturalisation of exotic plants has commonly been considered as a threat to native biodiversity, and most scientific studies have been centred around their negative impacts instead of considering the socio-ecological opportunities that the introduction of a new species could bring. During the last decades, numerous studies have highlighted the importance of control, monitoring, and managing of exotic species introduced in new areas [51–55]. In fact, many conservation policies have been implemented to conserve the native flora and sometimes, although being controversial, eradicate exotic invaders [56]. As an example, the Strategic Plan for Biodiversity called for urgent action by the parties to the Convention on Biological Diversity (CBD) to reduce the rate of biodiversity loss by 2020 [57]. To that end, they encountered the target of identifying, prioritising, and managing invasion pathways by 2020 to prevent the introduction of invasive exotic species [58]. Despite all this, it has been shown that human influence on the landscape is not always negative regarding the preservation of biodiversity assets. Traditional rural activities also represent important natural values and they maintain a positive relationship with diversity, at least in the Mediterranean biome [59].

2. The Mediterranean biome: historical relationships among the Mediterranean regions

The Mediterranean biome is defined because of its distinctive Mediterranean climate. It is mainly characterised by mild wet winters and warm to hot, dry summers and may occur on the west side of continents between about 30 and 40° latitude [60]. The summer drought period characteristic of the Mediterranean climate has become accentuated by the deforestation that has taken place around the Mediterranean regions during the last 2000 years and the subsequent loss of plant evapotranspiration and evaporation from soils [61–63].

Mediterranean regions with the above-described climatic characteristics are located in five different continents of the planet, including the Mediterranean basin, California, central Chile, Southern Australia, and South Africa. The international relationships and trade activities among these five Mediterranean regions started long time ago, during the colonialism. As an example, the Malaespina expedition (1789–1794) carried out by Spaniards involved four of the five Mediterranean regions; it departed from Spain and reached Chile, California, and Australia.

Although sharing climatic patterns, the five Mediterranean regions have had different biogeographical and environmental histories associated with the density of human populations as well as the time and intensity of the changes and land-use shifts that humans have caused in the territory. Despite being the Mediterranean biome known for its diverse flora, including five of the biodiversity hotspots [56] comparable to tropical rainforests or coral reefs, it has been subject to several impacts caused by humans but with different intensities and durations in each of the Mediterranean regions.

The use of fire and grazing and agricultural practises have been ancestral activities in the Mediterranean basin that play an essential role in shaping the current cultural landscape. The landscape, however, is also subjected to natural phenomena. Among anthropic impacts of important concern in the Mediterranean biome are alteration of land use, habitat fragmentation caused

http://dx.doi.org/10.5772/intechopen.69185

mainly by land clearing and urbanisation, climate change, alteration of fire regimes, and cross-introductions of alien species [64]. The cultural component generated a landscape maintained and cared for humans by means of reciprocal interactions and interdependencies involving natural processes and human activities. Thus, human culture and technology became integral parts of the ecosystems that underlie the Mediterranean cultural landscapes. In this sense, historical and societal characteristics of the Mediterranean basin, especially those associated with demographic pressure and millenary exploitation of land for agriculture and grazing, determined the patterns of land use change that the territory has presented since ancient times. Those anthropic effects on the environment led to the formation of seminatural systems or cultural landscapes [65, 66]. In fact, agro-silvo-pastoral systems (savanna-like formations characterised by a continuous herbaceous layer with scattered trees and land-use management defined by continuous extensive grazing with a low stocking rate in flatlands and rotation of grazing and cereal cropping in the better-drained hillsides) constitute a characteristic type of exploitation in the Mediterranean biome (e.g. *dehesas* in Spain, *montados* in Portugal). In these socio-ecological systems, plants have co-evolved with people over a long time [67, 68].

Similar cultural landscapes can also be observed in the other four Mediterranean regions worldwide, but the co-evolution of plants with humans has recently occurred, given that effective colonisation times vary depending on the region considered. Due to the fact that Mediterranean regions are enough far away from each other, which impedes the natural flow of exotic species among them, exotic species spreading is expected to follow the cultural landscapes associated with the main navigation routes from the colonialism period onwards (**Figure 2**).

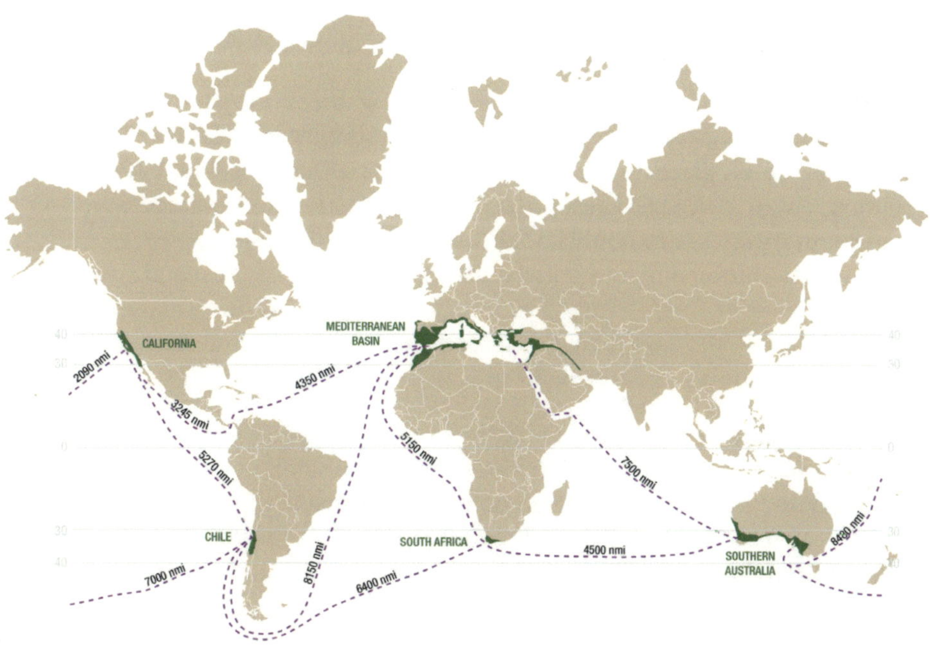

Figure 2. A map of the world showing the five Mediterranean regions and the navigation routes among them. Distances between the regions are shown in nautical miles (modified from Ref. [69]).

For example, Spaniard colonialism occurred in Chile during the sixteenth century due to the expedition led by Pedro de Valdivia that took place in 1541. In central Chile, sclerophyllous forests were cleared and displaced by modified woodlands dominated by introduced exotic species [70], and later those woodlands were opened for grazing and cropping, resulting into grasslands (i.e. called *espinales* in Chile) mostly dominated by species coming from the Mediterranean basin. The implementation of European livestock and agricultural culture led to big direct (ploughing, cropping, and grazing) and indirect (fire and deforestation employed as techniques for preparing the land for agriculture and livestock farming) changes, the extent of which are not well known yet [71]. Associated with the Spaniards arrival, the exotic species were introduced by exozoochory, coupled with merino sheep transported for wool trade, with hay for livestock fodder and with wool and cereals [72]. Several studies have associated the naturalisation (i.e. exotic plants that have been able to establish and self-perpetuate in a new area, according to the definition provided by Richardson et al. [48]) of those plants with processes of grazing by livestock [73–76]. Thus, although Chilean and Spanish plant communities have undergone different processes of invasion, previous reports highlighted the large number of species common to both regions (64% of Chile's non-native flora, [77]).

California was first sighted by Spaniards led by Juan Rodriguez Cabrillo in 1542, after several failed attempts to find a land that was famous for its gold and gems [78]. In the eighteenth century, the Spaniards supported the establishment of ranches granted free of charge, which covered large areas but did not have many inhabitants. The main aim was to support the agricultural and livestock development of the area. Again, the agrarian European culture was exported and implemented in this settlement. The *chaparral* and coastal grasslands of California constitute one of the principal plant formations in the region; they are characterised by having grazing activities and fire as the main agents of disturbance.

South Africa was first sighted by Europeans during the Portuguese expedition led by Bartolomé Díaz in 1488. After that, in the seventeenth century, disputes between Portuguese and Dutch settlers took place and South Africa became a Dutch colony. Finally, in the eighteenth century, South Africa became a British colony [79]. The Mediterranean-type ecosystem of Western South Africa is commonly known as *fynbos*. South African *fynbos* comprise mainly shrublands, grassy shrublands, and grasslands. As in other livestock-based economies, in the Mediterranean region of South Africa, a process of clearing for pasture and grazing intensification also took place [60]. Since then, Europe, particularly the Mediterranean basin, has been the source of 60% of the naturalised exotic grasses in southern Africa [80].

Regarding the Mediterranean region of Australia, Dutch expeditions along the Indic Ocean over the seventeenth century reached the Western Australia State. Nevertheless, British settlers were the ones that finally conquered both Mediterranean regions in Australia. Western Australia was colonised in 1829 and the South Australia State in 1836 [81]. Originally, *mallee* and *kwongan* were typical formations of shrublands and woodlands in this ecoregion. In the late 1800s, large-scale clearing of those formations began giving rise to opened woodlands and grasslands for agrarian and farming activities [82]. Although constituting a biodiversity hotspot, the Mediterranean region of Australia is dominated by naturalised annuals originating from the Mediterranean basin.

For all the above exposed, relationships among all the Mediterranean regions—but of particular importance relationships between the Mediterranean basin and each of the others Mediterranean regions worldwide—have been frequent over the last centuries. As a consequence, natural and ancestral cultural factors based on silvo-pastoral activity converge in the Mediterranean biome, involving the establishment of disturbance regimes that favoured the entry of alien species [73, 83]. Due to the historical-cultural context of the colonialism [19], Europe, and especially the Mediterranean basin, has constituted the main source area donor of exotic flora to the whole Mediterranean biome. Surprisingly, while in some Mediterranean-type ecosystems a displacement of the native flora caused by exotic invaders has occurred, this trend has barely been observed in others. Thus, Mediterranean-type ecosystems worldwide provide a great chance to compare the impact of plant species introduction [84, 85].

3. Exotic plant species naturalisation in the Mediterranean biome

As a result of the landscape transformation carried out by humans in all the Mediterranean regions, empty niches were created (notice that the notion of empty niche does not imply species extinction). That has constituted an opportunity window [86, 87] for the entry of exotic species [73, 83, 88] which were already adapted to disturbances and cultural landscapes [89, 90] in their region of origin for millennia (eco-evolutionary experience [91]). The long-term coexistence of Mediterranean agro-silvo-pastoral systems with anthropic management in the Mediterranean basin has determined processes of co-evolution between plants (crops, forages, etc.) and human practices [67, 76, 92, 93]. In fact, most of the species present in the Mediterranean biome originated in the Mediterranean basin. Plants from the Mediterranean basin presenting more advantageous traits in a context of livestock grazing and ploughing became selected [94]. In particular, among the exotic flora in the Mediterranean biome, the families most represented are Poaceae, Asteraceae, and Fabaceae, in accordance with the three most invasive families worldwide [95]. The rapid growth and high reproduction rates of annual plant species and their capacity to resist unfavourable periods in the form of seeds makes them more suitable to develop in disturbed open spaces by fire, ploughing, or grazing [96, 97]. Poaceae and Fabaceae are typical families associated with livestock grazing and crop cultivation practices while Asteraceae take advantage in spreading mechanisms and dispersal ability [98]. On the contrary, native flora in the rest of Mediterranean regions lacked adaptations to continuous grazing and other disturbances such as fire; thus, native species have resulted negatively affected with the introduction of livestock and crops and the alteration of fire regimes, which have favoured the establishment of exotic species [73, 75, 99, 100].

The factor determining which exotic species became naturalised in a new area depended on the scale of analysis. At a broad scale (i.e. continental), it has been reported that climate determined the possibilities of exotic species establishment because of the significant biogeographical association existing between the climates in the source and the recipient regions [101]. Therefore, the similarity of the climate in source and recipients areas played a crucial role in the current distribution of exotic species [102–105]. The species' climatic tolerance was essential to successfully establish in the new region which highlighted the importance of co-adaptive mechanisms.

Additionally, habitat characteristics of the source area (climate, soil nutrient status, propagule pressure, disturbance, and remarkably human activities) determined the communities' potential to act as main donors of exotic species [101]. However, the relative importance of those factors ultimately depended upon the climate of the recipient region and the distribution of the main land uses which highlighted the importance of human pressure as a driver of exotic species distribution [106].

Species that can potentially colonise new areas undergo environmental filtering. Regarding abiotic filters (climate- and edaphic-related) acting at a regional scale, previous reports suggested a filtering process in both the source and the recipient areas but acting with different intensity in each of them [107]. Existing literature stated that the influence of abiotic factors was stronger in the recipient area, where especially the climate determined the successful naturalisation of exotic species [107], agreeing with previous reports about invasibility [108–110]. For example, in the Mediterranean region of central Chile, the increase in species richness with precipitation, and with the shortening of summer drought, was greater for exotic than for native species [107]. Therefore, exotic species naturalisation appeared enhanced by an improvement in the main limiting resource (water) (supported by the 'resource availability hypothesis') [103, 111, 112]. In the recipient area, not only abiotic filters but also biotic ones can be acting as filters, for example, plant-plant interactions occurring along the processes of plant community's secondary succession [113–115]. Possible processes between exotic and native species in the invaded area could be competence, tolerance, or facilitation; as a consequence, the relationship observed between both species groups would be negative, neutral, or positive, respectively. However, sometimes, this relationship appears conspicuous; the 'invasion paradox' stated that the relationship between species richness of native and exotic flora depends on the scale of study (being negative at small spatial scales < 10 m^2 and positive at larger spatial scales > 10 m^2) [116]. At the ecological community level, a consensus on the general impact of exotic species diversity has not been reached yet.

4. Positive relationships between native and exotic species in the Mediterranean biome

Although the frequency of exotic species in Mediterranean ecosystems is considerable, it varies among different regions [117] and types of studied system. Habitats with a higher degree of invasion are usually those related to anthropogenic activities [118] in terms of both, human population density and human-mediated disturbances [119]. Disturbances open a window of opportunity that promotes exotic species success by altering the environmental and soil conditions as well as by establishing new interactions within the native plant community [120]. For example, exotic species appeared associated with disturbed biological soil crusts [121], agrarian practices [122], grazing land use (e.g. 87.5% of the species significantly associated with continuously grazed grasslands in Southern Australia were exotic [123]), and fire events—the latter especially in California [120, 124]. Thus, better understanding of the influence of exotic species on native diversity is highly relevant to the management of Mediterranean ecosystems. However, a publication bias towards studies focusing on biological invasions by exotic species with negative impacts [125], especially in ecological terms, has been common within the scientific community. For example, in Mediterranean grasslands, invasion by exotic species has been frequently cited as a key threat responsible for decreased native abundance, richness, growth and regeneration [126–128], and altered species composition [128]. Although the scientific community is aware of reports about serious negative impacts in ecology [129, 130], economy [34, 35], and society [131], studies focusing on the possible benefits that exotic species can provide have been overlooked receiving much less attention [132–134]. Fortunately, over the past years, the bias in the literature towards negative impacts from exotic species has started to revert. Positive associations between exotics and natives have recently been reported, especially in the Mediterranean biome. Here, the findings of some of the most outstanding studies reporting positive relationships are summarised.

In terms of richness, positive correlations between native and exotic species have been reported in all the Mediterranean regions. In central Chile, positive relationships were found in both shrublands [135] and grasslands [107, 122]. Similar relationships were also documented in the Californian shrublands [123] and for post-fire vegetation communities in the Californian *chaparral* [124, 136] and grasslands [137]. In the Mediterranean basin, positive correlations between native and exotic species were also found on different systems such as floodplains [138] and marine benthos communities [139].

Controlling for the environmental conditions and land uses has been proposed as key important steps when assessing relationships between natives and exotics at a broad scale. When including environmental variability, positive relationships were found in South Australian grasslands in terms of species richness, cover, and Shannon diversity index [140]. In the Mediteranean basin, common anthropic factors such as landscape heterogeneity and human pressure partially explained both native and exotic richness, but a significant percentage remained unexplained, revealing that biotic interactions between both species groups might be occurring [141].

Some studies that were not conducted in the Mediterranean biome stated that exotic species established in early successional stages impeded the re-establishment of native ones [142–144]. Contrarily, no apparent competition between both species groups has been reported along secondary succession in grasslands of central Chile and post-fire Californian plant communities [122, 124, 145]. Rather, in Ref. [122] a complementary role between native and exotic species possibly as a consequence of niche segregation at a local scale in earlier stages of succession was documented. Whereas time since the disturbing event had a positive effect on native species richness, exotic species richness and cover increased right after the disturbance and then richness remained stable until the end of the chronosequence while cover decreased [122, 140]. The increase in vegetation cover associated with the early colonisation by exotics seemed to create the appropriate conditions for the successful re-establishment of native species, which increased in number. Thus, exotic species acted as passengers [146] or even as facilitator species, playing a complementary role to the natives' one. In central Chile and Southern Australia, the coexistence under the conditions of extensive livestock grazing was achieved through two different strategies: alien species were mainly grazing-tolerant species, whereas native species were grazing-defensive species [94, 122, 140].

Other positive effects of exotic species in native communities have been shown in other systems or for other type of species. For example, an exotic nitrogen-fixing shrub was able to build up an island of fertility under its canopy by accumulating considerable stocks of C, N, and P in the soil and by improving the soil hydrological properties [147]. Likewise, the presence of another exotic species, *Lantana camara*, also improved soil quality [148].

Regarding effects of exotics on germination rates or seedling growth of native species in grasslands, the presence of the invasive species *Thymus vulgaris* originated in the Mediterranean basin that presents allelopathy showed no negative effect [149]. Moreover, facilitative effects on the reproductive success of co-flowering native plants have been reported in the presence of the invasive weed *Oxalis pescaprae* [150]. Likewise, it has been shown that litter of invasive exotic plants facilitated growth of the dominant native plants by altering soil moisture in Californian shrublands [151]. Additionally, in some cases, the presence of exotic plant species facilitated the visit of pollinators to native species [152]. In central Chile, alien European rabbits filled a role similar to the one played by native mammals, by dispersing native seeds [153].

Recent studies even showed that certain invasive species have become keystones for the survival of local endemisms. Therefore, eradication programmes to re-establish the original vegetation might provoke severe local extinction of endemic species [127, 154], population bottlenecks, and cascading effects across trophic levels [126, 155], as well as on pollinator communities [150]. Some exotic species can contribute to achieve native species conservation policies [133]; sometimes, even the employment of exotic species has been proposed as an effective action for ecological restoration [156, 157].

Many of the results reviewed here pointed out that exotic species coming from the Mediterranean basin were pre-adapted to the environmental conditions and land-use management in the recipient regions. Similarly, it has been shown that once they got naturalised in a non-native environment, they formed mixed native/exotic plant communities due to effective mechanisms of tolerance and facilitation that allow the coexistence between both species groups. This coexistence did not seem to be aggressive in opposition to the traditional beliefs that aliens' naturalisation always decreases native biodiversity [158] or displaces it by exclusion [49]. On the contrary, the coexistence between native and exotic species appears to be smooth as exotics seem to ameliorate the harsh environmental conditions created after a disturbance so that natives can re-establish in that area.

5. Towards a new paradigm

As reported here, many positive interactions between exotic and native assets have been described over the past decades. It seems therefore that it is time to rethink the traditional paradigm which, by default, considers exotic species as a threat for Mediterranean-type ecosystems. It would be desirable the employment of neutral terminology such as nonindigenous species to avoid negative predisposition to the effect of exotic species [159]. In fact, it has been reported that the perception of the consequences that exotic species have varies among different stakeholders [160]. Stakeholders with different socio-cultural contexts have very different

http://dx.doi.org/10.5772/intechopen.69185

opinions about exotic species; some of them recognise the benefits of exotic species not only on the native flora at local scale but also on other species they have established relationships with [154], on the ecosystem services [124], on the human wellbeing [161], on the local and global economy [127], and so on. Thus, the crucial importance of adopting a more balanced view of exotic species and understanding their presence in a new area as a holistic process [162] need to be highlighted.

Assessing the functional roles that exotics may have established in their new areas to avoid unexpected results from incorrect management and being critical and open minded to find any possible mutualistic interactions between exotics and natives are crucial. In the foreseeable future, it would be desirable to evaluate in detail other possible facilitation processes between natives and exotics by studying plant-plant interactions. Future outcomes could also include the combination of the current ecological knowledge in invasion processes with forthcoming global change scenarios [37]. Merging the prospection of future climate change [8], landscape heterogeneity, land-use shifts, and the subsequent modelled displacement of exotic plant species distribution with current knowledge on invasive processes would be a fruitful study for determining areas which are more prone to invasion.

Evaluating the factors involved in naturalisation and invasion processes is key to accomplish the objectives of the Millennium Ecosystem Assessment Program [163]. The species reshuffling and its impact on the native plant communities undoubtedly contribute to the emergence of new environmental scenarios. These new scenarios have ecological and socioeconomic repercussions that are difficult to evaluate short term [164] or at a determined spatial scale. As multi-scale patterns are determinant of naturalisation success, evaluations should be conducted from broader to finer geographical scales and in the short, medium, and especially in long term [165]. To that end, we should take a comprehensive and systemic approach, coming to understand that fundamental ecology is context dependent, being tightly bound up to the social-cultural history, the anthropic activity, the economic implications, and the social tradeoffs [124]. Thus, it is necessary to accomplish transdisciplinary decision-making processes [166] which take into account not only ecological consequences but also ecosystem services and human well-being [167].

Acknowledgements

I would like to thank Miguel Ángel Casado, Belén Acosta-Gallo, Isabel Castro, Miguel Brun, and Greg R. Guerin for their support.

Author details

Irene Martín-Forés

Address all correspondence to: imfores@pdi.ucm.es

Complutense University of Madrid, Madrid, Spain

References

[1] Steffen W, Andreae MO, Bolin B, Cox PM, Crutzen PJ, Cubasch U, Held H, Nakicenovic N, Scholes RJ, Talaue-McManus L, Turner II BL. The Achilles' heels of the Earth system. Environment. 2004;**46**:9-20

[2] Schellnhuber HJ, Crutzen PJ, Clark WC, Hunt J. Earth system analysis for sustainability. Environmental: Science and Policy for Sustainable Development. 2005;**47**:10-25

[3] Crutzen PJ. Geology of mankind. Nature. 2002;**415**:23-23

[4] Margalef R. Ecología. 9th ed. Omega: Barcelona; 1998. 951 p.

[5] Pineda FD, Schmitz MF, Acosta-Gallo B, Fernández Pastor M, Arnaiz C, Díaz P, Ruiz-Labourdette D. Informe y síntesis de la actividad investigadora desarrollada en el marco de la Red de Seguimiento del Cambio Global (RSCG) en los Parques Nacionales españoles (RPN). Madrid: Fundación Biodiversidad. Convenio Organismo Autónomo Parques Nacionales (OAPN) – Oficina Española del Cambio Climático (OECC) – Fundación Biodiversidad (FB) – Agencia Española de Meteorología (AEMT); 2014.

[6] Acosta-Gallo B. Comportamiento de los componentes aéreo y subterráneo de pastizales en diferentes condiciones ambientales [thesis]. Faculty of Biology, Complutense University of Madrid; 2005.

[7] Vitousek PM. Beyond global warming: Ecology and global change. Ecology. 1994;**75**: 1861-1876

[8] IPCC. Pachauri RK, Meyer LA, editors. Climate Change 2014: Synthesis Report. Contribution of Working Groups I, II and III to the Fifth Assessment Report of the Intergovernmental Panel on Climate Change. IPCC, Geneva, Switzerland; 2014. 151 p

[9] Malone TF, Roederer JG. Global Change. ICSU Press: Cambridge; 1984

[10] Vitousek PM, D'Antonio CM, Loope LL, Rejmánek M, Westbrooks R. Introduced species: A significant component of human-caused global change. New Zealand Journal of Ecology. 1997;**21**:1-1

[11] Palang H, Mander Ü, Naveh Z. Holistic landscape ecology in action. Landscape and Urban Planning. 2000;**50**:1-6

[12] Lev-Yadun S, Gopher A, Abbo S. The cradle of agriculture. Science. 2000;**288**:1602-1603

[13] Zeder MA. Domestication and early agriculture in the Mediterranean Basin: Origins, diffusion, and impact. Proceedings of the National Academy of Sciences of the United States of America. 2008;**105**:11597-11604

[14] Estévez J. Study of the faunal remains (original in Spanish). In: Fosca C, Olaria C, editors. Servico de Publicaciones Diputación de Castellón. Spain: Castellón; 1988. pp. 281-337

[15] Boessneck J, von den Driesch A. Finds of animal bones from four caves in southern Spain (original in German). In: Boessneck J, von den Driesch A, editors. Studies on Finds of Early Animal Bones from the Iberian Peninsula 7. Munich: Institut für Palaeoanatomir, Domestikationsforschung und Geschichte de Tiermedizin der Universität München;1980. pp. 1-83

[16] Brown TA, Jones AK, Powell W, Allaby RG. The complex origins of domesticated crops in the fertile Crescent. Trends in Ecology and Evolution. 2008;**24**:103-109

[17] Preston CD, Pearman DA, Hall AR. Archaeophytes in Britain. Botanical Journal of the Linnean Society. 2004;**145**:257-294

[18] Cronk QCB, Fuller JL. Plant invaders: The Threat to Natural Ecosystems. Chapman and Hall: London; 1995. 256 p

[19] National Geographic Society. 1491: America Before Columbus. William Reese Company;1991. 154 p

[20] Hulme PE. Trade, transport and trouble: Managing invasive species pathways in an era of globalization. Journal of Applied Ecology. 2009;**46**:10-18

[21] Baier SL, Bergstrand JH. The growth of world trade: Tariffs, transport costs, and income similarity. Journal of International Economics. 2001;**53**:1-27

[22] Hall CM. Tourism and biological exchange and invasions: A missing dimension in sustainable tourism? Tourism Recreation Research. 2015;**40**:81-94

[23] Pieterse JN. Globalization and Culture: Global Mélange. USA: Lanham MD, Rowman & Littlefield; 2015. 219 p

[24] Levine JM, D'Antonio CM. Forecasting biological invasions with increasing international trade. Conservation Biology. 2003;**17**:322-326

[25] Perrings C, Dehnen-Schmutz K, Touza J, Williamson M. How to manage biological invasions under globalization. Trends in Ecology and Evolution. 2005;**20**:212-215

[26] Meyerson LA, Mooney HA. Invasive alien species in an era of globalization. Frontiers in Ecology and the Environment. 2007;**5**:199-208

[27] Bradley BA, Blumenthal DM, Early R, Grosholz ED, Lawler JJ, Miller LP, Sorte CJB, D'Antonio CM, Diez JM, Dukes JS, Ibañez I, Olden JD. Global change, global trade, and the next wave of plant invasions. Frontiers in Ecology and the Environment. 2012;**10**:20-28

[28] Paskoff R, Manríquez H. Ecosystem and legal framework for coastal management in central Chile. Ocean and Coastal Management. 1999;**42**:105-117

[29] Rouget M, Richardson DM, Cowling RM, Lloyd JW, Lombard AT. Current patterns of habitat transformation and future threats to biodiversity in terrestrial ecosystems of the Cape Floristic Region, South Africa. Biological Conservation. 2003;**112**:63-85

[30] Dukes JS, Mooney HA. Disruption of ecosystem processes in western North America by invasive species. Revista Chilena de Historia Natural. 2004;**77**:411-437

[31] Schwartz MW, Thorne JH, Viers JH. Biotic homogenization of the California flora in urban and urbanizing regions. Biological Conservation. 2006;**187**:282-291

[32] Vitousek PM, D'Antonio CM, Loope LL, Westbrooks R. Biological invasions as global environmental change. American Scientist. 1996;**84**:468-478

[33] Pimentel D, Zuniga R, Morrison D. Update on the environmental and economic costs associated with alien-invasive species in the United States. Ecological Economics. 2005; **52**:273-288.

[34] Pimentel D, McNair S, Janecka J, Wightman J, Simmonds C, O'Connell C, Wong E, Russel L, Zern J, Aquino T, Tsomondo T. Economic and environmental threats of alien plant, animal, and microbe invasions. Agriculture, Ecosystems and Environment. 2001;**84**:1-20

[35] Pimentel D, Zuniga R, Morrison D. Update on the environmental and economic costs associated with alien-invasive species in the United States. Ecological Economics. 2005;**52**:273-288

[36] Mack RN, Simberloff D, Lonsdale WM, Evans H, Clout M, Bazzaz FA. Biotic invasions: Causes, epidemiology, global consequences, and control. Ecological Applications. 2000;**10**:689-710

[37] Sala OE, Chapin FS III, Armesto JJ, Berlow E, Bloomfield J, Dirzo R, Huber-Sanwald E, Huenneke LF, Jackson RB, Kinzig A, Leemans R, Lodge DM, Mooney HA, Oesterheld M, LeRoy Poff N, Sykes MT, Walker BH, Walker M, Wall DH. Global biodiversity scenarios of the year 2100. Science. 2000;**287**:1770-1774

[38] MA (Millennium Ecosystem Assessment). Ecosystems and Human Well-Being: A Framework for Assessment. Island Press: Washington DC; 2003

[39] Corvalan C, Hales S, McMichael A. Ecosystems and Human Well-being: Health Synthesis [a report for the Millennium Ecosystem Assessment]. Geneva: World Health Organization; 2005

[40] Díaz S, Fargione J, Chapin FS III, Tilman D. Biodiversity loss threatens human well-being. PLoS Biology. 2006;**4**:e277

[41] Martín-López B, Gómez-Baggethun E, González JA, Lomas PL, Montes C. The assessment of ecosystem services provided by biodiversity: Rethinking concepts and research needs. In: Aronoff JB, editor. Handbook of Nature Conservation: Global, Environmental and Economic Issues. Nova Science Publisher Inc. New York, USA; 2009. 479 p

[42] Cardinale BJ, Duffy JE, Gonzalez A, Hooper DU, Perrings C, Venail P, Narwani A, Mace GM, Tilman D, Wardle DA, Kinzig AP, Daily GC, Loreau M, Grace JB, Larigauderie A, Srivastava DS, Naeem S. Biodiversity loss and its impact on humanity. Nature. 2012;**486**:59-67

[43] Mace GM, Norris K, Fitter AH. Biodiversity and ecosystem services: A multilayered relationship. Trends in Ecology and Evolution 2012;**27**:19-26

[44] Brock JH, Wade M, Pyšek P, Green D. Plant Invasions: Studies from North America and Europe. Leiden: Backhuys; 1997

[45] Luken JO, Thieret JW. Assessment and Management of Plant Invasions. New York: Springer; 1997

[46] Dukes JS, Mooney HA. Does global change increase the success of biological invaders? Trends in Ecology and Evolution. 1999;**14**:135-139

[47] Higgins SI, Richardson DM, Cowling RM, Trinder-Smith TH. Predicting the landscape scale distribution of alien plants and their threat to plant diversity. Conservation Biology. 1999;**13**:303-313

[48] Richardson DM, Pyšek P, Rejmánek M, Barbour MG, Panetta FD, West CJ. Naturalization and invasion of naturalized plants: Concepts and definitions. Diversity and Distributions. 2000;**6**:93-107

[49] Winter M, Schweigera O, Klotza S, Nentwigc W, Andriopoulosd P, Arianoutsoud M, Basnoue C, Delipetrouf P, Didžiulisg V, Hejdah M, Hulme PE, Lambdonj PW, Perglh J, Pyšek P, Royl DB, Kühna I. Plant extinctions and introductions lead to phylogenetic and taxonomic homogenization of the European flora. Proceedings of the National Academy of Sciences of the United States of America. 2009;**106**:21721-21725

[50] Figueroa JA, Teillier S, Castro SA. Diversity patterns and composition of native and exotic floras in central Chile. Acta Oecologica. 2011;**37**:103-109

[51] Blossey B. Before, during and after: The need for long-term monitoring in invasive plant species management. Biological Invasions. 1999;**1**:301-311

[52] Zavaleta ES, Hobbs RJ, Mooney HA. Viewing invasive species removal in a whole-ecosystem context. Trends in Ecology and Evolution. 2001;**16**:454-459

[53] Pyšek P, Richardson DM. Invasive species, environmental change and management, and health. Annual Review of Environment and Resources. 2010;**35**:25-55

[54] Barbier EB, Knowler D, Gwatipedza J, Reichard SH, Hodges AR. Implementing policies to control invasive plant species. BioScience. 2013;**63**:132-138

[55] Essl F, Dullinger S, Rabitsch W, Hulme PE, Wilson JRU, Richardson DM. Delayed biodiversity change: no time to waste. Trends in Ecology and Evolution. 2015;**30**:375-378

[56] Myers N, Mittermeier RA, Mittermeier CG, da Fonseca GAB, Kent J. Biodiversity hotspots for conservation priorities. Nature. 2000;**403**:853-858

[57] CBD. Global Biodiversity Outlook 4. Montreal: Secretariat of the Convention on Biological Diversity; 2014

[58] Hulme PE. Invasion pathways at a crossroad: Policy and research challenges for managing alien species introductions. Journal of Applied Ecology. 2015;**52**:1418-1424.

[59] Balmford A, Moore JL, Brooks T, Burgess N, Hansen LA, Williams P, Rahbek C. Conservation conflicts across Africa. Science. 2001;**291**:2616-2619

[60] Lionello P, Malanotte-Rizzoli P, Boscolo R, Alpert P, Artale V, Li L, Luterbacher J, May W, Trigo R, Tsimplis M, Ulbrich U, Xoplaki E. The Mediterranean climate: An overview of the main characteristics and issues. Developments in Earth and Environmental Sciences. 2012;**4**:1-26

[61] Reale O, Dirmeyer P. Modeling the effects of vegetation on Mediterranean climate during the Roman classical period: Part I: Climate history and model sensitivity. Global Planet. 2000;**25**:163-184

[62] Reale O, Shukla J. Modeling the effects of vegetation on Mediterranean climate during the Roman classical period: Part II: Model simulation. Global Planet. 2000;**25**:185-214

[63] Dümenil-Gates L, Liess S. Impacts of deforestation and afforestation in the Mediterranean region as simulated by the MPI atmospheric GCM. Global and Planetary Change. 2001;**30**:309-328

[64] Wade TG, Riitters KH, Wickham JD, Jones KB. Distribution and causes of global forest fragmentation. Conservation Ecology. 2003;**7**:7

[65] Pineda FD, Schmitz MF, Hernández S. Interacciones entre infraestructuras y conectividad natural del paisaje. I Congreso Ingeniería Civil, Territorio y Medio Ambiente. Col. Ing. CC y Puertos, Madrid; 2002

[66] Schmitz MF, Pineda FD, Castro H, de Aranzabal I, Aguilera P. Cultural landscape and socioeconomic structure. Environmental value and demand for tourism in a Mediterranean territory. Sevilla: Junta de Andalucía; 2005

[67] Di Castri F. Mediterranean-type shrublands of the world. In: Di Castri F, Goodall DW, Specht RL, editors. Mediterranean-type Shrublands. Amsterdam: Elsevier; 1981. pp. 1-52

[68] Perevolotsky A, Seligman NG. Role of grazing in Mediterranean rangeland ecosystems–inversion of a paradigm. BioScience. 1998;**48**:1007-1017

[69] Fox MD. Mediterranean weeds: Exchanges of invasive plants between the five Mediterranean regions of the world. In: Di Castri F, Hansen AJ, Debussche M, editors. Biological Invasions in Europe and the Mediterranean Basin. The Netherlands: Springer; 1990. pp. 179-200

[70] Van de Wouw P, Echeverría C, Rey-Benayas JM, Holmgren M. Persistent Acacia savannas replace Mediterranean sclerophyllous forests in South America. Forest Ecology and Management. 2011;**262**:1100-1108

[71] Turner BL, Gómez Sal A, González Bernáldez F, Di Castri F, editors. Global Land Use Change. A Perspective from the Columbian Encounter. Madrid, Espana: CSIC; 1995

[72] Castro SA, Figueroa JA, Muñoz-Schick M, Jaksic FM. Minimum residence time, biogeographical origin, and life cycle as determinants of the geographical extent of naturalized plants in continental Chile. Diversity and Distributions. 2005;**11**:183-191

[73] Holmgren M, Aviles R, Sierralta L, Segura AM, Fuentes ER. Why have European herbs so successfully invaded the Chilean matorral? Effects of herbivory, soil nutrients, and fire. Journal of Arid Environment. 2000;**44**:197-211

[74] Pauchard A, Alaback PB. Influence of elevation, land use, and landscape context on patterns of alien plant invasions along roadsides in protected areas of south-central Chile. Conservation Biology. 2004;**18**:238-248

[75] Del Pozo A, Ovalle C, Casado MA, Acosta B, De Miguel JM. Effects of grazing intensity in grasslands of the Espinal of central Chile. Journal of Vegetation Science. 2006;**17**:791-798

[76] HilleRisLambers J, Yelenik SG, Colman BP, Levine JM. California annual grass invaders: The drivers or passengers of change? Journal of Ecology. 2010;**98**:1147-1156

[77] Pauchard A, Cavieres L, Bustamante R. Comparing alien plant invasions among regions with similar climates: where to from here? Diversity and Distributions. 2004;**10**:371-375

[78] Bouvier VM. Women and the conquest of California, 1542-1840: Codes of silence. University of Arizona Press. Tucson, Arizona, USA; 2004. 266p

[79] Chidester D. Religions of South Africa (Routledge Revivals). Routledge. London, UK; 2014. 306p

[80] Milton SJ. Grasses as invasive alien plants in South Africa: Working for water. South African Journal of Science. 2004;**100**:69-75

[81] Lines WJ. Taming the Great South Land: A History of the Conquest of Nature in Australia. University of California Press. Berkeley, California, USA; 1991

[82] The Australian National Botanic Gardens. Mallee Plants-surviving Harsh Conditions. Acton: Education Services Australian National Botanic Gardens; 2004

[83] Le Houérou NH. Plant invasion in the rangelands of the isoclimatic Mediterranean zone. In: Groves RH, Di Castri F, editors. Biogeography of Mediterranean Invasions. Camebridge: Cambridge University Press; 1991. pp. 393-404

[84] Kruger FJ, Breytenbach GJ, Macdonald IAW, Richardson DM. The characteristics of invaded Mediterranean-climate regions. In: Drake JA, Mooney, HA, Di Castri F, Groves RH, Kruger FJ, Rejmánek M, Williamson M, editors. Biological Invasions: A Global Perspective. New York, The USA: John Wiley & Sons; 1989. pp. 389-405

[85] Groves RH, Di Castri F, editors. Biogeography of Mediterranean Invasions. Camebridge (UK): Cambridge University Press; 1991. 485 p

[86] Tilman D. Community invasibility, recruitment limitation, and grassland biodiversity. Ecology. 1997;**78**:81-92

[87] Zobel M, Otsus M, Liira J, Moora M, Möls T. Is small-scale species richness limited by seed availability or microsite availability? Ecology. 2000;**81**:3274-3282

[88] Jauni M, Gripenberg S, Ramula S. Non-native plant species benefit from disturbance: A meta-analysis. Oikos. 2015;**124**:122-129

[89] Prinzing A, Durka W, Klotz S, Brandl R. Which species become aliens? Evolutionary Ecology Research. 2002;**4**:385-405

[90] Arianoutsou M, Delipetrou P, Vilà M, Dimitrakopoulos PG, Celesti-Grapow L, Wardell-Johnson G, Henderson L, Fuentes N, Ugarte-Mendes NE, Rundel PW. Comparative patterns of plant invasions in the Mediterranean Biome. Plos One. 2013;**8**:1-13

[91] Saul WC, Jeschke JM, Heger T. The role of eco-evolutionary experience in invasion success. NeoBiota. 2013;**17**:57-74

[92] Holmgren M. Exotic herbivores as drivers of plant invasions and switch to ecosystem alternative states. Biological Invasions. 2002;**4**:25-33

[93] Ricotta C, La Sorte FA, Pyšek P, Rapson GL, Celesti-Grapow L, Thompson K. Phyloecology of urban alien floras. Journal of Ecology. 2009;**97**:1243-1251

[94] De Miguel JM, Casado MA, Del Pozo A, Ovalle C, Moreno-Casasola P, Travieso-Bello AC, Barrera M, Ricardo N, Tecco PA, Acosta B. How reproductive, vegetative and defensive strategies of Mediterranean grassland species respond to a grazing intensity gradient. Plant Ecology. 2010;**210**:97-110

[95] Pyšek P. Is there a taxonomic pattern to plant invasions? Oikos. 1998;**82**:282-294

[96] Le Floc'h E. Invasive plants of the Mediterranean Basin. In: Groves RH, Di Castri F, editors. Biogeography of Mediterranean Invasions. Camebridge (UK): Cambridge University Press; 1991. pp. 67-80

[97] Gómez-González S, Torres-Díaz C, Valencia G, Torres-Morales P, Cavieres LA, Pausas JG. Anthropogenic fires increase alien and native annual species in the Chilean coastal matorral. Diversity and Distributions. 2010;**17**:58-67

[98] Venable DL. The evolutionary ecology of seed heteromorphism. The American Naturalist. 1985;**126**:577-595

[99] Milchunas DG, Lauenroth WK. Quantitative effects of grazing on vegetation and soils over a global range of environments. Ecological Monographs. 1993;**63**:327-366

[100] Ovalle C, Del Pozo A, Casado MA, Acosta B, De Miguel JM. Consequences of landscape heterogeneity on grassland diversity and productivity in the espinal agroforestry system of central Chile. Landscape Ecology. 2006;**21**:585-594

[101] Casado MA, Acosta-Gallo B, Sánchez-Jardón L, Martín-Forés I, Castro I, Ovalle C, del Pozo A, de Miguel JM. Interactive effects of source and recipient habitats on plant invasions: Distribution of exotic species in Chile. Diversity and Distributions. 2015;**21**:609-619

[102] Alpert P, Bone E, Holzapfel C. Invasiveness, invasibility and the role of environmental stress in the spread of non-native plants. Perspectives in Plant Ecology, Evolution and Systematics. 2000;**3**:52-66

[103] Davis MA, Grime J, Thompson K. Fluctuating resources in plant communities: A general theory of invasibility. Journal of Ecology. 2000;**88**:528-534

[104] Chytrý M, Maskell LC, Pino J, Pyšek P, Vilà M, Font X, Smart SM. Habitat invasions by alien plants: A quantitative comparison among Mediterranean, subcontinental and oceanic regions of Europe. Journal of Applied Ecology. 2008;**45**:448-458

[105] Chytrý M, Pyšek P, Wild J, Pino J, Maskell L, Vilà M. European map of alien plant invasions based on the quantitative assessment across habitats. Diversity and Distributions. 2009;**15**:98-107

[106] Marini L, Battisti A, Bona E, Federici G, Martini F, Pautasso M, Hulme PE. Alien and native plant life-forms respond differently to human and climate pressures. Global Ecology and Biogeography. 2012;**21**:534-544

[107] Martín-Forés I, Sánchez-Jardón L, Acosta-Gallo B, Del Pozo A, Castro I, de Miguel JM, Ovalle C, Casado MA. From Spain to Chile: Environmental filters and success of herbaceous species in Mediterranean-climate regions. Biological Invasions. 2015;**17**:1425-1438

[108] Kolar CS, Lodge DM. Progress in invasion biology predicting invaders. Trends in Ecology and Evolution. 2001;**16**:199-204

[109] Dawson W, Burslem DF, Hulme PE. Factors explaining alien plant invasion success in a tropical ecosystem differ at each stage of invasion. Journal of Ecology. 2009;**97**:657-665

[110] González-Moreno P, Diez JM, Ibáñez I, Font X, Vilà M. Plant invasions are context-dependent: Multiscale effects of climate, human activity and habitat. Diversity and Distributions. 2014;**20**:720-731

[111] Cleland EE, Smith MD, Andelman SJ, Bowles C, Carney KM, Horner-Devine MC, Drake JM, Emery SM, Gramling JM, Vandermast DB. Invasion in space and time: nonnative species richness and relative abundance respond to interannual variation in productivity and diversity. Ecology Letters. 2004;**7**:947-957

[112] Richardson DM, Pyšek P. Plant invasions: Merging the concepts of species invasiveness and community invasibility. Progress in Physical Geography. 2006;**30**:409-431

[113] Armas C, Pugnaire FI. Plant interactions govern population dynamics in a semi-arid plant community. Journal of Ecology. 2005;**93**:978-989

[114] Armas C, Schöb C, Gutiérrez JR. Modulating effects of ontogeny on the outcome of plant–plant interactions along stress gradients. New Phytologist. 2013;**200**:7-9

[115] Jeffers ES, Bonsall MB, Froyd CA, Brooks SJ, Willis KJ. The relative importance of biotic and abiotic processes for structuring plant communities through time. Journal of Ecology. 2015;**103**:459-472

[116] Fridley JD, Stachowicz JJ, Naeem S, Sax DF, Seabloom EW, Smith MD et al. The invasion paradox: Reconciling pattern and process in species invasions. Ecology. 2007;**88**:3-17

[117] Rejmánek M, Richardson DM, Pyšek P. Plant Invasions and Invasibility of Plant Communities in Vegetation Ecology. 2nd ed. Oxford (UK): John Wiley & Sons Ltd; 2013. pp. 387-424

[118] Montserrat V, Joan P, Xavier F. Regional assessment of plant invasions across different habitat types. Journal of Vegetation Science. 2007;**18**:35-42

[119] Estay SA, Navarrete SA, Toro SR. Effects of human mediated disturbances on exotic forest insect diversity in a Chilean Mediterranean ecosystem. Biodiversity and Conservation. 2012;**21**:3699-3710

[120] Carvalho LM, Antunes PM, Martins-Loução MA, Klironomos JN. Disturbance influences the outcome of plant–soil biota interactions in the invasive *Acacia longifolia* and in native species. Oikos. 2010;**119**:1172-1180

[121] Hernandez RR, Sandquist DR. Disturbance of biological soil crust increases emergence of exotic vascular plants in California sage shrub. Plant Ecology. 2011;**212**:1709

[122] Martín-Forés I, Castro I, Acosta-Gallo B, del Pozo A, Sánchez-Jardón L, de Miguel JM, Ovalle C, Casado MA. Alien plant species coexist over time with native ones in Chilean Mediterranean grasslands. Journal of Plant Ecology. 2016;**9**:682-691

[123] Price JN, Wong NK, Morgan JW. Recovery of understorey vegetation after release from a long history of sheep grazing in a herb-rich woodland. Austral Ecology. 2010;**35**:505-514

[124] Guo Q. Temporal changes in native-exotic richness correlations during early post-fire succession. Acta Oecologica. 2017; **80**:47-50

[125] Davis MA, Chew MK, Hobbs RJ, Lugo AE, Ewel JJ, Vermeij GJ, Brown JH, Rosenzweig ML, Gardener MR, Carroll SP, Thompson K, Pickett STA, Stromberg JC, Del Tredici P, Suding KN, Ehrenfeld JG, Grime JP, Mascaro J, Briggs JC. Don't judge species on their origins. Nature. 2011;**474**:153-154

[126] Lenz TI, Moyle-Croft JL, Facelli JM. Direct and indirect effects of exotic annual grasses on species composition of a South Australian grassland. Austral Ecology. 2003;**28**:23-32

[127] Gornish ES, Ambrozio dos Santos P. Invasive species cover, soil type, and grazing interact to predict long-term grassland restoration success. Restoration Ecology. 2015;**24**:222-229

[128] Flory SL, Clay K. Non-native grass invasion alters native plant composition in experimental communities. Biological Invasions. 2010;**12**:1285-1294

[129] Lowe S, Browne M, Boudjelas S, De Poorter M. 100 of the World's Worst Invasive Alien Species: A Selection from the Global Invasive Species Database. Auckland, New Zealand: Invasive Species Specialist Group; 2000. p. 12

[130] McGeoch MA, Butchart SHM, Spear D, Marais E, Kleynhans EJ, Symes A, Chanson J, Hoffmann M. Global indicators of biological invasion: species numbers, biodiversity impact and policy responses. Diversity and Distributions. 2010;**16**:95-108

[131] Pejchar L, Mooney HA. Invasive species, ecosystem services and human well-being. Trends in Ecology and Evolution. 2009;**24**:495-504

[132] Rodríguez LF. Can invasive species facilitate native species? Evidence of how, when, and why these impacts occur. Biological Invasions. 2006;**8**:927-939

[133] Schlaepfer MA, Sax DF, Olden JD. The potential conservation value of nonnative species. Conservation Biology. 2011;**25**:428-437

[134] Pienkowski T, Williams S, McLaren K, Wilson B, Hockley N. Alien invasions and livelihoods: Economic benefits of invasive Australian red claw crayfish in Jamaica. Ecological Economics. 2015;**112**:68-77

[135] Sax DF. Native and naturalized plant diversity are positively correlated in shrub communities of California and Chile. Diversity and Distributions. 2002;**8**:193-210

[136] Guo Q. No consistent small-scale native–exotic relationships. Plant Ecology. 2015;**216**: 1225-1230

[137] Seabloom EW, Harpole WS, Reichman OJ, Tilman D. Invasion, competitive dominance, and resource use by exotic and native California grassland species. Proceedings of the National Academy of Sciences. 2003;**100**:13384-13389

[138] Aguiar FC, Ferreira MT, Albuquerque A. Patterns of exotic and native plant species richness and cover along a semi-arid Iberian river and across its floodplain. Plant Ecology. 2006;**184**:189-202

[139] Çinar ME, Katagan T, Öztürk B, Dagli E, Açik S, Bitlis B, Bakir K, Dogan A. Spatio-temporal distributions of zoobenthos in Mersin Bay (Levantine Sea, eastern Mediterranean) and the importance of alien species in benthic communities. Marine Biology Research. 2012;**8**:954-968

[140] Martín-Forés I, Guerin GR, Lowe AJ. Weed abundance is positively correlated with native plant diversity in grasslands of southern Australia. PlosOne (accepted)

[141] Bartomeus I, Sol D, Pino J, Vicente P, Font X. Deconstructing the native-exotic richness relationship in plants. Global Ecology and Biogeography. 2012;**21**:524-533

[142] Kulmatiski A. Exotic plants establish persistent communities. Plant Ecology. 2006;**187**: 261-275

[143] Vilà M, Espinar JL, Hejda M, Hulme PE, Jarošík V, Maron JL, Perlg J, Schaffner U, Sun Y, Pyšek P. Ecological impacts of invasive alien plants: A metaanalysis of their effects on species, communities and ecosystems. Ecology Letters. 2011;**14**:702-708

[144] Tognetti PM, Chaneton EJ. Invasive exotic grasses and seed arrival limit native species establishment in an old-field grassland succession. Biological Invasions. 2012;**14**: 2531-2544

[145] Kimball S, Goulden ML, Suding KN, Parker S. Altered water and nitrogen input shifts succession in a southern California coastal sage community. Ecological Applications. 2014;**24**:1390-1404

[146] MacDougall AS, Turkington R. Are invasive species the drivers or passengers of change in degraded ecosystems? Ecology. 2005;**86**:42-55

[147] Stinca A, Chirico GB, Incerti G, Bonanomi G. Regime shift by an exotic nitrogen-fixing shrub mediates plant facilitation in primary succession. Plos One. 2005;**10**:e0123128

[148] Patel S. A weed with multiple utility: Lantana camara. Reviews in Environmental Science and Bio/Technology. 2011;**10**:341-351

[149] Nielsen JA, Frew RD, Whigam PA, Callaway RM, Dickinson KJM. Germination and growth responses of co-occurring grass species to soil from under invasive *Thymus vulgaris*. Allelopathy Journal. 2015;**35**:139-152

[150] Ferrero V, Castro S, Costa J, Acuna P, Navarro L, Loureiro J. Effect of invader removal: Pollinators stay but some native plants miss their new friend. Biological invasions. 2013;**15**:2347-2358

[151] Bartomeus I, Vilà M, Santamaría L. Contrasting effects of invasive plants in plant–pollinator networks. Oecologia. 2008;**155**:761-770

[152] Castro SA, Bozinovic F, Jaksic FM. Ecological efficiency and legitimacy in seed dispersal of an endemic shrub (*Lithrea caustica*) by the European rabbit (*Oryctolagus cuniculus*) in central Chile. Journal of Arid Environments. 2008;**72**:1164-1173

[153] Wolkovich EM, Bolger DT, Cottingham KL. Invasive grass litter facilitates native shrubs through abiotic effects. Journal of Vegetation Science. 2009;**20**:1121-1132

[154] Van Riel P, Jordaens K, Martins AMF, Backeljau T. Eradication of exotic species. Trends in Ecology and Evolution. 2000;**15**:515

[155] Courchamp F, Caut S, Bonnaud E, Bourgeois K, Angulo E, Watari Y. Eradication of alien invasive species: surprise effects and conservation successes. In: Veitch CR, Clout MN, Towns DR, editors. Island Invasives: Eradication and Management. Gland, Switzerland: IUCN; 2011. pp. 285-289

[156] Jeddi K, Chaieb M. Restoring degraded arid Mediterranean areas with exotic tree species: Influence of an age sequence of *Acacia salicina* on soil and vegetation dynamics. Flora-Morphology, Distribution, Functional Ecology of Plants. 2012;**207**:693-700

[157] Bussotti F, Pollastrini M, Holland V, Brüggemann W. Functional traits and adaptive capacity of European forests to climate change. Environmental and Experimental Botany. 2015;**111**:91-113

[158] Davis MA. Biotic globalization: does competition from introduced species threaten biodiversity? Bioscience. 2003;**53**:481-489

[159] Colautti RI, MacIsaac HJ. A neutral terminology to define 'invasive'species. Diversity and Distributions. 2004;**10**:135-141

[160] García-Llorente M, Martín-López B, González JA, Alcorlo P, Montes C. Social perceptions of the impacts and benefits of invasive alien species: Implications for management. Biological Conservation. 2008;**141**:2969-2983

[161] Shackleton S, Kirby D, Gambiza J. Invasive plants – friends or foes? Contribution of prickly pear (*Opuntia ficus-indica*) to livelihoods in Makana Municipality, Eastern Cape, South Africa. Development Southern Africa. 2011;**28**:177-193

[162] Schlaepfer MA, Sax DF, Olden JD. Toward a more balanced view of non-native species. Conservation Biology. 2012;**26**:1153-1158

[163] Carpenter S, Mooney HA, Agard J, Capistrano D, DeFries RS, Díaz S, Dietz T, Duraiappah AK, Oteng-Yeboah A, Pereira EM, Perrings C, Rield WV, Sarukhan J, Schies RJ, Whyte A. Science for managing ecosystem services: Beyond the millennium assessment. Proceedings of the National Academy of Science of the United States if America. 2009;**106**:1305-1312

[164] Rockstrom J, Steffen W, Noone K, Persson Å, Chapin I FS, Lambin EF, Lenton TM Scheffer M, Folke C, Schellnhuber HJ, Nykvist B, De Wit CA, Huges T, Van Der Leeuw S,

Rodhe H, Sörlin S, Snyder PK, Costanza R, Svedin U, Falkenmark M, Karlberg L, Corell RW, Fabry VJ, Hansen J, Walker B, Liverman D, Richardson K, Crutzen P, Foley JA. A safe operating space for humanity. Nature. 2009;**461**:472-475

[165] Strayer DL, Eviner VT, Jeschke JM, Pace ML. Understanding the long-term effects of species invasions. Trends in Ecology and Evolution. 2006;**21**:645-651

[166] Seastedt TR, Hobbs RJ, Suding KN. Management of novel ecosystems: Are novel approaches required? Frontiers in Ecology and the Environment. 2008;**6**:547-553

[167] Bennett EM, Cramer W, Begossi A, Cundill G, Díaz S, Egoh BN, Geijzendorffer IR, Krug CB, Lavorel S, Lazos E, Lebel L, Martín-López B, Meyfroidt P, Mooney HA, Nel JL, Pascual U, Payet K, Harguindeguy NP, Peterson GD, Prieur-Richard AH, Reyers B, Roebeling P, Seppelt R, Solan M, Tschakert P, Tscharntke T, Turner II BL, Verburg PH, Viglizzo EF, White PCL, Woodward G. Linking biodiversity, ecosystem services, and human well-being: Three challenges for designing research for sustainability. Current Opinion in Environmental Sustainability. 2015;**14**:76-85

Chapter 9

Amphibians and Reptiles of the Mediterranean Basin

Kerim Çiçek and Oğzukan Cumhuriyet

Additional information is available at the end of the chapter

http://dx.doi.org/10.5772/intechopen.70357

Abstract

The Mediterranean basin is one of the most geologically, biologically, and culturally complex region and the only case of a large sea surrounded by three continents. The chapter is focused on a diversity of Mediterranean amphibians and reptiles, discussing major threats to the species and its conservation status. There are 117 amphibians, of which 80 (68%) are endemic and 398 reptiles, of which 216 (54%) are endemic distributed throughout the Basin. While the species diversity increases in the north and west for amphibians, the reptile diversity increases from north to south and from west to east direction. Amphibians are almost twice as threatened (29%) as reptiles (14%). Habitat loss and degradation, pollution, invasive/alien species, unsustainable use, and persecution are major threats to the species. The important conservation actions should be directed to sustainable management measures and legal protection of endangered species and their habitats, all for the future of Mediterranean biodiversity.

Keywords: amphibians, conservation, Mediterranean basin, reptiles, threatened species

1. Introduction

The Mediterranean basin is one of the most geologically, biologically, and culturally complex region and the only case of a large sea surrounded by Europe, Asia and Africa. The Basin was shaped by the collision of the northward-moving African-Arabian continental plate with the Eurasian continental plate which occurred on a wide range of scales and time in the course of the past 250 mya [1].

The Basin stretches approx. 3800 km east to west from the tip of Portugal to the shores of Lebanon and 1000 km north to south from Italy to Morocco and Libya (**Figure 1**) [1, 2]. It covers the area surrounding the Mediterranean Sea, and includes partly or entirely 30 countries which are spread across 3 continents. It also includes 11,879 islands and islets [3].

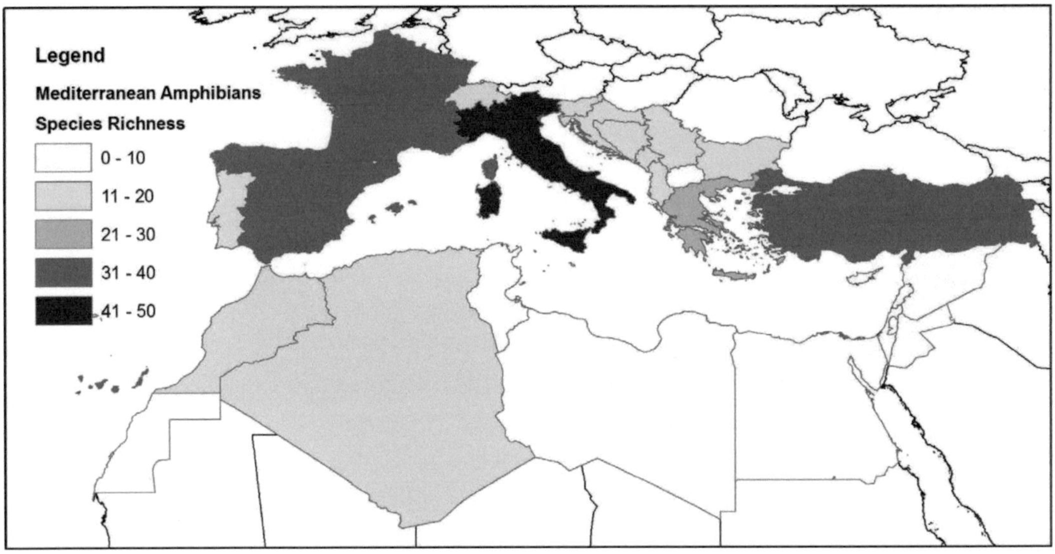

Figure 1. The amphibian richness of Mediterranean basin.

The Mediterranean region is considered to be 1 of 34 biodiversity hotspots due to its high level of floristic endemism [4] as well as the largest of the world's 5 Mediterranean-climate regions. The region flora includes more than 25,000 vascular plants while half of them are endemic [1, 2]—in other words, they are found nowhere else in the world.

The geographic structure of the Basin is an important factor in understanding its biodiversity. While coastal areas are extensive due to the presence of numerous archipelagos and islands, much of the area consists of mountainous terrain with many areas above 2000 m elevation and peaks as high as 4500 m [2, 3, 5]. The Mediterranean region consists of various landscapes such as high mountains, rocky shores, impenetrable scrub, semi-arid steppes, coastal wetlands, sandy beaches, and myriad islands of various shapes and sizes [1, 2].

The status and distribution of Mediterranean herptiles has been evaluated by Cox et al. [5] 9 years ago. The purpose of this chapter is to re-evaluate amphibian and reptile diversity and to discuss the major threats and conservation status of Mediterranean herptiles. The Amphibia Web [6] and The Reptile Database [7] were used for determining Mediterranean herptile list. Major threats and conservation status of species for the IUCN Red List of threatened species [8] are also addressed.

2. Amphibian and reptiles diversity

Amphibians (Amphibia) and reptiles (Reptilia) are two fascinating but poorly understood group of vertebrates, distributed around the world. For the time being, there are 7655 amphibian [6] and 10,450 reptilian [7] species recorded. Unfortunately, many amphibian and reptile species are threatened and declining all-around the world. Habitat loss and degradation, introduced invasive species, environmental pollution, disease and parasitism, unsustainable use, and global climate change are major threats on species [6, 7]. There are 117 amphibian species and 398 reptile species, and most of them are endemic distributed throughout the Basin (**Table 1**).

Order	Family	No. of species	No. of endemic species
Amphibians			
Caudata (newts and salamanders)	Plethodontidae	8	8 (100%)
Caudata	Proteidae	1	1 (100%)
Caudata	Salamandridae	40	23 (58%)
Total—Newts and salamanders		49	32 (65%)
Anura (frogs and toads)	Alytidae	12	11 (92%)
Anura	Bombinatoridae	3	1 (33%)
Anura	Bufonidae	12	7 (59%)
Anura	Dicroglossidae	1	0 (0%)
Anura	Hylidae	8	5 (63%)
Anura	Pelobatidae	3	2 (50%)
Anura	Pelodytidae	2	2 (100%)
Anura	Ranidae	27	20 (74%)
Total—Frogs and Toads		68	48 (70%)
Total—Amphibians		117	80 (68%)
Reptiles			
Testudines (turtles and tortoises)	Cheloniidae	3	0 (0%)
Testudines	Dermochelyidae	1	0 (0%)
Testudines	Emydidae	3	0 (0%)
Testudines	Geoemydidae	3	2 (66%)
Testudines	Testudinidae	4	3 (75%)
Testudines	Trionychidae	2	0 (0%)
Total—Turtles and Tortoises		16	5 (31%)
Sauria (lizards)	Agamidae	23	10 (43%)
Sauria	Anguidae	5	4 (80%)
Sauria	Blanidae	3	2 (66%)
Sauria	Chamaeleonidae	2	0 (0%)
Sauria	Eublepharidae	1	0 (0%)
Sauria	Gekkonidae	51	26 (51%)
Sauria	Lacertidae	132	86 (65%)

Order	Family	No. of species	No. of endemic species
Sauria	Phyllodactylidae	7	3 (42%)
Sauria	Scincidae	36	25 (70%)
Sauria	Varanidae	2	0 (0%)
Total—Lizards		262	160 (60%)
Ophidia (snakes)	Atractaspididae	3	2 (66%)
Ophidia	Boidae	2	0 (0%)
Ophidia	Colubridae	65	27 (42%)
Ophidia	Elapidae	5	0 (0%)
Ophidia	Lamprophiidae	1	0 (0%)
Ophidia	Leptotyphlopidae	3	0 (0%)
Ophidia	Natricidae	3	1 (33%)
Ophidia	Typhlopidae	3	2 (66%)
Ophidia	Viperidae	29	15 (52%)
Total—Snakes		114	47 (41%)
Amphisbaenia (amphisbaenians)	Amphisbaenidae	4	3 (75%)
Amphisbaenia	Trogonophiidae	1	1 (100%)
Total—Amphisbaenians		5	4 (80%)
Crocodylia (crocodilians)	Crocodylidae	1	0 (0%)
Total—Crocodilians		1	0 (0%)
Total—Reptiles		398	216 (54%)

Table 1. The amphibian and reptile diversity and endemism of Mediterranean basin.

2.1. Amphibian diversity

The amphibian fauna of the Mediterranean basin represents two orders: salamanders (Caudata) and anurans (Anura). A total of 117 amphibian species are found and 80 (68%) of them are endemic in the Basin (**Table 2**, **Figure 1**).

A total of 49 salamander species are present in this Region and 65% of them are endemic. The Salamandridae is the most diverse family. A total of 18 species with 7 genera (*Calotriton, Chioglossa, Euproctus, Ichthyosaura, Lyciasalamandra, Pleurodeles* and *Salamandrina*) are endemic to the Basin (**Table 2**). The only single member of Proteidae, *Proteus anguinus,* is present in the Balkan Peninsula and is endemic to the Basin. The other six members of the family are found in eastern North America.

The anurans have 68 species and 70% of them are endemic to the Region. The families Alytidae, Bufonidae and Ranidae consist of 75% of the group. A fascinating species of midwife

http://dx.doi.org/10.5772/intechopen.70357

	Number of amphibians				Number of reptiles						
	Urodela	Anura	Total	Endemics	Amphisbaenia	Crocodilia	Testudines	Sauria	Serpentes	Total	Endemics
Albania	5	10	15	5	0	0	6	13	17	36	11
Algeria	9	4	13	12	1	0	4	70	32	107	50
Andorra	2	2	4	3	0	0	0	8	0	8	4
Bosnia and Herzegovina	8	8	16	2	0	0	3	6	8	17	5
Bulgaria	6	12	17	1	0	0	5	11	19	35	7
Canary Islands	0	2	2	2	0	0	0	16	1	17	15
Croatia	7	9	16	3	0	0	6	15	17	38	11
Cyprus	0	3	3	1	0	0	5	12	12	29	7
Egypt	0	8	8	1	0	1	7	65	37	110	23
France	13	26	39	20	0	0	9	19	14	42	22
Greece	7	17	24	8	1	0	9	33	24	67	29
Israel/Palestine	2	8	10	5	1	0	9	43	46	99	29
Italy	15	28	43	27	0	0	9	26	23	58	29
Jordan	1	3	4	2	0	0	7	61	44	112	27
Lebanon	2	4	6	2	2	0	5	26	26	59	19
Libyan Arab Jamahiriya	0	2	2	2	0	0	4	47	25	76	19
Macedonia	4	6	10	1	0	0	5	12	16	33	7
Malta	0	2	2	2	0	0	1	5	4	10	5
Monaco	1	1	2	2	0	0	2	1	0	3	0
Montenegro	4	11	15	3	0	0	4	10	6	20	8
Morocco	2	11	13	12	4	0	4	70	30	108	58
Portugal	7	13	20	14	2	0	7	17	12	38	26

	Number of amphibians				Number of reptiles						
	Urodela	Anura	Total	Endemics	Amphisbaenia	Crocodilia	Testudines	Sauria	Serpentes	Total	Endemics
Serbia	7	12	19	1	0	0	4	6	8	18	4
Slovenia	4	12	16	2	0	0	3	4	14	21	5
Spain	8	28	36	23	2	0	6	45	14	67	45
Switzerland	5	12	17	1	0	0	2	6	8	16	4
Syrian Arab Republic	2	3	5	2	2	0	7	46	40	95	22
Tunisia	1	4	5	4	1	0	4	44	24	73	34
Turkey (except for NE Anatolia)	15	13	28	12	3	0	11	56	49	119	39
Western Sahara	0	2	2	1	0	0	4	39	20	63	16

Table 2. The number of amphibians and reptiles in the Mediterranean countries.

toads (*Alytes*) have five species which are found across western Europe, northern Africa and Majorca.

The amphibian diversity is highest in Europe, especially in areas of higher rainfall, notably in northern Italy, France, western and northern Spain, Portugal, Slovenia and Croatia (**Figure 1**, **Table 3**) [5]. On the contrary, the diversity is much lower in the eastern and southern parts of the Basin where there are large arid and semiarid habitats. The higher amphibian diversity is observed in European countries of the western Mediterranean, especially in Italy, France and Spain [5]. The amphibian richness increases from south to north and from east to west of the Basin [1]. The reason lies in larger areas of humid habitats in the north and west of the Basin, which are an ideal habitat for amphibians.

Country	EX	CR	EN	VU	NT	LC	DD	NE
Albania	0	0	1	1	0	12	0	1
Algeria	0	0	1	2	2	9	0	0
Andorra	0	0	0	0	1	3	0	0
Bosnia and Herzegovina	0	0	0	1	1	14	0	0
Bulgaria	0	0	0	0	1	16	0	1
Canary Islands	0	0	0	0	0	2	0	0
Croatia	0	0	0	2	1	13	0	0
Cyprus	0	0	0	0	0	3	0	0
Egypt	0	0	0	0	0	8	0	1
France	0	0	1	2	4	32	0	3
Greece	0	1	1	3	0	18	1	0
Israel/Palestine	0	3	0	0	1	5	1	1
Italy	0	0	3	6	4	30	0	5
Jordan	0	0	0	0	0	2	1	2
Lebanon	0	0	0	0	1	4	1	1
Libyan Arab Jamahiriya	0	0	0	0	0	2	0	1
Macedonia	0	0	0	0	0	10	0	1
Malta	0	0	0	0	0	2	0	0
Monaco	0	0	0	0	1	1	0	0
Montenegro	0	0	0	0	0	14	0	2
Morocco	0	0	1	1	3	8	0	1
Portugal	0	0	0	1	5	13	0	2
Serbia	0	0	0	0	0	17	0	3
Slovenia	0	0	0	2	0	14	0	1

Country	EX	CR	EN	VU	NT	LC	DD	NE
Spain	0	1	1	3	7	22	0	4
Switzerland	0	0	0	0	0	17	0	3
Syrian Arab Republic	0	0	0	0	1	4	0	2
Tunisia	0	0	0	1	0	4	0	0
Turkey (except for NE Anatolia)	0	2	5	2	2	12	1	5
Western Sahara	0	0	0	0	1	1	0	0

Table 3. The conservation status of amphibians in Mediterranean countries.

2.2. Reptile diversity

The reptiles of the region represent five orders: Crocodylia (crocodilians), Testudines (turtles and tortoises), Amphisbaenia (amphisbaenians), Sauria (lizards) and Ophidia (snakes). The great majority of the species are lizards (262 species, 66%) and snakes (114 species, 29%) (**Figure 2**, **Table 4**). About 54% of the reptiles are endemic to the Basin. The most diverse families are Lacertidae (132 species), Gekkonidae (51 species) and Scincidae (36 species) for lizards; and Colubridae (65 species) and Viperidae (29 species) for snakes.

The reptile diversity is the highest in the eastern part of the Basin, particularly in southern Turkey, Lebanon, south-western Syria, Israel/Palestine, Jordan and parts of northern Egypt [5].

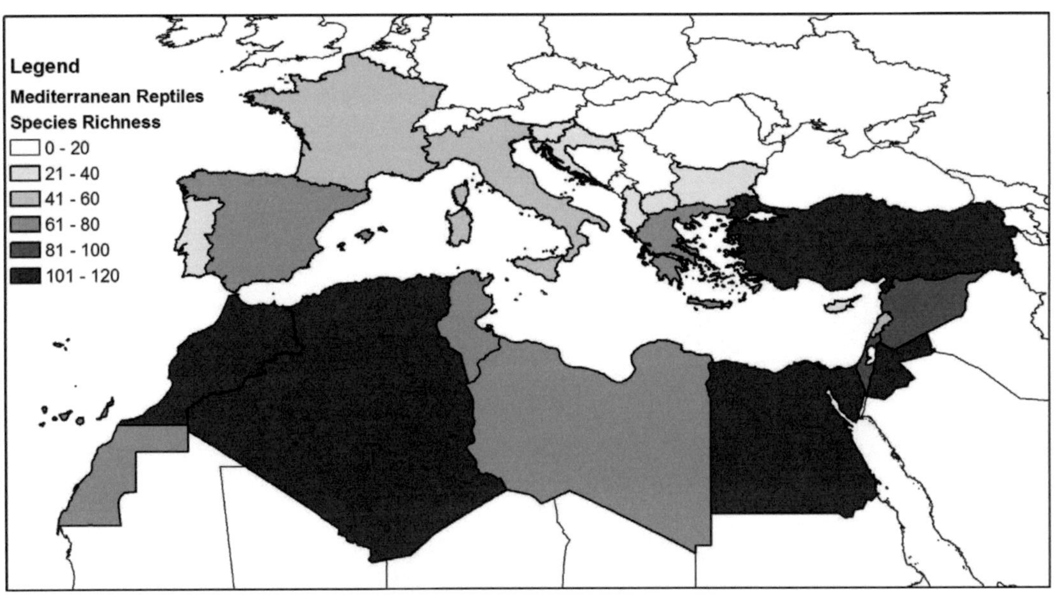

Figure 2. The reptile richness in Mediterranean basin.

http://dx.doi.org/10.5772/intechopen.70357

Country	EX	CR	EN	VU	NT	LC	DD	NE
Albania	0	0	0	3	3	27	0	4
Algeria	0	1	4	2	11	75	6	9
Andorra	0	0	1	0	1	5	0	1
Bosnia and Herzegovina	0	0	0	1	2	13	0	2
Bulgaria	0	0	0	1	4	28	0	5
Canary Islands	0	3	0	0	0	14	0	1
Croatia	0	0	1	3	4	29	0	2
Cyprus	0	0	3	1	1	22	0	2
Egypt	0	2	2	3	4	80	3	16
France	0	1	1	4	6	26	0	5
Greece	0	0	2	4	8	48	0	8
Israel/Palestine	0	4	5	4	2	76	1	7
Italy	0	1	1	4	6	40	2	5
Jordan	0	1	4	4	1	88	1	13
Lebanon	0	0	4	2	0	42	1	10
Libyan Arab Jamahiriya	0	1	1	2	4	39	1	28
Macedonia	0	0	0	1	3	27	0	3
Malta	0	0	0	1	0	9	0	0
Monaco	0	0	0	0	1	2	0	0
Montenegro	0	0	0	3	2	13	0	2
Morocco	0	1	3	4	13	74	4	9
Portugal	0	1	2	3	6	21	0	7
Serbia	0	0	0	1	4	13	0	2
Slovenia	0	0	0	2	2	17	0	1
Spain	0	2	5	2	10	40	0	8
Switzerland	0	0	0	0	1	14	0	1
Syrian Arab Republic	0	0	3	2	3	78	1	11
Tunisia	0	0	1	1	8	75	2	14
Turkey (except for NE Anatolia)	0	1	5	2	5	91	3	38
Western Sahara	0	1	1	1	3	49	4	4

Table 4. The conservation status of reptiles in Mediterranean countries.

The species diversity is much higher in North Africa than in western Europe. The reptile diversity of North Africa is the highest in the mountainous area, in semi-arid regions along the northern margins of the Sahara and in the Nile Valley. The Balkans has much higher reptile diversity than elsewhere in Europe. At the other hand, the diversity is very low in northern Europe [5]. In contrast to amphibians, the species diversity of the reptiles increases from north to south and from west to east, along with gradients of the extent to which arid and semi-arid habitats are present [1, 5].

3. Conservation status

3.1. Conservation status of Mediterranean amphibians

About 29% of Mediterranean amphibians are globally threatened, while 5% are critically endangered, 11% endangered and 13% vulnerable (**Figures 3–5**). Rest of the species are evaluated as near threatened (15%), least concerned (49%), data deficient (<1%) and 7% is not evaluated. The salamanders and newts have higher share of threatened species (20 species, 17%). Among frogs and toads, 13 species (11%) are globally threatened. One of the endangered species is the Hula painted frog, *Latonia nigriventer,* from Israel/Palestine that is listed as extinct up to 2004. The species is restricted to an area under 2 km^2 due to heavy predation pressure by the waterbird populations [13]. The newts and salamanders have higher number of threatened species than frogs and toads (**Table 5**) [5].

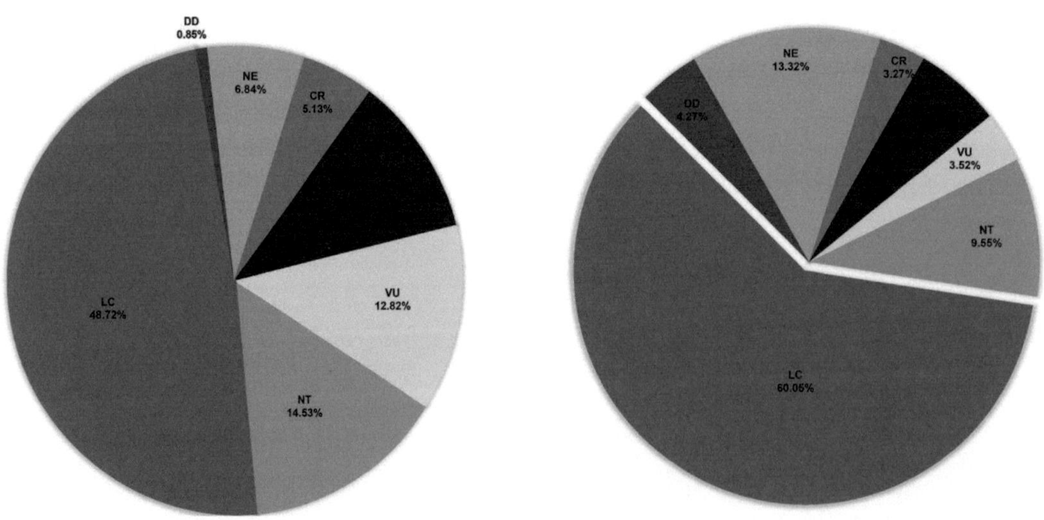

Figure 3. Summary of conservation status for Mediterranean amphibians (left) and reptiles (right). EX: extinct, EW: extinct in the wild, CR: critically endangered, EN: endangered, VU: vulnerable, NT: near threatened, LC: least concern, DD: data deficient, NE: not evaluated.

http://dx.doi.org/10.5772/intechopen.70357

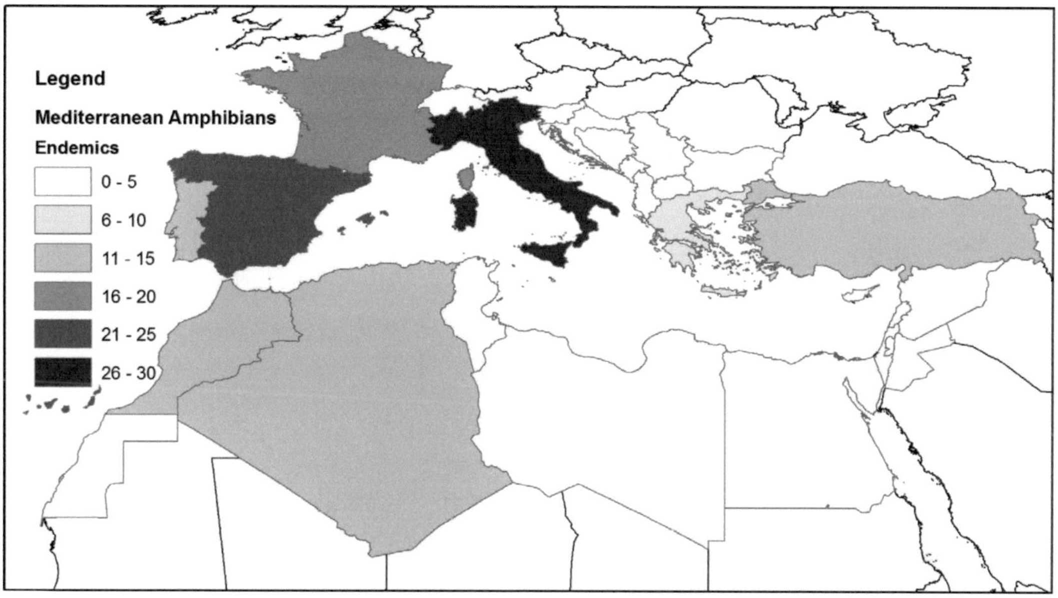

Figure 4. The species richness of endemic amphibians in the Mediterranean basin.

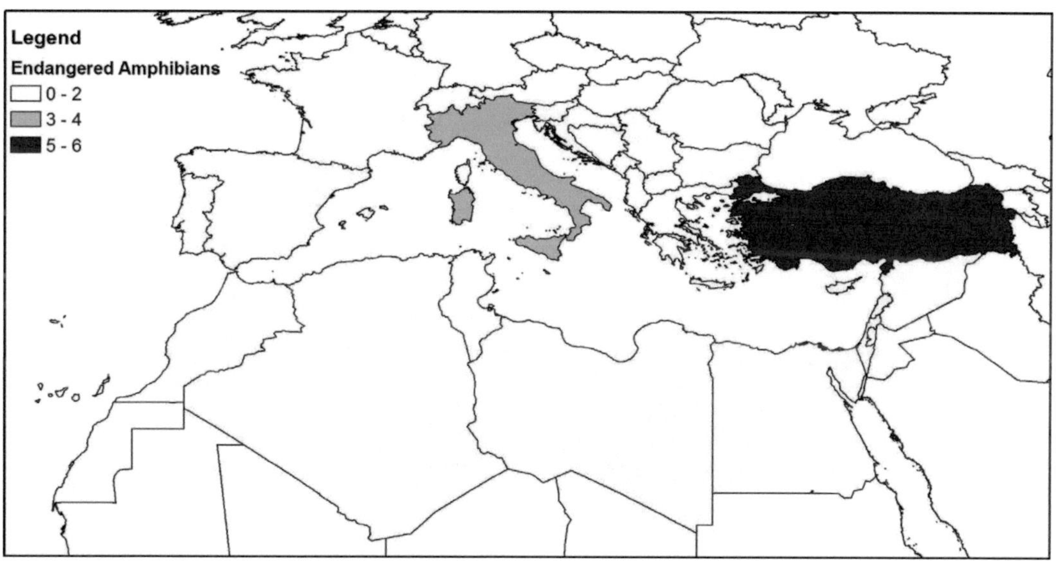

Figure 5. The species richness of threatened amphibians in the Mediterranean basin.

3.2. Conservation status of Mediterranean reptiles

About 13% of Mediterranean reptiles are globally threatened (51 species), out of which 3% is critically endangered, 6% endangered and 4% vulnerable. A total of 347 species are assessed

IUCN Red List categories	Caudata	Anura	Total	Crocodilia	Testudines	Amphisbaenia	Sauria	Ophidia	Total
Extinct									
Critically endangered	2	4	7	0	2	0	11	0	13
Endangered	7	6	13	0	2	0	16	6	26
Vulnerable	11	4	15	0	2	0	11	1	15
Near threatened	9	8	16	0	2	0	28	8	39
Least concern	16	41	57	1	7	5	155	71	242
Data deficient	0	1	1	0	1	0	8	8	18
Not evaluated	4	4	8	0	0	0	33	20	56
Total	49	68	117	1	16	5	262	114	398
Endemic	37 (76%)	48 (71%)	83 (63%)	0(%)	5 (31%)	4 (80%)	158 (60%)	47 (41%)	214 (54%)

Table 5. The IUCN red List categories and endemis of amphibians and reptiles in Mediterranean countries.

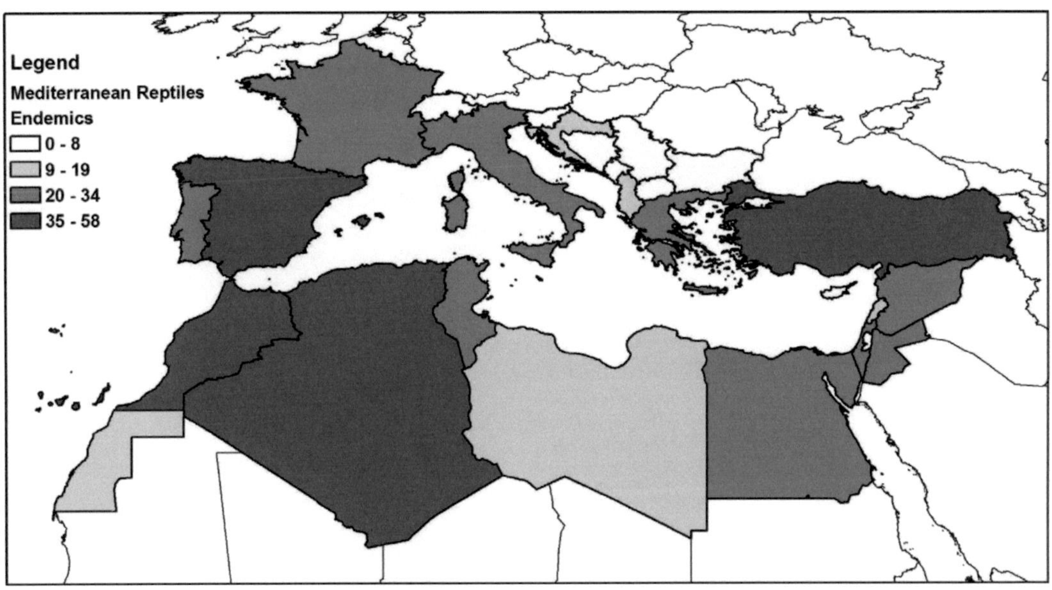

Figure 6. The species richness of endemic reptiles in the Mediterranean basin.

as near threatened (10%), least concerned (60%), data deficient (4%) and 13% is not evaluated (**Figures 3**, **6** and **7**).

Amphisbaenians and crocodilians are not considered threatened species in the Region. At the other side, the chelonians have six threatened species. Among the lizards, there are

http://dx.doi.org/10.5772/intechopen.70357

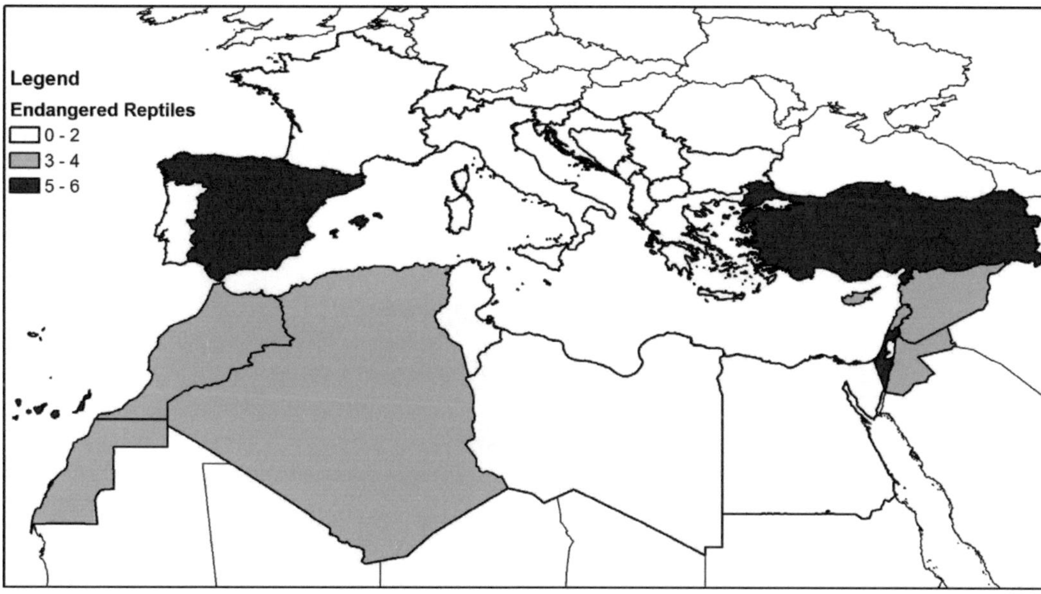

Figure 7. The species richness of threatened reptiles in the Mediterranean basin.

38 species considered threatened. Snakes have only seven threatened species (2%). The endemic lizard genus *Gallotia* occurs only on the Canary Islands and consist of eight species. The genus has evolved there almost 20 mya, ever since the first islands emerged from the sea [11, 12]. They are adapted to eating significant quantities of plants. The overall share of threatened amphibians in the Mediterranean basin is as twice higher (29%) than that for reptiles (14%).

4. Major threats

The Mediterranean basin is the second largest biodiversity hotspot in the world. It covers more than 2 million km^2 . The Basin stretches west to east from Portugal to Lebanon and north to south from Italy to Morocco and Libya [2]. The Region is home to approx. 455 million people, from a wide variety of countries and cultures for some 8000 years [2, 10] The Gross National Income per capita in the Mediterranean EU countries being 10 times that of the north African ones [10]. The poor countries mostly depend on natural resources and this threatens natural resources at high levels. Besides, economic development increases the pressures on natural resources, the conservation challenges and options of the Basin are driven by these economic inequalities [10]. Species provide us with essential services as not only food, fuel, clothes and medicine, but also purification of water and air, prevention of soil erosion, regulation of climate, pollination of crops by insects and much more [10]. Many threats come up thanks to these entries. The human-induced factors threaten the Mediterranean biodiversity and nature more than any other biological 'hotspot' [10].

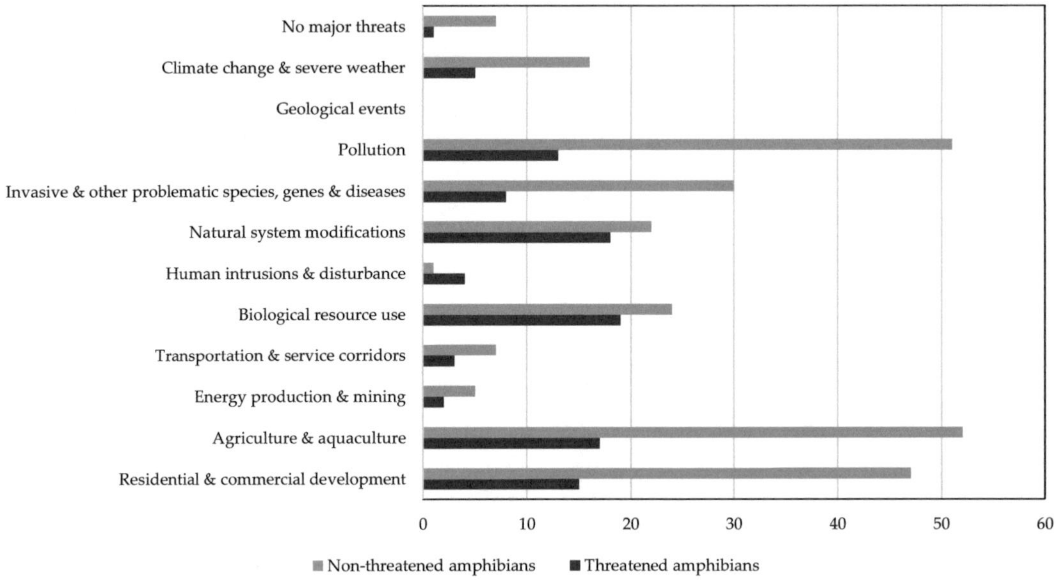

Figure 8. The major threats to Mediterranean amphibians.

Fifty-three percent of amphibians and 20% of reptiles are suffering from "residential & commercial developments" (**Figure 8**). Urbanization, industrial areas, tourism and recreation areas negatively affected the herptile species. Another important factor is "agriculture and aquaculture" activities and almost half of the species (59% for amphibians and 25% for reptiles) are affected by such activities. Along with the increase of human population, the food needs are also increasing day by day. The expansion of agricultural areas, livestock farming, overgrazing, aquaculture and mariculture activities are causing habitat loss and degradation and intervening in the food webs.

One of the biggest contemporary concerns is the growing need for energy as well as the need for nutrients and technological developments. The most basic resource used to meet the growing energy needs is still natural resource. Among natural resources, fossil fuels and mines are used most commonly. Use of renewable energy sources as alternative energy sources are not reached desired level. The "energy production and mining" activities have low effect on Mediterranean herptiles (6% for amphibians and 5% for reptiles).

Another problem brought by urbanization and population increase is in the construction of roads, especially narrow transport corridors cause wildlife mortality. Besides, these corridors create specific stress to biodiversity by fragmentation of the habitats and lead to other threats including farms, invasive species and poachers. The "transportation and service corridors" activities have low effect on amphibians (9%) and reptiles (6%).

Unsustainable harvesting, hunting and fishing activities are directly or indirectly affecting the amphibians and reptiles. Some species are used in traditional medicine, food and pet trade.

The threats are driven by destroying or declining natural populations [10]. The amphibians (37%) and reptiles (22%) are densely used as "biological resource use" for many purposes. Almost half of salamanders and snakes are suffering from commercial purpose and persecution. The Mediterranean marine turtle species are severely affected by accidental capture in fishing gear, also called as "bycatch" [10].

Besides, "human intrusions and disturbance" have low pressure on amphibians (4%) and reptiles (4%). While "natural system modifications" severely affect the amphibians (32%), it has low pressure on reptiles (9%). The dam construction, for water management or use, and other ecosystem modifications make significant pressure on natural herptile populations. Forest fires are deliberately excluded to open such areas, especially in the Mediterranean region in Turkey, it is observed that these activities have been carried out in the summer. The endemic *Lyciasalamandra* species living in this Region are highly affected by forest fires. In addition to the destruction of the area for the construction of the dams, the alteration of the water flow direction of the rivers disturbs the natural habitat areas, especially amphibians are highly affected due to degradation and reducing habitat quality.

Non-indigenous animal species, pathogens and genes are appearing as major threatening factors to biodiversity being the process that is expected to continue in the future. Mediterranean amphibians (34%) are more affected by "invasive and/or problematic species, pathogens, and genes" than reptiles (4%). The American bullfrog (*Lithobates catesbeianus* or *Rana catesbeiana*) is one of the invasive species in western Europe. Another invasive species, *Trachemys scripta,* is popular in the pet trade and has been introduced into the Mediterranean region by people releasing it to the wild.

The amphibians (59%) are more sensitive to "pollution" than reptiles (4%). Many chemical pollutants are increasing sensitivity to illness and mortality rates and reducing the reproductive success [10]. Domestic/industrial waste carries pollution to the sea and rural areas through rivers and sewage systems, in particular. Pollutants that cause water pollution from agricultural, silvicultural and aquaculture systems containing foodstuffs, toxic chemicals and sediments also pollute natural habitats as well as agricultural areas. Apart from these pollutants, trash and soil pollutants and even atmospheric pollutants are serious threat to species.

Today, global "climate change" emerges as a factor that affects the changing nature of natural habitats. Temperature fluctuations (changing in temperature extremes, increasing average summer temperatures and reducing winter/spring temperatures) cause the alteration of habitats, breeding phenology and host-parasite relationship of herptile species. Mediterranean amphibians (18%) are more affected by global climate change than reptiles (3%).

Mediterranean amphibians and reptiles are affected by these major threats (habitat loss and degradation, invasive alien species, harvesting, pollution natural disasters, disease, human disturbance, vehicle collusion and persecution) (**Figures 8** and **9**) [5]. While the most common threats for amphibians are habitat loss and degradation, pollution and invasive alien species,

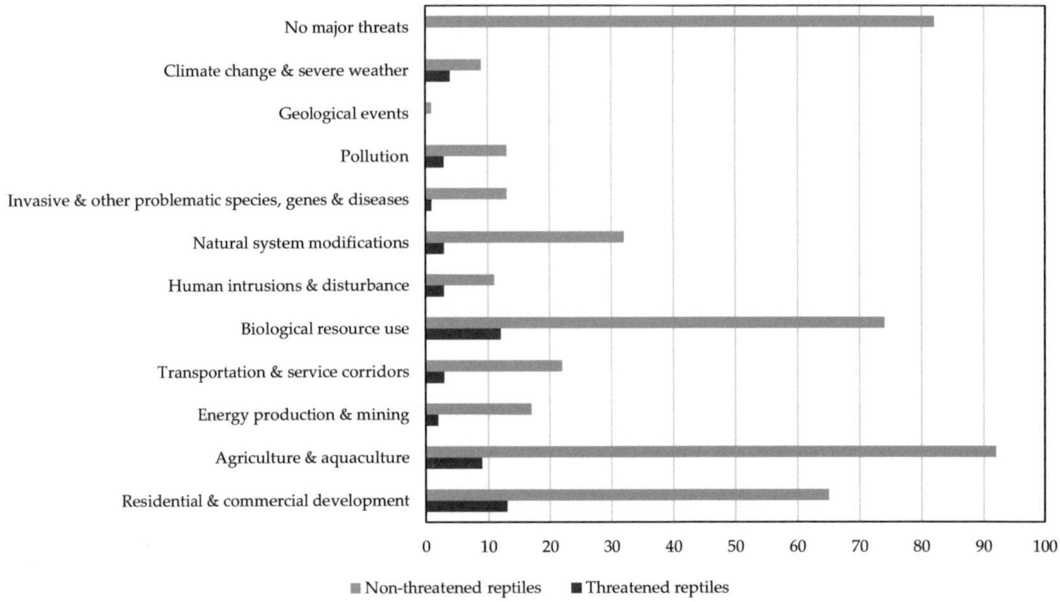

Figure 9. The major threats on to Mediterranean reptiles.

the most common ones for Mediterranean reptiles are habitat loss and degradation, harvesting and persecution [5]. On the other hand, there is no major threat for about 10% amphibians and 21% reptiles in the Mediterranean.

5. Conservation

The major threats to amphibians and reptiles in the Mediterranean are quite different from each other [5]. Therefore, each group needs specific conservation activities. Island species particularly need urgent conservation studies. Although amphibians (especially salamanders) have a high tendency to be threatened, and reptiles much less so, there are many more reptile species on the edge of extinction in the Region than amphibians [5, 9]. Several methods can be applied by scientists in order to protect species. Land/water protection and management, species management, education and raising awareness, and monitoring and research are major actions for Mediterranean herptiles [10, 14].

5.1. Land/water protection and management

The Mediterranean region is densely populated and more than 30% of all international tourists visit its coastal areas [15], thus direct disturbance by humans is an important threat to natural resources [10]. The Region is also considered as the cradle of Europe's civilization and one of the most important centers of crop plants origin [2]. However, the traditional farming practices have been abandoned in recent years in favor of intensive and industrial-scale farming methods [2].

The area conservation and management are important for endemic and threatened species with high risk status. "Land/water management" include many different types of actions such as conserving or restoring habitats and controlling invasive/problematic species. The tourism, urbanization, deforestation, intensive farming, overgrazing and fires are causing habitat loss for many threatened species. Therefore, site protection and management has crucial importance for sustainability of the threatened amphibians and reptiles.

5.2. Species protection and management

Improvement and enforcement of legal protection for threatened species and their habitats is the most urgent conservation action to be taken at both regional and national levels [5]. Species Action Plans can be an effective means for determining the specific conservation actions that are needed and for promoting coordinated activities. The primary goal of species conservation is the preservation of viable populations of wild species in their original native range [10]. Another solution could be captive breeding studies for endangered species close to extinction as part of intensive management activities. Besides, measures to be taken in conjunction with legal regulations are essential for the sustainability of protected areas. All countries should have endangered species red list database along with IUCN Red List to determine conservation priorities.

5.3. Education and awareness raising activities

There is no way of protecting a species or effective conservation without support of local people. The education and raising awareness have important role for an effective conservation activity. Collaboration between regional actors such as locals, farmers, landowners, NGOs and policy-makers should enhance conservation efforts to prevent biodiversity loss [10].

An official undergraduate program could enhance the knowledge and skills of students for environmental conservation. Additionally, creating a high school environmental course could be useful in terms of raising awareness. Increasing the exchange of knowledge, skills and knowledge in structured settings outside their undergraduate programs could be an effective way to reach outside of normal learning for practitioners, stakeholders and other interested people.

5.4. Monitoring and researches

Monitoring and inventory surveys on the endangered amphibians and reptiles will be helpful for identifying threats and create key activities for protection of the species. The main topics could be determining population/community trends, habitat quality, modeling climate change impacts and attitudes of local populations.

6. Conclusion

The Mediterranean basin's biodiversity are facing many pressures and urgent action is required to preserve its future. Fortunately, many stakeholders such as regional and governmental organizations, NGOs, scientists and conservation practitioners are cooperating to preserve Mediterranean natural resources. The key conservation actions should be focused on sustainable management and legal protection of endangered species and their habitats [10]. Besides, it is not possible to deny importance of education, awareness-raising activities and monitoring studies for sustainability of Mediterranean amphibians and reptiles.

Acknowledgements

We thank Dilara Arslan, Çağdaş Yaşar, Sevde Karagöz and Onur Obut for their help to preparing data set and Ahmet Burak Kaya for reviewing English style.

Appendix 1.

Amphibians and reptiles of the Mediterranean basin

Class	Order	Family	Species	IUCN Red List category	Endemic (Yes/No)
Amphibia	Caudata	Plethodontidae	*Hydromantes ambrosii*	NT	Y
Amphibia	Caudata	Plethodontidae	*Hydromantes flavus*	VU	Y
Amphibia	Caudata	Plethodontidae	*Hydromantes genei*	VU	Y
Amphibia	Caudata	Plethodontidae	*Hydromantes imperialis*	NT	Y
Amphibia	Caudata	Plethodontidae	*Hydromantes italicus*	NT	Y
Amphibia	Caudata	Plethodontidae	*Hydromantes sarrabusensis*	VU	Y
Amphibia	Caudata	Plethodontidae	*Hydromantes strinatii*	NT	Y
Amphibia	Caudata	Plethodontidae	*Hydromantes supramontis*	EN	Y
Amphibia	Caudata	Proteidae	*Proteus anguinus*	VU	Y
Amphibia	Caudata	Salamandridae	*Calotriton arnoldi*	CR	Y
Amphibia	Caudata	Salamandridae	*Calotriton asper*	NT	Y
Amphibia	Caudata	Salamandridae	*Chioglossa lusitanica*	VU	Y
Amphibia	Caudata	Salamandridae	*Euproctus montanus*	LC	Y
Amphibia	Caudata	Salamandridae	*Euproctus platycephalus*	EN	Y

Class	Order	Family	Species	IUCN Red List category	Endemic (Yes/No)
Amphibia	Caudata	Salamandridae	*Ichthyosaura alpestris*	LC	N
Amphibia	Caudata	Salamandridae	*Lissotriton boscai*	LC	Y
Amphibia	Caudata	Salamandridae	*Lissotriton helveticus*	LC	N
Amphibia	Caudata	Salamandridae	*Lissotriton italicus*	LC	Y
Amphibia	Caudata	Salamandridae	*Lissotriton kosswigi*	NE	N
Amphibia	Caudata	Salamandridae	*Lissotriton vulgaris*	LC	N
Amphibia	Caudata	Salamandridae	*Lyciasalamandra antalyana*	EN	Y
Amphibia	Caudata	Salamandridae	*Lyciasalamandra atifi*	EN	Y
Amphibia	Caudata	Salamandridae	*Lyciasalamandra billae*	CR	Y
Amphibia	Caudata	Salamandridae	*Lyciasalamandra fazilae*	EN	Y
Amphibia	Caudata	Salamandridae	*Lyciasalamandra flavimembris*	EN	Y
Amphibia	Caudata	Salamandridae	*Lyciasalamandra helverseni*	VU	Y
Amphibia	Caudata	Salamandridae	*Lyciasalamandra luschani*	VU	Y
Amphibia	Caudata	Salamandridae	*Neurergus strauchii*	VU	Y
Amphibia	Caudata	Salamandridae	*Ommatotriton vittatus*	LC	Y
Amphibia	Caudata	Salamandridae	*Pleurodeles nebulosus*	VU	Y
Amphibia	Caudata	Salamandridae	*Pleurodeles poireti*	EN	Y
Amphibia	Caudata	Salamandridae	*Pleurodeles waltl*	NT	Y
Amphibia	Caudata	Salamandridae	*Salamandra algira*	VU	Y
Amphibia	Caudata	Salamandridae	*Salamandra atra*	LC	N
Amphibia	Caudata	Salamandridae	*Salamandra corsica*	LC	Y
Amphibia	Caudata	Salamandridae	*Salamandra infraimmaculata*	NT	N
Amphibia	Caudata	Salamandridae	*Salamandra lanzai*	VU	Y
Amphibia	Caudata	Salamandridae	*Salamandra salamandra*	LC	N
Amphibia	Caudata	Salamandridae	*Salamandrina perspicillata*	LC	Y

Class	Order	Family	Species	IUCN Red List category	Endemic (Yes/No)
Amphibia	Caudata	Salamandridae	*Salamandrina terdigitata*	LC	Y
Amphibia	Caudata	Salamandridae	*Triturus anatolicus*	NE	Y
Amphibia	Caudata	Salamandridae	*Triturus carnifex*	LC	N
Amphibia	Caudata	Salamandridae	*Triturus cristatus*	LC	N
Amphibia	Caudata	Salamandridae	*Triturus dobrogicus*	NT	N
Amphibia	Caudata	Salamandridae	*Triturus ivanbureschi*	NE	N
Amphibia	Caudata	Salamandridae	*Triturus karelinii*	LC	N
Amphibia	Caudata	Salamandridae	*Triturus macedonicus*	NE	Y
Amphibia	Caudata	Salamandridae	*Triturus marmoratus*	LC	Y
Amphibia	Caudata	Salamandridae	*Triturus pygmaeus*	NT	Y
Amphibia	Anura	Alytidae	*Alytes cisternasii*	NT	Y
Amphibia	Anura	Alytidae	*Alytes dickhilleni*	VU	Y
Amphibia	Anura	Alytidae	*Alytes maurus*	NT	Y
Amphibia	Anura	Alytidae	*Alytes muletensis*	VU	Y
Amphibia	Anura	Alytidae	*Alytes obstetricans*	LC	N
Amphibia	Anura	Alytidae	*Discoglossus jeanneae*	NT	Y
Amphibia	Anura	Alytidae	*Discoglossus galganoi*	LC	Y
Amphibia	Anura	Alytidae	*Discoglossus montalentii*	NT	Y
Amphibia	Anura	Alytidae	*Discoglossus pictus*	LC	Y
Amphibia	Anura	Alytidae	*Discoglossus sardus*	LC	Y
Amphibia	Anura	Alytidae	*Discoglossus scovazzi*	LC	Y
Amphibia	Anura	Alytidae	*Latonia nigriventer*	CR	Y
Amphibia	Anura	Bombinatoridae	*Bombina bombina*	LC	N
Amphibia	Anura	Bombinatoridae	*Bombina pachypus*	EN	Y
Amphibia	Anura	Bombinatoridae	*Bombina variegata*	LC	N
Amphibia	Anura	Bufonidae	*Barbarophryne brongersmai*	NT	Y
Amphibia	Anura	Bufonidae	*Bufo bufo*	LC	N
Amphibia	Anura	Bufonidae	*Bufo spinosus*	NE	Y
Amphibia	Anura	Bufonidae	*Bufotes balearicus*	LC	Y
Amphibia	Anura	Bufonidae	*Bufotes boulengeri*	LC	Y
Amphibia	Anura	Bufonidae	*Bufotes siculus*	LC	Y
Amphibia	Anura	Bufonidae	*Bufotes variabilis*	DD	N
Amphibia	Anura	Bufonidae	*Bufotes viridis*	LC	N

http://dx.doi.org/10.5772/intechopen.70357

Class	Order	Family	Species	IUCN Red List category	Endemic (Yes/No)
Amphibia	Anura	Bufonidae	*Epidalea calamita*	LC	N
Amphibia	Anura	Bufonidae	*Sclerophrys kassasii*	LC	Y
Amphibia	Anura	Bufonidae	*Sclerophrys mauritanica*	LC	Y
Amphibia	Anura	Bufonidae	*Sclerophrys regularis*	LC	N
Amphibia	Anura	Dicroglossidae	*Hoplobatrachus occipitalis*	LC	N
Amphibia	Anura	Hylidae	*Hyla arborea*	LC	N
Amphibia	Anura	Hylidae	*Hyla heinzsteinitzi*	CR	Y
Amphibia	Anura	Hylidae	*Hyla intermedia*	LC	Y
Amphibia	Anura	Hylidae	*Hyla meridionalis*	LC	Y
Amphibia	Anura	Hylidae	*Hyla molleri*	NE	Y
Amphibia	Anura	Hylidae	*Hyla orientalis*	NE	N
Amphibia	Anura	Hylidae	*Hyla sarda*	LC	Y
Amphibia	Anura	Hylidae	*Hyla savignyi*	LC	N
Amphibia	Anura	Pelobatidae	*Pelobates cultripes*	NT	Y
Amphibia	Anura	Pelobatidae	*Pelobates syriacus*	LC	N
Amphibia	Anura	Pelobatidae	*Pelobates varaldii*	EN	Y
Amphibia	Anura	Pelodytidae	*Pelodytes ibericus*	LC	Y
Amphibia	Anura	Pelodytidae	*Pelodytes punctatus*	LC	Y
Amphibia	Anura	Ranidae	*Pelophylax bedriagae*	LC	Y
Amphibia	Anura	Ranidae	*Pelophylax bergeri*	LC	Y
Amphibia	Anura	Ranidae	*Pelophylax caralitanus*	NT	Y
Amphibia	Anura	Ranidae	*Pelophylax cerigensis*	CR	Y
Amphibia	Anura	Ranidae	*Pelophylax cretensis*	EN	Y
Amphibia	Anura	Ranidae	*Pelophylax cypriensis*	NE	Y
Amphibia	Anura	Ranidae	*Pelophylax epeiroticus*	VU	Y
Amphibia	Anura	Ranidae	*Pelophylax esculentus*	LC	N
Amphibia	Anura	Ranidae	*Pelophylax hispanicus*	LC	Y
Amphibia	Anura	Ranidae	*Pelophylax kurtmuelleri*	LC	Y
Amphibia	Anura	Ranidae	*Pelophylax lessonae*	LC	N
Amphibia	Anura	Ranidae	*Pelophylax perezi*	LC	Y
Amphibia	Anura	Ranidae	*Pelophylax ridibundus*	LC	N
Amphibia	Anura	Ranidae	*Pelophylax saharicus*	LC	Y
Amphibia	Anura	Ranidae	*Pelophylax shqipericus*	EN	Y

Class	Order	Family	Species	IUCN Red List category	Endemic (Yes/No)
Amphibia	Anura	Ranidae	*Rana catesbeiana*	LC	N
Amphibia	Anura	Ranidae	*Rana dalmatina*	LC	N
Amphibia	Anura	Ranidae	*Rana graeca*	LC	Y
Amphibia	Anura	Ranidae	*Rana holtzi*	CR	Y
Amphibia	Anura	Ranidae	*Rana iberica*	NT	Y
Amphibia	Anura	Ranidae	*Rana italica*	LC	Y
Amphibia	Anura	Ranidae	*Rana latastei*	VU	Y
Amphibia	Anura	Ranidae	*Rana macrocnemis*	LC	N
Amphibia	Anura	Ranidae	*Rana perezii*	LC	Y
Amphibia	Anura	Ranidae	*Rana pyrenaica*	EN	Y
Amphibia	Anura	Ranidae	*Rana tavasensis*	EN	Y
Amphibia	Anura	Ranidae	*Rana temporaria*	LC	N
Reptilia	Testudines	Cheloniidae	*Caretta caretta*	VU	N
Reptilia	Testudines	Cheloniidae	*Chelonia mydas*	EN	N
Reptilia	Testudines	Cheloniidae	*Eretmochelys imbricata*	CR	N
Reptilia	Testudines	Dermochelyidae	*Dermochelys coriacea*	VU	N
Reptilia	Testudines	Emydidae	*Emys orbicularis*	NT	N
Reptilia	Testudines	Emydidae	*Emys trinacris*	DD	N
Reptilia	Testudines	Emydidae	*Trachemys scripta*	LC	N
Reptilia	Testudines	Geoemydidae	*Mauremys caspica*	LC	N
Reptilia	Testudines	Geoemydidae	*Mauremys leprosa*	LC	Y
Reptilia	Testudines	Geoemydidae	*Mauremys rivulata*	LC	Y
Reptilia	Testudines	Testudinidae	*Testudo graeca*	LC	N
Reptilia	Testudines	Testudinidae	*Testudo hermanni*	NT	Y
Reptilia	Testudines	Testudinidae	*Testudo kleinmanni*	CR	Y
Reptilia	Testudines	Testudinidae	*Testudo marginata*	LC	Y
Reptilia	Testudines	Trionychidae	*Rafetus euphraticus*	EN	N
Reptilia	Testudines	Trionychidae	*Trionyx triunguis*	LC	N
Reptilia	Amphisbaenia	Amphisbaenidae	*Blanus cinereus*	LC	Y
Reptilia	Amphisbaenia	Amphisbaenidae	*Blanus mettetali*	LC	Y
Reptilia	Amphisbaenia	Amphisbaenidae	*Blanus strauchi*	LC	N
Reptilia	Amphisbaenia	Amphisbaenidae	*Blanus tingitanus*	LC	Y
Reptilia	Amphisbaenia	Trogonophiidae	*Trogonophis wiegmanni*	LC	Y

http://dx.doi.org/10.5772/intechopen.70357

Class	Order	Family	Species	IUCN Red List category	Endemic (Yes/No)
Reptilia	Sauria	Agamidae	*Agama impalearis*	LC	Y
Reptilia	Sauria	Agamidae	*Agama spinosa*	LC	N
Reptilia	Sauria	Agamidae	*Phrynocephalus arabicus*	LC	Y
Reptilia	Sauria	Agamidae	*Phrynocephalus maculatus*	LC	N
Reptilia	Sauria	Agamidae	*Pseudotrapelus aqabensis*	NE	Y
Reptilia	Sauria	Agamidae	*Pseudotrapelus sinaitus*	LC	N
Reptilia	Sauria	Agamidae	*Stellagama stellio*	LC	N
Reptilia	Sauria	Agamidae	*Trapelus agnetae*	LC	N
Reptilia	Sauria	Agamidae	*Trapelus boehmei*	LC	Y
Reptilia	Sauria	Agamidae	*Trapelus lessonae*	LC	N
Reptilia	Sauria	Agamidae	*Trapelus mutabilis*	LC	N
Reptilia	Sauria	Agamidae	*Trapelus ruderatus*	LC	N
Reptilia	Sauria	Agamidae	*Trapelus savignii*	VU	Y
Reptilia	Sauria	Agamidae	*Trapelus schmitzi*	DD	N
Reptilia	Sauria	Agamidae	*Trapelus tournevillei*	LC	Y
Reptilia	Sauria	Agamidae	*Uromastyx acanthinura*	NT	N
Reptilia	Sauria	Agamidae	*Uromastyx aegyptia*	NT	N
Reptilia	Sauria	Agamidae	*Uromastyx alfredschmidti*	NT	Y
Reptilia	Sauria	Agamidae	*Uromastyx dispar*	NT	Y
Reptilia	Sauria	Agamidae	*Uromastyx geyri*	NT	Y
Reptilia	Sauria	Agamidae	*Uromastyx nigriventris*	NE	Y
Reptilia	Sauria	Agamidae	*Uromastyx ocellata*	NT	N
Reptilia	Sauria	Agamidae	*Uromastyx ornata*	NT	N
Reptilia	Sauria	Anguidae	*Anguis cephallonica*	NT	Y
Reptilia	Sauria	Anguidae	*Anguis graeca*	NE	Y
Reptilia	Sauria	Anguidae	*Anguis veronensis*	NE	Y
Reptilia	Sauria	Anguidae	*Hyalosaurus koellikeri*	LC	Y
Reptilia	Sauria	Anguidae	*Pseudopus apodus*	LC	N
Reptilia	Sauria	Blanidae	*Blanus alexandri*	NE	Y
Reptilia	Sauria	Blanidae	*Blanus aporus*	NE	Y
Reptilia	Sauria	Blanidae	*Blanus mariae*	NE	Y
Reptilia	Sauria	Chamaeleonidae	*Chamaeleo africanus*	LC	N
Reptilia	Sauria	Chamaeleonidae	*Chamaeleo chamaeleon*	LC	N

Class	Order	Family	Species	IUCN Red List category	Endemic (Yes/No)
Reptilia	Sauria	Eublepharidae	*Eublepharis angramainyu*	LC	N
Reptilia	Sauria	Gekkonidae	*Asaccus elisae*	LC	N
Reptilia	Sauria	Gekkonidae	*Bunopus blanfordii*	NE	Y
Reptilia	Sauria	Gekkonidae	*Bunopus tuberculatus*	LC	N
Reptilia	Sauria	Gekkonidae	*Cyrtopodion scabrum*	LC	N
Reptilia	Sauria	Gekkonidae	*Euleptes europaea*	NT	Y
Reptilia	Sauria	Gekkonidae	*Hemidactylus dawudazraqi*	NE	N
Reptilia	Sauria	Gekkonidae	*Hemidactylus flaviviridis*	LC	N
Reptilia	Sauria	Gekkonidae	*Hemidactylus foudaii*	LC	Y
Reptilia	Sauria	Gekkonidae	*Hemidactylus granosus*	NE	N
Reptilia	Sauria	Gekkonidae	*Hemidactylus lavadeserticus*	NE	Y
Reptilia	Sauria	Gekkonidae	*Hemidactylus mindiae*	LC	Y
Reptilia	Sauria	Gekkonidae	*Hemidactylus robustus*	LC	N
Reptilia	Sauria	Gekkonidae	*Hemidactylus sinaitus*	LC	N
Reptilia	Sauria	Gekkonidae	*Hemidactylus turcicus*	LC	N
Reptilia	Sauria	Gekkonidae	*Mediodactylus amictopholis*	EN	Y
Reptilia	Sauria	Gekkonidae	*Mediodactylus heterocercus*	LC	N
Reptilia	Sauria	Gekkonidae	*Mediodactylus kotschyi*	LC	N
Reptilia	Sauria	Gekkonidae	*Pristurus flavipunctatus*	LC	N
Reptilia	Sauria	Gekkonidae	*Pristurus rupestris*	LC	N
Reptilia	Sauria	Gekkonidae	*Quedenfeldtia moerens*	LC	Y
Reptilia	Sauria	Gekkonidae	*Quedenfeldtia trachyblepharus*	NT	Y
Reptilia	Sauria	Gekkonidae	*Saurodactylus brosseti*	LC	Y
Reptilia	Sauria	Gekkonidae	*Saurodactylus fasciatus*	VU	Y
Reptilia	Sauria	Gekkonidae	*Saurodactylus mauritanicus*	LC	Y
Reptilia	Sauria	Gekkonidae	*Stenodactylus doriae*	LC	N
Reptilia	Sauria	Gekkonidae	*Stenodactylus grandiceps*	LC	N
Reptilia	Sauria	Gekkonidae	*Stenodactylus mauritanicus*	NE	Y
Reptilia	Sauria	Gekkonidae	*Stenodactylus petrii*	LC	N
Reptilia	Sauria	Gekkonidae	*Stenodactylus slevini*	LC	N
Reptilia	Sauria	Gekkonidae	*Stenodactylus stenurus*	NE	Y
Reptilia	Sauria	Gekkonidae	*Stenodactylus sthenodactylus*	LC	N
Reptilia	Sauria	Gekkonidae	*Tarentola angustimentalis*	LC	Y

http://dx.doi.org/10.5772/intechopen.70357

Class	Order	Family	Species	IUCN Red List category	Endemic (Yes/No)
Reptilia	Sauria	Gekkonidae	*Tarentola annularis*	LC	N
Reptilia	Sauria	Gekkonidae	*Tarentola bischoffi*	NE	Y
Reptilia	Sauria	Gekkonidae	*Tarentola boehmei*	LC	Y
Reptilia	Sauria	Gekkonidae	*Tarentola boettgeri*	LC	Y
Reptilia	Sauria	Gekkonidae	*Tarentola chazaliae*	LC	N
Reptilia	Sauria	Gekkonidae	*Tarentola delalandii*	LC	Y
Reptilia	Sauria	Gekkonidae	*Tarentola deserti*	LC	Y
Reptilia	Sauria	Gekkonidae	*Tarentola ephippiata*	LC	N
Reptilia	Sauria	Gekkonidae	*Tarentola fascicularis*	NE	Y
Reptilia	Sauria	Gekkonidae	*Tarentola gomerensis*	LC	Y
Reptilia	Sauria	Gekkonidae	*Tarentola mauritanica*	LC	Y
Reptilia	Sauria	Gekkonidae	*Tarentola mindiae*	LC	Y
Reptilia	Sauria	Gekkonidae	*Tarentola neglecta*	LC	Y
Reptilia	Sauria	Gekkonidae	*Tropiocolotes algericus*	LC	N
Reptilia	Sauria	Gekkonidae	*Tropiocolotes bisharicus*	LC	N
Reptilia	Sauria	Gekkonidae	*Tropiocolotes nattereri*	LC	Y
Reptilia	Sauria	Gekkonidae	*Tropiocolotes nubicus*	DD	N
Reptilia	Sauria	Gekkonidae	*Tropiocolotes steudneri*	LC	N
Reptilia	Sauria	Gekkonidae	*Tropiocolotes tripolitanus*	LC	Y
Reptilia	Sauria	Lacertidae	*Acanthodactylus aegyptius*	NE	Y
Reptilia	Sauria	Lacertidae	*Acanthodactylus blanci*	EN	Y
Reptilia	Sauria	Lacertidae	*Acanthodactylus boskianus*	LC	N
Reptilia	Sauria	Lacertidae	*Acanthodactylus busacki*	LC	N
Reptilia	Sauria	Lacertidae	*Acanthodactylus dumerilii*	LC	N
Reptilia	Sauria	Lacertidae	*Acanthodactylus erythrurus*	LC	Y
Reptilia	Sauria	Lacertidae	*Acanthodactylus grandis*	LC	N
Reptilia	Sauria	Lacertidae	*Acanthodactylus hardyi*	NE	N
Reptilia	Sauria	Lacertidae	*Acanthodactylus harranensis*	CR	Y
Reptilia	Sauria	Lacertidae	*Acanthodactylus longipes*	LC	N
Reptilia	Sauria	Lacertidae	*Acanthodactylus maculatus*	LC	Y
Reptilia	Sauria	Lacertidae	*Acanthodactylus opheodurus*	LC	N
Reptilia	Sauria	Lacertidae	*Acanthodactylus orientalis*	LC	N
Reptilia	Sauria	Lacertidae	*Acanthodactylus pardalis*	VU	Y

Class	Order	Family	Species	IUCN Red List category	Endemic (Yes/No)
Reptilia	Sauria	Lacertidae	*Acanthodactylus robustus*	LC	N
Reptilia	Sauria	Lacertidae	*Acanthodactylus savignyi*	NT	Y
Reptilia	Sauria	Lacertidae	*Acanthodactylus schmidti*	LC	Y
Reptilia	Sauria	Lacertidae	*Acanthodactylus schreiberi*	EN	Y
Reptilia	Sauria	Lacertidae	*Acanthodactylus scutellatus*	LC	N
Reptilia	Sauria	Lacertidae	*Acanthodactylus spinicauda*	CR	Y
Reptilia	Sauria	Lacertidae	*Acanthodactylus taghitensis*	DD	Y
Reptilia	Sauria	Lacertidae	*Acanthodactylus tilburyi*	NE	N
Reptilia	Sauria	Lacertidae	*Acanthodactylus tristrami*	LC	N
Reptilia	Sauria	Lacertidae	*Algyroides fitzingeri*	LC	Y
Reptilia	Sauria	Lacertidae	*Algyroides marchi*	EN	Y
Reptilia	Sauria	Lacertidae	*Algyroides moreoticus*	NT	Y
Reptilia	Sauria	Lacertidae	*Algyroides nigropunctatus*	LC	Y
Reptilia	Sauria	Lacertidae	*Anatololacerta anatolica*	LC	Y
Reptilia	Sauria	Lacertidae	*Anatololacerta budaki*	NE	Y
Reptilia	Sauria	Lacertidae	*Anatololacerta danfordi*	LC	Y
Reptilia	Sauria	Lacertidae	*Anatololacerta pelasgiana*	NE	Y
Reptilia	Sauria	Lacertidae	*Apathya cappadocica*	LC	Y
Reptilia	Sauria	Lacertidae	*Archaeolacerta bedriagae*	VU	Y
Reptilia	Sauria	Lacertidae	*Atlantolacerta andreanskyi*	NT	Y
Reptilia	Sauria	Lacertidae	*Dalmatolacerta oxycephala*	LC	Y
Reptilia	Sauria	Lacertidae	*Darevskia praticola*	NT	N
Reptilia	Sauria	Lacertidae	*Darevskia rudis*	LC	N
Reptilia	Sauria	Lacertidae	*Darevskia valentini*	LC	N
Reptilia	Sauria	Lacertidae	*Dinarolacerta mosorensis*	VU	Y
Reptilia	Sauria	Lacertidae	*Dinarolacerta montenegrina*	LC	Y
Reptilia	Sauria	Lacertidae	*Gallotia atlantica*	LC	Y
Reptilia	Sauria	Lacertidae	*Gallotia auaritae*	CR	Y
Reptilia	Sauria	Lacertidae	*Gallotia bravoana*	CR	Y
Reptilia	Sauria	Lacertidae	*Gallotia caesaris*	LC	Y
Reptilia	Sauria	Lacertidae	*Gallotia galloti*	LC	Y
Reptilia	Sauria	Lacertidae	*Gallotia intermedia*	CR	Y
Reptilia	Sauria	Lacertidae	*Gallotia simonyi*	CR	Y

http://dx.doi.org/10.5772/intechopen.70357

Class	Order	Family	Species	IUCN Red List category	Endemic (Yes/No)
Reptilia	Sauria	Lacertidae	*Gallotia stehlini*	LC	Y
Reptilia	Sauria	Lacertidae	*Hellenolacerta graeca*	NT	Y
Reptilia	Sauria	Lacertidae	*Iberolacerta aranica*	CR	Y
Reptilia	Sauria	Lacertidae	*Iberolacerta aurelioi*	EN	Y
Reptilia	Sauria	Lacertidae	*Iberolacerta bonnali*	NT	Y
Reptilia	Sauria	Lacertidae	*Iberolacerta cyreni*	EN	Y
Reptilia	Sauria	Lacertidae	*Iberolacerta galani*	NT	Y
Reptilia	Sauria	Lacertidae	*Iberolacerta horvathi*	NT	N
Reptilia	Sauria	Lacertidae	*Iberolacerta martinezricai*	CR	Y
Reptilia	Sauria	Lacertidae	*Iberolacerta monticola*	VU	Y
Reptilia	Sauria	Lacertidae	*Lacerta agilis*	LC	N
Reptilia	Sauria	Lacertidae	*Lacerta bilineata*	LC	N
Reptilia	Sauria	Lacertidae	*Lacerta media*	LC	N
Reptilia	Sauria	Lacertidae	*Lacerta pamphylica*	LC	Y
Reptilia	Sauria	Lacertidae	*Lacerta schreiberi*	NT	Y
Reptilia	Sauria	Lacertidae	*Lacerta trilineata*	LC	N
Reptilia	Sauria	Lacertidae	*Lacerta viridis*	LC	N
Reptilia	Sauria	Lacertidae	*Latastia longicaudata*	LC	N
Reptilia	Sauria	Lacertidae	*Mesalina bahaeldini*	LC	Y
Reptilia	Sauria	Lacertidae	*Mesalina brevirostris*	LC	N
Reptilia	Sauria	Lacertidae	*Mesalina guttulata*	LC	N
Reptilia	Sauria	Lacertidae	*Mesalina martini*	LC	N
Reptilia	Sauria	Lacertidae	*Mesalina olivieri*	LC	N
Reptilia	Sauria	Lacertidae	*Mesalina pasteuri*	DD	N
Reptilia	Sauria	Lacertidae	*Mesalina rubropunctata*	LC	N
Reptilia	Sauria	Lacertidae	*Mesalina simoni*	LC	Y
Reptilia	Sauria	Lacertidae	*Ophisops elbaensis*	DD	N
Reptilia	Sauria	Lacertidae	*Ophisops elegans*	LC	N
Reptilia	Sauria	Lacertidae	*Ophisops occidentalis*	LC	Y
Reptilia	Sauria	Lacertidae	*Parvilacerta fraasii*	EN	Y
Reptilia	Sauria	Lacertidae	*Parvilacerta parva*	LC	N
Reptilia	Sauria	Lacertidae	*Philochortus zolii*	EN	N
Reptilia	Sauria	Lacertidae	*Phoenicolacerta cyanisparsa*	LC	Y

Class	Order	Family	Species	IUCN Red List category	Endemic (Yes/No)
Reptilia	Sauria	Lacertidae	*Phoenicolacerta kulzeri*	EN	Y
Reptilia	Sauria	Lacertidae	*Phoenicolacerta laevis*	LC	Y
Reptilia	Sauria	Lacertidae	*Phoenicolacerta troodica*	LC	Y
Reptilia	Sauria	Lacertidae	*Podarcis bocagei*	LC	Y
Reptilia	Sauria	Lacertidae	*Podarcis carbonelli*	EN	Y
Reptilia	Sauria	Lacertidae	*Podarcis cretensis*	EN	Y
Reptilia	Sauria	Lacertidae	*Podarcis erhardii*	LC	Y
Reptilia	Sauria	Lacertidae	*Podarcis filfolensis*	LC	Y
Reptilia	Sauria	Lacertidae	*Podarcis gaigeae*	VU	Y
Reptilia	Sauria	Lacertidae	*Podarcis guadarramae*	NE	Y
Reptilia	Sauria	Lacertidae	*Podarcis hispanicus*	LC	Y
Reptilia	Sauria	Lacertidae	*Podarcis levendis*	VU	Y
Reptilia	Sauria	Lacertidae	*Podarcis lilfordi*	EN	Y
Reptilia	Sauria	Lacertidae	*Podarcis liolepis*	NE	N
Reptilia	Sauria	Lacertidae	*Podarcis melisellensis*	LC	Y
Reptilia	Sauria	Lacertidae	*Podarcis milensis*	NT	T
Reptilia	Sauria	Lacertidae	*Podarcis muralis*	LC	Y
Reptilia	Sauria	Lacertidae	*Podarcis peloponnesiacus*	LC	Y
Reptilia	Sauria	Lacertidae	*Podarcis pityusensis*	NT	Y
Reptilia	Sauria	Lacertidae	*Podarcis raffonei*	CR	Y
Reptilia	Sauria	Lacertidae	*Podarcis siculus*	LC	Y
Reptilia	Sauria	Lacertidae	*Podarcis tauricus*	LC	Y
Reptilia	Sauria	Lacertidae	*Podarcis tiliguerta*	LC	Y
Reptilia	Sauria	Lacertidae	*Podarcis vaucheri*	LC	Y
Reptilia	Sauria	Lacertidae	*Podarcis virescens*	NE	Y
Reptilia	Sauria	Lacertidae	*Podarcis waglerianus*	LC	Y
Reptilia	Sauria	Lacertidae	*Psammodromus algirus*	LC	Y
Reptilia	Sauria	Lacertidae	*Psammodromus blanci*	NT	Y
Reptilia	Sauria	Lacertidae	*Psammodromus edwarsianus*	NE	N
Reptilia	Sauria	Lacertidae	*Psammodromus hispanicus*	LC	Y
Reptilia	Sauria	Lacertidae	*Psammodromus microdactylus*	EN	Y
Reptilia	Sauria	Lacertidae	*Psammodromus occidentalis*	NE	Y
Reptilia	Sauria	Lacertidae	*Psammophis aegyptius*	LC	N

http://dx.doi.org/10.5772/intechopen.70357

Class	Order	Family	Species	IUCN Red List category	Endemic (Yes/No)
Reptilia	Sauria	Lacertidae	*Psammophis biseriatus*	NE	N
Reptilia	Sauria	Lacertidae	*Psammophis rukwae*	LC	N
Reptilia	Sauria	Lacertidae	*Psammophis schokari*	LC	N
Reptilia	Sauria	Lacertidae	*Psammophis sibilans*	LC	N
Reptilia	Sauria	Lacertidae	*Psammophis tanganicus*	NE	N
Reptilia	Sauria	Lacertidae	*Pseuderemias mucronata*	DD	N
Reptilia	Sauria	Lacertidae	*Scelarcis perspicillata*	LC	Y
Reptilia	Sauria	Lacertidae	*Teira dugesii*	LC	Y
Reptilia	Sauria	Lacertidae	*Timon kurdistanicus*	LC	N
Reptilia	Sauria	Lacertidae	*Timon lepidus*	NT	Y
Reptilia	Sauria	Lacertidae	*Timon nevadensis*	NE	Y
Reptilia	Sauria	Lacertidae	*Timon pater*	LC	Y
Reptilia	Sauria	Lacertidae	*Timon princeps*	LC	Y
Reptilia	Sauria	Lacertidae	*Timon tangitanus*	LC	Y
Reptilia	Sauria	Lacertidae	*Zootoca vivipara*	LC	N
Reptilia	Sauria	Lacertidae	*Acanthodactylus ahmaddisii*	EN	N
Reptilia	Sauria	Lacertidae	*Acanthodactylus aureus*	LC	N
Reptilia	Sauria	Lacertidae	*Acanthodactylus bedriagai*	NT	Y
Reptilia	Sauria	Lacertidae	*Acanthodactylus beershebensis*	CR	Y
Reptilia	Sauria	Phyllodactylidae	*Asaccus barani*	NE	Y
Reptilia	Sauria	Phyllodactylidae	*Ptyodactylus ananjevae*	NE	Y
Reptilia	Sauria	Phyllodactylidae	*Ptyodactylus guttatus*	LC	N
Reptilia	Sauria	Phyllodactylidae	*Ptyodactylus hasselquistii*	LC	N
Reptilia	Sauria	Phyllodactylidae	*Ptyodactylus oudrii*	LC	Y
Reptilia	Sauria	Phyllodactylidae	*Ptyodactylus puiseuxi*	LC	N
Reptilia	Sauria	Phyllodactylidae	*Ptyodactylus ragazzii*	LC	N
Reptilia	Sauria	Scincidae	*Chalcides bedriagai*	NT	Y
Reptilia	Sauria	Scincidae	*Chalcides boulengeri*	NE	N
Reptilia	Sauria	Scincidae	*Chalcides chalcides*	LC	Y
Reptilia	Sauria	Scincidae	*Chalcides colosii*	LC	Y
Reptilia	Sauria	Scincidae	*Chalcides ebneri*	CR	Y
Reptilia	Sauria	Scincidae	*Chalcides guentheri*	VU	Y
Reptilia	Sauria	Scincidae	*Chalcides lanzai*	NT	Y

Class	Order	Family	Species	IUCN Red List category	Endemic (Yes/No)
Reptilia	Sauria	Scincidae	*Chalcides manueli*	VU	Y
Reptilia	Sauria	Scincidae	*Chalcides mauritanicus*	EN	Y
Reptilia	Sauria	Scincidae	*Chalcides mertensi*	LC	Y
Reptilia	Sauria	Scincidae	*Chalcides minutus*	VU	Y
Reptilia	Sauria	Scincidae	*Chalcides mionecton*	LC	Y
Reptilia	Sauria	Scincidae	*Chalcides montanus*	NT	Y
Reptilia	Sauria	Scincidae	*Chalcides ocellatus*	LC	Y
Reptilia	Sauria	Scincidae	*Chalcides parallelus*	EN	Y
Reptilia	Sauria	Scincidae	*Chalcides polylepis*	LC	Y
Reptilia	Sauria	Scincidae	*Chalcides pseudostriatus*	NT	Y
Reptilia	Sauria	Scincidae	*Chalcides sepsoides*	LC	Y
Reptilia	Sauria	Scincidae	*Chalcides sexlineatus*	LC	Y
Reptilia	Sauria	Scincidae	*Chalcides sphenopsiformis*	LC	N
Reptilia	Sauria	Scincidae	*Chalcides striatus*	LC	Y
Reptilia	Sauria	Scincidae	*Chalcides viridanus*	LC	Y
Reptilia	Sauria	Scincidae	*Eumeces algeriensis*	LC	Y
Reptilia	Sauria	Scincidae	*Eumeces schneideri*	LC	N
Reptilia	Sauria	Scincidae	*Eurylepis taeniolata*	NE	N
Reptilia	Sauria	Scincidae	*Heremites auratus*	NE	N
Reptilia	Sauria	Scincidae	*Ophiomorus latastii*	DD	Y
Reptilia	Sauria	Scincidae	*Ophiomorus punctatissimus*	LC	Y
Reptilia	Sauria	Scincidae	*Scincopus fasciatus*	DD	N
Reptilia	Sauria	Scincidae	*Scincus albifasciatus*	LC	N
Reptilia	Sauria	Scincidae	*Scincus scincus*	LC	N
Reptilia	Sauria	Scincidae	*Trachylepis quinquetaeniata*	LC	N
Reptilia	Sauria	Scincidae	*Ablepharus budaki*	LC	Y
Reptilia	Sauria	Scincidae	*Ablepharus chernovi*	LC	N
Reptilia	Sauria	Scincidae	*Ablepharus kitaibelii*	LC	N
Reptilia	Sauria	Scincidae	*Ablepharus rueppellii*	LC	Y
Reptilia	Sauria	Varanidae	*Varanus griseus*	LC	N
Reptilia	Sauria	Varanidae	*Varanus niloticus*	LC	N
Reptilia	Ophidia	Atractaspididae	*Atractaspis engaddensis*	LC	N
Reptilia	Ophidia	Atractaspididae	*Micrelaps muelleri*	LC	Y

http://dx.doi.org/10.5772/intechopen.70357

Class	Order	Family	Species	IUCN Red List category	Endemic (Yes/No)
Reptilia	Ophidia	Atractaspididae	*Micrelaps tchernovi*	NE	Y
Reptilia	Ophidia	Boidae	*Eryx colubrinus*	NE	N
Reptilia	Ophidia	Boidae	*Eryx jaculus*	LC	N
Reptilia	Ophidia	Colubridae	*Coronella austriaca*	LC	N
Reptilia	Ophidia	Colubridae	*Coronella girondica*	LC	Y
Reptilia	Ophidia	Colubridae	*Dasypeltis scabra*	LC	N
Reptilia	Ophidia	Colubridae	*Dolichophis caspius*	LC	N
Reptilia	Ophidia	Colubridae	*Dolichophis cypriensis*	EN	Y
Reptilia	Ophidia	Colubridae	*Dolichophis jugularis*	LC	N
Reptilia	Ophidia	Colubridae	*Dolichophis schmidti*	LC	N
Reptilia	Ophidia	Colubridae	*Eirenis aurolineatus*	LC	Y
Reptilia	Ophidia	Colubridae	*Eirenis barani*	LC	Y
Reptilia	Ophidia	Colubridae	*Eirenis collaris*	LC	N
Reptilia	Ophidia	Colubridae	*Eirenis coronella*	LC	N
Reptilia	Ophidia	Colubridae	*Eirenis coronelloides*	LC	N
Reptilia	Ophidia	Colubridae	*Eirenis decemlineatus*	LC	Y
Reptilia	Ophidia	Colubridae	*Eirenis eiselti*	LC	Y
Reptilia	Ophidia	Colubridae	*Eirenis hakkariensis*	DD	N
Reptilia	Ophidia	Colubridae	*Eirenis levantinus*	LC	Y
Reptilia	Ophidia	Colubridae	*Eirenis lineomaculatus*	LC	Y
Reptilia	Ophidia	Colubridae	*Eirenis modestus*	LC	N
Reptilia	Ophidia	Colubridae	*Eirenis occidentalis*	NE	Y
Reptilia	Ophidia	Colubridae	*Eirenis persicus*	LC	N
Reptilia	Ophidia	Colubridae	*Eirenis punctatolineatus*	LC	N
Reptilia	Ophidia	Colubridae	*Eirenis rothii*	LC	Y
Reptilia	Ophidia	Colubridae	*Eirenis thospitis*	DD	Y
Reptilia	Ophidia	Colubridae	*Elaphe quatuorlineata*	NT	Y
Reptilia	Ophidia	Colubridae	*Elaphe sauromates*	LC	N
Reptilia	Ophidia	Colubridae	*Hemorrhois algirus*	LC	Y
Reptilia	Ophidia	Colubridae	*Hemorrhois hippocrepis*	LC	N
Reptilia	Ophidia	Colubridae	*Hemorrhois nummifer*	LC	N
Reptilia	Ophidia	Colubridae	*Hemorrhois ravergieri*	LC	N
Reptilia	Ophidia	Colubridae	*Hierophis gemonensis*	LC	Y

Class	Order	Family	Species	IUCN Red List category	Endemic (Yes/No)
Reptilia	Ophidia	Colubridae	*Hierophis viridiflavus*	LC	Y
Reptilia	Ophidia	Colubridae	*Lycophidion capense*	LC	N
Reptilia	Ophidia	Colubridae	*Lytorhynchus diadema*	LC	N
Reptilia	Ophidia	Colubridae	*Macroprotodon abubakeri*	DD	Y
Reptilia	Ophidia	Colubridae	*Macroprotodon brevis*	NT	Y
Reptilia	Ophidia	Colubridae	*Macroprotodon cucullatus*	LC	Y
Reptilia	Ophidia	Colubridae	*Macroprotodon mauritanicus*	NE	Y
Reptilia	Ophidia	Colubridae	*Malpolon insignitus*	NE	N
Reptilia	Ophidia	Colubridae	*Malpolon monspessulanus*	LC	N
Reptilia	Ophidia	Colubridae	*Muhtarophis barani*	NE	Y
Reptilia	Ophidia	Colubridae	*Platyceps collaris*	LC	Y
Reptilia	Ophidia	Colubridae	*Platyceps elegantissimus*	DD	N
Reptilia	Ophidia	Colubridae	*Platyceps florulentus*	LC	N
Reptilia	Ophidia	Colubridae	*Platyceps najadum*	LC	N
Reptilia	Ophidia	Colubridae	*Platyceps rhodorachis*	LC	N
Reptilia	Ophidia	Colubridae	*Platyceps rogersi*	LC	N
Reptilia	Ophidia	Colubridae	*Platyceps sinai*	DD	Y
Reptilia	Ophidia	Colubridae	*Platyceps tessellata*	NE	Y
Reptilia	Ophidia	Colubridae	*Platyceps ventromaculatus*	LC	N
Reptilia	Ophidia	Colubridae	*Rhagerhis moilensis*	NE	N
Reptilia	Ophidia	Colubridae	*Rhynchocalamus melanocephalus*	LC	N
Reptilia	Ophidia	Colubridae	*Rhynchocalamus satunini*	NE	N
Reptilia	Ophidia	Colubridae	*Spalerosophis diadema*	LC	N
Reptilia	Ophidia	Colubridae	*Spalerosophis dolichospilus*	DD	Y
Reptilia	Ophidia	Colubridae	*Telescopus dhara*	LC	N
Reptilia	Ophidia	Colubridae	*Telescopus fallax*	LC	N
Reptilia	Ophidia	Colubridae	*Telescopus hoogstraali*	EN	Y
Reptilia	Ophidia	Colubridae	*Telescopus nigriceps*	LC	N
Reptilia	Ophidia	Colubridae	*Telescopus obtusus*	NE	N
Reptilia	Ophidia	Colubridae	*Telescopus tripolitanus*	NE	N
Reptilia	Ophidia	Colubridae	*Zamenis hohenackeri*	LC	N
Reptilia	Ophidia	Colubridae	*Zamenis lineatus*	DD	Y
Reptilia	Ophidia	Colubridae	*Zamenis longissimus*	LC	N

http://dx.doi.org/10.5772/intechopen.70357

Class	Order	Family	Species	IUCN Red List category	Endemic (Yes/No)
Reptilia	Ophidia	Colubridae	*Zamenis scalaris*	NE	Y
Reptilia	Ophidia	Colubridae	*Zamenis situla*	LC	N
Reptilia	Ophidia	Elapidae	*Naja haje*	LC	N
Reptilia	Ophidia	Elapidae	*Naja nubiae*	LC	N
Reptilia	Ophidia	Elapidae	*Naja pallida*	NE	N
Reptilia	Ophidia	Elapidae	*Walterinnesia aegyptia*	LC	N
Reptilia	Ophidia	Elapidae	*Walterinnesia morgani*	NE	N
Reptilia	Ophidia	Lamprophiidae	*Boaedon fuliginosus*	NE	N
Reptilia	Ophidia	Leptotyphlopidae	*Myriopholis algeriensis*	LC	N
Reptilia	Ophidia	Leptotyphlopidae	*Myriopholis cairi*	NE	N
Reptilia	Ophidia	Leptotyphlopidae	*Myriopholis macrorhyncha*	NE	N
Reptilia	Ophidia	Natricidae	*Natrix maura*	LC	Y
Reptilia	Ophidia	Natricidae	*Natrix natrix*	LC	N
Reptilia	Ophidia	Natricidae	*Natrix tessellata*	LC	N
Reptilia	Ophidia	Typhlopidae	*Letheobia episcopus*	DD	Y
Reptilia	Ophidia	Typhlopidae	*Letheobia simonii*	LC	Y
Reptilia	Ophidia	Typhlopidae	*Xerotyphlops vermicularis*	NE	N
Reptilia	Ophidia	Viperidae	*Bitis arietans*	LC	N
Reptilia	Ophidia	Viperidae	*Cerastes boehmei*	NE	Y
Reptilia	Ophidia	Viperidae	*Cerastes cerastes*	LC	N
Reptilia	Ophidia	Viperidae	*Cerastes gasperettii*	LC	N
Reptilia	Ophidia	Viperidae	*Cerastes vipera*	LC	N
Reptilia	Ophidia	Viperidae	*Daboia deserti*	NT	Y
Reptilia	Ophidia	Viperidae	*Daboia mauritanica*	NT	Y
Reptilia	Ophidia	Viperidae	*Daboia palaestinae*	LC	Y
Reptilia	Ophidia	Viperidae	*Echis coloratus*	LC	N
Reptilia	Ophidia	Viperidae	*Echis leucogaster*	LC	N
Reptilia	Ophidia	Viperidae	*Echis pyramidum*	LC	N
Reptilia	Ophidia	Viperidae	*Macrovipera lebetina*	LC	N
Reptilia	Ophidia	Viperidae	*Macrovipera schweizeri*	EN	Y
Reptilia	Ophidia	Viperidae	*Montivipera albizona*	EN	Y
Reptilia	Ophidia	Viperidae	*Montivipera bornmuelleri*	EN	Y
Reptilia	Ophidia	Viperidae	*Montivipera bulgardaghica*	LC	Y

Class	Order	Family	Species	IUCN Red List category	Endemic (Yes/No)
Reptilia	Ophidia	Viperidae	*Montivipera raddei*	NT	N
Reptilia	Ophidia	Viperidae	*Montivipera xanthina*	LC	Y
Reptilia	Ophidia	Viperidae	*Pseudocerastes fieldi*	LC	N
Reptilia	Ophidia	Viperidae	*Vipera ammodytes*	LC	N
Reptilia	Ophidia	Viperidae	*Vipera anatolica*	EN	Y
Reptilia	Ophidia	Viperidae	*Vipera aspis*	LC	N
Reptilia	Ophidia	Viperidae	*Vipera barani*	NT	Y
Reptilia	Ophidia	Viperidae	*Vipera berus*	LC	N
Reptilia	Ophidia	Viperidae	*Vipera latastei*	NT	Y
Reptilia	Ophidia	Viperidae	*Vipera monticola*	NT	Y
Reptilia	Ophidia	Viperidae	*Vipera seoanei*	LC	Y
Reptilia	Ophidia	Viperidae	*Vipera ursinii*	VU	N
Reptilia	Ophidia	Viperidae	*Vipera walser*	NE	Y
Crocodilia	Crocodilia	Crocodylidae	*Crocodylus niloticus*	LC	N

Author details

Kerim Çiçek* and Oğzukan Cumhuriyet

*Address all correspondence to: kerim.cicek@ege.edu.tr

Zoology Section, Department of Biology, Faculty of Science, Ege University, Izmir, Turkey

References

[1] Blondel J, Aronson J, Bodiou J-Y, Boeuf G. The Mediterranean Region: Biological Diversity in Space and Time. 1st ed. New York: Oxford University Press Inc; 2010. 376 p. DOI: 978-0-19-955798-1

[2] Communities E. Natura 2000 in the Mediterranean Region. 1st ed. Luxembourg: Office for Official Publications of the European Communities; 2009. 12 p. DOI: 10.2779/77695

[3] Arnold C. Mediterranean Islands, 1st ed. London: Survival Books; 2008. 416 p. DOI: 978-0955648915

[4] Mittermeier RA, Turner WR, Larsen FW, Brooks TM, Gascon C. Global biodiversity conservation: The critical role of hotspots. In: Zachos FE, Habel JC, editors. Biodiversity

Hotspots: Distribution and Protection of Conservation Priority Areas. 1st ed. Heidelberg: Springer; 2011. p. 3-22. DOI: 783642209918

[5] Cox N, Chanson J, Stuart S. The Status and Distribution of Reptiles and Amphibians of the Mediterranean Basin, 1st ed. Switzerland and Cambridge: IUCN; 2006. 42 p. DOI: 978-2-8317-0912-3

[6] AmphibiaWeb. AmphibiaWeb [Internet]. 2017 [Updated: 2017]. Available from: http://amphibiaweb.org [Accessed: March 27, 2017]

[7] Uetz P, Freed P, Hošek J. The Reptile Database [Internet]. 2017 [Updated: 2017]. Available from: http://reptile-database.org [Accessed: March 27, 2017]

[8] IUCN. The IUCN Red List of Threatened Species [Internet]. 2017 [Updated: 2017]. Available from: http://www.iucnredlist.org/ [Accessed: March 27, 2017]

[9] Whitfield SM, de ScottRyan TJ, Buhlmann KA, Tuberville TD, Metts BS, Greene JL, Mills T, Leiden Y, Poppy S. The global decline of reptiles, déjà amphibians. BioScience. 2000;**50**(8):53-666

[10] Cuttelod A, García N, Abdul Malak D, Temple H, Katariya V. The Mediterranean: A biodiversity hotspot under threat. In: Vié J-C, Hilton-Taylor C, Stuart SN, editors. The 2008 Review of The IUCN Red List of Threatened Species. 1st ed. Switzerland: IUCN Gland; 2008. p. 1. DOI: 978-2-8317-1063-1

[11] Barahona F, Evans SE, Mateo JA, García-Márquez M, López-Jurado LF. Endemism, gigantism and extinction in island lizards: the genus Gallotia on the Canary Islands. Journal of Zoology. 2000;**250**(3):373-388

[12] Cox SC, Carranza S, Brown RP. Divergence times and colonization of the Canary Islands by Gallotia lizards. Molecular Phylogenetics & Evolution. 2010;**56**:747-757

[13] IUCN SSC Amphibian Specialist Group. Latonia nigriventer. The IUCN Red List of Threatened Species 2012: e.T6715A13339841 [Internet]. [Updated: 2012] Available from: http://dx.doi.org/10.2305/IUCN.UK.2012-1.RLTS.T6715A13339841.en [Accessed: 08 April 2017]

[14] Salafsky N, Salzer D, Stattersfield AJ, Hilton-Taylor C, Neugarten R, Butchart SHM, Collen B, Cox N, Master LL, O'Connor S, Wilkie D. A standard lexicon for biodiversity conservation: Unified classifications of threats and actions. Conservation Biology. 2007;**22**:897-911. DOI: 10.1111/j.1523-1739.2008.00937.x

[15] Blue Plan. The Blue Plan's Sustainable Development Outlook for the Mediterranean. 1st ed. Sophia Antipolis, France: UNEP Blue Plan Activity Centre; 2008. 26 p

Section 3

Risks and Hazards

Chapter 10

Vulnerability of Soil and Water in Mediterranean Agro-Forestry Systems

António Canatário Duarte

Additional information is available at the end of the chapter

http://dx.doi.org/10.5772/intechopen.70094

Abstract

The study watershed is located within the Idanha Irrigation Scheme, Portugal. A hydrological and water quality station was installed at the outlet of the catchment. The AnnAGNPS model was applied in this study, and afterwards it was calibrated and validated to the conditions in the study catchment. The antecedent soil moisture conditions play an important role for rapid runoff and flash flooding. Relative Water Supply (RWS) is always bellow the unity value and sometimes below the value 0.5. Sometimes in very dry years, like the year 2004-2005 (302 mm), a runoff coefficient is equal to 0.40. Spatial distribution of runoff was primarily influenced by topography and soil management, which is common in Mediterranean agricultural systems, namely in grain crop systems such as oats and wheat. The simulation of spatial distribution of nitrate load shows a dependence of the spatial distribution of runoff, due to its high solubility. Spatial distribution of soil erosion by water indicates that the process does not depend directly on the runoff distribution in the catchment. Therefore, soil erosion is greatly influenced by deficient land cover whenever erosivity of rainfall is strong. Phosphorus losses were less than nitrate losses, due to their lower water solubility and mobility in soil.

Keywords: Mediterranean environment, agro-forestry systems, soil and water conservation, soil erosion, water scarcity, water pollution, Portugal

1. Introduction

Farming activities, as a part of natural resource management practice, impact both soil and water quality at the watershed level. The nonpoint source (NPS) pollution with nitrates and phosphates from agricultural areas, sometimes associated to soil erosion by water, is an important environmental problem [1]. For instance, the source of nitrates and phosphates from agriculture activities is reported as responsible for 46–87 and 20–40%, respectively, into European continental waters [2]. Soil and water conservation practices also help to reduce the loss of chemicals in runoff and maintain water quality [3]. Increases in nutrient losses and riverine nutrient loads have caused eutrophication of many coastal and freshwater ecosystems in many Mediterranean regions (**Figure 1**). Nonpoint source (NPS) pollution is an important environmental and water quality problem closely related to hydrologic behavior of watersheds as a basic management territorial unit [4]. However, watershed being a geographically dynamic unit, its behavior varies both spatially and temporarily. Nonpoint source pollution has become a global environmental issue and it has been a critical issue concerning environmental degradation caused in recent years.

The Mediterranean climate plays an important role in these environmental issues, like soil erosion and degradation of quality water bodies, given the irregularity and uncertainty of climatic patterns. The Mediterranean climate is characterized by irregular inter-annual and intra-annual precipitations (wet years mixed with recurrent droughts), and high concentrations of rainfall over a few months, many times over only a few days, and low rainfall during the summer [5, 6]. On the other hand, many of the conventional Mediterranean arable land cropping systems have a soil surface, which remains uncovered during long periods of time, corresponding to the soil preparation and crop establishment phases, which frequently occur during the periods of high rainfall concentration (**Figure 1**) [7, 8]. The runoff regime associated to this climatic condition determines in many years a low capacity for harvesting a convenient volume of water for agricultural activity and the implementation of some strategies for coping with water scarcity [9]. Both water scarcity and Mediterranean forestry systems create preferential conditions for wildfires, which can considerably change the hydrological processes and the landscape's vulnerability in major peak flow and erosion events [10]. Thus, to have a better understanding of the processes of soil erosion and nonpoint source pollution at the basin scale, to better actuate on their prevention, monitoring studies conducted at a watershed scale are difficult to replicate in the way that traditional plot-scale research is designed, in order to compare responses of alternative management practices using only field observations [11]. At plot scale, it is easier to conduct research actions to study those processes, but it is impossible to obtain accurate valid information for large areas (**Figure 2**). So, computer simulation models provide an efficient and effective alternative for evaluating the effects of agricultural practices on soil and water quality at the watershed level [12].

Figure 1. Soil erosion by water and consequences downstream, namely eutrophication of the water bodies.

http://dx.doi.org/10.5772/intechopen.70094

Figure 2. Plots to evaluate soil erosion by water under some crop rotations (Experimental Station in Polytechnic Institute of Castelo Branco/School of Agriculture).

Erosion is mainly due to climatic conditions, namely the precipitation patterns, and the inadequate use of soil by agriculture and forestry, but also through building constructions and uncontrolled water runoff from roads and other sealed surfaces. According to this, the regions of the Mediterranean basin where the higher rates of soil erosion are mapped are coincident with the regions where the highest values of erosivity density are verified (**Figure 3**). In more than one-third of the total land of the Mediterranean basin, average yearly soil losses can

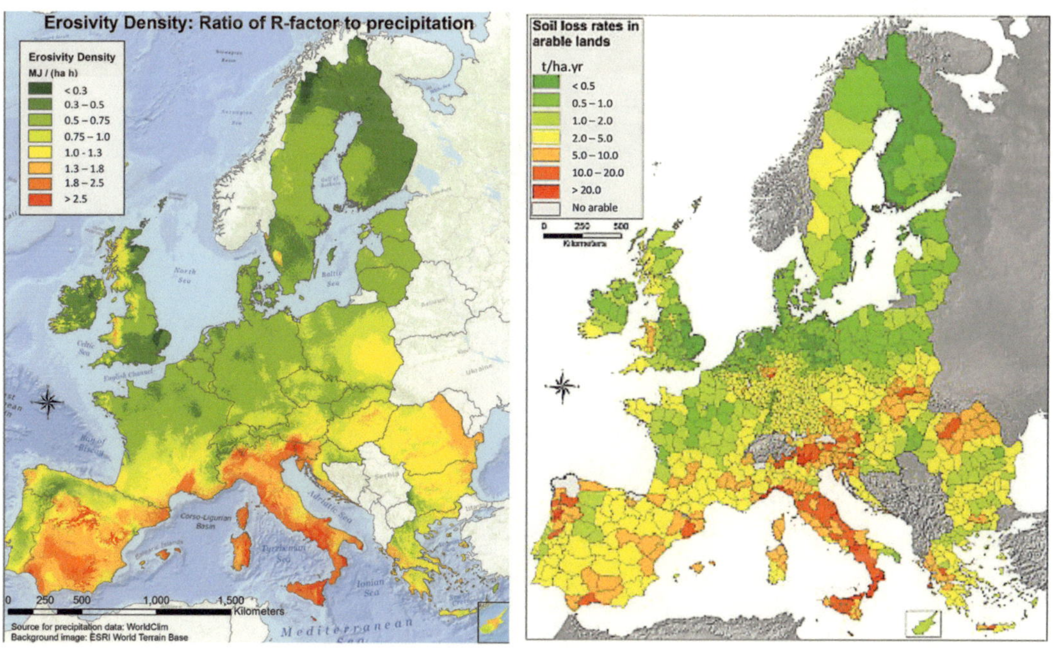

Figure 3. Erosivity density (erosivity factor per precipitation unit) and soil erosion by water in Europe [13].

exceed 15 tons ha^{-1}. On the other hand, loss of organic matter is mainly due to intensive use of the land by agriculture, especially when organic residues are not sufficiently produced or recycled to soil. Agronomists consider soil with less than 1.7% organic matter to be in predesertification stage [13].

The main objective of this study is to understand the dynamics of pollutant sediments, nitrates, and phosphorus in close relation to hydrologic behavior of a small basin in typical Mediterranean environmental conditions.

2. Methodology

The study watershed is located within the Idanha Irrigation Scheme, Idanha-a-Nova County, Portugal, near the border with Spain and just north of the Tagus River (**Figure 4**). The study area is close to the International Tagus Natural Park, where we can find many terrestrial and aquatic unique ecosystems, like a Mediterranean forest very well adapted to these conditions (holm oak forest—*Quercus ilex L.*), and some communities of rare birds of prey (imperial eagle) and black stork. Therefore, this region is very sensitive from an environmental point of view. It covers an area of 189 ha and a perimeter of 6510 m, and presents a third-order hierarchy stream. The main natural stream is 2300 m long and runs north-southwest. The drainage density of the perennial streams is 12.2 m ha^{-1} [14].

Altitude varies from 212 m at the outlet of the basin to 248 m, and the slopes range from 0 to 4%; thus, the topography is flat to gently undulating (**Figure 5**). The limits of the study catchment are well defined.

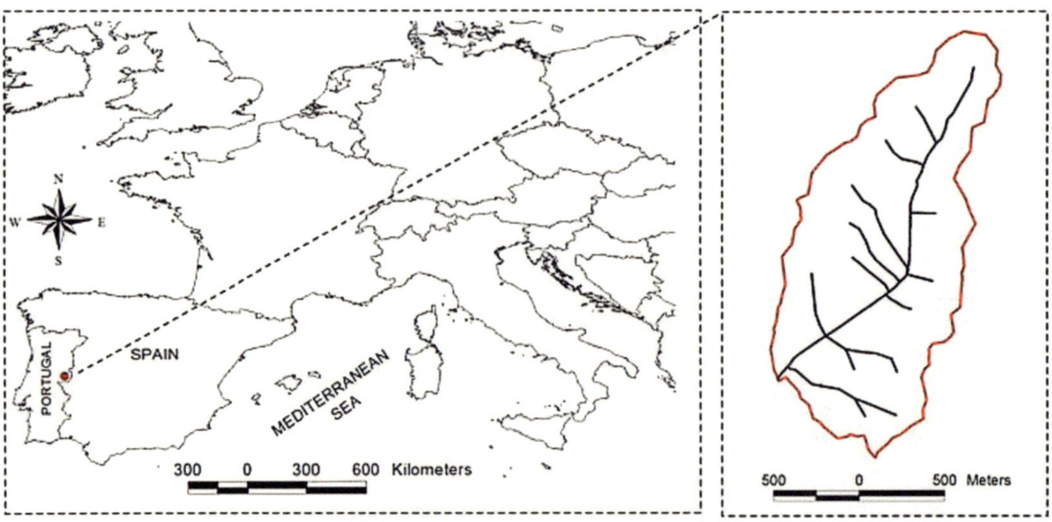

Figure 4. Location of the study catchment in Portugal.

http://dx.doi.org/10.5772/intechopen.70094

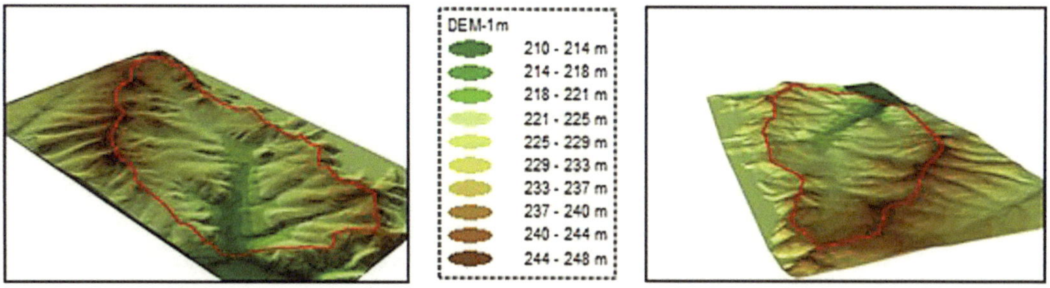

Figure 5. Topography of the study catchment and a downstream and upstream view.

The climate is typically Mediterranean continental. Average annual rainfall is 638 mm, with a rainless summer. The average temperature varies from 8.1°C in January to 25.3°C in August; the average reference evapotranspiration (ET_0) ranges from 0.5 mm day^{-1} in January to 9.0 mm day^{-1} in July (**Figure 6**) [14]. According to the Köppen climate classification, the study catchment enters a Mediterranean Csa type climate (humid temperate with a dry and hot summer).

The study catchment covers an area of 189 ha and it is subdivided into 18 fields. About one-third (31%) of the catchment is not irrigable and is now devoted to a young cork (*Quercus suber* L.) and oak tree (*Quercus rotundifolia* Lam.) forest that was planted in 2001. Three distinct areas can be identified in terms of agricultural practices: uncultivated zone (area with cork and oak young forest), an area with intensive agricultural use in monoculture (maize and tobacco), that in some recent years has remained fallow, and an area with diversity of agricultural uses with smaller fields. The area with young cork and oak forest, located in the eastern part of the basin, can be seen in long fallow as it has not been cultivated for several years [14]. The irrigated fields use sprinklers with either center pivot units or stationary systems. As a consequence of the Common Agriculture Policy, that was put in practice at the beginning of this century, namely almost disappearing with tobacco, the irrigated area decreased to less than half after 2005 irrigation season, with a corresponding increase in fallow area. This reduction has clear implications on the hydrologic behavior of the basin during the irrigation season and subsequent rainfall season [15].

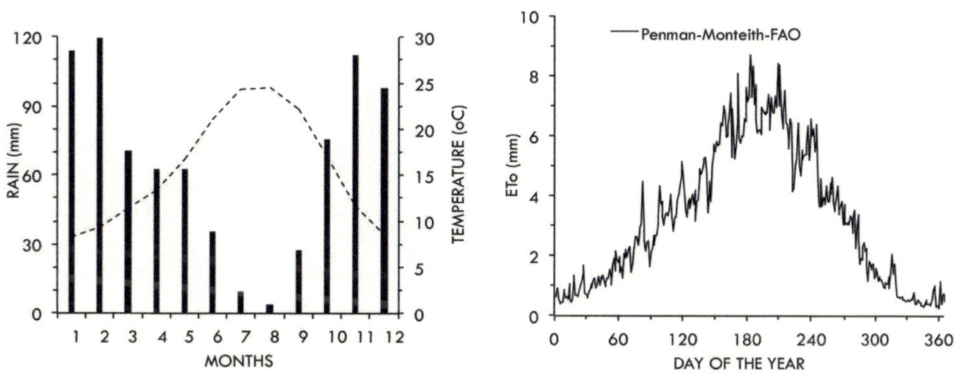

Figure 6. Characterization of climate in the study catchment through parameters of rainfall, temperature, and evapo transpiration.

Soils in the catchment were mapped from preexisting unpublished studies, field inspections, photo-interpretation techniques, and detailed characterization of one soil profile per soil group. According to the FAO classification system [16, 17], the predominant soil groups in the watershed are Luvisols and Cambisol, these ones originated from fluvial deposits associated with the tributaries of the Tagus River (**Figure 7**). Fluvisols are also present in the catchment area, these originated from alluvial deposits associated with the main creek. An impermeable, fractured soil layer underlies the Luvisols classes at a depth of approximately 0.4 m. Luvisols[1] are soils with a subsurface horizon of high activity clay accumulation and high base saturation and show marked textural differences within the profile. The surface horizon is depleted in clay while the subsurface argic horizon has accumulated clay (Bt horizon). A wide range of parent materials and environmental conditions lead to a great diversity of soils in this Reference Soil Group. Other names used for this soil type include sols lessivés (France), Alfisols (Soil Taxonomy), and Mediterranean soils (Portugal). The Luvisols of the Mediterranean region are widely distributed throughout Portugal, Spain, Italy, Greece, Albania, Croatia, Turkey, and Cyprus, which represent a significant percentage of the total area of soils in these countries. Luvisol is a fertile soil suitable for a wide range of agricultural uses. In the Mediterranean, it is commonly used for cereals. On sloping land, it requires measures such as man-made terraces to control erosion, and it is best suited for fruit trees, vineyards, olives, and grazing [13].

A hydrological station was constructed and installed in 2004 at the outlet of a catchment (39°50′48″ N, 7°10′00″ W). The station consisted of (i) a long-throated flume, with a triangular control section for small water depths and triangular/trapezoidal control section for large water depths [18] and (ii) an ultrasonic sensor ("The Probe," manufactured by Milltronics, Siemens Milltronics Process Instruments Inc., Ontario, Canada) connected to a data logger continuously measuring and recording the water level at the flume (**Figure 8**). Pollutants, nitrates, sediments, and salts, were evaluated with a multiparameter probe for monitoring and logging water quality (TROLL 9500 Water Quality Instruments, manufactured by In-Situ Inc.,

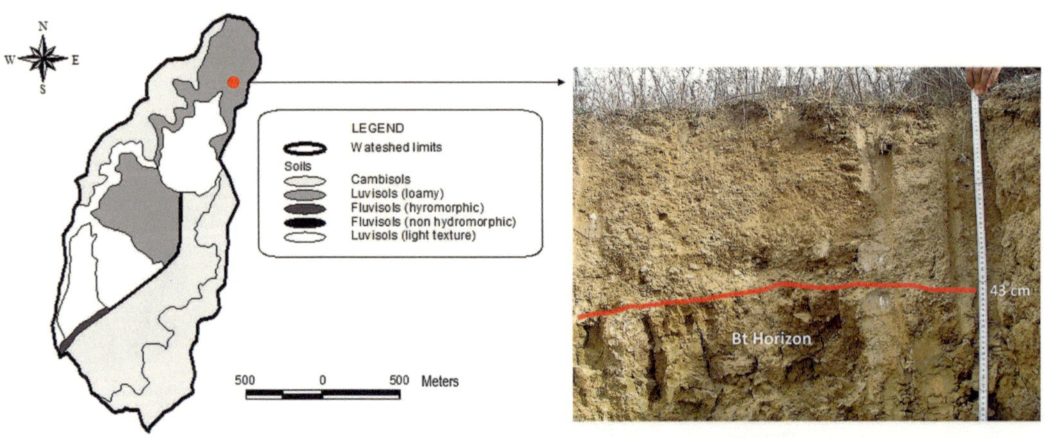

Figure 7. Soils in the study catchment (FAO nomenclature), and a soil profile from a Luvisol.

[1]From the Latin, luere, meaning to wash.

http://dx.doi.org/10.5772/intechopen.70094

Figure 8. Hydrometric and water quality station located at the outlet of the study catchment.

Fort Collins, Colorado, USA). Farming practices, related to Mediterranean crops rotations, were recorded by farmers and verified by direct observations during visits to the catchment to apply to a computer model. The irrigation methods used in the study watershed are sprinkler center pivot and stationary sprinkler (in the areas not covered by the pivot machines).

Several available hydrologic models were evaluated and the Annualized **Ag**ricultural **N**on-**P**oint **S**ource (AnnAGNPS) pollution model was selected as the simulation tool to be used in this study. AGNPS is a suite of computer models resulting from the joint effort between the Agricultural Research Service (ARS) and Natural Resources Conservation Service (NRCS) agencies and it has been developed to evaluate farming and conservation practices through prediction of nonpoint source pollutant loadings within agricultural watersheds. Within AGNPS, AnnAGNPS is a continuous simulation, mixed-land use, watershed-scale computer model designed to predict the origin and movement of water, sediment, and chemicals at any location in agricultural watersheds [19]. The model estimates erosion caused by different processes such as sheet and rill, tillage-induced gullies, classical gullies, and streambed and bank sources [20]. AnnAGNPS has been calibrated, validated, and applied for runoff and sediment yield losses from watersheds in different geographic locations, conditions, and management practices [21, 22]. The AnnAGNPS model was applied in this study and afterward was calibrated and validated to the conditions in the study catchment [23]. **Table 1** lists the main input parameters commonly used in AnnAGNPS simulations.

Irrigation performance and catchment water usage were quantified for various irrigation and rainfed seasons using the following indicators [35, 36]:

$$\text{Runoff coefficient:} \quad ROC = \frac{RO}{R+I} \tag{1}$$

$$\text{Relative irrigation supply:} \quad RIS = \frac{I}{I_{\text{required}}} \tag{2}$$

where ROC is the fraction of rainfall and irrigation that contribute to runoff verified at the outlet of the basin, RO (mm) is the runoff verified at the outlet of the study catchment, R (mm) is rainfall, I is irrigation applied on the irrigated fields in the study basin, and I_{required} is the required irrigation calculated by FAO methodology, as the water requirements to satisfy the full water needs of the crop [37].

Group of parameters	Input variables	Methods
Climate	Daily rainfall	Measurements at the Ladoeiro [24] and Ribeiro de Freixo stations (data not published)
	Daily maximum and minimum temperatures, wind direction and speed, daily percentage cloud cover and dew point temperature	Measurements at the Ribeiro de Freixo station (data not published)
	Annual distribution EI_{30}	Calculated with measured data, from Ladoeiro data [24] and the methodology described by Wischmeier and Smith [25]
	Type of rainfall distribution (TR-55)	Comparison with the 24 h rainfall distribution curve calculated with the data from Ladoeiro station [24]
	Two year 24 h precipitation	Gumbel method [26] applied to the data from Ladoeiro station [24]
Topography	Drainage area and limit of the catchment, cells area, reaches length, mean slope of cells and reaches, RUSLE LS factors	Application of TopAGNPS and AgFlow [27]—ArcView interface 3.2, using a DEM with 1 m vertical resolution
	Topographic and hydrologic configuration	Manipulation of the CSA and MSCL parameters, and comparison to observed and simulated natural stable reaches
Soils	Depth (horizons)	Field observations, cleaning of profiles
	Texture	Method of Robison pipette [28]
	Saturated hydraulic conductivity	Rawls and Brakensiek [29]
	Bulk density	Mass/volume of clods with wax to measure their submerged weight
	Field capacity and wilting point	Richard's methodology
	Percentage of organic matter	Methodology of Walkley-Black [30]
	pH	ISO [31]
	Coarse elements (%)	Field observations
Operations and Management	Crop data	Farmer's information and some bibliography related with crops in the catchment
	Crop operations, included irrigation	Farmer's information and some observation in the fields
	Irrigation application rate	Measured in field [32, 33]
	RUSLE-factors C and P	Wischmeier and Smith [25]
Others	CN	TR-55 [34]
	Manning's n	Wischmeier and Smith [25]

Note: DEM, digital elevation model; CSA, critical source area; MSCL, minimum source critical length; CN, curve number.

Table 1. AnnAGNPS Input parameters and methods used in their evaluation.

3. Results

In Mediterranean climatic conditions, the accumulated erosivity curve shows a stationary phase in the summer months (June, July, August, and September), because it has not practically rained, and two phases of growth, one in the first five months of the year and another, more intense, in the last months of the year. This phase is particularly important, specifically for winter crops such as cereals, since the surface of the soil is unprotected [38, 39]. In contrast, the annual erosive rain is less than 50% of the total annual rain, according to the results of 14 years (2001–2014). The adjustment of linear tendency shows that the larger the total rainfall, the less is the erosive rain (**Figure 9**). Therefore, the analysis of rainfall and erosivity (EI30 methodology, as a compound index of kinetic energy of the rain, E, and the maximum 30-min intensity) [25] distribution over many years (2001–2014) shows typically Mediterranean climate conditions.

Case of a single event (**Figure 10**) shows that the superficial runoff dominates the hydrological response of this basin during the most significant events, known as Hortonian model [40]. This behavior is in part responsible for an almost impermeable layer (horizon Bt), characteristic of Luvisols. The significant variability in runoff is justified not only by differences in characteristics of storms and surface soil moisture content, but also by preceding long dry periods which were likely to ease the effects of rainfall [41, 42]. So, the event occurred on October 20, 2004, in dry soil conditions, reached a discharge equal to 8289 m^3 in sequence of 39.9 mm of precipitation (**Figure 9**). In a subsequent event, on October 27, 2004, in wet soil conditions, the total discharge was 80,457 m^3 derived from 42.3 mm of precipitation (**Figure 10**). This significant difference between two runoff events caused by almost the same precipitation illustrates the decisive influence of the antecedent soil moisture conditions in the magnitude of the flash floods at the small basin scale. It is considered by some authors (e.g., [43]) as the most important soil factor for rapid runoff and flash flooding, with a tendency to produce in Mediterranean region more extreme floods than other European regions [44].

The relation between the water entry in the catchment and the runoff coefficient is completely different under irrigation and rainy seasons. In fact, it is possible to establish a good positive correlation in irrigation, while in the rainy season no relation is possible (**Figure 11**). As is comprehensible, in irrigation conditions, the runoff coefficient is proportional to the volume of irrigation water provided to the fields, given that the water is supplied by a regular behavior. Nevertheless, only from a minimum value of water supplied to the irrigation fields, start to have runoff at the outlet of the catchment. For example, in the 2007 and 2009 irrigations seasons,

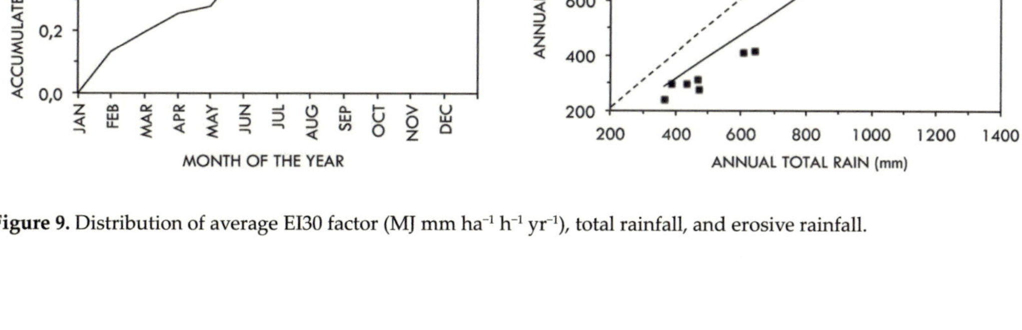

Figure 9. Distribution of average EI30 factor (MJ mm ha^{-1} h^{-1} yr^{-1}), total rainfall, and erosive rainfall.

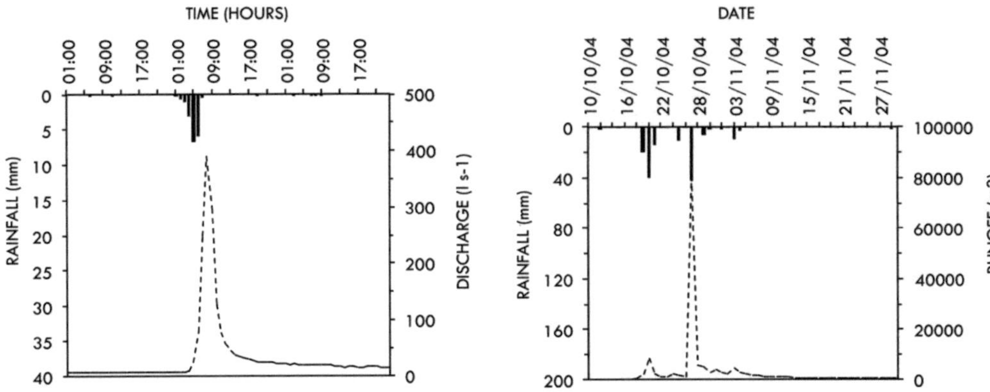

Figure 10. Precipitation and hydrograph (dotted line), recorded between November 30 and December 2, 2004, showing a hortonian behavior [40], and influence of the antecedent soil moisture conditions on the magnitude of hydrologic events.

the water supplied to the irrigations fields was 56 mm and 54 mm respectively **(Table 2)**, and the runoff observed in the outlet of the study basin was 0 mm in both irrigation seasons (**Figure 11**, left chart). On the other hand, as was concluded earlier [14], the reduction of the runoff coefficient is more than proportional relatively to the reduction of water supplied to the catchment. Therefore, it is necessary to make difference on the concept irrigation efficiency at different territorial scales; performance indicators of irrigation at field scale or at catchment scale are not the same [45]. During usual rainy seasons, as in the irrigation seasons, it is possible to obtain a good positive relation between rainfall and runoff coefficient, but in some unusual years the relation is completely shuffled. That is characteristic for Mediterranean climatic conditions, namely precipitation patterns (**Figure 11**, right chart). The highest value of runoff coefficient verified in the study catchment (0.36) is obtained for the rainfall of only 302 mm in a very dry season (2004/2005).

In many regions in the Mediterranean basin, like in the study catchment, the indicator cropping intensity (*CI*), defined as the ratio of arable to cultivated area, is low due to some climatic,

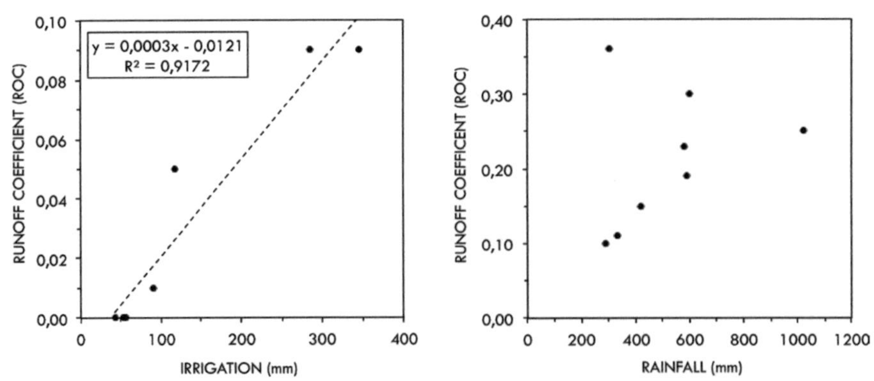

Figure 11. Runoff coefficient (*ROC*) in irrigation and rainfed seasons evaluated in the study catchment.

institutional, or socio-economic factors. This is one of the threats to the region, as a result of the higher susceptibility to environmental damage, like wild fires, and the consequences of soil erosion [46]. In the last century, many rural regions in the Mediterranean basin had significant demographic and socio-economic changes, which determined a substantial decrease in population, abandonment of agricultural land and increase of the human-induced desertification process [47]. Water scarcity and more frequent droughts are other serious problems in the Mediterranean region, which require an enhancement in water management and more efficient water allocation, distribution, and use [48]. Relative water supply (RWS) is always below the unity value, except in the irrigation seasons of 2004 and 2008 in the study catchment, and sometimes even below the value 0.5, clearly denoting water scarcity in most of the irrigation seasons (**Table 2**). Both, deficient water distribution in many irrigation districts and/ or low water storage in the reservoirs, are the most frequent reasons that determine water scarcity conditions on the irrigation fields.

Mediterranean climate, as referred earlier, presents a strong intra- and inter-annual irregularity, being especially evident in the rainfall patterns [49]. Additionally, studies of hydrological behavior in Mediterranean climate using precipitation and discharge averages are not relevant, since this kind of climate is dominated by extreme events [50]. The two graphics in

Cropping season	Year	Rain (mm)	Irrigation (mm)	Runoff (mm)	*CI*	*ROC*	*RIS*
Irrigation	2004		346	32	0.81	0.09	1.11
	2005		118	6	0.39	0.05	0.66
	2006		90	1	0.38	0.01	0.62
	2007		56	0	0.24	0.00	0.60
	2008		286	25	0.64	0.09	1.03
	2009		54	0	0.16	0.00	0.42
	2010						
	2011						
	2012		43	0	0.21	0	0.49
Rainfed	2004–2005	302		109	0.23	0.36	
	2005–2006	591		112	0.14	0.19	
	2006–2007	1021		255	0.14	0.25	
	2007–2008	422		62	0.06	0.15	
	2008–2009	333		36	0.16	0.11	
	2009–2010	601		180	0.32	0.30	
	2010–2011	579		134	0.24	0.23	
	2011–2012	288		28	0.24	0.10	

Table 2. Level of agricultural intensification (*CI*), runoff coefficient (*ROC*), and relative irrigation supply (*RIS*), in the study catchment, during seven irrigation seasons and eight rainfed seasons.

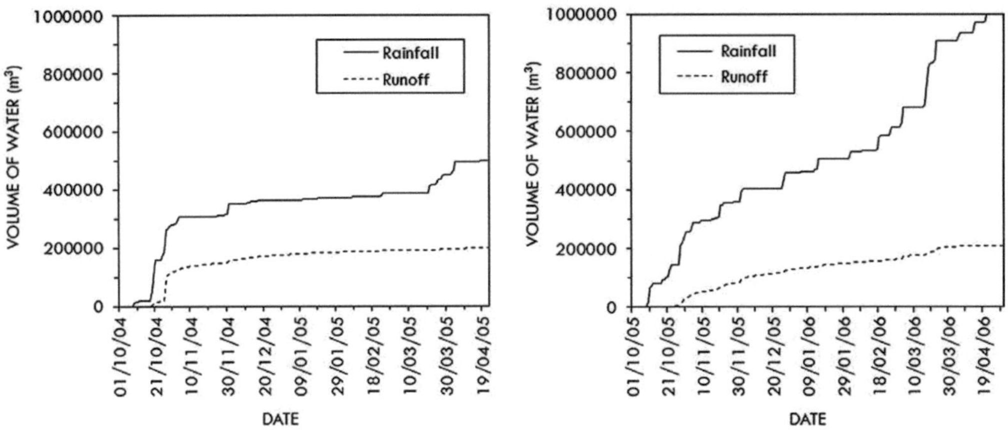

Figure 12. Rainfall and runoff verified in the rainfed seasons 2004–2005, as a dry year, and 2005–2006, as a meteorologically normal year.

Figure 12 show exactly what is a very abnormal (2004–2005) and normal year (2005–2006), as to the quantity and the distribution of rainfall and runoff. In fact, in 2004–2005, the amount of precipitation was 302 mm, distributed mainly into five events and giving a few more intense runoff events, highlighting the first event with a great magnitude. In the same period of time, the amount of runoff was 109 mm, determining a runoff coefficient equal to 0.36, a value which is uncommon in these conditions. In the hydrological year 2005–2006, the amount of rainfall (591 mm) was close to the average value in this region (638 mm) and was distributed by many more rainfall events. It should be noted that the amount of runoff (112 mm) was almost the same as that of the previous year, but giving a lower runoff coefficient (0.19). In the first year of analysis (2004–2005), the runoff started on October 19 and finished on April 20, without any gap in this period (**Figure 12**). Considering that, between December 13 and March 20 the amount of precipitation was 15.3 mm, and the continuous runoff was possible only because the natural channels' network was not disconnected from the aquifers or groundwater table [51].

Mediterranean rainfall can reach, sometimes with very high intensity, very important runoff events, like this one illustrated in **Figure 13**, which occurred in the hydrological year 2004–2005. This event occurred in the conditions of saturated soil, due to the amount of rainfall in the previous days (October 21–23; 77.9 mm), but, as noted before [52], big events can be reached when the intensity of precipitation exceeds the initial capacity of soil infiltration, or the maximum intensity of precipitation occurs in a favorable moment of the storm. In the present event, verified in a very dry year (302 mm), the runoff coefficient was equal to 0.85, characteristic to streams with torrential patterns, and the erosion annual rate in this event was 79.1%; with the potential to carry coarse elements outside the basin. Among the two different types of events identified by Camarasa-Belmonte [53], this one is related to the type with high intensity, occurring at the beginning of the storm and generating hydrographs that are very similar to the hyetographs. The other type of events, according to Camarasa-Belmonte [53], occur in the conditions of low intensity of precipitation and rain peaks at the end of the

http://dx.doi.org/10.5772/intechopen.70094

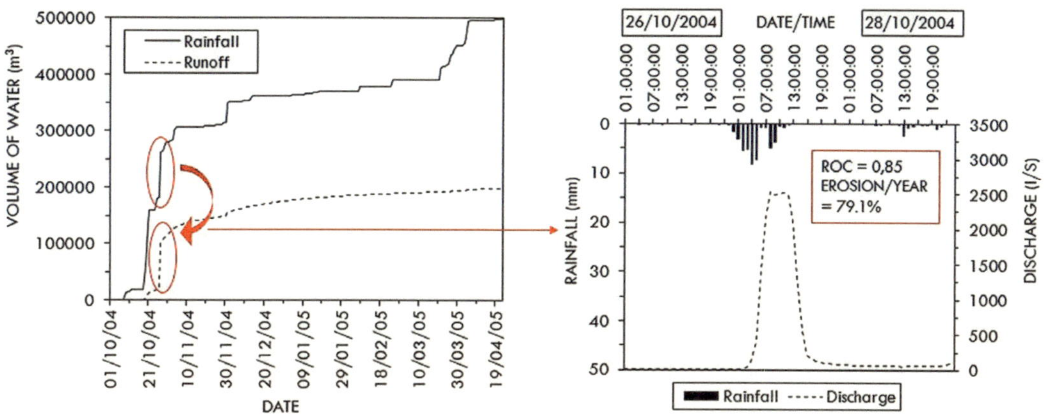

Figure 13. Runoff and erosion rates verified in a very strong event in October 2004.

event, generating hydrographs quite different from hyetographs. Both rising time and recession time are reached relatively quickly, due to the characteristics of this storm and the physical conditions of the catchment. Since this catchment does not have important water storage in the channels, soil and surface, rapid recession time is verified, when the storm stops [54].

The effects of surface runoff and soil erosion, as well as nonpoint source pollution at watershed scale, can be predicted with hydrological models. They are effective to simulate various combinations and scenarios, and indicate the best management practices to minimize these processes [55]. Several hydrological models were evaluated to simulate runoff and soil erosion, as well as nonpoint source pollution by nitrates and phosphorus. The selection indicates the AnnAGNPS model (Annualized Agricultural NonPoint Source), as the most suitable for this task. As for spatial distribution of runoff volume and the ratio of erosion in the basin, the results based on observations (**Figure 14**) seem to indicate that there is a distinction between

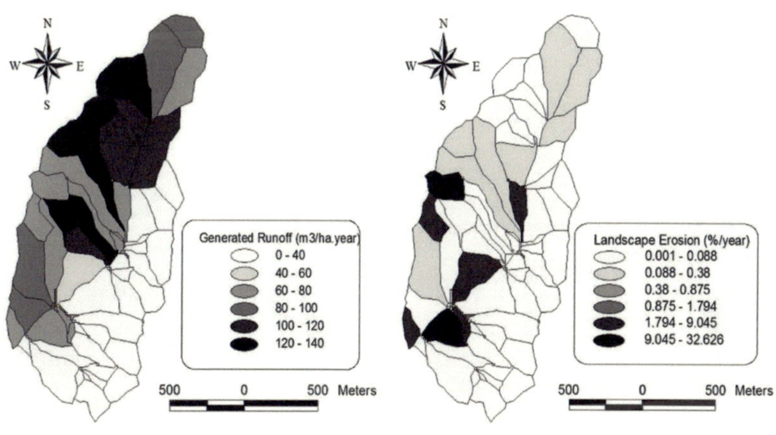

Figure 14. Average simulated runoff and erosion rates, generated in each cell and subcatchment, by AnnAGNPS model (period 2003–2005).

the two runoff-producing areas in the basin: one that generates low runoff (0–40 m^3 ha^{-1} yr^{-1}), which corresponds to the uncultivated area in the basin (58.6 ha), and another, which corresponds to the remaining area of the watershed (130.4 ha) occupied by various crops, producing higher runoff (41–140 m^3 ha^{-1} yr^{-1}). The variation of runoff in the cultivated part of the basin depends on the topography, the type of soil, and their intrinsic characteristics. The relations showed in **Figure 6** enable to conclude that there is no correlation between the average sediment production and the average runoff volume in most of the cells in the basin [38]. Although having an influence on the erosion process, the average volume of runoff is not the determining factor. Other factors such as land cover throughout the year, especially during the concentration of rainfall, and the occurrence of more or less erosive storms appear to have more significant influence [56, 39].

The massive use of fertilizers is common to all the agricultural systems around the world. Nitrogen losses, from the agricultural nonpoint sources, has become one the most significant threats to the quality of water bodies [57]. Due to the solubility of this contaminant (inorganic-N), there is a close relationship between hydrological processes and the loss of nitrogen, with the superficial and subsuperficial runoff. Organic-N is adsorbed in the fine soil particles, while their loss occurs preferably through soil erosion [58]. The loss of nitrates is not affected by rainfall variability as the sediment loss. According to some earlier studies, the base flow is the main source of nitrates at the surface of water bodies [59]. Regarding the nonpoint source pollution of nitrates (**Figure 15**), the research has shown that the process occurs when there is significant runoff, given the solubility of nitrates, and when there is availability of this nutrient in the soil, in sequence of more or less intense fertilization [60]. The values simulated by AnnAGNPS are similar to the reference values obtained in other studies: pristine watersheds and organic agriculture (0.76–10.85 kg NO_3-N ha^{-1} yr^{-1}) [61]; nonirrigated agriculture (26.10 kg NO_3-N ha^{-1} yr^{-1}) [59]; irrigated agriculture (59.00 kg NO_3-N ha^{-1} yr^{-1}) [60]. Moreover, only about 0.3 mg l^{-1} of nitrates is needed for growth of algal and to promote the eutrophication [62]. The critical value was

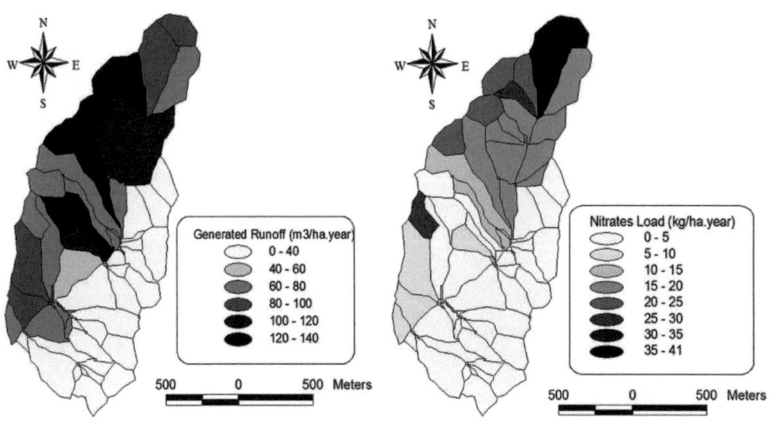

Figure 15. Average simulated runoff and nitrates load, generated in each cell and subcatchment, by AnnAGNPS model (period 2003–2005).

largely surpassed in the period of simulation, as shown in **Figure 15**. According to the efficiency of nitrogen fertilization practices, the loss of nitrates can reach high values (30–80%) in intensive cropping systems [63], like maize or tobacco in the study catchment. So, the riparian buffers and vegetation strips on stream water can be a very good measure to improve water quality downstream of the agricultural fields, as observed and analyzed in some earlier studies [64, 65].

The continuous use of phosphorus in agriculture, many times with low P-use efficiency by crops, increases in phosphorus concentration in the soil, and increasing the risk of phosphorus in runoff along with leaching to water bodies, contribute to the eutrophication [66, 67]. Phosphorus is an element with little mobility in the soil, and can be transferred from agricultural systems to water bodies dissolved in the superficial runoff, leaching to deeper soil layers, or in conjunction with mineral and organic sediment in the water erosion process. Similar to this study, Olness et al. [68] observed losses of P to agricultural watersheds lower than 5% of the most recent P fertilization, with a total P discharge in runoff waters ranging from 1.0 to 11.5 kg P ha^{-1}. Considering the volume of runoff verified in our basin, the value highlighted by some authors as the level to prevent the risk of eutrophication (0.05 mg L^{-1}) [69] is probably often exceeded. Most of the phosphorus used in the agricultural fields, as in our basin, is for winter cereals crops (oat, wheat, rye), most common in Mediterranean agricultural systems. Phosphorus can load in solution outside the basin in surface or subsurface runoff, depending on the nature of P-forms (mineral or organic) and in the soil texture, or attached to the finest soil particles [70, 56]. The analysis of the relation showed in **Figure 16**, referent to the simulated phosphorus load by AnnAGNPS model, suggests that most of the phosphorus is lost in solution and not with the soil colloidal particles loaded outside the watershed, given the subcatchments where high values of P-load occurred (10–12 kg ha^{-1}.yr), and relatively low rates of soil erosion were verified. The phosphorus loss clearly occurred mostly in the agricultural fields and almost null in the forested area, while the high loss values were verified in the fields occupied by winter cereals.

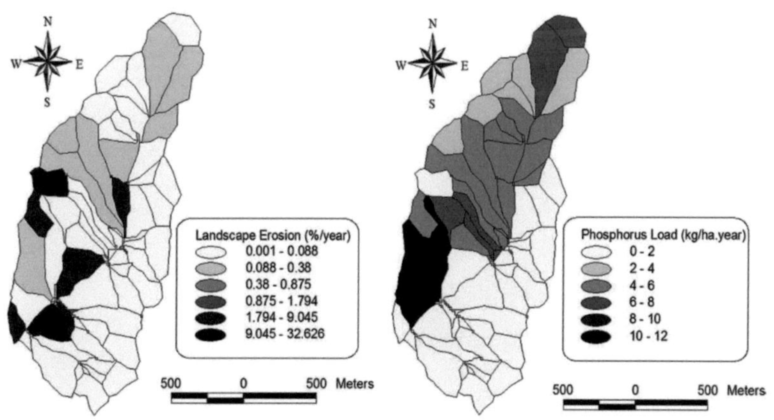

Figure 16. Simulated average phosphorus load by AnnAGNPS model, generated in each subcatchment (period 2003–2005).

4. Conclusions

Both soil and water resources in the Mediterranean environmental conditions are exposed to physical, chemical, and biological degradation. The phase of pronounced erosivity in the Mediterranean climatic condition corresponds to the last 3 months of the year and it is particularly important, mainly for winter crops such as cereals, since the surface of the soil is unprotected. The superficial runoff dominates the hydrological response of the study basin during the most significant events. The antecedent soil moisture condition is considered a factor of greater importance. The irregularity of intra- and inter- annual precipitation is a pattern specific to the Mediterranean climate, that usually determines the grade of soil cover and the soil erosion process. The climate change can additionally aggravate this situation. Water scarcity is another threat in the irrigated agricultural systems of the Mediterranean region. It is verified in many irrigation seasons in this study, where the indicator RWS was much higher. Mediterranean rainfall, sometimes with very high intensity and verified in dry years, can reach very important runoff events like the one that occurred at the end of October in the hydrological year 2004–2005. Runoff coefficient was equal to 0.85 and an annual erosion rate equal to 79.1%. Spatial distribution of runoff was primarily influenced by topography and soil management, which is common to Mediterranean agricultural systems, namely in grain crops systems such as oats and wheat. The simulation of spatial distribution of nitrate loads shows a dependence of the spatial distribution of runoff, due to its high solubility. Despite the close relation between soil erosion, water and runoff, the simulation using the AnnAGNPS model of spatial distribution of soil erosion by water indicates that the process does not directly depend on it. Therefore, soil erosion is mostly influenced by deficient land cover whenever erosivity of rainfall is strong. Phosphorus losses were less than nitrate losses, due to their lower water solubility and mobility in soil.

Acknowledgements

The author gratefully acknowledges funding from Portuguese Foundation for Science and Technology (FCT) for financially supporting this research through the project GeoBioSciences, GeoTechnologies, and GeoEngineering (UID/GEO/04035/2013).

Author details

António Canatário Duarte[1,2]*

*Address all correspondence to: acduarte@ipcb.pt

1 Polytechnic Institute of Castelo Branco, School of Agriculture, Castelo Branco, Portugal

2 GEOBIOTEC—Center for GeoBioSciences, GeoTechnologies and GeoEngineering (Project funding by Portuguese Foundation for Science and Technology UID/GEO/04035/2013), Aveiro, Portugal

References

[1] Thornton JA, Rast W, Holland MM, Jolankai G, Ryding SO. Man and the Biosphere Series. Paris: UNESCO; 1999. p. 478

[2] EEA—European Environment Agency. Nutrients in Europe Ecosystems. Environmental Assessment Report No. 4 EEA. 1999

[3] Berry JK, Delgado JA, Khosla R, Pierce FJ. Precise conservation for environmental sustainability. Journal of Soil and Water Conservation. 2003;**58**(6):332-339

[4] European Community. Directive 2000/60/EC of October 23 2000 of the European Parliament and of the Council establishing a framework for community action in the field of water policy. Official Journal of the European Communities. 2000;**L327**:1-72

[5] García-Barron L, Aguilar M, Sousa A. Evolution of annual rainfall irregularity in the southwest of the Iberian Peninsula. Theoretical and Applied Climatology. 2011;**103**(1-2):13-26

[6] Lionello P, Malanotte P, Boscolo R. Mediterranean Climate Variability. Amsterdam: Elsevier; 2006

[7] Poesen J, Nachtergaele J, Verstraeten G, Valentin C. Gully erosion and environmental change: Importance and research needs. Catena. 2003;**50**(2-4):91-133

[8] DeSantisteban L, Casalí J, López JJ. Assessing soil erosion rates in cultivated areas of Navarre (Spain). Earth Surface Processes and Landforms. 2006;**31**(4):487-506

[9] Pereira LS, Cordery I, Lacovides I. Coping with Water Scarcity-Addressing the Challenges. Springer; Dordrecht (Netherlands), 2009. p. 382. DOI: 10.1007/978-1-4020-9579-5

[10] Stoof CR, Vervoort RW, Iwema J, Elsen E, Ferreira AJ, Ritsema CJ. Hydrological response of a small catchment burned by experimental fire. Hydrology and Earth System Sciences. 2012;**16**(267-285):1027-5606

[11] Mohammed H, Yohannes F, Zeleke G. Validation of agricultural non-point source (AGNPS) pollution model in Kori watershed, South Wollo, Ethiopia. International Journal of Applied Erath Observation and Geoinformation. 2004;**6**:97-109

[12] He C. Integration of geographic information systems and simulation model for watershed management. Environmental Modelling & Software. 2003;**18**:809-813

[13] EC, European Commission, Joint Research Centre & European Soil Bureau, editors. Soil Atlas of Europe. Luxembourg: European Commission; 2005

[14] Duarte AC. Non-point source pollution caused by agricultural activity, at watershed scale [thesis]. Córdoba (Spain): University of Córdoba; 2006. p. 192

[15] Duarte AC. Non-point source pollution in an agricultural catchment and the quality of return flows, under Mediterranean conditions. Fresenius Environmental Bulletin. 2014;**23/No12a/2014**:3237-3247

[16] FAO, Food and Agriculture Organization of the United Nations, editors. World Reference Base for Soil Resources-FAO World Soil Resources Report 84. Rome; 1998

[17] FAO, Food and Agriculture Organization of the United Nations, editors. World reference base for soil resources 2014. International soil classification system for naming soils and creating legends for soil maps. Update 2015. World Soil Resources Reports No. 106. Rome; 2016

[18] Bos MG, Replogle JA, Clemmens AJ. Flow measuring flumes for open channel systems. New York: John Wiley & Sons; 1991. p. 321

[19] Cronshey RG, Theurer FG. AnnAGNPS—Non-point pollutant loading model. In: Proceedings of First Federal Interagency Hydrologic Modelling Conference; 19-23 April; Las Vegas, NV, ASAE (American Society of Agricultural Engineers), St. Joseph-MI (USA); 1998

[20] Bingner RL, Wells RR, Momm HG, Theurer FD, Frees LD. Development and application of gully erosion components within the USDA AnnAGNPS watershed model for precision agriculture. In: Proceedings of the International Conference on Precision Agriculture; July 18-21; Denver, Colorado, International Society of Precisition Agriculture (ISPA), Monticello-IL (USA); 2010

[21] Sadeghi A, Kwang Y, Graff C, McCarty G, McConnell L, Shirmohammadi A, Hively D, Sefton KA. Assessing the performance of SWAT and AnnAGNPS Models in a Coastal Plain Watershed, Choptank River, Maryland, USA. In: ASABE Meeting. St. Joseph, MI: American Society of Agricultural and Biological Engineers; 2007

[22] Licciardello F, Zema DA, Zimbone SM, Bingner RL. Runoff and soil erosion evaluation by AnnAGNPS model in a small Mediterranean watershed. Transactions of the ASABE. 2007;**59**(5):1585-1593

[23] Duarte AC, Mateos L, Fereres E. Modelling runoff with AnnAGNPS model in a small agricultural catchment, in Mediterranean environment. In: Proceedings of the 21st Century Watershed Technology Conference and Workshop; May 27-June 1, 2012; Bari (Italy). St. Joseph MI: ASABE; 2012

[24] INAG-Serviço Nacional de Informação de Recursos Hídricos. Boletim de Precipitação Anual—Estação do Ladoeiro-14N/02U [Internet]. 2008. Available from: http//snirh.inag.pt/snirhwww.php?main_id=18item=4.3 [Accessed: January 25, 2008]

[25] Wischmeier WH, Smith DD. Predicting Rainfall Erosion Losses—Agricultural Handbook No.537. Washington, DC: United States Department of Agriculture; 1978

[26] Gumbel EJ. Statistics of Extremes. New York: Columbia University Press; 1958

[27] Garbrecht J, Martz LW. TOPAGNPS, an automated digital landscape analysis tool for topographic evaluation, drainage identification watershed segmentation and subcatchment parametrization for AGNPS 2001. In: Agricultural Research Service, editor. Washington, DC: Watershed Modelling Technology; 1999

[28] USDA-SCS, United States Department of Agriculture-Soil Conservation Service. Soil Survey Laboratory Methods and Procedures for Collecting Soil Samples-Soil Survey Report 1. Washington, DC; 1982

[29] Rawls WJ, Brakensiek DL. Estimation of soil and water retention and hydraulic properties. In: Morel-Seytoux H.J., editor. Unsatured Flow in Hydrologic Modelling Theory and Practice. Beltsville, MD: Kluwer Academic Publishers; 1989. pp. 275-300

[30] Nelson DW, Sommers LE. Total carbon, organic carbon, and organic matter. In: Page AL, Miller H, Keeney DR, editors. Methods of Soil Analysis: Chemical and Microbiological Properties. 2nd ed. Madison, WI: Soil Science Society of America; 1982. pp. 539-577

[31] International Organization for Standardization. Soil Quality—Determination of pH—Standard ISO 10390:2005. Geneva, Switzerland: Standards ISO; 2005

[32] ASAE, American Society of Agricultural Engineers, editor. ASAE Standards 2005. St. Joseph, MI, American Society of Agricultural Engineers (ASAE); 2005

[33] Merriam JL, Keller J. Farm Irrigation Systems Evaluation: A Guide for Management. Logan, UT: Utah State University; 1978

[34] USDA-NRCS, United States Department of Agriculture-Natural Resources Conservation Service, editor. Urban Hydrology for Small Watersheds-Technical Release 55. Washington, DC; 1986

[35] Molden DJ, Gates TK. Performance measures for evaluation of irrigation-water-delivery systems. Journal of Irrigation and Drainage Engineering. 1990;**116**:804-823

[36] Burt CM, Clemmens AJ, Strelkoff TS, Solomon KH, Bliesner RD, Hardy LA, Howell TA, Eisenhauer DE. Irrigation performance measures: Efficiency and uniformity. Journal of Irrigation and Drainage. 1997;**123**:423-442

[37] Allen RG, Pereira LS, Raes D, Smith M. Crop Evapotranspiration. Guidelines for Computing Crop Water Requirements-FAO Irrigation and Drainage Paper No 56. Rome: FAO, Food and Agriculture Organization of the United Nations; 1998

[38] Pathak P, Wani SP, Singh P, Sudi R. Sediment flow behaviour from small agricultural watershed. Agricultural Water Management. 2004;**67**:105-117. DOI: 10.1016/j.agwat.2004.02.005

[39] Duarte AC, Mateos L, Fereres E. Application of the AnnAGNPS model to a small agricultural watershed, to analyse the spatial and temporal distribution of the pollutants and runoff. In: Proceedings of the 5th International Congress of the European Society for Soil Conservation; 25-28 June; Palermo (Italy), European Society for Soil Conservation (ESSC); 2007

[40] Horton RE. Erosional development of streams and their drainage basins: Hydrophysical approach to quantitative morphology. Bulletin of the Geological Society of America. 1945;**56**:275-370

[41] Taguas EV, Ayuso JL, Peña A, Yuan Y, Pérez R. Evaluating and modelling the hidrological and erosive behaviour of an olive orchard microcatchment under no-tillage with bare soil in Spain. Earth Surface Processes and Landforms. 2009;**34**:738-751

[42] Beven KJ. Rainfall-Runoff Modelling—The Primer. 2nd ed. Chichester: Wiley-Blackwell; 2012. p. 488

[43] Grillakis MG, Koutroulis AG, Kommab J, Tsanis IK, Wagner W, Bloschl G. Initial soil moisture effects on flash flood generation—A comparison between basins of contrasting hydro-climatic conditions. Journal of Hydrology. 2016;**541**:206-217

[44] Gaume E, Bain V, Bernardara P, Newinger O, Barbuc M, Bateman A, Blaskovicová L, Bloschl G, Borga M, Dumitrescu A, Daliakopoulos I, Garcia J, Irimescu A, Kohnova S, Koutroulis A, Marchi L, Matreata S, Medina V, Preciso E, Sempere-Torres D, Stancalie G, Szolgay J, Tsanis I, Velasco D, Viglione A. A compilation of data on European flash floods. Journal of Hydrology. 2009;**367**:70-78

[45] Lorite IJ, Mateos L, Fereres E. Evaluating irrigation performance in a Mediterranean environment. II. Variability among crops and farmers. Irrigation Science. 2004;**23**:85-92

[46] Lourenço L, Nunes AN, Bento-Gonçalves A, Vieira A. Soil erosion after wildfires in Portugal: What happens when heavy rainfall events occurs. In: IN-TECH, editor. Research on Soil Erosion. IN-TECH, Rijeka (Croatia); 2012. pp. 65-88. DOI: http://dx.doi.org/10.5772/50447

[47] Moreira F, Viedma O, Arianoutsou M, Curt T, Koutsias N, Rigolot E. Landscape and wild fire interactions in southern Europe: Implications for landscape management. Journal of Environmental Management. 2011;**92**:2389-2402

[48] Esteve P, Varela-Ortega C, Blanco-Gutierrez I, Downing TE. A hydro-economic model for the assessment of climate change impacts and adaptation in irrigated agriculture. Ecological Economics. 2015;**120**:49-58

[49] García-Barrón L, Camarillo JM, Morales J, Sousa A. Temporal analysis (1940-2010) of rainfall aggressiveness in the Iberian Peninsula basins. Journal of Hydrology. 2015;**525**:747-759

[50] Brakenridge GR. River flood regimen and floodplain stratigrafy. In: Baker VR, Kochel RC, Patton PC, editors. Flood Geomorphology. Wiley and Sons, New York (USA); 1988. pp. 139-155

[51] Longobardi A, Villani P, Grayson RB, Western AW. On the relationship between runoff coefficient and catchment initial conditions. In: Proceeding of MODSIM 2003; 2003. pp. 867-872

[52] Cerdà A, Doerr SH. Soil wettability, runoff and erodibility of major dry-Mediterranean land uses types on calcareous soils. Hydrological Processes. 2007;**21**:2325-2336

[53] Camarasa-Belmonte AM. Flash floods in Mediterranean ephemeral streams in Valencia Region (Spain). Journal of Hydrology. 2016;**541**:99-115

[54] Bras R. Hydrology. An Introduction to Hydrologic Science. New York: Addison-Wesley Pub; 1990. p. 633

[55] Zema DA, Bombino G, Denisi P, Licciardello F, Zimbone SM. Prediction of surface runoff and soil erosion at watershed scale: Analisys of the AnnAGNPS model in different environmental conditions. In: Godone D, Stanchi S, editor. Research on Soil Erosion. IN-TECH, Rijeka (Croatia); 2012. pp. 4-31. DOI: http://dx.doi.org/10.5772/50427

[56] Lemunyon JL, Daniel TC. Quantifying phosphorus losses from the agricultural system. Journal of Soil and Water Conservation. 2002;**57**(6):399-401

[57] Carpenter SR, Caraco NF, Correl DL, Howarth RW, Sharpley AN, Smith VH. Non-point pollution of surface waters with phosphorus and nitrogen. Ecological Applications. 1998;**8**(3):559-568

[58] Dekun S. Characteristics of nitrogen loss on slope under simulated rainfall. Bulletin of Soil and Water Conservation. 2009;**29**(5):98-101

[59] Schilling K, Zhang YK. Baseflow conntribution to nitrate-nitrogen export from a large agricultural watershed, USA. Journal of Hydrology. 2004;**295**:305-316

[60] Causapé J, Quilez D, Aragues R. Salt and nitrate concentrations in the surface waters of the CR-V irrigation district (Bardenas I, Spain): diagnosis and prescriptions for reducing off-site contamination. Journal of Hydrology. 2004;**295**(1-4):87-100

[61] Honisch M, Hellmeier C, Weiss K. Response of surface and subsurface water. Geoderma. 2002;**105**(3-4):277-298

[62] Brooks KN, Folliott PF, Gregersen HM, DeBano LF. Hydrology and the Management of Watersheds. 2nd ed. Ames: Iowa State University Press; 1997. p. 502

[63] Angás P, Lampurlané J, Cantero-Martinez C. Tillage and N fertilization effects on N dynamics and barley yield under semiarid Mediterranean conditions. Soil &Tillage Research. 2006;**87**:59-71

[64] Broadmeadow S, Niesbet TR. The effects of riparian forest management on the freshwater environment: A literature review of best management practice. Hydrology and Earth System Science. 2004;**8**(3):286-305

[65] Fiener P, Auerswald K. Seasonal variation of grassed waterway effectiveness in reducing runoff and sediment delivery from agricultural watersheds in temperate Europe. Soil & Tillage Research. 2006;**87**:48-58

[66] Kleinman PJA, Sharpley AN, McDowell RW, Flaten DN, Buda AR, Tao L, Bergstrom L, Zhu Q. Managing agricultural phosphurus for water quality protection: Principles for progress. Plant and Soil. 2011;**349**:169-182. DOI: http//dx.doi.org/10.1007/s11104-011-0832-9

[67] Sharpley AN, Daniel TC, Gibson G, Bundy L, Cabrera M, Sims T, Stevens R, Lemunyon J, Kleinman PJA, Parry R. Brest Management Practices to Minimize Agricultural Phosphurus Impacts on Water Quality. USDA-ARS Publication 163. Washington DC: U.S. Gov. Printing Office; 2006

[68] Olness A, Smith SJ, Rhoades ED, Menzel RG. Nutrient and sediment discharge from agricultural watersheds in Oklahoma. Journal of Environmental Quality. 1975;**4**:331-336

[69] Golterman HL, Oude NT. Eutrophication of lakes, rivers and coastal seas. In: Hutzinger O, editor. The Handbook of Environmental Chemistry, Vol. 5 Part A. Berlin: Springer-Verlag; 1991. pp. 80-124

[70] Kronvang B, Bechmann M, Lundekvam H, Behrendt H, Rubaek GH, Schoumans OF, Syversen N, Andersen HE, Hoffman CC. Phosphorus losses from agricultural areas in river basins: Effects and uncertainties of target mitigation measures. Journal of Environmental Quality. 2005;**34**:2129-2145. DOI: 10.2134/jeq2004.0439

Chapter 11

Resilience of Mediterranean Forests to Climate Change

Sferlazza Sebastiano, Maetzke Federico Guglielmo,
Miozzo Marcello and
La Mela Veca Donato Salvatore

Additional information is available at the end of the chapter

http://dx.doi.org/10.5772/intechopen.68943

Abstract

In the Mediterranean region, forests play a key role in the welfare of urban and rural peoples, by providing highly appreciated marketed goods as well as high value but nonmarket services. Anthropogenic climate change entails a significant impact on Mediterranean forests, such as the reduced species diversity, high density, pest, and diseases. In this chapter, we present the achievements of the project LIFE ResilForMed (Resilience of Mediterranean Forests to climate change), as Sicilian responds on adaptation and mitigation to climate change: (i) map of Sensitivity Forest Areas to desertification of the Sicilian region; (ii) list of bird indicators showing sensitivity to desertification in Sicily; (iii) set of resilience indicators aimed to assess the resilience of Mediterranean forests to climate changes; (iv) list of sporadic and endemic tree species of Sicilian forests; (v) development of optimal management models suitable to improve or consolidate the resilience of forest and preforest ecosystems. This work is at the forefront of developing knowhow and transferable best management practices that may help Mediterranean forests to adapt to climate change and thereby, safeguard their multifunctional benefits for future generations.

Keywords: climate change, Mediterranean forests, adaptation strategy, resilience indicators, best management practices

1. Introduction

Resilience is defined as the degree, manner, and pace of recovery of an ecosystem properties following natural or human disturbance [1, 2]. As such, it is a quantitative, process-specific measure (ecosystems are more or less resilient for a given variable), and not a generic, qualitative property (ecosystems are not resilient or unresilient). Resilience is a quantitative property describing return to equilibrium and is hence only applicable to stable systems [3]. Mediterranean-type ecosystems have been documented as being highly resilient to disturbances [2, 4]. Recovery of ground cover after a disturbance of vegetation can be relatively fast, while time taken to return to a species composition similar to the initial one might be considerably longer [3]. For example, in different Mediterranean-type ecosystems, it has been observed that both grasslands [5] and shrublands [6, 7] return to their previous state in just a few years. On the other hand, other studies [8, 9] have found a slower tree recovery rate. The entire Mediterranean basin is characterized by landscape patterns with compositions resulting from countless, long, and complicated cultural and historical processes that developed in complex and varied environments [10–12]. In the Mediterranean region, a long history of natural (fire, floods, extreme droughts) and anthropogenic influences (overgrazing, forest harvesting, land use changes) have been the determining factors in shaping a mosaic of vegetation types with high ecological diversity and resilience.

In an environment such as the Mediterranean, forests play a key role in the welfare of urban and rural Mediterranean peoples, by providing highly appreciated marketed goods (e.g., firewood, cork, mushrooms) as well as high value but nonmarket services (e.g., soil protection and conservation, water regulation, recreation and landscape enhancement). The forests have been developing over thousands of years during which the climate has undergone significant changes. Phytocoenoses adapt to slow and constant changes, but anthropogenic climate change is faster and is not likely to allow ecosystems to adapt effectively: observed temperature increases and the occurrence of many serious problems for forest and preforest ecosystems, such as the reduced species diversity, high density, pest and diseases, threaten to make the adaptation more difficult. Forests have a dual function within climate change: they contribute to the mitigation of climate change, and their adaptation to climate change effects ensures their sustainable development under future site conditions [13].

The projections of climate change for 2016–2035 in Mediterranean Europe from the Intergovernmental Panel on Climatic Change (IPCC), estimate an increase in mean seasonal temperature up to 1.2°C in summer and 0.9°C in winter: mean seasonal precipitation is expected to increase by up to 25% in summer, but drop by up to 20% in winter [14]. This will cause a change in the distribution of the vegetation belts; moreover, extreme events such as flash floods, snowstorms, windstorms, fires, droughts, and heat waves are expected with increased frequency and intensity. The projections of the effects of climate change on European forests are complex: in Mediterranean basin, where water availability is a limiting factor, the most frequent summer droughts may lead to reduced productivity and lower resilience of forest.

Looking after the long-term future of forest resources, climate adaptation measures for forests became a high priority for the EU and its Member States. The European Union's new Forest Strategy noted the importance of both policy and practical actions suitable to improve or consolidate the resilience and adaptive capacity of forests. Particular attention in this respect is paid in setting up actions proposed by the European Union Strategy on Adaptation to Climate Change [15]. In view of the specific and wide ranging nature of climate change impacts on the EU territory, adaptation measures need to be taken at all levels, from local to regional and national levels. A more climate-resilient environment is the overall purpose of the European Union Adaptation Strategy. This means improving the preparation and capacity to respond to the effects of climate change at local, regional, national, and continental levels, developing a coherent approach and improving coordination [15].

In this chapter, we present the achievements of the project LIFE ResilForMed (Resilience of Mediterranean Forests to Climate Change) co-financed by the EU, as Sicilian responds on adaptation and mitigation to climate change. The project has mapped Sicily's forests to identify the most sensible areas for the desertification of risk, where urgent application of forest management techniques aimed at increasing resilience are needed. The bird communities susceptible to the desertification risk in Sicily were individuated too. ResilForMed has developed optimal management models for forest resilience, summarized in five forestry Best Management Practices (BMPs), which may be applied to other Mediterranean ecosystems as well as to other biomes. Finally, the project has involved local communities: (i) in forest protection measures and establishing indicators to define the role of communities and ecosystems in adapting to climate change; (ii) implementing the forest management models for fostering forest resilience through participatory tables; (iii) developing a forest governance model that consider the participation of lower level governments and the rural populations. The project ResilForMed is at the forefront of developing knowhow and transferable best management practices that may help Mediterranean forests to adapt to climate change and thereby safeguard their multifunctional benefits for future generations.

2. Forest resources in Sicily (Italy)

Sicily is the widest island of the Mediterranean basin and the broader Italian region (25711 km^2). The region also includes smaller islands around the main one: the Aeolian archipelago, the Pelagian archipelago, the Aegadian archipelago, the islands of Pantelleria and Ustica. The island is characterized by huge variability in geological, climatic, and biological assets, and it is an expression of secular anthropogenic disturbances performed by various cultures and people that shaped the local landscapes and preserved an extraordinary biodiversity. From a geomorphological viewpoint, about 62% of the Sicilian surface is characterized by hills, about 24% can be ascribed to mountains, and about 14% to plains. Regional average annual precipitation is 679 mm, ranges from a minimum of about 450–500 mm in southern Sicily and subcoastal plains to about 1500–1700 mm in north-eastern mountain areas facing the sea. Regional average annual temperature is 16.4°C, ranges from 18–20°C near shorelines to 8–10°C in the highest mountain areas [16].

The agricultural systems characterize the Sicilian traditional landscape [17], including extensive active or abandoned farms, shrublands, pastures, and grasslands. Sicilian forests cover the slopes and the top of the main northern and north-eastern mountain ranges [18]. Sicily also has a large percentage of its forest surface protected by Regional Natural Parks (Mt. Etna, Madonie, Nebrodi, Sicani Mts. and Alcantara River), National Park "Isola di Pantelleria", and

Natura 2000 network sites. Thus, forest and shrubland species composition and stand structure are typical of Mediterranean landscapes and exhibit a very high variability due to natural and anthropogenic influences [18–20]. In addition to these forest and preforest ecosystems (i.e., shrublands and woodlands), there are widespread reforested lands that are generally comprised of nonnative species, such as conifer and eucalypt plantations [18]. Sicily with its 274,454 ha of forests is one of the Italian regions with a lower coefficient of woodiness, about 10.6%, this forest surface increases to 512,121 ha if other wooded lands are taken into account [21]. It is known that historically forested areas were wider, with a loss of forest areas that lasted until the beginning of the last century, when this trend stopped and huge investments were done to reafforestate wide cleared surfaces. The Sicilian forest and preforest hierarchical classification system includes 14 forest categories (level 1) and 58 forest types (level 2) divided in subtypes and/or variants (level 3) [22]. Categories are characterized by one or few dominant species (e.g., beech forest, downy oak forest, etc.) and/or main physiognomic type (i.e., forest, shrubland). Types are the reference unit homogeneous in floristic, ecological, management features. Subtypes identify a specific meso- and pedo-climatic conditions and/or silviculture practices, while the variants identify a specific species composition. A general view on the Sicilian forest and preforest classification system at the level of category is provided:

Beech (*Fagus sylvatica* L.) forests: the Sicilian beech forests cover 15,964 ha [22] and are of particular phytogeographic and ecological interest because they are located at the southernmost limit of the entire European distribution area of the species [23]. Beech grows from 1200 up to 2000 m a.s.l. on the ridges and high slopes of the Madonie and Nebrodi mountains, as well as on the slopes of Mount Etna. The category includes four forest types due to the ecological amplitude of this species in colonizing different substrates (i.e., calcareous, siliceous, volcanic). Traditionally, beech forests in Sicily were mostly managed as coppice for charcoal production. Today, these stands are used for occasional grazing and wood harvesting, resulting in highly diversified structures which cannot be easily classified as standard structural managing models (simple coppice, coppice with standards, or selection coppice).

Chestnut (*Castanea sativa* Mill.) forests: these stands cover 11,520 ha [22] and find their optimum habitat between 1000 and 1400 m a.s.l. by colonizing siliceous and volcanic substrates. However, chestnut forests extend to lower altitudes (400–500 m a.s.l.) on the slopes of north-eastern Sicily (Madonie, Nebrodi, and Peloritani Mts.). On the south-eastern slopes of Mt. Etna, chestnut spreads over the upper montane region where it reaches its highest altitude. The category includes a thermophile type and a mesophile one. Coppicing was the traditional management system to provide firewood, small and medium poles used in agriculture. Today, chestnut fruit production is dramatically reduced and is rather oriented toward high quality products in limited areas.

Cork oak (*Quercus suber* L.) forests: these stands cover 18,830 ha from sea level up to 500 m a.s.l. on the north (Madonie and Nebrodi Mts.) and south-east (Iblei Mts.) of Sicily, by colonizing siliceous and volcanic substrates [22]. Cork oaks are usually dominant in silvopastoral systems in many areas of Sicily, where trees were exploited alternately for cork production or firewood, the shrubby, and herbaceous layer were used for grazing and occasionally mushroom picking [24, 25].

Holm oak (*Quercus ilex* L.) forests: the evergreen holm oak forests represent the potential vegetation in more than half of Sicilian coastal and sub-coastal hills. Their extension is nowadays much more limited due to the historical spread of agriculture and forest harvesting. These stands cover 28,650 ha, include four forest types due to the ecological amplitude of this species in colonizing different environments with different plant communities [22]. On the island, holm oak is a key species in many primary and secondary formations from the sea level up to 1500 m a.s.l., forms pure and mixed forests. Regular and frequent coppicing of these stands was the traditional management system to provide firewood and charcoal. Today, following the cessation of coppicing and with stand aging, the stumps transformation into more homogeneous stand structures is increasing. Charcoal production, very important in the past, is practically disappeared.

Downy oak forests: these stands cover 84,753 ha, about 17% of the Sicilian forest surface, from sea level up to 1200 m a.s.l. on the slopes of north Sicily (Sicani, Madonie, Nebrodi and Peloritani Mts.), as well as on the slopes of east Sicily (Mt. Etna and Iblei Mts.) [22]. These stands include two main species: one is downy oak (*Quercus pubescens* Willd.), the other is sessile oak (*Quercus petraea* L.). The category include five forest types due to the ecological amplitude and maximized morphologic variability of downy oak driven by different climatic and edaphic conditions, ranging from xerophile to meso-xerophile coenoses and from calcareous to volcanic substrates. Traditionally, downy oak forests were managed as coppice for charcoal and firewood production, resulting in highly diversified structures which cannot be easily classified as standard structural managing models (simple coppice, coppice with standards, or selection coppice).

Turkey oak (*Quercus cerris* L.) forests: these stands cover 25,289 ha from 400 up to 1500 m a.s.l. on the slopes of Nebrodi Mts., only a few small nuclei are on Etna Mt. and Ficuzza Natural Reserve [22]. The category includes a thermophile type and a mesophile one: with regard to the first type, it is a Sicilian endemism (*Quercus gussonei* (Borzì) Brullo) exclusive to the Nebrodi Mts. and Ficuzza Natural Reserve. These stands are managed in two ways: coppice and high forests. The presence of uncontrolled grazing cattle and the competition with shrubs represent risk factors to natural regeneration of Turkey oak [26].

Orno-ostrietum forests: these stands include two main species, manna ash (*Fraxinus ornus* L.) and European hop-hornbeam (*Ostrya carpinifolia* Scop.), which form limited relict coenoses, about 100 ha, on the slopes of north-eastern Sicily (Nebrodi Mts. and Mt. Etna) [22]. The category includes a xerophile type and a meso-xerophile one.

Riparian vegetation: the category represents riparian formations along streams and rivers, which are characterized by highly variable intra- and inter-annual flow mainly resulting from variable in-flowing runoff. The category covers 19,177 ha [22], encompassing five forest types due to high variability in species richness, composition, and density. The most widespread species are African tamarisk (*Tamarix africana* Poir.), narrow-leafed ash (*Fraxinus oxycarpa* Bieb.), various willow species (i.e., *Salix alba* L., *Salix pedicellata* Desf., *Salix purpure* L., *Salix gussonei* Brullo & Spamp.), black poplar (*Populus nigra* L.), and oriental plane (*Platanus orientalis* L.).

Pioneer vegetation: these stands cover 4470 ha, from the sea level up to montane region on shallow and degraded soils, abandoned croplands, and rocky slopes. The category includes seven forest types for the high variability in specific species composition, structure, and evolutionary dynamics [22]. A Sicilian endemic species, Etna birch (*Betula aetnensis* Rafin), constitutes one type; other species are European aspen (*Populus tremula* L.), field elm (*Ulmus minor* Mill.), manna ash (*Fraxinus ornus* L.), black locust (*Robinia pseudoacacia* L.), ailanthus (*Ailanthus altissima* Mill.), Italian alder (*Alnus cordata* Loisel.), and acacia ssp.

Corsican pine (*Pinus nigra* ssp. *laricio* (Poir.) Mair) forests: these stands cover 4316 ha from 1000 up to 2000 m a.s.l. on the slopes of Mount Etna [22]. The category includes three forest types, ranging from xerophile to mesophile coenoses, depending to ecological drivers (elevation, soil type, and bioclimate) and anthropogenic factors. The ancillary species associated to Corsican pine at elevation up to 1500 m a.s.l are downy oaks, Turkey oak, and chestnut, while, at higher altitudes, beech, Etna birch, and other pioneer species (e.g., *Genista aetnensis* Raf., *Astragalus siculus* Biv., *Berberis aetnensis* C. Presl, *Juniperus hemisphaerica* J. Presl & C. Presl.

Mediterranean pine forests: it groups Mediterranean pine stands with high naturalness degree. These stands occupy limited extensions of the Sicilian forest landscape, above 2240 ha [22]. The category includes four forest types by colonizing different substrates (i.e., calcareous, siliceous, volcanic): the most common species are Aleppo pine (*Pinus halepensis* Mill.), stone pine (*Pinus pinea* L.), and maritime pine (*Pinus pinaster* Ait.).

Plantations: these stands cover 105,460 ha, about 21% of the Sicilian forest surface, from sea level up to the upper montane region [22], due to extensive afforestation program conducted by regional forest service after World War II. In terms of distribution, plantations are main constituted of *Eucalyptus* ssp., Mediterranean conifers (e.g., *Pinus halepensis* Mill., *Pinus pinea* L., *Pinus pinaster* Ait., *Cupressus* ssp., *Cedrus* ssp.) and mountain-Mediterranean conifers (e.g., *Pinus* ssp., *Cedrus* ssp., *Abies* ssp., *Pseudotsuga menziesii*). Afforestation has been based mainly on *Eucalyptus* ssp. and conifers because they are fast-growing species, and also because it was believed that this would lead to rapid restoration of soil hydrological processes and protective functions against soil erosion.

Mediterranean shrublands: these are large systemic units for the high variability of Mediterranean maquis, mainly grouped according to the most representative dominant species [27]. These are primary formations and/or developed from secondary successional processes on degraded preforest communities due to frequent fires or other anthropogenic actions. The category covers 108,570 ha, about 21% of the Sicilian forest surface and includes eight forest types, ranging from xerophile to meso-xerophile coenoses, colonizing different substrates (i.e., calcareous, siliceous, volcanic) [22]. The most widespread species are *Spartium junceum* L., *Pistacia lentiscus* L., *Calicotome infesta* (C. Presl) Guss., *Rhus* ssp., *Euphorbia* ssp., *Juniperus* ssp., *Cistus* ssp., *Quercus ilex* L., *Quercus calliprinos* Webb., *Olea europaea* var. *sylvestris* (Miller) Lehr, *Chamaerops humilis* L.

Supra-Mediterranean shrublands: these stands cover 30,730 ha in the supra- and oro-Mediterranean belt of island. These are primary formations and/or developed from secondary successional processes on shallow and degraded soils. The category comprises five forest types, ranging from xerophile to mesophile coenoses, colonizing different substrates (i.e., calcareous, siliceous, volcanic) [22]. The most widespread species are *Erica arborea* L., *Ilex aquifolium* L., *Prunus* spp., *Rosa* spp., *Crataegus* spp., *Pyrus* spp., *Genista aetnensis* Raf., *Cytisus scoparius* (L.) Link.

The Sicilian forests can be considered at high risk of degradation for the growing phenomenon of forest fires and even more for the effects of climate change (e.g., flash floods, snowstorms, windstorms, droughts, and heat waves). Thus, climate changes may weaken or reduce forest and semi-natural ecosystems as well as exacerbate the potential desertification risk in Sicily, one of the Italian regions most threatened by this land degradation.

3. Climate resilience and adaptation strategy

Effective approach aimed at mitigating the effects of climate change to forest and preforest ecosystems of Sicily has involved a portfolio of different actions:

i identification of areas at high risk to climate change at regional and landscape scale;

ii analysis, assessment, and quantification of impacts of climate change;

iii definition of the role of communities and ecosystems in the strategies of adapting to climate change;

iv development of optimal management models suitable to improve or consolidate the resilience of forest and preforest ecosystems.

In order to identify the areas at high risk to climate change at regional and landscape scale, the environmental sensitivity areas (ESAs) to desertification of Sicilian region [28] and the Regional Forest [21] maps were overlaid, using the spatial analysis tools in a GIS environment. By this way, we developed sensitivity forest areas to desertification of the Sicilian region at a scale of 1:25,000. The different classes and sub-classes of sensitivity to desertification are distributed in a different way (**Table 1**). Regarding the results, the largest part of the Sicilian forest areas (49% of the area) was classified as "fragile" (mainly F2); 29% of the area was classified as "critical" (mainly C2), 17% as "potential", and only 5% as "not affected." More detailed results are shown in **Table 1**. The forest categories were also classified according to the level of sensitivity to desertification (**Figure 1**). The highest observed level of sensitivity (i.e., critical) was found in six forest categories (in decreasing percentage of land covered): Mediterranean shrublands, Riparian vegetation, Mediterranean pine forests, Pioneer vegetation, and Supra-Mediterranean shrublands that are, most likely, degraded successional stages of former Mediterranean forest types exposed to natural and anthropogenic stressors; on the other hand, the Turkey oak forests and beech forests showed the lowest level of sensitivity to desertification, suggesting the positive contribute of relatively stable forests for preventing desertification in Mediterranean region. Effective priority actions can be better planned if based on the knowledge of the degree of risk to desertification and the scale of the spatial domains. The map of forest areas sensitive to desertification highlighted the relationships between conditioning factors as well as the identification of the priority areas of action, to implement the best forest management models. The priority areas of action were chosen using the following criteria [18]: (1) high level of sensitivity to desertification; (2) encompassed in a protected area, parks and/or Natura 2000 sites; (3) articulate samples of regional forest types; and (4) includes the presence of selected silvicultural practices. On this basis, one or two priority areas were chosen in each of the main forest territories (Sicani Mts., Madonie Mts., Nebrodi Mts., Mount Etna, the Calatino area, and Pantelleria Island) that are also considered to be the representative of main ecological and sociocultural variations of the entire island.

In order to assess and quantify the main effects of climate change, we performed a diachronic survey on identified priority areas through the photos/images interpretation of collected materials over a 57-year period (1955–2012), exactly in four ages: 1955, 1968, 1988, and 2012. The 1955 and 1968 surveys were performed by the Italian Military Geographic Institute (IGMI) using black and white film; the 1988 survey was performed by the Sicily Regional Territory Service with orthoimages; and for 2012, free satellite remote sensing images were exploited. For methods and data processing used, see Ref. [18]. Land use/land cover (LULC) changes were identified and quantified in the priority areas over time by considering the whole period (1955–2012) and sub-periods (1955–1968, 1968–1988, 1988–2012). Land use (LU) changes were ranked according to three classes [18]: (i) "unvaried" was given to the forest surfaces where no change had happened over time; (ii) "evolution" was given to the forest surfaces where successional dynamics resulted in a more complex structure or composition of the forest and preforest ecosystem; and (iii) the rank "degradation" included regressive dynamic cases with a clear simplification of structure and composition of the forest and preforest ecosystem. Land cover (LC) changes were also investigated and four classes of change were identified [18]: "unvaried"; "less significant," where the change in observed LC was less than or equal to 20%; "increase" or "decrease," where a progressive or regressive change, respectively, of over 20%

Sensitivity class	Sensitivity sub-class	Description	Forest areas (ha)	Forest areas (%)
Not affected		Areas non threatened	25,368.13	5.2
Potential		Areas threatened under climate and land use/land cover changes	81,908.48	16.7
Fragile	F1	Areas in which any changes in the delicate balance of natural and human activities is likely to bring	82,015.64	16.7
	F2		10,3607.96	21.1
	F3		53,825.26	11.0
Critical	C1	Areas already degraded through past actions, showing a threat to the environment of the surrounding lands	39,097.44	8.0
	C2		89,053.35	18.1
	C3		16,141.68	3.3

Table 1. Classification and distribution of Sicilian forest areas in terms of sensitivity to desertification.

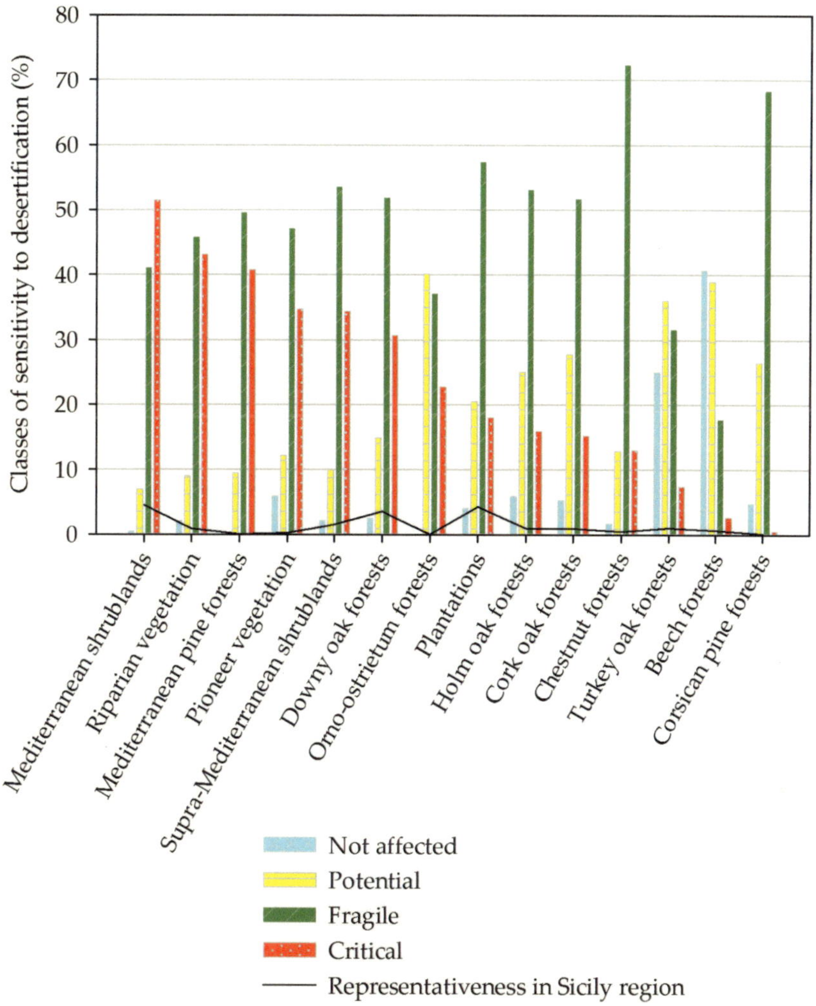

Figure 1. Classifications of forest categories according to the level of sensitivity to desertification in Sicily region.

was recorded for LC. The 20% threshold is considered the minimum significant value that can detect regressive or progressive LC changes: it was chosen on the basis of a general overview of this type of phenomenon observed in Sicilian forests, together with the general knowledge of forest stand cover dynamics according to the mean Mediterranean climatic and ecological driving factors [29, 30].

The role of bird community and forest ecosystems in the strategies of adapting to climate change was analyzed on identified priority areas. Focusing on the bird communities, a generalized linear modeling (GLM) approach has been used. This model allows to analyze the effect of different variables with a high elasticity and is therefore widely used in the ecological analysis [31]. According to the approach chosen, 37 bird indicators were tested to assess sensitivity to desertification. For more details of methods used, see Ref. [32]. As result, a total of 22 indicators (3 community indicators and 19 bird species) were highlighted as presenting sensitivity to desertification (**Table 2**).

With regard to forest ecosystems, we suggested a set of qualitative and quantitative indicators with the aim to assess the resilience of Mediterranean forests to climate changes. Thus, resilience indicators are of crucial importance because they can be used for a variety of purposes, such as: (i) describe and assess the state of the forests; (ii) identify pressure conditions within forests; (iii) assess the effects of forest management on forest composition, structure and functioning, discriminating among alternative policies, forecast future trends. An extensive field campaign on the unvaried areas, where no LULC change had occurred in the period 1955–2012, was set up to measure and calculate the resilience indicators. In particular, 12 forest types and 8 forest categories of Sicily were selected to define effective resilience indicators

EURING code	Indicator	P
	Species richness *sensu lato*	0.000
	Species richness *sensu stricto*	0.000
	Woodiness Bird Community Index	0.000
6700	*Columba palumbus*	0.001
8760	*Dendrocopos major*	0.000
10660	*Troglodytes troglodytes*	0.002
10990	*Erithacus rubecula*	0.000
12020	*Turdus viscivorus*	0.002
12650	*Sylvia cantillans*	0.000
12770	*Sylvia atricapilla*	0.012
13110	*Phylloscopus collybita*	0.064
13150	*Regulus ignicapilla*	0.000
14610	*Periparus ater*	0.060
14620	*Cyanistes caeruleus*	0.000
14790	*Sitta europaea*	0.030
14870	*Certhia brachydactyla*	0.000
15080	*Oriolus oriolus*	0.010
16360	*Fringilla coelebs*	0.000
16490	*Carduelis chloris*	0.001
16530	*Carduelis carduelis*	0.000
18580	*Emberiza cirlus*	0.000
18600	*Emberiza cia*	0.005

Table 2. List of bird indicators showing sensitivity to desertification in Sicily and the observed significance level (P-value).

and quantify the local threshold values of some indicators such as dendrometric parameter (i.e., the minimum level required to maintain the resilience and adaptive capacity of forests). For more details of methods used, see Ref. [33]. The following resilience indicators were selected and deemed effective:

1. Tree composition.
2. Forest crown cover.
3. Dendrometric parameters. The following parameters were measured: tree density (n ha^{-1}), basal area (G, in m^2 ha^{-1}), mean tree diameter (D_m, in cm), mean tree height (H_m, in m), stem volume (V, m^3 ha^{-1}). **Table 3** offers an overview of the minimum level required to maintain the resilience and adaptive capacity of forests of every parameter for each forest type investigated.
4. Structural diversity: vertical and horizontal distribution.
5. Presence/absence of sporadic and/or endemic tree species.
6. Presence/absence of old-growth trees.
7. Presence/absence of natural regeneration: composition, seedlings and saplings density, limiting factors.
8. Presence/absence of deadwood (woody debris, standing dead trees, and stumps woody debris): volume and decay class.
9. Presence/absence of hydrogeological instability phenomena.

In addition, the field campaign allowed to identify and lists the sporadic and endemic tree species as an important element of the biodiversity and resilience of Sicilian forests. A sporadic tree species is a kind of species that is seldom found in the forest ecosystem in a well-defined region. This specific definition contains an important aspect: it is always important to take in consideration the contest where such species live to evaluate their sporadic nature, both in the absolute and relative meaning of the term. On the other hand, an endemic tree species is a kind of species that live only in one geographic region closely adapted to its particular environment (e.g., large or small areas, islands). Nevertheless these species, being sporadic and/or endemic, are less competitive than the dominant ones. For these reasons, we must protect and widespread them as well as enhance their value and presence. The list of sporadic and endemic tree species is given in **Table 4**.

Data collecting, analyzing, and monitoring of selected resilience indicators were needed to develop optimal management models for improving or strengthening the resilience of Mediterranean forest ecosystems. The test was restricted to 12 forest types listed in **Table 3** on sample plots. It was adopted that a rigorous methodological approach to evaluate objectively the current condition and the optimum of the investigated forest types (**Figure 2**): taking into account the optimal conditions to ensure good forest resilience, we evaluated as the current condition deviates from the optimum. Then, we estimated the forest dynamics in the short

Forest category	Forest type	Species	Tree density (n ha^{-1})	Basal area (m^2 ha^{-1})	D_m (cm)	H_m (m)	V (m^3 ha^{-1})
Downy oak forests	*Quercus pubescens* forest of xeric environments	*Quercus pubescens*	1241	19	14	7	98
Cork oak forests	*Quercus suber* forest of xeric environments	*Quercus suber*	573	9	14	5	29
Holm oak forests	Mountain *Quercus ilex* forest of carbonatic substrata	*Quercus ilex*	608	30	25	13	169
	Quercus ilex forest of xeric environments, variant of volcanic substrata	*Quercus ilex*	477	13	19	11	70
Turkey oak forests	*Quercus cerris* forest tipica	*Quercus cerris*	1050	28	18	14	168
Beech forests	*Fagus sylvatica* forest tipica on calcareous substratum	*Fagus sylvatica*	4042	36	11	10	207
	Fagus sylvatica forest tipica on siliceous substratum	*Fagus sylvatica*	1750	36	16	10	220
Corsican pine forests	*Pinus laricio* forest tipica	*Pinus nigra* ssp. *laricio*	859	38	24	14	287
Mediterranean pine forests	*Pinus pinaster* forest	*Pinus pinaster*	2896	55	16	10	356
Plantations	*Eucalyptus* plantation tipica	*Eucalyptus camaldulensis*	827	19	17	12	105
	Eucalyptus plantation, variant mixed with Mediterranean evergreen species	*Eucalyptus camaldulensis*	198	7	21	12	36
	Pinus halepensis plantation	*Pinus halepensis*	198	28	42	16	257

Table 3. Minimum level of main dendrometric parameters required to maintain the resilience and adaptive capacity of 12 Sicilian forest types.

and medium term in the absence of interventions and, in case of stagnation or regression, it was defined the most appropriate adaptation strategy.

Under this approach, we developed the optimal management models to be applied in relation to forest category (or type), summarized in five forestry Best Management Practices (BMPs) suitable to improve or consolidate the resilience of Sicilian forest and preforest ecosystems:

BMP1. Actions favoring mixing of species and hydrogeological stability of forests: practices in order to favor of species improving soil quality (nutrients, texture, and structure),

Scientific name	Common name (in Italian)	Endemic	IUCN Red List
Acer monospessulanum L.	Acero minore		
Acer obtusatum Willd et K.	Acero etneo		
Acer pseudoplatanus L.	Acero di monte		
Betula aetnensis Raf.	Betulla dell'Etna	Sicily, Mt. Etna	
Carpinus orientalis Miller	Carpino orientale		
Celtis tournefortii subsp. asperrima (Lojac.) Raimondo & Schicchi	Bagolaro dell'Etna	Sicily, Mt. Etna	Vulnerable (VU)
Celtis australis L.	Bagolaro		
Ceratonia siliqua L.	Carrubbo		
Cercis siliquastrum L.	Albero di Giuda		
Fraxinus angustifolia subsp. *oxycarpa* (Willd.) Franco & Rocha Afonso	Frassino meridionale		
Genista aetnensis (Raf. ex Biv.) DC.	Ginestra dell'Etna	Sicily, Mt. Etna; Sardinia	
Ilex aquifolium L.	Agrifoglio		
Malus sylvestris Miller	Melo selvatico		
Ostrya carpinifolia Scop.	Carpino nero		
Pinus nigra subsp. *laricio* Poiret	Pino laricio	Sicily, Mt. Etna; Calabria	Vulnerable (VU)
Pistacia terebinthus L.	Terebinto		
Prunus avium L.	Ciliegio		
Prunus mahaleb L. subsp. *cupaniana* (Guss.) Arc.	Ciliegio canino di Cupani	Sicily	Vulnerable (VU)
Pyrus spinosa Forssk.	Pero mandorlino		
Pyrus castribonensis Raimondo, Schicchi & Mazzola	Pero di castelbuono	Sicily, Madonie Mts.	Vulnerable (VU)
Pyrus pyraster (L.) Burgsd	Pero selvatico		
Pyrus sicanorum Raimondo, Schicchi & Marino	Pero sicano	Sicily, Sicani Mts.	Endangered (EN)
Pyrus vallis-demonis Raimondo & Schicchi	Pero di Valdemone	Sicily, Nebrodi Mts.	Endangered (EN)
Sorbus aria (L.) Crantz	Sorbo montano		
Sorbus aucuparia L.	Sorbo selvatico		
Sorbus aucuparia L. subsp. *praemorsa* (Guss.) Nyman	Sorbo degli uccellatori	Sicily; Sardinia; Corsica	Vulnerable (VU)
Sorbus domestica L.	Sorbo domestico		
Sorbus graeca (Spach) Kotschy	Sorbo meridionale		

Scientific name	Common name (in Italian)	Endemic	IUCN Red List
Sorbus torminalis L. (Crant)	Ciavardello		
Taxus baccata L.	Tasso		
Tilia platyphyllos Scop.	Tiglio nostrale		
Ulmus glabra Hudson	Olmo montano		
Ulmus minor Mill.	Olmo minore		
Ulmus minor subsp. *canescens* (Melville) Browicz & Ziel.	Olmo canescente		
Zelkova sicula Di Pasquale, Garfi & Quézel	Zelkova siciliana	Sicily, Iblei Mts.	Critically Endangered (CR)

Table 4. List of sporadic and endemic tree species of Sicilian forests.

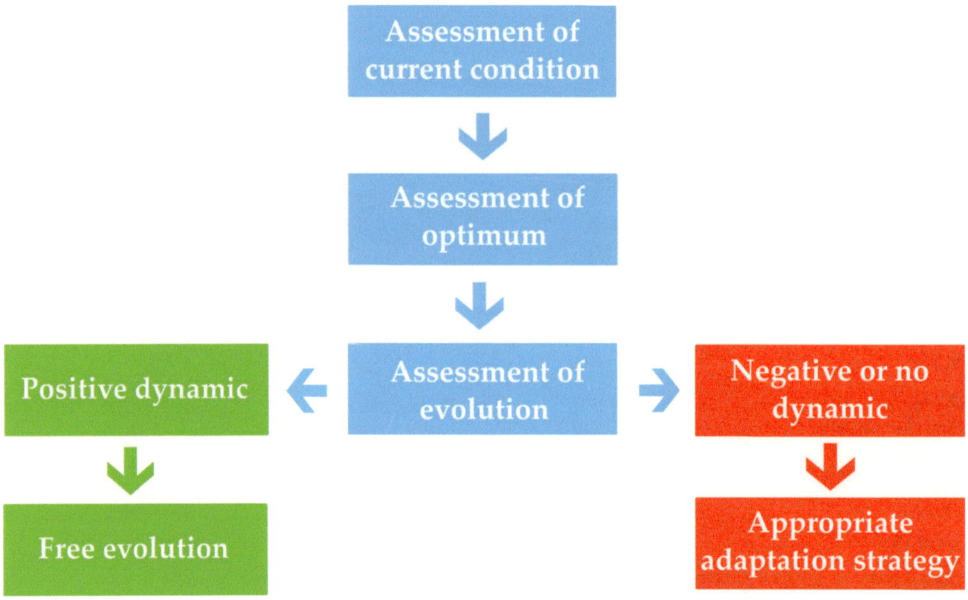

Figure 2. Flowchart of applied methodology to evaluate the current condition and the optimum of Sicilian forest types for defining adaptation strategies to climate changes.

which at the same time ensure or maintain adequate levels of organic matter in the soil, ensuring the improvement of water supply and the same conditions of light and temperature favorable to the acceleration of progressive successional forest dynamics.

BMP2. Renaturalization of forest plantations: interventions aimed to pander renaturalization processes with the aim of increasing the stability and thus the efficiency of ecological-structural ecosystem. Interventions to reintegrate native species in forests free of seed-bearing trees were also foreseen.

http://dx.doi.org/10.5772/intechopen.68943

BMP3. Remedial measures and restoration of degraded forests: silvicultural actions on degraded woods due to fire, extreme droughts, overgrazing, erosion, land use changes, aimed at recovering the forest ecosystems.

BMP4. Actions aimed at enhancing complex structural forests: (i) conversion of coppices into high forests within the protected areas, in order to create communities with greater productivity and ecosystem stability; (ii) to create favorable conditions for sporadic tree species, the application of methods of management system that take care of the global needs of the forest and of each single species were foreseen, such as the tree-oriented silviculture techniques. Tree-oriented silviculture is based on targeted interventions aimed to advantage only some selected trees (target trees) and it makes possible focusing the efforts mainly on such aspects as the species mixture, stand structure, regeneration, and intra/inter specific competition dynamics.

BMP5. Actions favoring connectivity in agroforestry systems: forest interventions aimed at reducing the fragmentation of forest areas in order to increase connectivity and reduce the influence of anthropogenic activities in the surrounding areas (agricultural lands, pastures).

In particular, the BMPs were implemented in a set of forest interventions for nine forest categories on 120 ha of forest areas in Sicily (**Table 5**).

Forest category	Forest type	BMP	Forest interventions
Downy oak forests	*Quercus pubescens* forest of xeric environments	03	(a) Salvage felling of fire degraded downy oak forests
			(b) Introduction of native oak species in stands lacking of seed-bearing trees
			(c) Reducing forest fuel loads to decrease wildfire risk
			(d) Grazing exclusion
Cork oak forests	*Quercus suber* forest of xeric environments	01	Bio-engineering technique for soil erosion control
		02	Removal of exotic species
		03	(a) Salvage felling of fire degraded downy oak forests
			(b) Introduction of native oak species in stands lacking of seed-bearing trees
			(c) Reducing forest fuel loads to decrease wildfire risk
			(d) Grazing exclusion
Holm oak forests	Mountain *Quercus ilex* forest of carbonatic substrata	04	(a) Conversion of coppices into high forests
			(b) Tree-oriented silviculture techniques
	Quercus ilex forest of xeric environments, variant of volcanic substrata	02	(a) Removal of exotic plantations
			(b) Introduction of native oak species in stands lacking of seed-bearing trees
			(c) Grazing exclusion

Forest category	Forest type	BMP	Forest interventions
Turkey oak forests	*Quercus cerris* forest tipica	04	(a) Conversion of coppices into high forests (b) Tree-oriented silviculture techniques
Beech forests	*Fagus sylvatica* forest tipica on calcareous substratum	04	(a) Conversion of coppices into high forests (b) Tree-oriented silviculture techniques
	Fagus sylvatica forest tipica on siliceous substratum	01	(a) Introduction of native tree species (b) Bio-engineering technique for soil erosion control (c) Grazing exclusion
Corsican pine forests	*Pinus laricio* forest tipica	02	Removal of exotic plantations
Mediterranean pine forests	*Pinus pinaster* forest	04	Selective thinning on *Pinus pinaster* to reduce interspecific competition
Plantations	*Eucalyptus* ssp. plantation tipica	02	(a) Selective thinning on *Eucalyptus* spp. (b) Cutting all resprouts of *Eucalyptus* spp. stumps (c) Introduction of native tree and shrub species
		05	Establishment of wooded strips with native tree and shrub species as ecological corridors
	Eucalyptus ssp. plantation, variant mixed with Mediterranean evergreen species	02	(a) Selective thinning on *Eucalyptus* spp. (b) Cutting all resprouts of *Eucalyptus* spp. stumps (c) Introduction of native oak species in stands lacking of seed-bearing trees
	Pinus halepensis plantation	02	(a) Selective thinning on *Pinus halepensis* to reduce interspecific competition (b) Introduction of native tree species
		03	Reducing forest fuel loads to decrease wildfire risk
		05	Establishment of wooded strips with native tree and shrub species as ecological corridors
	Mixed coniferous plantation	02	Removal of exotic species
Mediterranean shrublands		03	(a) Salvage felling of fire degraded maquis (b) Introduction of native tree and shrub species (c) Reducing forest fuel loads to decrease wildfire risk

Table 5. Set of forest management interventions for each forest category (or type) suitable to improve or consolidate the resilience of 9 forest categories in Sicily.

http://dx.doi.org/10.5772/intechopen.68943

All these developments have been addressed in drafting of the "Guidelines for assessing the Resilience of Mediterranean Forests to Climate Change", forthcoming publication. This publication sets out the specific actions that policymakers, forest managers, and other stakeholders should take to improve or consolidate the resilience of forest and preforest ecosystems in Sicily and in Mediterranean Region. In this way, we suggested an approach to conserve biodiversity, to maintain resilience, and to give Mediterranean forest systems the best possible chance of adapting to climate changes.

Acknowledgements

This work was supported by grants of the Project LIFE11+ ENV IT 000215 RESILFORMED—RESILienza al cambiamento climatico delle FOReste MEDiterranee (Resilience of Mediterranean forests to climate change). http://www.resilformed.eu

Author details

Sferlazza Sebastiano[1]*, Maetzke Federico Guglielmo[1], Miozzo Marcello[2] and La Mela Veca Donato Salvatore[1]

*Address all correspondence to: sebastiano.sferlazza@unipa.it

1 Department of Agricultural and Forest Sciences, University of Palermo, Palermo, Italy

2 D.R.E.Am Italia Soc. Coop. Agr., Pratovecchio Stia (AR), Italy

References

[1] Westman WE. Measuring the inertia and resilience of ecological systems. BioScience. 1978;**28**:705-710

[2] Westman WE. Resilience: Concepts and measures. In: Dell B, Hopkins AJM, Lamont BB, editors. Resilience in Mediterranean-type Ecosystems. Dordrecht: Dr W. Junk Publishers; 1986. pp. 5-19. DOI: 10.1007/978-94-009-4822-8

[3] Lavorel S. Ecological diversity and resilience of Mediterranean vegetation to disturbance. Diversity and Distributions. 1999;**5**:3-13.

[4] Malanson GP, Trabaud L. Ordination analysis of components of resilience of *Quercus coccifera* garrigue. Ecology. 1987;**68**:463-472.

[5] Caturla RN, Raventos J, Guardia R, Vallejo VR. Early post-fire regeneration dynamics of Brachypodium retusum Pers. (Beauv.) in old fields of the Valencia region (eastern Spain). Acta Oecologica. 2000;**21**:1-12

[6] Malanson GP, Trabaud L. Vigour of post-fire resprouting by *Quercus coccifera* L. Journal of Ecology. 1988;**76**(2):351-365

[7] Seligman NG, Henkin Z. Regeneration of a dominant Mediterranean dwarf-shrub after fire. Journal of Vegetation Science. 2000;**11**:893-902

[8] Lloret F, Calvo E, Pons X, Diaz-Delgado R. Wildfires and landscape patterns in the Eastern Iberian Peninsula. Landscape Ecology. 2002;**17**:745-759.

[9] Baeza MJ, Valdecantos A, Alloza JA, Vallejo VR. Human disturbance and environmental factors as drivers of long-term post-fire regeneration patterns in Mediterranean forests. Journal of Vegetation Science. 2007;**18**:243-252

[10] Braudel F. Civiltà e Imperi del Mediterraneo nell'età di Filippo II. Torino, Italy: Einaudi; 1986

[11] Grove AT, Rackham O. The Nature of Mediterranean Europe: An Ecological History. New Haven (CT), USA: Yale University Press; 2001. p. 384

[12] Barbera G, Cullotta S. The Halaesa landscape (III B.C.) as ancient example of the complex and bio-diverse traditional Mediterranean polycultural landscape. Landscape History. 2014;**35**(2):53-66. DOI: 10.1080/01433768.2014.981395

[13] Jandl R, Cerbu G, Hanewinkel M, Berger F, Gerosa G, Schüler S. Management strategies to adapt Alpine space forests to climate change risks–An introduction to the Manfred project. In: Cerbu G., Hanewinkel M., Gerosa G., Jandl R., editors. Management Strategies to Adapt Alpine Space Forests to Climate Change Risks. InTech; August 28, 2013 under CC BY 3.0 license. p. 1-12. DOI: 10.5772/56267

[14] IPCC. Near-term Climate Change: Projections and Predictability. In: Climate Change 2013 – The Physical Science Basis: Working Group I Contribution to the Fifth Assessment Report of the Intergovernmental Panel on Climate Change. Cambridge, United Kingdom, and New York, USA: Cambridge University Press; 2014. pp. 953-1028. https://doi.org/10.1017/CBO9781107415324

[15] European Commission. COM(2013)216. An EU Strategy on Adaptation to Climate Change. www.eur-lex.europa.eu [Internet]. [Update: 27/01/2017]. Available from: http://ec.europa.eu/transparency/regdoc/rep/1/2013/EN/1-2013-216-EN-F1-1.Pdf [Accessed: 27-01-2017]

[16] Servizio Informativo Agrometeorologico Siciliano. www.sias.regione.sicilia.it [Internet]. [Updated: 27/01/2017]. Available from: http://www.sias.regione.sicilia.it/frameset_dati.htm [Accessed: 27-01-2017]

[17] Barbera G, Cullotta S. An inventory approach to the assessment of main traditional landscapes in Sicily (central Mediterranean basin). Landscape Research. 2012;**37**(5):539-569. http://dx.doi.org/10.1080/01426397.2011.607925

[18] La Mela Veca DS, Cullotta S, Sferlazza S, Maetzke FG. Anthropogenic influences in land use/land cover changes in Mediterranean forest landscapes in Sicily. Land. 2016;**5**(3):1-13. DOI: 10.3390/land5010003

[19] Mazzoleni S, Di Pasquale G, Mulligan M, Di Martino P, Rego F. Recent dynamics of the Mediterranean vegetation and landscape. West Sussex, UK: John Wiley & Sons Ltd; 2004. 320 p.

[20] Ruhl J, Chiavetta U, La Mantia T, La Mela Veca DS, Pasta S Land cover change in the Nature Reserve "Sughereta di Niscemi" (SE Sicily) in the 20th century. In: Erasmi S, Cyffka B, Kappas M, editors. Remote Sensing & GIS for Environmental Studies. Gottingen, Germany: Gottinger Geographische Abhandlungen; 2005. pp. 54-62.

[21] Hofmann A, Cibella R, Bertani R, Miozzo M, Fantoni I, Luppi S. Strumenti conoscitivi per la gestione delle risorse forestali della Sicilia. Sistema Informativo Forestale Regionale. Città di Castello. Perugia, Italy: Regione Siciliana; 2011. 208 p.

[22] Camerano P, Cullotta S, Varese P. Strumenti conoscitivi per la gestione delle risorse forestali della Sicilia. Tipi Forestali. Città di Castello, Perugia, Italy: Regione Siciliana; 2011. 192 p.

[23] Pignatti S. I boschi d'Italia. Sinecologia e biodiversità. Torino: UTET; 1998. 680 p.

[24] Pizzurro GM, Maetzke FG Influenza del clima sulla crescita del sughero in soprassuoli produttivi siciliani. Forest. 2009;**6**:107-119. DOI: 10.3832/efor0562-006

[25] Pizzurro GM, Maetzke FG, La Mela Veca DS. Differences of raw cork quality in productive cork oak woods in Sicily in relation to stand density. Forest Ecology and Management. 2010;**260**:923-929. DOI: 10.1016/j.foreco.2010.06.013

[26] Sala G, Giardina G, La Mantia T. I fattori di rischio per la biodiversità forestale in Sicilia: il caso studio del cerro di Gussone. L'Italia Forestale e Montana. 2011;**66**(1):71-80. DOI: 10.4129/ifm.2011.1.06

[27] Cullotta S, Marchetti M. Forest types for biodiversity assessment at regional level: The case study of Sicily (Italy). European Journal of Forest Research. 2007;**126**:431-447. DOI: 10.1007/s10342-006-0166-y

[28] Regione Siciliana. Environmental Sensitive Areas to Desertification, ESAs [Internet]. 2011 [Updated: 23/09/2013]. Available from: http://www.sitr.regione.sicilia.it/?p=569 [Accessed: 23-09-2013]

[29] Mazzoleni S, Di Pasquale G, Mulligan M, Di Martino P, Rego F, editors. Recent Dynamics of the Mediterranean Vegetation and Landscape. West Sussex, UK: John Wiley & Sons Ltd.; 2004. p.320. DOI: 10.1002/0470093714

[30] Quézel P, Médail F. Ecologie et biogéographie des forêts du bassin méditérranéen. Paris, France: Elsevier, Collection Environnment; 2003. 573 p.

[31] Rushton SP, Ormerod SJ, Kerby G. New paradigms for modelling species distributions? Journal of Applied Ecology. 2004;**41**:193-200.

[32] Londi G, Tellini Florenzano G, Campedelli T, Cutini S Azione A4–Impiego degli uccelli come degli indicatori dello stato delle foreste [Report]. LIFE Resilformed; 2014. p. 42 Available from: http://www.resilformed.eu/

[33] Maetzke FG, Cullotta S, La Mela Veca DS, Sferlazza S, Clementi G. Azione A4–Definizione, attraverso indicatori, del ruolo di comunità e degli ecosistemi nelle misure di adattamento ai cambiamenti climatici. Componente forestale. [Report]. LIFE Resilformed; 2014. p. 49. Available from: http://www.resilformed.eu/

Chapter 12

Drought-Forest Fire Relationships

Tuğrul Varol, Mertol Ertuğrul and Halil Barış Özel

Additional information is available at the end of the chapter

http://dx.doi.org/10.5772/intechopen.68487

Abstract

This study was carried out to determine the methods that bear the most realistic results in predicting the number of fires and burned area under the climate conditions in future. Different indices and statistical methods were used in predicting the burned area and the number of fires. With this aim, in addition to the indices used in estimating the climate, Machine Learning and multivariate adaptive regression spline (MARS) models are also used in predicting these factors. According to the results obtained in several studies, the relationship between the drought and fire indices burned area and the number of fires changes from region to region. While better results are obtained in predicting the burned area and the number of fires via the drought indices being used in this study and the MARS models that the combinations of these indices use, it is seen that a 30–39% success was achieved for predicting the amount of burned area via Machine Learning methods (Kernel Nearest Neighbor (kNN), Recursive Partitioning and Regression Trees (RPART), Support Vector Machine (SVM) and RF), and this success ranges widely from 8 to 41% in terms of the number of fires. RPART, of these four algorithms, performed the best in fire prediction, but kNN was the worst.

Keywords: drought, forest fire, number of fires, burned area, machine learning, silviculture

1. Introduction

The fires that occurred in the Mediterranean Region between 2001 and 2014 are as follow: 3250 in France, 1425 in Greece, 6525 in Italy, 16,000 in Spain, 21,800 in Portugal and 2200 in Turkey, 205 of which were seen in Antalya. These statistics emphasize how much important it is to understand the factors which have impact on the burned areas and the number of fires and their consequences in terms of the sustainable management of forests [1]. The amount of burned areas and the number of fires are affected by numerous variables such as the size of the fire-sensitive forest area, topography, landscape (e.g., road, creek, lake, and agricultural areas), flammable materials characteristics, fire season, altitude, firefighting policy and the efficiency of organization, the number of concurrent fires and climate conditions [2]. For this reason, although the relationship between these factors and climate variables was examined first for the estimation of the number of fires and the amount of burned areas, it was seen in the research in Canada that the majority of variance could not be explained. Second, therefore, the relationships were examined by developing several indexes such as KBDI (Keetch-Byram Drought Index), SPEI (The Standardized Precipitation-Evapotranspiration Index), FWI (Fire Weather Index) and PDSI (Palmer Drought Severity Index). In the studies conducted across the world, various results were obtained in the explanation of the variance depending on the study areas and variables utilized such as 12, 35, and 66% [2–9]. In the third phase, descriptive regression models were used. It was stated by Viegas et al. [10] in France that ISI (Initial Spread Index), as a FWI component, is successful under extreme fire conditions. Jong et al. [11] state that the 75, 90 and 99% uses of at least one of the FWI components for the UK is of a significant advantage in calibration studies. Bedia et al. [12] emphasized DSR (Daily Severity Ratio) and FWIP90 as FWI components for a better understanding of the spatial and temporal distribution of fires in Spain and identified an increase in the amount of burned areas. Venalainen et al. [13] ascertained a 99% relationship for Central and South Europe and 95% relationship for Eastern Europe between FWI and burned areas in the estimation of forest fires in Europe from 1960 to 2012. When Urbieta et al. [14] examined the fire activities for Europe (Portugal, Spain, Southern France, Italy and Greece) and the Pacific coasts of the USA (Oregon and California) through FWI components, it was found $R^2 > 0.70$ for the models obtained for Europe and $R^2 > 0.50$ for the USA. New approaches have been developed in order to look into the relationship of the variables having effects on the burned forest areas and the number of fires. Some of these approaches are the Machine Learning methods (MLM) used in ecological applications such as kNN (Kernel Nearest Neighbor), SVM (Support Vector Machine), RPART (Recursive Partitioning and Regression Trees), and RF (Random Forest) [15, 16]. T_{max} and relative humidity were determined to be more successful in explaining fire activity in linear regression than RF algorithm along with all effective climate factors [15]. Moreover, RPART algorithm is made use of in determining the need for saplings after fire [17].

This study was carried out in three phases in Antalya region. In the first phase, it was investigated the relationship between the amount of burned area and the number of fires through the variables in **Table 1**. In the second phase, the descriptive regression equations that are used in such countries as Portugal, Spain, Canada and the USA were tried to be achieved in order to obtain the nearest estimations for the monitored numbers of fire and amount of burned area. In the third phase, the eligibility of MLM such as kNN, SVM, RPART and RF was investigated in estimating the burned areas and the number of fires. The results obtained in each phase were discussed in terms of the models' predicting abilities and the variables. The data sets comprised of either meteorological data or fire statistics cover the period between 2001 and 2014.

http://dx.doi.org/10.5772/intechopen.68487

KBDI	Keetch-Byram drought index
KBDIX	Maximum Keetch-Byram drought index
KBDIP90	90th percentile Keetch-Byram drought index
SPEI	The standardized precipitation-evapotranspiration index
FFMC	Fine fuel moisture code
FFMCX	Maximum fine fuel moisture code
FFMCP90	90th percentile fine fuel moisture code
DC	Drought code
DCX	Maximum drought code
DCP90	90th percentile drought code
DMC	Duff moisture code
DMCX	Maximum duff moisture code
DMCP90	90th percentile duff moisture code
BUI	Build-up index
BUIX	Maximum build-up index
BUIP90	90th percentile build-up index
ISI	Initial spread index
ISIX	Maximum initial spread index
ISIP90	90th percentile initial spread index
FWI	Fire weather index
FWIX	Maximum fire weather index
FWIP90	90th percentile fire weather index
DSR	Daily severity ratio
DSRX	Maximum daily severity ratio
DSRP90	90th percentile daily severity ratio
TX	Maximum daily temperature
TXX	Maximum of maximum daily temperature
TXP90	90th percentile
RH	Relative humidity
TPREC	Total precipitation

Table 1. Components related to the drought index and meteorological variables.

2. Materials and methods

2.1. Study area

Antalya region, which is of the largest forest area in Turkey after Amasya Region, includes a forest area of 1,146,062 ha, a rate of 60.43% [18]. The dominant tree species of the region is *Pinus brutia* Ten. (65%), and it is, respectively, succeeded by *Cedrus* sp. (16%), *Pinus nigra* (8%), *Abies* sp. (5%), *Juniperus* sp. (4%) and other leafed species (2%). Such forests are between 600 and 2100 m, and maquis shrubs are mainly seen up to this altitude. Antalya Regional Directorate of Forestry is located between 36° 00′ 45″ north latitudes and 29° 16′ 15″ east longitudes. Antalya Region has a hot-summer mediterranean climate (according to Köppen-Geiger climate classification system Csa), which is dry and hot in summers, warm and rainy in winter (**Figure 1**). The lowest precipitation and the highest temperature for a five-month period, from May to September, in 2001–2014 were obtained relatively in 2007 and 2003. Seeing the history of regional fires, it is seen that a total of 24,390 ha has burned in 2097 fires between 2001 and 2014. This value is above the national average (based on the data obtained from 28 Regional Directorates of Forestry) considering both the areal (~4145 ha) and numerical average (~1097). For the same period, the annual average burned area and number of fires are, respectively, ~1742 ha and ~150 [14]. August is of the highest average month in terms of both the number of fires and burned area in Antalya's 14-year fire records (BA). The forest areas in the region encountered the highest damage by the burning of 16,890 ha in 2008 and the most of the fires (214) occurred in 2013.

The daily, monthly and seasonal fire data and daily, monthly and seasonal meteorological data covering 2001–2014 for Antalya region were used. The fire statistics were obtained from General Directorate of Forestry [18], and meteorological data were obtained from GDM (General Directorate of Meteorology). The fire season is accepted to be May 1 to September 30. April and October were not taken into consideration because they, respectively, cover 0.26 and 4.28% of the yearly burned areas. Meteorological data are comprised of daily highest temperature, total precipitation, wind speed and relative humidity.

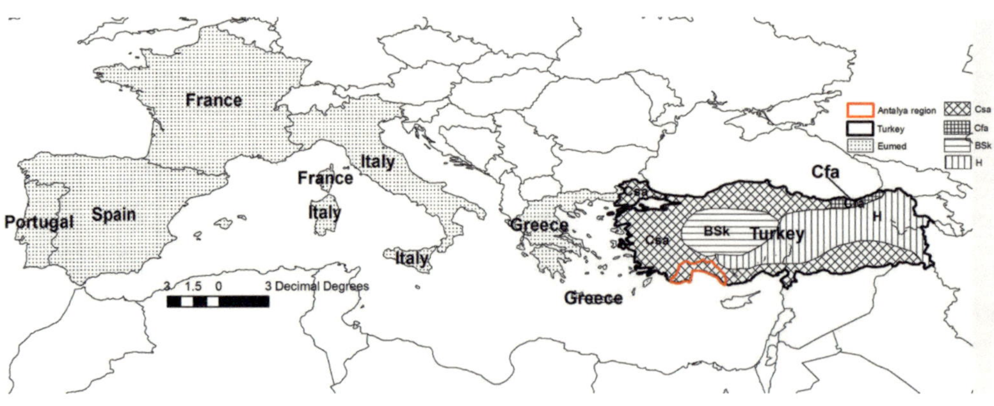

Figure 1. Location of Antalya Regional Directorate of Forestry.

http://dx.doi.org/10.5772/intechopen.68487

The annual average KBDI, FWI and SPEI values of the study area are seen in **Figure 2**. When compared to FWI values in EUMED region given by Urbeita et al. [14], FWI values are seen to be more stable. The highest and lowest values were encountered twice for the average FWI between 1995 and 2010. For the period 2001–2014, a decrease is widely seen to be in the annual average number of fires and burned areas to the eastwards of EUMED region from the west. Turkey itself, facing with 30,724 fires, changed the distribution of fires to some extent by outnumbering Greece (20,002 fires). When the comparison is made by taking forest areas into consideration, it is seen that the countries facing with fires the most are put in order relatively as follow: Portugal, Greece, Italy, Spain, France and Turkey.

2.2. KBDI, FWI, SPEI

KBDI involves a range of values from 0 to 800. Eight hundred signify the extreme drought, while 0 represents saturated soil. In the cases that daily rainfall data causing the change of index values reach significant amounts, KBDI value is generally needed to start from 0 [19]. Such rainfalls are frequently encountered in spring and winter seasons in Antalya region. KBDI (Q) is measured depending on average annual rainfall (R—inches) and daily maximum temperature (T—degrees Fahrenheit). The rate of change (dQ) in index is calculated via the following formula. In the formula, dT signifies a temporal change.

$$dQ = \frac{[800 - Q][0.968\, exp(0.0486T) - 8.30]dT}{1 + 10.88\, exp(-0.0441R)} x\, 10^{-3} \quad (1)$$

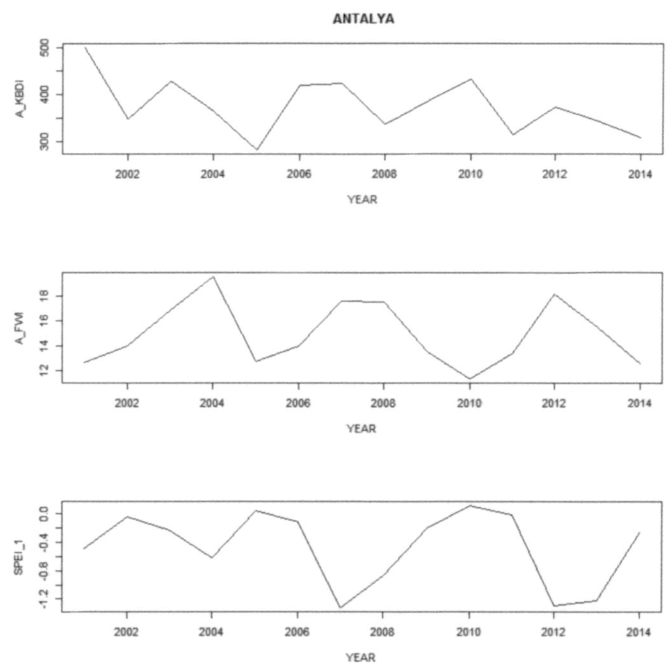

Figure 2. The average KBDI, FWI and SPEI values belonging to Antalya region.

The structure of FWI system is shown in **Figure 3**. Temperature, relative humidity, wind speed and rainfalls in 24 h are taken into consideration in calculating the components. Seven standard components of FWI allow numerical values for a possibility of fire [20].

SPEI, using the distinction between rainfall and evapotranspiration (P-ET_0) as data, is calculated in a similar way as SPI. Climatic water balance compares current quantity of water (P) with atmospheric steam deficit (ET_0). Therefore, the data used in SPEI are more suitable in measuring drought severity rather than considering only the rainfall [21]. The data used in calculating SPEI in the study area were obtained from the link [22] in the grid cell size by entering longitude and latitude values.

2.3. kNN, SVM, RPART, RF

Learning methods via artificial intelligence has become more usable in modeling the complex relationships and interactions without restricting assumptions of parametric statistics [23, 24]. Machine learning methods generate controllable approaches that try to model the relationships [23]. Some of these methods are artificial neural networks [25], classification and regression trees [26], support vector machines [27], random forest [28], fuzzy logic [29], maximum entropy [30] and kNN [15].

K-nearest neighbors (kNNs) are one of the oldest and simplest methods used in model classifications among machine learning methods. kNN labels each of unlabeled samples according to their closest neighbors in the data set. Therefore, its performance depends on the distance metrics (Euclidean distance, Minkowski distance, Mahalanobis distance) used for calculating its closest neighbors [31].

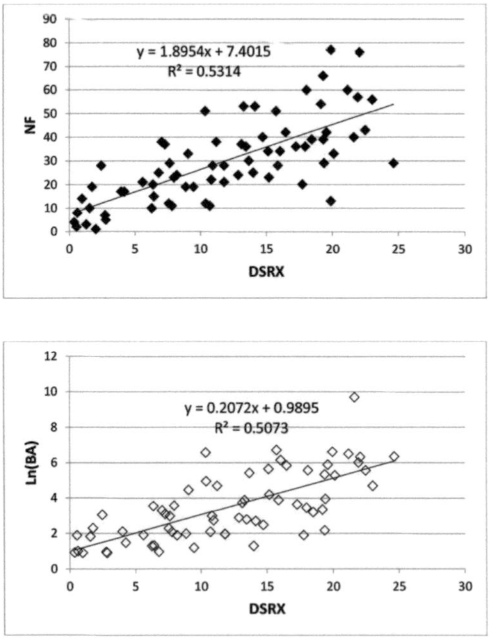

Figure 3. The relationship between monthly maximum DSR and monthly NF and BA's natural logarithm values between 2001 and 2014.

SVM offers some theoretical advantages since local minimums do not exist in the optimization phase of the model compared to other machine learning methods. The inputs in SVM are transformed through non-linear methods in the m-dimensional feature space. In this way, SVM finds the best linear classifier hyperplane in the feature space [32].

$$f(x) = w_0 + \sum_{i=1}^{m} \left(w_i \varnothing_i(x) \right) \quad (2)$$

Here, $\varnothing_i(x)$ represents the non-linear transformation according to the Kernel function ($K(x, x') = \sum_{i=1}^{m} \varnothing_i(x) \varnothing_i(x')$) [33].

In order to analyze complex ecological data sets, RPART comes up as the powerful statistical tools. One of the biggest reasons for this is that it offers useful alternatives, while it models non-linear data, which includes independent variables interacting with each other [29]. Regression trees were used in numerous ecological applications such as the relationship between the severity and frequency of forest fires [34].

Breiman [35] suggests, instead of producing a single decision tree, unifying the multivariate decisions trees ($\varnothing_k$), each of which is trained through different training data sets. RF (Random forest) is a learning algorithm that generates multi-classifiers instead of a single classifier and then classifies the new data (x) with the votes from the predictions ($h(x,\varnothing_k)$, k=1,...) [36]. In the performance, evaluation of the algorithm of these four different classifications was used four factors which were also preffered by Hong et al. [37]. These factors can be defined as follows:

Precision or positive predictive value (PPV) is the proportion of predicted positive samples that were correctly classified: $\mathrm{PPV} = \frac{\mathrm{TP}}{\mathrm{TP+FP}}$; sensitivity, recall, hit rate or true positive rate (TPR) is the proportion of positive samples that were correctly identified: $\mathrm{TPR} = \frac{\mathrm{TP}}{\mathrm{TP+FN}}$; F1 score is the harmonic mean of precision and sensitivity: $\mathrm{F1} = \frac{2\mathrm{TP}}{2\mathrm{TP+FP+FN}}$; finally, accuracy $\mathrm{ACC} = \frac{\mathrm{TP+TN}}{\mathrm{P+N}}$.

Here, TP is the number of true positive, FP is the number of false negative, FN is the number of false negative, P is the number of real positive cases in data, and N is the number of real negative cases in the data.

2.4. Statistical analysis

The statistical program, R x 64 3.2.3, was used in order to carry out statistical analysis in the study. The statistical analyses were utilized in order to determine both the components of drought indices and the machine learning methods as well as determining the relationship between KBDI, SPEI and FWI and the number of fires, the data of burned areas. All data for drought indices and components were measured as daily, monthly and seasonal data (May 1 to September 30). Since most of the forest fires arise under extreme air conditions [8], maximum and 90% extreme variables of drought indices and their own components were calculated.

Because of the uneven distribution of the raw data for burned area used in the study, their natural logarithm equivalents were used instead of such data. It was examined the correlation between the natural logarithm values of the data for the number of fires and burned area and all the components of all data for drought indices. Each component was later used in the step-wise regression analysis.

3. Results and discussion

In the first phase, no relationship between the drought indices and either the number of fires or the data of burned areas was found. The variance values that can be explained by daily data range from 2 to 18%. The best correlation results were achieved for the natural logarithm of the data for the burned area and the number of fires. As a result of the correlation analyses by using monthly data, variance results between natural logarithm and the data for the number of fires and burned area are, respectively, 53 and 51% through maximum DSR values (**Figure 3**).

When the corrected R^2 values were calculated for all the variables in **Table 1**, it was determined that the highest correlation (56%) for the number of fires belongs to the maximum DSR and FWI. The highest correlation with 50% among the corrected R^2 values calculated for all the variables in terms of the burned area still belongs to the maximum DSR. Urbieta et al. [14] laid stress on a strong relationship between FWI and the burned area in EUMED for the last 30 years and said that 60% of this relationship could be explained through the R^2 values. In the same study, the R^2 values are able to explain the variance ranging from 20 to 55% for the number of fires and number of large fires. It is of high importance for estimating the activity in fire season that the R^2 values bear similar results for the both areas where the variables differ such as the differences in the study periods, socio-economic conditions, fire extinguishing activities, fuel component accumulations.

Canadian FWI system is one of the most commonly used fire weather indices all over the world thanks to its success in determining the fire risk and flexibility although it was originally developed for Canada [38–40]. Beside this, six components forming the system can individually estimate the fire risk in a succesful way in different conditions. In particular, FWI and FFMC among these components stand out in explaining the fire activity [41]. FWI component itself is used not only for generally determining the fire severity of many fuel types, but also for explaining the fire danger [42, 43]. As for FFMC component, it is generally used as the indicator of fire outbreaks and the potential human-induced fire outbreak danger [43–49]. In addition to these two indices, it was also determined that there is a close relationship between ISI and DSR and the weather conditions in another study carried out in New Zealand. DSR, obtained with the calculation of FWI, as well as FFMC and FWI, was found as a component to be of the highest relationship with high temperature and high west winds in Canterbury, New Zealand. In a different research area of the same study, ISI was also added to these indices in this relationship with different weather conditions [41]. The increase in DSR leads to an interpretation that extinguishing fires needs more efforts and time [50–52].

The data on the forest fires in recent years in EUMED countries display that the forest fires have decreased [14, 53]. Although Turkey and EUMED countries have different tendencies, it is seen that the relationship between forest fires and climates is of vital importance. The more the number and variety of fire and climate data increase, the clearer the relationship between them will come out to be [14]. When considered the other studies carried out in the Mediterranean region, it is seen that the areas, the drought indices (FWI, DSR) of which are of low long-term averages such as those of Turkey, are able to tolerate the climate changes better

[54, 55]. For that reason, even though there is an increase in the number of fires in the Antalya region, it is thought to be of a downtrend in the amount of burned area and the frequency of large fires.

Along with the assumption that a comprehensive model would be more determinant and easy to generalize [56, 57], a two-sided (forward–backward) step-wise regression analysis was made by using all the variables in **Table 1** in the second phase of the study, and the results are presented in **Table 2**.

Extreme values (90th percentile) of monthly BUIX and FWI, DSR and TX are the components that could be used for estimating both the number of fires and the burned area in Antalya region. The components in the both regression equations are statistically significant ($P < 0.01$). When the equations were examined, it is seen that the extreme values are more satisfactory than the components in terms of both NF and BA for Antalya region. The shared predicting factors in our study and that of Amatulli et al. [58] are the maximum values of BUI and DSR. The regression equation found by Amatulli et al. [58] can explain about 75% of the variance in estimating the amount of the burned area, similarly 71% of it (**Figure 4**) can be explained in our study with the equation in **Table 2**. When these results are assessed on the basis of countries, the explained variance is lower because large fires occur intensely in some regions like in Turkey. While a strong relationship between the amount of burned areas and FWI, BUI, ISI and SSR values in Balıkesir region was stated by Ertuğrul and Varol [6], in Muğla region was found a relationship only between SSR and the amount of burned area. Similarly, the regression equation found by Balshi et al. [59] for the west of Canada explains 80% of the variance, whereas the one for the east of Canada is able to explain only 43% of the variance. Antalya region comes in second after Muğla region in terms of the burned area and the three regions (Antalya, Muğla and Balıkesir) are of approximately 61% of the burned areas in Turkey (for the period 1977–2014).

It is shown in **Figure 4**, the monthly monitored and estimated values graphic of the burned area and number of fires. Once **Figure 4** was examined, the estimated and monitored values for both BA and NF are seen to be distributed equally on the trend line. The high and low estimates are considered to be caused by climate and index components, responsible for the unexplained part of variance.

Estimated equivalent values of NF and Ln (BA) data monitored in Antalya region for the period 2001–2014 as in **Figures 5** and **6** are shown monthly. While there are low estimations for the years 2005–2007 in terms of the number of fires, high estimates are seen for the years 2008–2011. The period in which high estimates are made in terms of the number of fires is succeeded by another period of time with low estimates. While there are low estimates for 2001–2011 in terms of burned area, there are high estimates for the years 2011–2012.

When examined **Figures 5** and **6**, an above average loss of field for 2005–2009 causes low estimates in the given period. Even only the years 2007 and 2008 lead to an increase by 3.82 in the average for the period 2001–2014. Considering **Figures 5** and **6**, the number of fires goes up the average by 1.4 in terms of number of fires especially in 2005, 2006 and 2007, and it was understood that this caused low estimates.

	Regression model NF		***P value***
	-35.54–0.39DMCX+0.44BUIX-2.08FWIP90+5.1DSRP90+1.7TXP90		**9.086e-14**
Variables	***P value***	**Variables**	***P value***
DMCX	0.000565***	DSRP90	0.002219**
BUIX	0.000160***	TXP90	0.004018**
FWIP90	0.034976*		
	Regression model Ln(BA)		***P value***
	-5.31+0.01BUIX-0.30FWIP90+0.57DSRP90+0.25TXP90		9.911e-12
Variables	***P value***	**Variables**	***P value***
BUIX	6.58e-05***	DSRP90	0.002314**
FWIP90	0.007374**	TXP90	0.000395***

Table 2. Multivariate adaptive regression spline models (MARS) selected through step-wise regression analysis by using monthly NF and BA.

http://dx.doi.org/10.5772/intechopen.68487

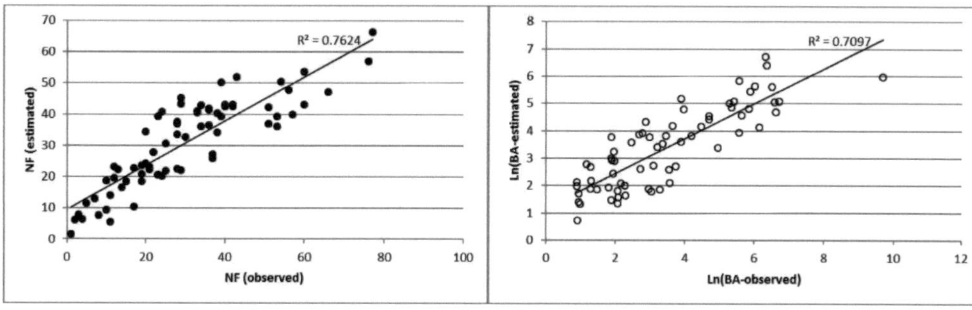

Figure 4. The relationship between the monitored values and estimated values of monthly NF and BA natural logarithm values for the period 2001–2014.

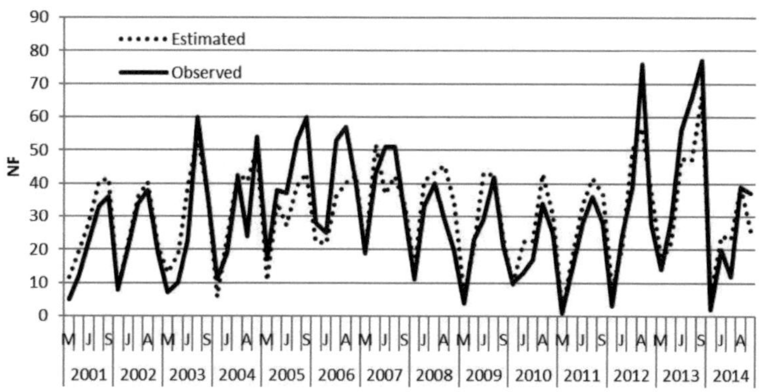

Figure 5. Monthly monitored and estimated NF values for 2001–2014.

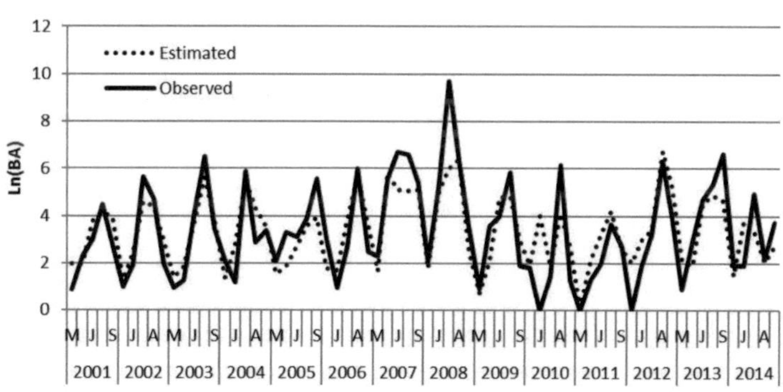

Figure 6. Natural logarithm values of monitored and estimated monthly values of BA for 2001–2014.

When it is considered that approximately 42% of the fires that occurred in Turkey between 2004 and 2014 stemmed from the negligence and carelessness, the fires caused by either physical or human induced factors in EUMED region, thus resulting in non-linear trends, makes us consider that non-parametric fire modelling would be more convenient. Syphard et al. [60]

and Vilar et al. [57] also expressed the existence of a non-linear relationship between the fires and independent variables. Therefore, Machine Learning algorithms were used in the third phase of the study.

These four algorithms (kNN, SVM, RPART and RF) managed to predict of the amount of burned areas with a success rate ranging 30–39%. In terms of the number of fires, its success widely ranges from 8 to 41%. The unsuccessful algorithms failing to predict both the number of fires and the amount of burned areas were able to be very successful in predicting the possibility of the destruction of areas greater than 300 ha. Also, the prediction results for the possibility of fire outbreaks according to drought value are seen in **Table 3**.

In the study carried out by Cortez and Morais [32] in the Montesinho National Park in Portugal, the most successful results were obtained with SVM model, while the performance of kNN and SVM models in Antalya (79.21% and 30.81) was seen to be relatively lower than those of other models. The difference is thought to stem from that the study in Portugal was carried out in a national park and also the direct weather inputs were used instead of drought indices as an input. The drought indices take into consideration not only weather inputs but also the factors such as fine fuel moisture and duff moisture. The fire possibility is of poor prediction accuracy for all models compared to large fires. RPART displays the best performance in predicting both the existence of fires and large fires. kNN, on the other hand, displays the worst performance for the both classifications.

The predictions trying to understand the relationship between the fires and droughts qualitatively were tested via the analyses in three phases. The results of these studies show that the use of MARS models bears better results at predicting both the amount of burned areas and the number of fires rather than using several drought indices. Because much of the area burned occurs during extreme fire weather condition [8], the extreme values (maximum and 90th percentile) of drought indices take place in MARS models. As well as stating that MARS and Machine Learning methods are more successful because they take the complex relationships in the data into consideration, it is seen that MARS models perform better in both the

	PPV (%)	TPR (%)	F1 (%)	ACC. (%)	PPV (%)	TPR (%)	F1 (%)	ACC. (%)
	kNN				**SVM**			
Fire possibility	60.00	59.02	59.50	86.27	67.33	58.62	62.67	88.66
BA≥ 300 ha	97.86	100.00	98.92	99.30	99.16	98.74	98.95	99.30
Average	78.93	79.51	79.21	92.79	83.24	78.68	80.81	93.38
	RPART				**RF**			
Fire possibility	100.00	77.01	87.01	97.20	64.38	67.92	66.10	88.75
BA≥ 300 ha	100.00	98.89	99.44	99.58	98.82	99.87	99.34	99.53
Average	100.00	87.95	93.23	98.39	81.60	83.89	82.72	94.14

Table 3. The performance results of different classification algorithms in predicting forest fires.

article by Amatulli et al. [58] and our study. It is stated that it is successful because MARS models constitute functions in each hyper-region by taking regression coefficients into consideration [61]. Contrary to the expectations, Machine Learning methods such as kNN, SVM, RPART and RF were unable to perform as successfully as MARS and linear regression models in order to predict the amount of burned areas and the number of fires. The performances of Machine Learning methods vary from geographical region to region, and these methods perform better in larger areas (like EUMED) compared to its performance in smaller areas [62]. Trigo et al. [63] emphasized that through Machine Learning methods were obtained poor results in Portugal where the burner areas hit the top in 2003, and large fires became intense in a small region. It is considered that a single fire that destroyed an area of 16,890 ha in 2008 in Antalya region as the study area bore similar results. Similarly, Prasad et al. [64] also stated that the same variables would vary depending on the spatial and analysis scale because environmental and social conditions differ from region to region. SVM and RF stand out compared to Machine Learning methods in terms of prediction performance [65] and RPART algorithm produced more successful results in our study. Besides the statements of both Trigo et al. [63] and Prasad et al. [64], it is thought that RPART algorithm performed better for Antalya region as it used decision tree mechanism along with regression equations in a similar way to MARS.

4. Conclusion

The relationship between factors such as meteorological factors, FWI System components, KBDI and SPEI and both the amount of burned area and the number of fires was investigated for Antalya region. The results show that the extreme values of the given factors are more effective on the amount of burned area and the number of fires. DSRX explains the majority of the variance in the number of fires and the amount of burned area. Normal or extreme values of KBDI and SPEI could not take place in the equations obtained as a result of step-wise regression. In a study in Antalya carried out by Varol and Ertuğrul [6], it was found that there was a significant relationship between KBDI values and the burned area only in the years when big fires arose but there was no relationship between KBDI values and the number of fires. As a result of step-wise regression analysis, the extreme ones of FWI, DSR, T components for both the burned area and the number of fires were selected as significant variables. As a result of the obtained equations, approximately 65% of the variance in the number of fires and 62% of the burned area could be explained through 90th percentiles of these components.

This study will be able to be used as base work in order to estimate the number of fires and the amount of burned area in Antalya region for future climate scenarios. Moreover, when the extent of the study is expanded in a way that it covers fire sensitive forests, the differences and similarities between regions will be pointed out as well as a general view for Turkey which could be reached.

In a study carried out in Canada by Harrington et al. [3], the mean and extreme values of the FWI components were compared with each other, and the extreme values in six out of nine different regions were seen to be more successful in the explanation of the variance. In another study carried out by Flannigan and Harrington [2], the use of meteorological data as a satisfactory variable explained that it could explain 30% of the variance in Canada, and this value was a similar result to FWI components. It was pointed out that maximum temperature and low humidity have a strong relationship with the dry conditions, and therefore, this is a frequently preferred variable. In our study, the existence of maximum temperature in the equations of either the number of fires or burned area promotes this assertion as a result of step-wise regression.

Urbeita et al. [14] stated that SPEI8 produced better results in the period from autumn to spring. In our study, however, SPEI is seen to be ineffective compared to other indices and SPEI1 produced better results for the forest fires in Antalya region. It is thought that the analyses for investigating the relationship between SPEI and forest fires should be made in longer periods, and the studies, as in Antalya subregion, should be expanded to the worldwide; thus, it would bear better results [14].

What determine in fire statistics are surely not factors used only in calculating drought indices. However, it is a crucial data source that is not of a certain relationship between fire statistics and meteorological data. So, it is needed a detailed and foolproof record of both fire statistics and meteorological data records. In terms of forest fire statistics, much more certain data are existent for the period after 2001, while there are no data for earlier periods. Similarly, the lack of daily meteorological data led to the determination of the study period between 2001 and 2014 for Antalya region.

Many researchers also state that a number of social factors like unemployment or arson are of influence on forest fires [66]. Moreover, in some studies, it is pointed out that socio-economic and landscape factors would be more efficient on the burned area than climate conditions in local scale [67]. So, it will be necessary to take some social factors such as unemployment, population as well as road intensity into consideration in order that the number of fires and the amount of burned areas can be predicted more successfully [15]. Therefore, MARS is thought to be more efficient after it has been added more variables. In defining complex relationships, simple and multiple regression techniques are not productive for investigating the relationships in high-dimensional data sets, while MARS can be a solution to overcome these obstacles. Moreover, it is usually simpler to compare the models developed using MARS approach with other modeling and mathematical techniques. MARS approach is considered to be more suitable than other approaches for representing the temporal variability in the factors such as temperature, precipitation, relative humidity and wind. The effects of predictors on the distribution, the model, intervals, explained variance and the extrapolation skills [58] also support the aforementioned statements. It should not be forgotten that even the best model chosen presents the maximum variance explained in the observation period, and it should be taken into account that the model would exceed this period in the future climate conditions.

Determining the factors effective on the reasons for fires is of high importance in determining the way, quality and density degree of technical interference with natural and artificial forest stands that fire sensitive plant species generate especially in the regions of ecological conditions, which are also fire sensitive. Silvicultural activities closely affect the amount of fuel [68]. The plantation and restoration activities, which aim at establishing artificial forests, it is necessary to form fire resistant and especially fire retardant forests in the Mediterranean Region climate zone, which is fire sensitive in terms of especially ecological conditions. Thus, it is so important to choose the suitable origins and clones of those as well as choosing the fire resistant and indigenous plant species [69].

These results displayed that climate conditions and stand structure have important effects on the forest fires in Antalya region as the reason of their outbreaks. Similarly, the majority of the fires occurred in the Calabrian pine and cedar forests whose closure was high (71–100%) and where middle sized and thick trees were seen in the studies, which were carried out on the forest fires in Antalya climate zone and stand dynamics and silvicultural techniques. When analyzed the current fire statistics, most of the forest fires are seen to occur in the Calabrian pine and cedar forests which are of high closure and middle sized and thick trees near especially the residential areas, recreation areas and main arterial roads. Also, taking the current ecological conditions into consideration for silvicultural objectives and aims, it is quite important to enable the suitable mesophyte species to take place in the area in order to reduce the effects of fire danger in pure coniferous monoculture Calabrian pine and cedar plantations, where especially important silvicultural interventions are compulsory such as opening necessary maintenance paths [70].

Acknowledgements

The first results of this study were presented in the 1st International Mediterranean Science and Engineering Congress.

Author details

Tuğrul Varol*, Mertol Ertuğrul and Halil Barış Özel

*Address all correspondence to: tvarol@bartin.edu.tr

Department of Forest Engineering, Faculty of Forestry, Bartın University, Turkey

References

[1] EUROSTAT. Regions in the European Union Nomenclature of Territorial Units for Statistics NUTS 2010/EU-27 [Internet]. 2011. Available from: http://ec.europa.eu/eurostat/documents/3859598/5916917/KS-RA-11-011-EN.PDF [Accessed: 21 December 2016]

[2] Flannigan MD, Harrington J. A study of the relation of meteorological variables to monthly provincial area burned by wildfire in Canada (1953-80). Journal of Applied Meteorology. 1988; 247(4):441-452

[3] Harrington JB, Flannigan MD, Van Wagner CE. A study of the relation of components of the Fire Weather Index to monthly provincial area burned by wildfire in Canada 1953-80. Information Report PI-X-25. Ontario: Canadian Forestry Service, Petawawa National Forestry Institute; 1983. p. 65

[4] Viegas DX, Viegas MTSP, Ferreira AD. Moisture content of fine forest fuels and fire occurrence in central Portugal. International Journal of Wildland Fire. 1992;**2**(2):69-86

[5] Pausas JG. Changes in fire and climate in the eastern Iberian Peninsula (Mediterranean basin). Climatic Change. 2004;**63**(3):337-350

[6] Varol T, Ertuğrul M. Analysis of the forest fires in the Antalya region of Turkey using the Keetch–Byram drought index. Journal of Forestry Research. 2016;**27**(4):811-819. DOI: 10.1007/s11676-016-0235-0

[7] Ertuğrul M, Varol T. Evaluation of fire activity in some regions of Aegean Coasts of Turkey via Canadian Forest Fire Weather Index System (CFFWIS). Applied Ecology and Environmental Research. 2016;**14**(2):93-105

[8] Carvalho A, Flannigan MD, Logan K, Miranda AI, Borrego C. Fire activity in Portugal and its relationship to weather and the Canadian Fire Weather Index System. International Journal of Wildland Fire. 2008;**17**(3):328-338

[9] Flannigan MD, Krawchuk MA, de Groot WJ, Wotton BM, Gowman LM. Implications of changing climate for global wildland fire. International Journal of Wildland Fire. 2009;**18**(5):483-507

[10] Viegas DX, Bovio G, Ferreira A, Nosenzo A, Sol B. Comparative study of various methods of fire danger evaluation in southern Europe. International Journal of Wildland Fire. 2000;**9**(4):235-246

[11] Jong MCD, Wooster MJ, Kitchen K, Manley C, Gazzard R, McCall FF. Calibration and evaluation of the Canadian Forest Fire Weather Index (FWI) System for improved wildland fire danger rating in the United Kingdom. Natural Hazards and Earth System Sciences. 2016;**16**(5):1217-1237

[12] Bedia J, Herrera S, Camia A, Moreno JM, Gutiérrez JM. Forest fire danger projections in the Mediterranean using ENSEMBLES regional climate change scenarios. Climatic Change. 2014;**122**(1-2):185-199

[13] Venäläinen A, Korhonen N, Hyvärinen O, Koutsias N, Xystrakis F, Urbieta IR, Moreno JM. Temporal variations and change in forest fire danger in Europe for 1960-2012. Natural Hazards and Earth System Sciences. 2014;**14**(6):1477-1490

[14] Urbieta IR, Zavala G, Bedia J, Gutiérrez JM, San Miguel-Ayanz J, Camia A, Keeley JE, Moreno JM. Fire activity as a function of fire–weather seasonal severity and antecedent climate across spatial scales in southern Europe and Pacific western USA. Environmental Research Letters. 2015;**10**(11):1-11

[15] Oliveira S, Oehler F, San-Miguel-Ayanz J, Camia A, Pereira JM. Modeling spatial patterns of fire occurrence in Mediterranean Europe using Multiple Regression and Random Forest. Forest Ecology and Management. 2012;**275**:117-129

[16] Cutler DR, Edwards TC, Beard KH, Cutler A, Hess KT, Gibson J, Lawler JJ. Random forests for classification in ecology. Ecology. 2007;**88**(11):2783-2792

[17] Johnstone JF, Hollingsworth TN, Chapin FS, Mack MC. Changes in fire regime break the legacy lock on successional trajectories in Alaskan boreal forest. Global Change Biology. 2010;**16**(4):1281-1295

[18] GDF. Turkish Forest Statistics [Internet]. 2015. Available from: http://www.ogm.gov.tr/ekutuphane/Yayinlar/T%C3%BCrkiye%20Orman%20Varl%C4%B1%C4%9F%C4%B1-2015.pdf [Accessed: 15 October 2016]

[19] Fujioka FM. Staring up the Keetch–Byram drought index. In: Andrews PL, Potts DF, editors. 11th Conference on Fire and Forest Meteorology; 16-19 April 1991; Missoula. Montana: Society of American Foresters; 1991. pp. 74-80

[20] De Groot WJ. Interpreting the Canadian Forest Fire Weather Index (FWI) system. In: Fourth Central Region Fire Weather Committee Scientific and Technical Seminar; 2 April 1987; Winnipeg, Manitoba. Edmonton: Canadian Forestry Service, Northern Forestry Centre; 1987. pp. 3-14

[21] Beguería S, Vicente-Serrano SM, Reig F, Latorre B. Standardized precipitation evapotranspiration index (SPEI) revisited: Parameter fitting, evapotranspiration models, tools, datasets and drought monitoring. International Journal of Climatology. 2014;**34**(10): 3001-3023

[22] Vicente-Serrano SM, Begueria S. [Internet] 2014. Available from: http://sac.csic.es/spei/database.html [Accessed: 12 July 2016]

[23] Olden JD, Lawler JJ, Poff NL. Machine learning methods without tears: A primer for ecologists. The Quarterly Review of Biology. 2008;**83**(2):171-193

[24] De'Ath G. Boosted trees for ecological modeling and prediction. Ecology. 2007;**88**(1): 243-251

[25] Lek S, Delacoste M, Baran P, Dimopoulos I, Lauga J, Aulagnier S. Application of neural networks to modelling nonlinear relationships in ecology. Ecological Modelling. 1996;**90**(1):39-52

[26] De'ath G, Fabricius KE. Classification and regression trees: A powerful yet simple technique for ecological data analysis. Ecology. 2000;**81**(11):3178-3192

[27] Drake JM, Randin C, Guisan A. Modelling ecological niches with support vector machines. Journal of Applied Ecology. 2006;**43**(3):424-432

[28] Stojanova D, Panov P, Kobler A, Džeroski S, Taškova K. Learning to predict forest fires with different data mining techniques. In: Conference on Data Mining and Data Warehouses (SIKDD 2006); 10 October; Ljubljana. SIKDD. p. 255

[29] Iliadis LS, Papastavrou AK, Lefakis PD. A computer-system that classifies the prefectures of Greece in forest fire risk zones using fuzzy sets. Forest Policy and Economics. 2002;**4**(1):43-54

[30] Phillips SJ, Anderson RP, Schapire RE. Maximum entropy modeling of species geographic distributions. Ecological Modelling. 2006;**190**(3):231-259

[31] Weinberger KQ, Blitzer J, Saul L. Distance metric learning for large margin nearest neighbor classification. Journal of Machine Learning Research. 2009;**10**: 207-244

[32] Cortez P, Morais ADJR. A Data Mining Approach to Predict Forest Fires using Meteorological Data. [Internet]. 2007. Available from: http://www.dsi.uminho.pt/~pcortez [Accessed: 10 November 2016]

[33] Pontil M, Verri A. Properties of support vector machines. Neural Computation. 1998;**10** (4):955-974

[34] Rollins MG, Keane RE, Parsons RA. Mapping fuels and fire regimes using remote sensing, ecosystem simulation, and gradient modelling. Ecological Applications. 2004;**14**(1): 75-95

[35] Breiman L. Random forests. Machine Learning. 2001;**45**(1):5-32

[36] Hong PTT, Hai TTT, Hoang VT, Hai V, Nguyen TT. Comparative study on vision based rice seed varieties identification. In: Seventh International Conference on Knowledge and Systems Engineering (KSE); 8-10 October 2015; Ho Chi Minh, Vietnam. Vietnam National University; 2015. pp. 377-382

[37] Montillo AA. Random Forests, Guest Lecture: Statistical Foundations of Data Analysis. [Internet] 2009. Available from: (http://www.dabi.temple.edu/~hbling/8590.002/Montillo RandomForests_4-2-2009.pdf). [Accessed: 30 January 2017]

[38] De Jong MC, Wooster MJ, McCall FF. Calibration and evaluation of the Canadian Forest Fire Weather Index (FWI) System for improved wildland fire danger rating in the United Kingdom. Natural Hazards and Earth System Sciences. 2016;**16**(5):1217

[39] Taylor SW, Alexander ME. Science, technology and human factors in fire danger rating: The Canadian experience. International Journal of Wildland Fire. 2006;**15**(1):121-135

[40] Lawson BD, Armitage OB. Weather guide for the Canadian Forest Fire Danger Rating System. Natural Resources Canada. Edmonton, Alberta: Canadian Forest Service; 2008. p. 87

[41] Renwick J, Salinger J, Zheng X, Pearce G. Prediction of fire weather and fire. NIWA Client Report, AKI-2007-034; New Zealand Fire Service Commission; 2007. p. 44

[42] Ullah MR, Liu XD, Al-Amin M. Spatial-temporal distribution of forest fires and fire weather index calculation from 2000 to 2009 in China. Journal of Forest Science. 2013;**59**(7):279-287

[43] Stocks BJ, Lawson BD, Alexander ME, Van Wagner CE, McAlpine RS, Lynham TJ, Dubé DE. The Canadian system of forest fire danger rating. In: A Conference on Bushfire Modelling and Fire Danger Rating Systems; 11-12 July 1988; Canberra, Australia. Northern Forestry Center, Edmonton; 1991. pp. 9-18

[44] Wotton BM. Interpreting and using outputs from the Canadian Forest Fire Danger Rating System in research applications. Environmental and Ecological Statistics. 2009; **16**(2):107-131

[45] Chu T, Guo X. An assessment of fire occurrence regime and performance of Canadian fire weather index in south central Siberian boreal region. Natural Hazards and Earth System Sciences Discussions. 2014;**2**(7):4711-4742

[46] Stocks BJ. Wildfires and the fire weather indexes system in Ontario. Sault Ste. Marie, Ontario: Canadian Forestry Service, Great Lakes Forestry Centre. Information Report O-X-213; 1974. p. 17

[47] Kiil AD, Miyagawa RS, Quintilio D. Calibration and performance of the Canadian fire weather indexes in Alberta. Edmonton, Alberta: Canadian Forestry Service, Northern Forest Research Centre. Information Report NOR-X-173; 1977. p. 45

[48] Martell DL, Otukol S, Stocks BJ. A logistic model for predicting daily people-caused forest fire occurrence in Ontario. Canadian Journal of Forest Research. 1987;**17**(5):394-401

[49] Anderson KR, Englefield P. Quantile characteristics of forest fires in Saskatchewan. In: 4th Symposium on Fire and Forest Meteorology; 13-15 November 2001; Massachusetts, USA. Northern Forestry Centre, Edmonton; 2001. p. 9-16

[50] Van Wagner C. Development and structure of the Canadian forest fire weather index system. Ottawa, ON, Canada: Canadian Forest Service. For. Tech. Rep. 35; 1987. p. 48

[51] Whitman E, Sherren K, Rapaport E. Increasing daily wildfire risk in the Acadian Forest Region of Nova Scotia, Canada, under future climate change. Regional Environmental Change. 2015;**15**(7):1447-1459

[52] Flannigan M, Cantin AS, de Groot WJ, Wotton M, Newbery A, Gowman LM. Global wildland fire season severity in the 21st century. Forest Ecology and Management. 2013;**294**:54-61

[53] San-Miguel-Ayanz J, Moreno JM, Camia A. Analysis of large fires in European Mediterranean landscapes: Lessons learned and perspectives. Forest Ecology and Management. 2013;**294**:11-22

[54] Dimitrakopoulos AP, Vlahou M, Anagnostopoulou CG, Mitsopoulos ID. Impact of drought on wildland fires in Greece: Implications of climatic change? Climatic Change. 2011;**109**(3):331-347

[55] Pausas JG, Ribeiro E. The global fire–Productivity relationship. Global Ecology and Biogeography. 2013;**22**(6):728-736

[56] Catry FX, Rego FC, Bação FL, Moreira F. Modeling and mapping wildfire ignition risk in Portugal. International Journal of Wildland Fire. 2010;**18**(8):921-931

[57] Vilar L, Woolford DG, Martell DL, Martín MP. A model for predicting human-caused wildfire occurrence in the region of Madrid, Spain. International Journal of Wildland Fire. 2010;**19**(3):325-337

[58] Amatulli G, Camia A, San-Miguel-Ayanz J. Estimating future burned areas under changing climate in the EU-Mediterranean countries. Science of the Total Environment. 2013;**450-451**:209-222

[59] Balshi MS, McGuire AD, Duffy P, Flannigan M, Walsh J, Melillo J. Assessing the response of area burned to changing climate in western boreal North America using a Multivariate Adaptive Regression Splines (MARS) approach. Global Change Biology. 2009;**15**(3):578-600

[60] Syphard AD, Radeloff VC, Keeley JE, Hawbaker TJ, Clayton MK, Stewart SI, Hammer RB. Human influence on California fire regimes. Ecological applications. 2007;**17** (5):1388-1402

[61] Morissette-Thomas V, Cohen AA, Fülöp T, Riesco E, Legault V, Li Q, Milot E, Dusseault-Belanger F, Ferrucci L. Inflamm-aging does not simply reflect increases in pro-inflammatory markers. Mechanisms of Ageing and Development. 2014;**139**:49-57

[62] Flannigan MD, Logan KA, Amiro BD, Skinner WR, Stocks BJ. Future area burned in Canada. Climatic Change. 2005;**72**(1):1-16

[63] Trigo RM, Pereira J, Pereira MG, Mota B, Calado TJ, Dacamara CC, Santo FE. Atmospheric conditions associated with the exceptional fire season of 2003 in Portugal. International Journal of Climatology. 2006;**26**(13):1741-1757

[64] Prasad AM, Iverson LR, Liaw A. Newer classification and regression tree techniques: Bagging and random forests for ecological prediction. Ecosystems. 2006;**9**(2):181-199

[65] Witten IH, Frank E, Hall MA, Pal CJ. Data Mining: Practical Machine Learning Tools and. Second edition. Morgan Kauffmann Publishers. 2005; p. 558

[66] Leone V, Lovreglio R, Martín MP, Martínez J, Vilar L. Human factors of fire occurrence in the. In: Earth Observation of Wildland Fires in Mediterranean Ecosystems, Springer Berlin Heidelberg. 2009; pp.149-170.

[67] Costa L, Thonicke K, Poulter B, Badeck FW. Sensitivity of Portuguese forest fires to climatic, human, and landscape variables: subnational differences between fire drivers in extreme fire years and decadal averages. Regional Environmental Change. 2011;**11** (3):543-551

[68] Agee JK, Skinner CN. Basic principles of forest fuel reduction treatments. Forest Ecology and Management. 2005;**211**(1):83-96

[69] Baker LG, Spaine P, Covert SF. Effect of surface wettability on germination and gene expression in *Cronartium quercuum* f. *sp. fusiforme basidiospores*. Physiological and Molecular Plant Pathology. 2006;**68**(4):168-175

[70] Trabaud L. Natural and prescribed fire: survival strategies of plants and equilibrium in Mediterranean ecosystems. In: Tenhunen JD, Catarino FM, Lange OL, Oechel WC, editors. Plant Response to Stress-Functional Analysis in Mediterranean Ecosystem. NATO Advanced Science Institute. New York: Springer; 1987. pp. 607-621

Chapter 13

The Fire in the Mediterranean Region: A Case Study of Forest Fires in Portugal

Ana Cristina Gonçalves and Adélia M.O. Sousa

Additional information is available at the end of the chapter

http://dx.doi.org/10.5772/intechopen.69410

Abstract

Forest fires are a common disturbance in many forest systems in the world and in particular in the Mediterranean region. Their origins can be either natural or anthropogenic. The effects in regard to the time trends, vegetation, and soil will be reflected in the species distribution, forest composition, and soil potential productivity. In general, it can be said that the larger the fire and the shorter the time between two consecutive occurrences, the higher the probability to originate shifts in vegetation and soil degradation. In the Mediterranean region, the number of fire ignitions does not reflect the burnt area due to the occurrence of very large fires. The latter occur in a very small proportion of the number of ignitions, but result in very large burnt areas. Also there seems to be an increasing trend toward larger fires in the Mediterranean region due mainly to climatic and land use changes. This case study highlights the importance of vegetation regrowth a short time after the fire to maintain both forest systems and soil conservation.

Keywords: Portugal, burnt area, number of fires, spatial dynamic, temporal dynamic, vegetation

1. Introduction

Mediterranean forest types can be characterized by their heterogeneity, whether climatic, edaphic, geomorphologic, floristic, biogeographic or historical, and instability and vulnerability, consequences of the former and due to genetic and ecological factors as well as to the anthropogenic actions [1]. The Mediterranean flora is composed of a wide variety of tree, shrub, and herbaceous species, and their distribution depends on the edaphoclimatic conditions and on human intervention [1, 2]. The climatic conditions that are most influential on the species distribution and growth are the temperature and the precipitation, as well as their interactions. In the Mediterranean basin the mean annual temperatures vary between 5 and 8°C and the mean annual precipitation ranges from 300 mm to more than 2500 mm [1] with a marked seasonal and annual variability. As a result, droughts are frequent and vary between a couple of weeks and more than six months [1, 2]. In addition, rain falls frequently in torrential short-duration events mainly in autumn and in winter. Thus, growth is limited to mainly spring and autumn, and when the moisture balance is favorable, the growth can be luxurious [2]. Nonurban land use in the Mediterranean basin is distributed in a mosaic pattern with the three most frequent uses being forest, pasture, and agricultural cultures. This spatial arrangement is composed of very small areas with diffuse edges [1–3].

Fire was, and still is, a common feature in the Mediterranean landscape. Historically, fire has been described as having two main forces of ignition: natural, to a lesser extent, caused usually by summer lightning storms, not considered to be significant as frequently is followed by torrential rainfall [2]; and anthropogenic, in earlier times to clear areas for pasture, hunting, grazing, and/or agriculture [1, 2], though in ancient times it was also considered a war weapon [2].

Fires are common in many forest systems around the world, and in particular in the Mediterranean region. Forest fires can be a minor or major disturbance in forest ecosystems depending on their intensity and result in stand's species composition and structure changes [4, 5]. Thus the analysis of the spatial distribution and temporal frequency of forest fires and their relation to vegetation—trees, shrubs and herbaceous plants—is of utmost importance. In this chapter, a review of the origins (Section 2); of the effects of the time trends, on vegetation and on soil (Section 3); of the evaluation of pre- and post-fire vegetation dynamics with remote sensing (Section 4); and of the analysis of the fires, number, and burnt area, for the Mediterranean region and for Portugal (Section 5) is made. A case study is also presented to compare the vegetation prior the fire with that of regeneration after the fire (Section 6).

2. Origins of forest fires in the Mediterranean region

Forest fires are a driving force in the evolution, distribution, and organization of the forests around the world [4–6]. Fires may have a strong effect both on the vegetation and on the soil carbon sequestration and sinks as they may reduce regeneration and, consequently, the potential biomass accumulation [7–9], carbon stocks [10], and timber [4, 5]. Forest fires are one of the primordial factors affecting the Mediterranean-type ecosystems [11], causing more destruction to trees than all the other hazards, such as diseases, insects, wind throws, and frosts [12].

Humans have been using fire since ancient times [13, 14] and since 10,000 years ago have influenced their regime [15]. While in earlier times fire was linked to land management, currently other causes prevail. Since a couple of decades, an intensification in the fire occurrences has been observed, due to the combined effects of climate change [16–21] and alterations in the land use, such as the abandonment of agricultural lands, decrease in grazing, urban areas increase, and the interaction between the urban and the wildland areas [11, 22–36]. As a result, change in the vegetation structure and species composition occurred

[30, 36]. Forest fires in the Mediterranean ecosystems have complex effects on the ecological processes due to the differential responses to the vegetation structures and fire regimes [37–39]. The post-fire evolution of vegetation community is a consequence of its resilience [40, 41] which can be linked to biodiversity (e.g. [30, 42]), vegetation structure (e.g. [43]), and ecosystem functions (e.g. [44]).

Fire damages can be evaluated by their direct impacts that are the physical changes in the flora and fauna of the ecosystems affected; and the indirect losses, consequence of the former and which relate to the impacts on the economy and environment of both the burnt and the neighboring areas. Noteworthy are also the tangible losses that can be expressed in monetary terms, and the intangible losses of difficult quantification though affecting both the environment and the economy [45].

Any fire regime can be described as the function of a suite of variables. The most commonly used ones to describe a fire regime are the nature, pattern (size), season, intensity (energy released), and recurrence. Of notice are the features regarding fire recurrence, namely frequency, which is characterized by the number of fires within a life span in an area; and the mean fire return interval, that is the time interval between successive fires [46]. Fire frequency has a major role in the vegetation due to its effects on the regeneration and recruitment of plants, especially of trees, after a fire. In the Mediterranean basin, many fire regimes have a mean fire return interval of 15–20 years [36, 47], which can reduce the number of species because sexual maturity is not reached, and consequently no seed is available to enable plant regeneration and recruitment [48, 49]. Fire effects can be evaluated by fire intensity, fire severity, and burn severity [50, 51]. The latter is the more frequently used [52–55].

3. Effects of fires on the Mediterranean stands and forests

3.1. Time trends on forest fires

Most of the detected fires are small and can be seen as minor disturbances. However, large fires, though in a much smaller number, can be considered as major disturbances and can cause considerable impacts both at the landscape level with the destruction of large areas of vegetation and at the economic and social levels [56, 57]. According to the affected vegetation strata, fires can be classified as ground, surface, and crown fires [58]. The surface fires are recognized as having less impact on the vegetation communities and are sometimes used to control vegetation as prescribed burnings. They can be more easily controlled and extinguished [5]. On the contrary, when a fire evolves to a crown fire, its suppression is almost always impossible, causing also major impacts on the vegetation [59]. The transition from surface to crown fire is related to the vertical and horizontal arrangement of the vegetation. While a vegetation community with continuous horizontal and vertical spatial distribution can enhance fire spread into crowns, their discontinuity can reduce it [60–63]. The effects of heterogeneous vegetation structures in the fire behavior [64–68], as well as the "ladder effect" [69], result from the vegetation vertical and horizontal connectivity.

The term "megafire" is often used to classify very large fires, though no definition exists. They are frequently derived from several large fires, resulting in very large burnt areas and high damage levels, both in the vegetation and at the socioeconomic level. Megafires can be classified using three criteria: (i) fire behavior, corresponding to their intensity and spread rate [70]; (ii) resistance to control by suppression activities [71]; and (iii) fire severity, corresponding to the affected area, fatalities, burn severity, and economic losses [72]. San-Miguel-Ayanz et al. [11] analyzed megafires in Portugal, Spain, and Greece, and refer that each megafire was a singular event, with a set of large fires concentrated in time and space in one fire season. These events seem to be linked with extreme meteorological conditions, for example very high temperatures, very low humidity, and lightning storms. The same authors refer that the number of ignitions and the fast spread of fire (due to vegetation and topographic conditions) can be of major importance. The behavior of megafires makes it difficult to control their spread and suppress them [71, 73], and their control and extinction is only possible when they spread to lower fuel load vegetation or when the weather turns cooler and wetter [74].

3.2. Vegetation

Vegetation can be characterized according to their fire proneness. The typical vegetation of the Mediterranean regions is one of the most fire prone [14, 75]. In these vegetation types, fire had, and still has, a key role in its dynamics and structure. The role of fire dates back to the early Holocene [76] with a continuous role onward [14, 77, 78]. Fire has been recognized as affecting the landscape at the long term; however, its effects vary as function of the regeneration patterns of the vegetation, topography, and fire history [24, 79]. Vegetation community resilience has been verified as a significant proportion of these communities were able to maintain their characteristics [24].

Several authors refer that the fire risk is expected to increase in a frame of climate change in the Mediterranean region, especially in the European part (e.g. [80, 81]) as a consequence of the increase in drought events [82]. Thus the knowledge of soils and plant communities' behavior after fire is of utmost importance [78, 83–85]. Also, the foreseen longer and more frequent drought events along with the higher temperatures might enhance the expansion of plants better adapted to those climatic conditions [86, 87].

The plants' post-fire regeneration depends on their adaptive traits, which can be divided into: (i) resprouters, plants that have protected buds that are able to develop after the fire; and (ii) seeders, which correspond to plants that are affected by fires but have seed banks either in the canopy and/or in the soil that are able to maintain the germination ability [4, 9]. Additionally, there is a suite of factors controlling the development of the post-fire vegetation development: (i) the pre-fire plant species composition [88–92]; (ii) the severity and intensity of fire [6]; (iii) the season of fire occurrence [93]; (iv) the regeneration ability, by resprouting or seed bank in the canopy, and soil [9, 94, 95].

The post-fire dynamics of vegetation community is a very slow process that can be divided into three stages: initial, transition, and advanced. The initial stage is characterized by an intensive regeneration of the vegetation existing prior to the fire, including trees, shrubs, climbers, and herbaceous plants [88–92, 96] either by resprouting or by seed germination. The growing space let available by the fire is rapidly occupied, and about two years after the fire the species richness reaches a maximum [89, 90]. However, the spatial arrangement of the plants shows a high heterogeneity, characteristic of the Mediterranean ecosystems [89, 97]. This variability is also related to the site characteristics, with low species diversity been found in the sites where water is the main limiting factor [89, 98]. With time, competition for growing space increases gradually, limiting the regeneration of new seedlings, while some saplings tend to disappear, resulting in a decrease of species diversity [89]. The transition stage is the consequence of the increasing competition with the less-competitive plants (frequently herbaceous plants) disappearing while others gradually increase their frequency, such as shrubs, climbers, and some trees species [89]. The advanced stage is the consequence of the species characteristics (especially growth rate and life span) and site characteristics (especially temperature, water, and nutrient availability). At this stage, species with low growth rates and long life cycles are found (mainly trees but also some shrubs and climbers), while species of fast growth and shorter life spans tend to disappear [77, 89].

Short-time fire intervals may have a strong effect on the presence and abundance of species [36, 99–102] and may originate shifts in the predominant species in the Mediterranean ecosystems [103, 104]. As a result, infrequent fires or fires with medium and long return intervals tend to promote tree plant communities, while short return interval fires originate shrub- and herbaceous-dominant communities and also an intensification of the fire regime [8, 9, 46, 81, 99, 105].

3.3. Soil

The effects of fire on vegetation are easily perceived as they can be seen. However, the effects of fire on the soil are more difficult to recognize. Several authors [55, 106–108] refer the impacts of fire on the chemical, physical, and biological properties of the soils.

After a fire the soil properties can be affected by precipitation: total precipitation, intensity, and its temporal patterns [108]. Torrential rainfalls, characteristic to the Mediterranean region, are of special importance due to their negative effect on soil fertility as it increases the risk of runoff, erosion, and nutrient losses [32, 109–111]. As a result of the burning of the vegetation, some nutrients increase their availability immediately after the fire, such as nitrogen [106, 112], phosphorous [55, 106–108], and potassium [108]. Nonetheless, all three nutrients decrease about one year after the fire, explained, at least partially, by the losses originated by the rainfall. However, these effects are partially minimized under drought conditions [106]. Contrarily, soil carbon is partially or almost completely destroyed during the fire. As fires frequently occur during the hot dry months, the soil carbon after the fire will be recovered at a slow rate [106, 108, 113, 114], and similarly the carbon mineralization rate is also quite low under drought [106, 115–118].

The effect of vegetation on soil conservation is well known [119–121], and soil erosion and land degradation processes depend on it [122]. Also, erosive processes tend to increase after the occurrence of fires as soil is exposed [123], and the risk increases with the increase in the time needed for the vegetation to develop to a minimum ground cover threshold [122–124]. Several authors referred that most of the post-fire sediments are observed in the first year [123, 125, 126] and that the highest susceptibility to soil erosion occurs 4–6 months after the fire [125]. The protective effects of vegetation on soils [119–121] are due to both the ground cover and the improvement of the hydrological properties of the soil [125]. Contrarily, repeated fires in the same area, due to the direct effect of the fire on soil properties and the loss of vegetation, originate higher risk of water erosion and consequently increase soil degradation processes [125].

Fires are frequently followed by salvage cuttings in order to attain some timber economic return. These cuttings have two disadvantages in the short term; one is that it might increase soil erosion risk [127, 128]; and the other is that the damaged wood will not decompose *in situ* and consequently does not restore or increase the soil carbon stocks [129]. In the medium and long terms, it may further increase the erosion hazard [130] and reduce the soil seed bank, the establishment of seedlings, and the belowground resprouting organs [131–133].

In studies in *Pinus pinaster* stands, several authors [129, 134–136] stress that in unburnt stands the aboveground biomass, and thus the carbon stock, is considerably higher than in those that have burnt one or more times. On the contrary, in the latter the most important carbon stocks seem to be the soil organic carbon [134, 137, 138]. These findings are important in forest management, as after the fire the risk of soil erosion increases [139] and considerable soil losses may occur [140–143].

Fire occurrence, spread, and burnt area can be minimized with management practices that go toward the heterogeneity of the landscape [144]. Fuel loads and vertical and horizontal spatial distribution of the vegetation communities can promote or reduce the fire risk. Oak-dominated stands or forests may enhance the opportunities to reduce the burnt area [145] or reduce the fire risk [18]. However, the opposite is encountered in pine stands or forests and shrublands [146] where the vegetation structure has a higher degree of continuity [147]. Gutierrez and Lozano [148] state that according to their analysis carried out from 1980 to 2010, about 35% of the damages caused by forest fires could have been avoided if appropriated practices had been set out. The same authors refer also that there are significant differences between southern European countries (France, Greece, Italy, Portugal, and Spain) that were more efficient in controlling the fire damage in 2007 and 2009–2010 than other European countries (Austria, Bulgaria, Croatia, Cyprus, Estonia, and Finland).

4. Remote sensing evaluation of pre- and post-fire vegetation dynamics

Remote sensing has a primary role in the assessment [149–153] of pre- and post-fire vegetation [154] as it can be used even in the inaccessible areas or where the costs of fieldwork are prohibitive [155]. Time series of satellite images that can be used in monitoring the dynamics of vegetation are also of major importance [154, 156].

Pre- and post-fire vegetation have different spectral and spatial responses which allow their dynamics evaluation with remote sensing [157]. Though many methodologies and techniques exist, the most frequently used are image classification, vegetation indices, and spectral mixture analysis. Image classification enables the transformation of remotely sensed data into land cover/use classes, using either supervised or unsupervised techniques [154].

Vegetation indices have a strong relation with biomass and leaf area index, thus suitable for vegetation evaluation both prior and after fire events, whether spatially or temporally [52, 154, 158–166]. One of the most frequently used indices is Normalized Difference Vegetation Index (NDVI) [52, 167–169]. Other vegetation indices are also widely used, such as Soil Advanced Vegetation Index (SAVI) and Transformed Soil Advanced Vegetation Index (TSAVI) [52, 154, 157, 170]. Spectral mixture analysis enables the discrimination of the different fraction in each pixel. It is especially important in low and medium spatial resolution satellite images where each pixel is almost always composed of vegetation, soil, or other land cover/use. Thus it is suited for the post-fire analysis of the vegetation regrowth as instead of one spectral signature, it detects the cover fraction within each pixel [49, 158, 171]

5. Dynamics of fire

5.1. Mediterranean region

The statistics of forest fires in the Mediterranean basin countries had started in 1980s for France, Greece, Italy, Portugal, and Spain, while for Croatia, Morocco, and Turkey, the first data are from the 1990s; for Cyprus and Slovenia, from the 2000s; and for Algeria and the former Yugoslav republic of Macedonia around 2010s. Lebanon data are available for two years (2012 and 2015) and no data are available for Albania, Egypt, Israel, Libya, Montenegro, Syria, and Tunisia [172]. The larger number of fires per year is found in Portugal and Spain, more than 10,000 per year, followed by Italy, between 5,000 and 10,000 (**Figure 1**, top). These three countries have more than 100,000 ha of burnt area annually (**Figure 1**, center). There seems to be a trend toward larger number of fires and burnt area for the Iberian Peninsula. When analyzing the mean burnt area per fire and per country (**Figure 1**, bottom), it can be seen that for Slovenia, France, Turkey, Portugal, and Morocco, areas are smaller than 10 ha; in Algeria, Spain, Italy, and Cyprus, the mean burnt area per fire between 11 ha and 15 ha is found, while for former Yugoslav republic of Macedonia, it is 23 ha, for Greece, 31 ha, and for Croatia, 47 ha. These results point to the rather large number of small fires and a small number of fire events with very large burnt areas or megafires, as referred in other studies (e.g. [11, 74]). Another study [23] refers that for many European countries the more frequently affected areas are the wildland-urban interface, though regional variability is observed.

The temporal analysis of the number of fires and burnt area per country will be focused on the five European Mediterranean countries which have data available from 1980 onward. For this evaluation, from 1980 to 2015, four classes of five-year period and one of six-year period were considered. There seems to be a different trend for the number of fires and for the burnt area. France and Greece have the lowest number of fires with a rather small annual fluctuation values while Portugal and Spain have the highest values and an increasing trend up to 1995–1999 with a decreasing trend onward. Italy has an increasing trend up to 1990–1994 and decreasing afterward (**Figure 2**). The burnt area variability per country is larger than that of the number of fires. Similar to the number of fires, France and Greece have the lowest values of burnt area, but the variability between all time periods is larger, with an ascending trend during the 1980s, decreasing in the 1990s onward, except during 2000–2004 for France and irregular for Greece. Italy has predominantly descending trend in the burnt area values, especially noticeable from 1990–1994 onward. Portugal has an increasing trend of burnt area values up to 2000–2004, where the highest value was attained, and a rather constant value in the next two time periods. Spain has burnt area value increases from the first to the second time periods and a decreasing trend in the following two time periods and a rather constant trend afterward. The highest aforementioned values seem to be linked to the occurrence of megafires, namely in Italy in 1989; in Spain in 1989, 1994, and 2006; in Portugal in 2003 and 2005; and in Greece in 2007 [11, 74]. Furthermore, the changes in land use, in particular the decrease in or abandonment of agricultural, pastoral, and forestry activities, seem to have promoted the increase in the number of fires and burnt area [11, 22–30, 33, 34]

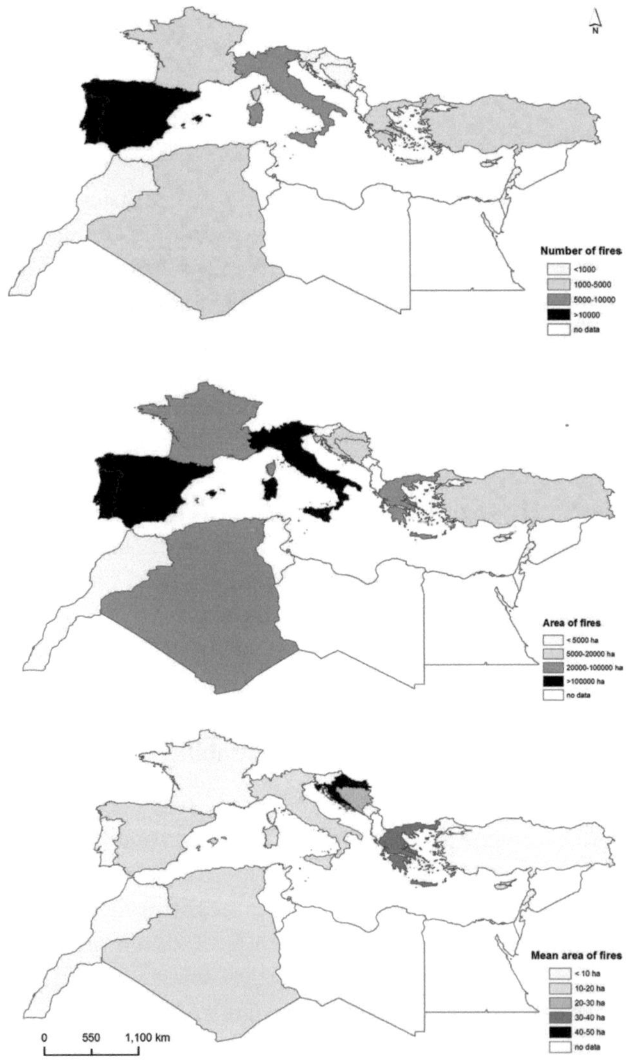

Figure 1. Average number of fires (top), average area of fires (center) and mean dimension of burnt area per fire per year (bottom), per country, in the period 1980–2015.

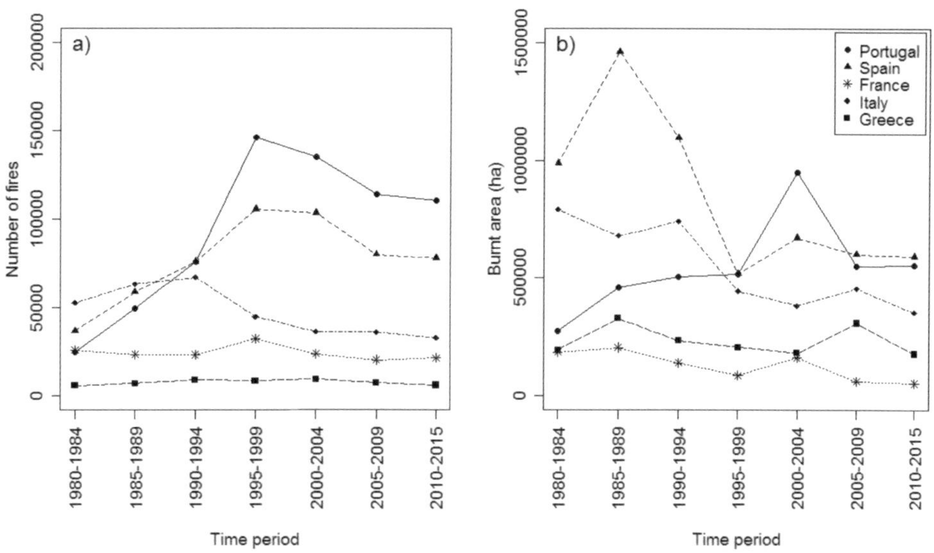

Figure 2. Number of fires (a) and burnt area (b) per country per five-year periods.

5.2. Portugal

A more detailed analysis for the dynamics of forest fires for Portugal for the time series data of 1980–2015 [173] was done, aggregating the burnt area per fire in the classes: <1, 1–25, 25–100, 100–500, and >500 ha. Most fires (66.5%) have burnt areas of less than 1 ha, corresponding to the lowest proportion of the total burnt area (3%). Contrarily, fires with burnt areas of more than 500 ha have the smallest number of events (0.5%), but have the largest proportion of burnt area (50%). The observed relation for the remaining classes is the following: the smaller the number of fires, the larger the burnt area (29.3, 2.5, 1.2% and 14, 12, 22%, respectively for burnt areas of 1–25, 25–100, and 100–500 ha). Other research describes similar trends [11, 174].

When the number of fires and burnt area is analyzed per NUT3 territorial units, it can be seen that the highest values are in northern and central Portugal (**Figure 3**). Though with different methodologies, several authors attained similar results [22, 174, 175]. In addition, a similar trend of inverse proportionality between the number of fires and burnt area is observed for all 23 NUT3 regions, with some variability for the five aforementioned classes of burnt area. Yet, variability is also observed within each of the NUT3 regions, if a smaller territorial unit is considered, for example the municipality (**Figure 4**). There seems to be more homogeneity in the municipalities of the northeastern and central eastern Portugal and more heterogeneity for those of coastal Portugal. This variability can be related to the occurrence of large fires, such as those referred by Fernandes et al. [73] and Tedim et al. [74]. And also with land use, especially the composition, structure and fuel loads of the forest systems [22, 174, 175]. The lower number of fires and burnt area corresponds to the agroforestry systems, which are characterized frequently by a tree cover of evergreen oaks (*Quercus suber* and *Quercus rotundifolia*), and pasture with extensive grazing [176, 177] thus with a horizontal and vertical discontinuity [174, 178, 179]. At the other edge are the forest systems

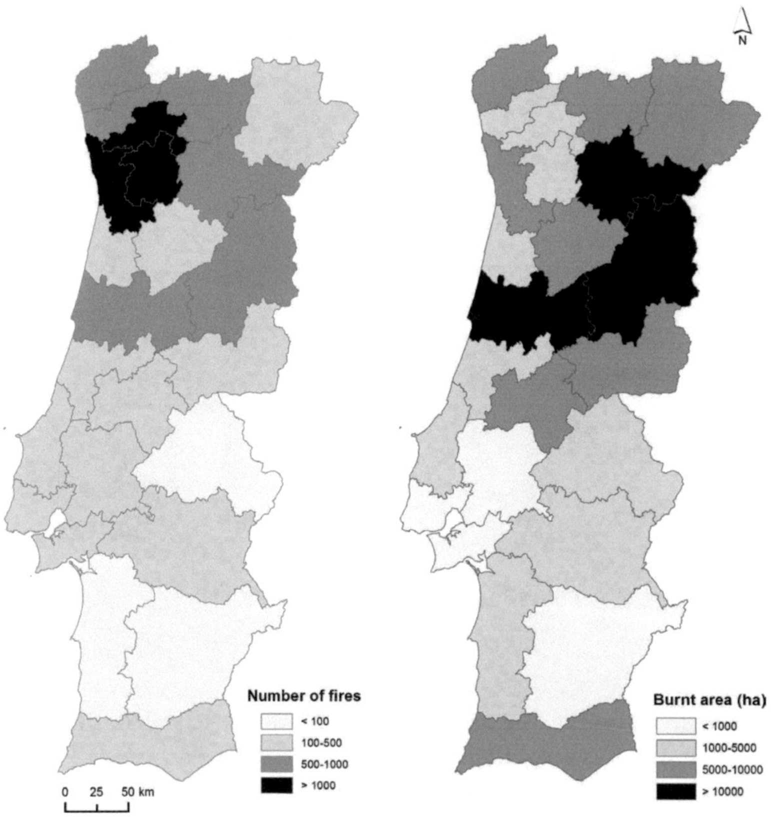

Figure 3. Average number of fires (left) and burnt area (right) per NUT3 region of Portugal in the period 1980–2015.

composed mainly of maritime pine (*Pinus pinaster*) and/or Tasmanian blue gum (*Eucalyptus globulus*), especially those with a well-developed shrub layer [178] corresponding to systems with higher horizontal and vertical continuity, and larger fuel loads, where the higher number of fires and larger burnt area occurs [22, 174, 175]. A similar trend was described for *Pinus halepensis* in Spain [60]. The temporal dynamics of the burnt area presents a rather large variability, increasing from the smallest to the largest class of burnt area (**Figure 5**), showing a similar trend to the observed by San-Miguel-Ayanz et al. [11]. There seems to be an inverse relation between the burnt area of the fires larger than 500 ha and that of the remaining classes, that is smallest burnt areas per year for the class larger than 500 ha correspond to the highest burnt areas per year in the other five classes (**Figure 5**). Noteworthy are the peaks of 2003 and 2005, as already referred, which are linked with the occurrence of megafires [11, 74].

It seems that the temporal and spatial patterns of fires are linked with the vegetation community structure both in density of ignitions [22] and burnt area [22, 175] with a positive trend for the former in Portugal since the 1980s [106, 174, 180]. The shrublands are among the vegetation communities mostly affected by fire due to the high rate of spread and low firefighting priority [27, 28, 181–185].

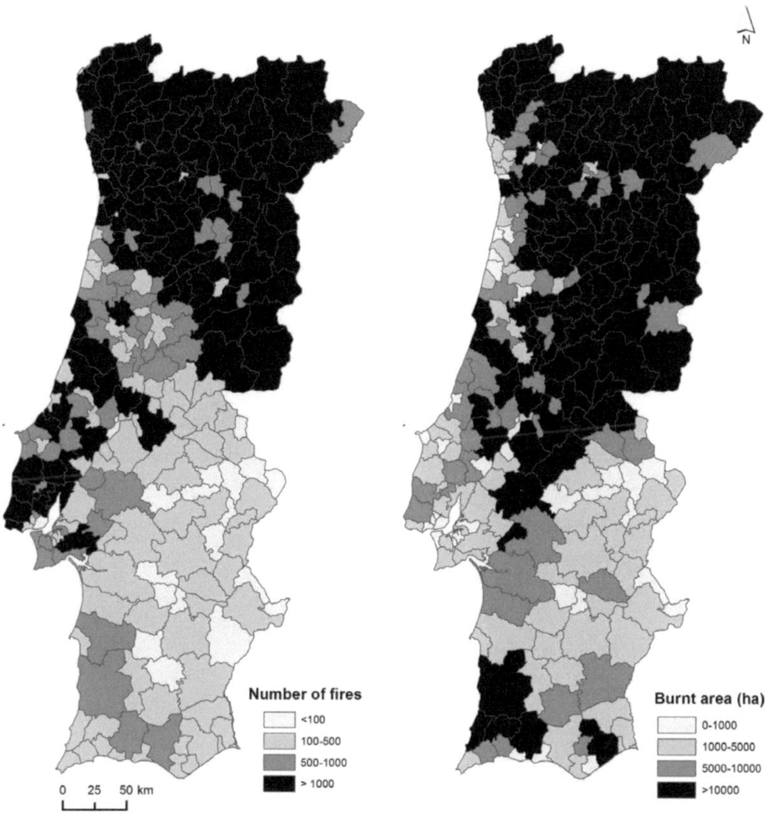

Figure 4. Average number of fires (left) and burnt area (right) per municipality region of Portugal in the period 1980–2015.

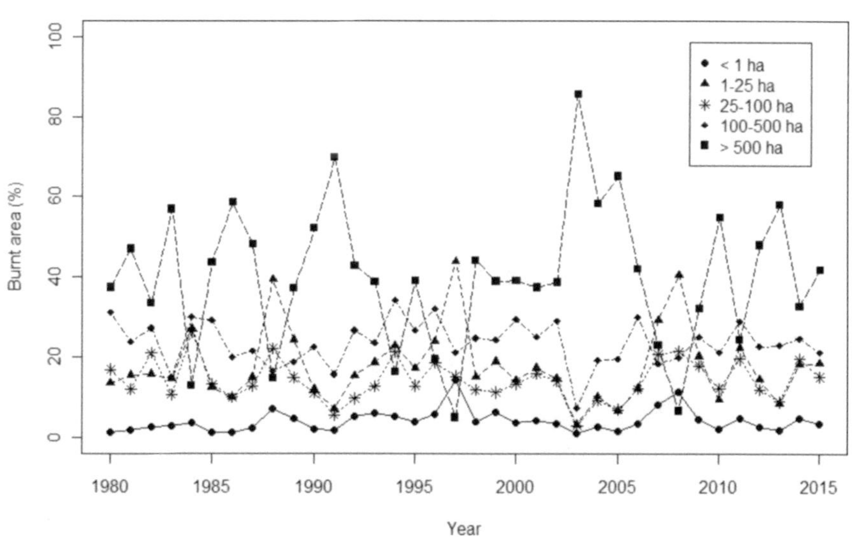

Figure 5. Burnt area per class of burnt area and per year.

6. Case study

6.1. Study area and remote sensing data

Mapped burnt areas in 2013 by the Portuguese Forest Services were used to study the vegetation recovery, which also encompass the affected main land cover/use types. The analyses of the vegetation recovery for large fire scars were carried out, using the vegetation index NDVI calculated with Landsat 8 images for the four years: 2013, 2014, 2015, and 2016.

The study area is located in northern central Portugal (**Figure 6**), one of the most affected regions by forest fires (cf. **Figures 3** and **4**) in the summer during dry season (June to September). The climate is Mediterranean, with a mean precipitation per month of 120 mm in the rainy season (December and May), decreasing in summer to 30 mm per month [186]. According to the land use map of Portugal (COS06), the studied areas are mainly composed of *Pinus pinaster* (37%), pastures (20%), *Eucalyptus globulus* (19%), annual crops (20%), oak (5%), and other broadleaved species (6%).

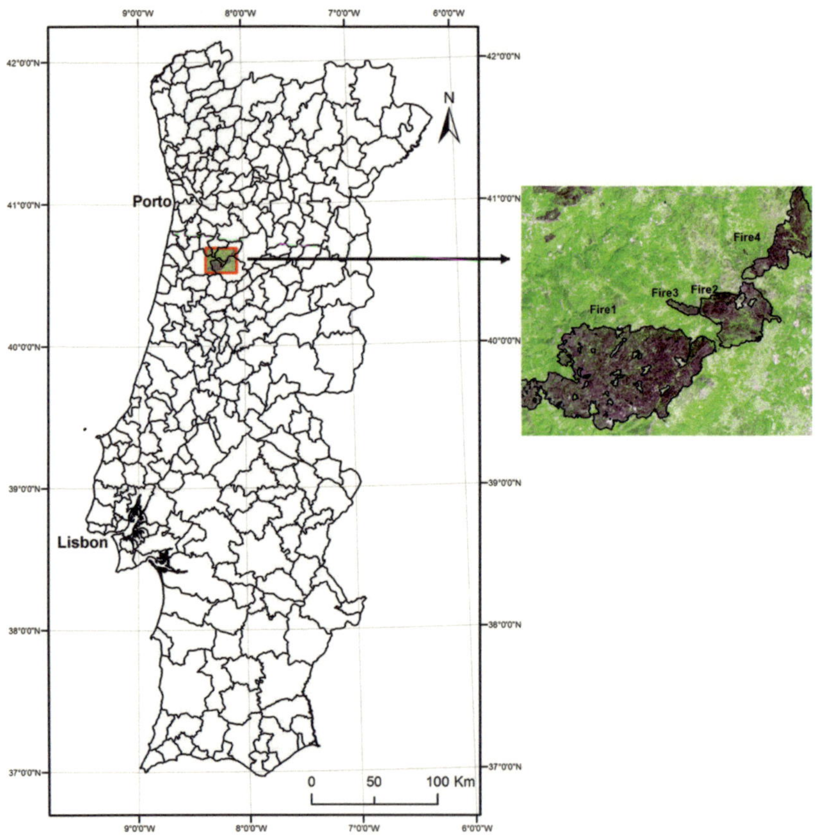

Figure 6. Study areas location (a) and false color composite Landsat image (RGB—SWIR, NIR, Red) (b).

Remote sensing data was available from Landsat 8 satellite images (**Table 1**) obtained from the United States Geological Survey (USGS), Glovis Visualization Viewer (GLOVIS) platform. A temporal time series of four years, of 2013 (fires' occurrence year), 2014, 2015, and 2016, was used (**Table 2**). The images were selected considering a similar acquisition date, to minimize cloud cover and phenological stage effects. Four fires were considered. The image of 2013 (6 july, 2013), covering Fire1, Fire2, and Fire4, corresponds to pre-fire image, and the other images to the post-fire, one, two, and three after the fire. As to Fire3 that occurred at the beginning of June of 2013 all images are post-fire. Thus, the time series enables the temporal vegetation analyses for a time frame of three years after the fire event. The satellite images were geometrically adjusted, image to image, considering that of 2013 as reference, to ensure a minimum geometric pixel deviation.

Landsat 8 OLI and TIR bands			
Bands		**(μm)**	**Spatial resolution (m)**
b1	Coastal	0.433–0.453	30
b2	Blue	0.450–0.515	30
b3	Green	0.525–0.600	30
b4	Red	0.630–0.680	30
b5	Near Infrared (NIR)	0.845–0.885	30
b6	Short Wave Infrared (SWIR1)	1.560–1.660	30
b10	Thermal Infrared I (TIR1)	10.6–11.19	100
b11	Thermal Infrared II (TIR2)	11.5–12.51	100
b7	Cirrus	0.500–0.680	30
b8	Short Wave Infrared (SWIR2)	2.100–2.300	30
b9	Pan	0.503–0.676	15

Table 1. Band characterization of the Landsat 8 Operational Land Imager (OLI) satellite.

Satellite/sensor	**Acquisition date**	**Fire**	**Month of fire event**	**Area (ha)**
Landsat 8 OLI	06/07/2013	Fire1	July	6391
		Fire2	August	1496
		Fire3	June	123
		Fire4	August	1348
	09/07/2014			
	26/06/2015			
	28/06/2016			

Table 2. Date of Landsat 8 images acquisition (WRS-2 204/32) and burnt area of the four fires.

6.2. Methodology

For this study four burnt areas were extracted using the fire perimeter provided by the Portuguese Forest Services [173]. **Figure 7** shows the fire scars with false-color composite images for the four years and an image with two active fires (Fire2 and Fire4).

The vegetation indices based on red (RED) and near infrared (NIR) bands are directly related to the vegetation photosynthetic activity [187], which is highly reflective in the NIR region and strongly absorbing in the RED [188]. As already referred, one of the mostly used indices to evaluate the vegetation changes, especially after fire, is NDVI [188]. It is particularly used to assess vegetation regeneration after the fire [52, 157, 170, 189]. This index is rather sensitive to vegetation activity presenting a strong decrease after fire due to the partial or total destruction of the vegetation, thus it is indicative of the fire damage [52].

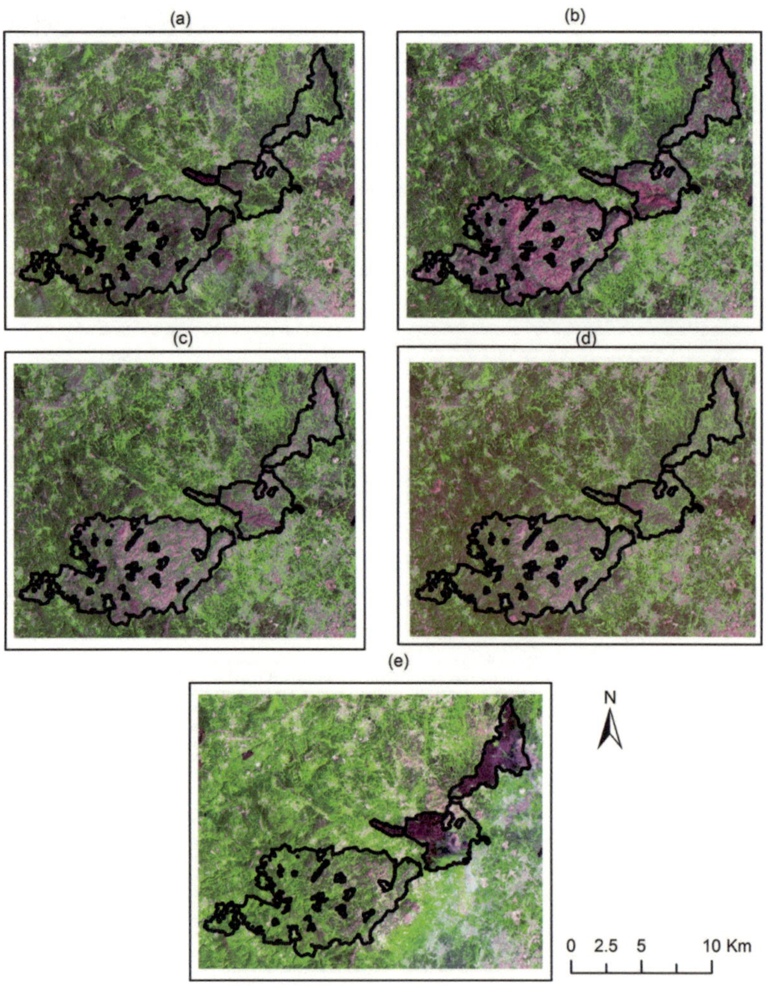

Figure 7. False-color composite Landsat 8 images (RGB—SWIR, NIR, Red) for 2013 (a), 2014 (b), 2015 (c), and 2016 (d) and an image with active Fire2 and Fire4 (e), including the burnt area perimeter (black line).

In this study, the NDVI was used to calculate on the basis of the normalized difference between NIR and RED bands (NDVI = ((NIR−RED)/(NIR+RED)). It was calculated both for pre- ($NDVI_{pre}$) and post-fire ($NDVI_{post}$) periods, as the mean value of NDVI before and after (for the 2014, 2015, and 2016 images) the fire event. The dates of satellite image acquisition were deliberately chosen to be a few days prior to the fire event and afterward annually in approximately the same dates, to eliminate the possible variations originated by the different vegetation phenological stage. The mean NDVI was obtained with the pixels located inside the selected burnt areas. The temporal vegetation variation was analyzed with the difference between the NDVI pre- and post-fire.

6.3. Post-fire analysis

In **Figure 7**, a strong decrease in NIR (green color tones) and increase in Red (rose color tones) reflectance within the fire perimeter from pre-fire image (**Figure 7a**) and the one acquired one year later (**Figure 7b**) for Fire1, Fire2, and Fire4 can be observed. For the two following years, 2015 (**Figure 7c**) and 2016 (**Figure 7d**), the fire scars are less evident. These changes are confirmed by the NDVI values (**Figure 8**), with the highest values of NDVI in the pre-fire image (2013) indicating the presence of dense vegetation and a relevant decrease in these values in the post-fire image (2014), where the fire scar is present. In the following two years, 2015 and 2016, an increase in the NDVI values can be observed, demonstrating a gradual recovery of vegetation, and even more accentuated in 2016. For Fire3 there is no pre-fire image, **Figure 8** shows a low NDVI value for 2013 (image acquired a few days after the fire event) and a gradual

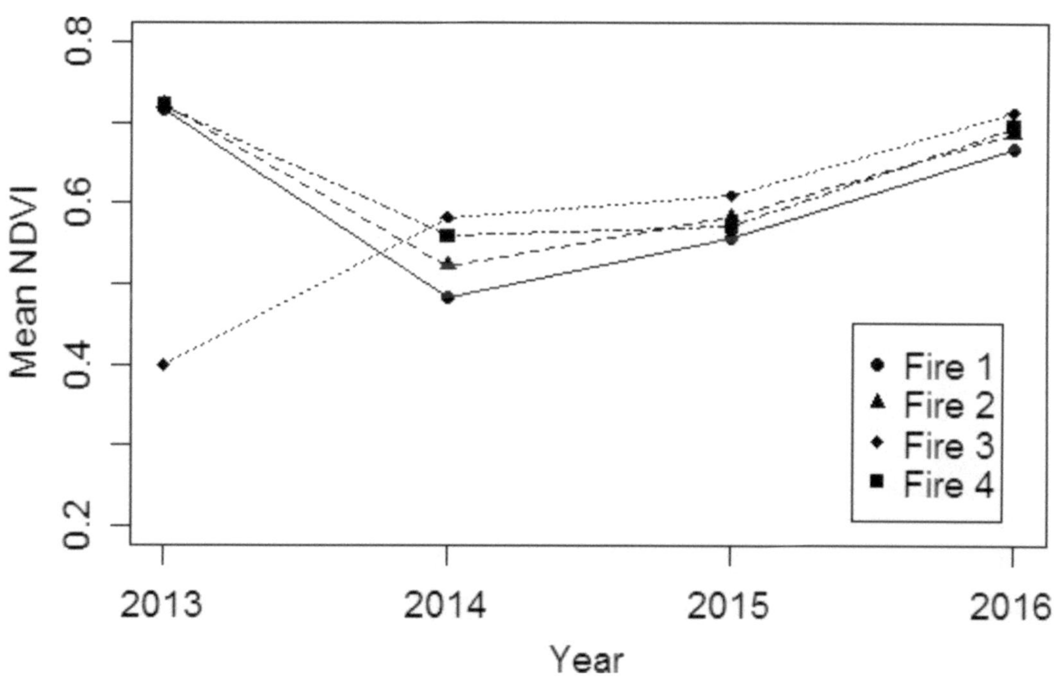

Figure 8. Evolution of mean NDVI values of the fire events from 2013 to 2016.

increase for next years in analysis. The increasing values of NDVI are indicative of the vegetation regeneration either by seed or by sprouts [4, 9, 94, 95], shortly after the fire [88–92, 96]. Nevertheless, for these fires, more than three years are necessary to achieve the pre-fire NDVI values, values, depending on several factors such as the pre-fire plant species composition [89–92], season of fire occurrence [93], and severity and intensity of fire [6]. The vegetation recovery will have also the advantage of reducing substantially the erosion risk [119–121, 125].

The interannual difference of the NDVI (**Figure 9**), the pre- and post-fire, shows also a gradual vegetation recovery, in the affected areas, by different land cover vegetation types. For all different land cover types the vegetation regeneration is similar. Differences of mean NDVI are between 0.2 and 0.25 for Fire1 and Fire2, and between 0.1 and 0.15 for Fire4. Lower mean NDVI points to lower severity, that is the maintenance of some vegetation or some live tree crowns after the fire event, which enables seed regeneration [9, 94, 95]. Therefore small positive differences were encountered for the *Pinus pinaster* and *Eucalyptus globulus* (Fire1), where the fire scar is accentuated one year after the fire event, as can be observed in **Figure 7**. The regeneration of these species is not sufficiently fast to increase the NDVI values.

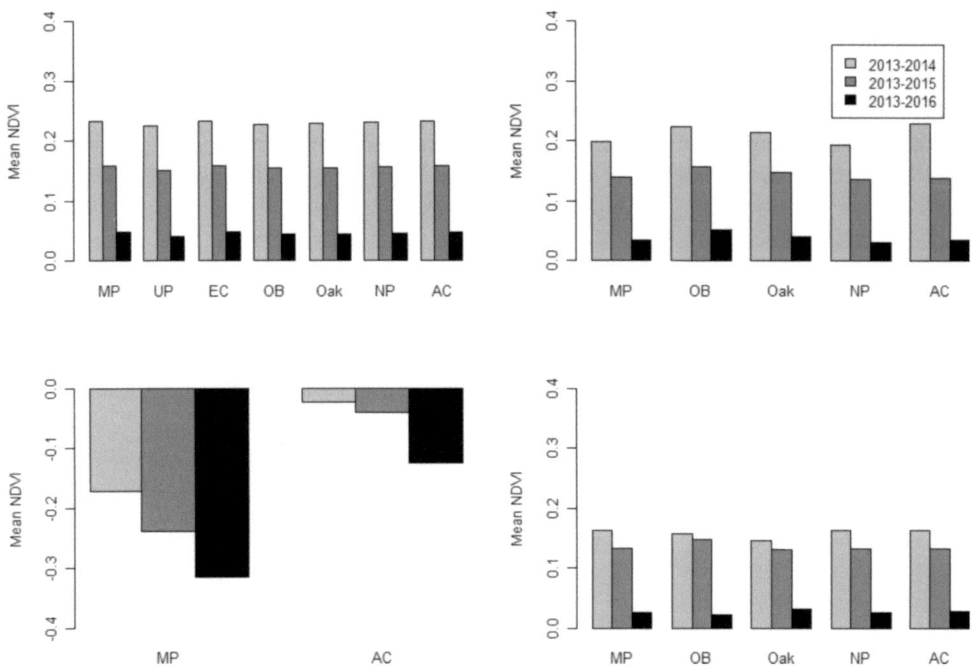

Figure 9. Difference of mean NDVI values per vegetation type (MP – Maritime pine, UP – Umbrella pine, EC – Tasmanian blue gum, OB – Other broadleaf species, Oak, NP – Natural pastures, AC – Annual culture).

7. Conclusions

Forest fires are a frequent feature in forest ecosystems. In the Mediterranean and, in particular in Portugal, they occur on regular annual basis. Nonetheless a wide range of variability, whether spatial or temporal, exists, partially explained by the type of forest system and the climate, especially in relation to the high temperatures and low humidity. The most affected systems are those with high spatial, both vertical and horizontal continuity, and large fuel loads. In Portugal the most affected areas are those covered by *Pinus pinaster* and *Eucalyptus globulus*, which are also areas with larger burnt areas. The species characteristics, high flammability, as well as the large continuous areas when compared to other species, high stand density of trees and shrubs can explain, at least partially, the fire propagation. The spatial analysis detected regional differences. In northern and central Portugal, the number of fires and the burnt area as well as the frequency of fires were much larger than those in southern Portugal. This seems to be linked with vegetation type. While the forest areas of northern and central Portugal are predominantly of *Pinus pinaster* and *Eucalyptus globulus*, in the southern Portugal the evergreen oaks (*Quercus suber* and *Quercus rotundifolia*) dominate the forest landscape. The NDVI based on a time series Landsat 8 satellite images allows the monitoring and evaluation of post-fire vegetation regeneration. The interannual difference of NDVI enables better understanding of the temporal variability of the recovery vegetation after the fire event, and a reduction of the potential soil erosion risk in the Mediterranean ecosystems. The differences between fires and land cover types can be an indicator of the fire severity.

Acknowledgements

This work is funded by National Funds through FCT—Foundation for Science and Technology under the Project UID/AGR/00115/2013.

Author details

Ana Cristina Gonçalves* and Adélia M.O. Sousa

*Address all correspondence to: acag@uevora.pt

Department of Rural Engineering, School of Sciences and Technology, Institute of Mediterranean Agricultural and Environmental Sciences (ICAAM), Institute of Research and Advanced Information (IIFA), University of Évora, Évora, Portugal

References

[1] Quezel P, Tomaselli R, Morandini R. Mediterranean forests and maquis: ecology, conservation and management. MAB Technical notes 2. Paris: UNESCO; 1977. p. 94

[2] Thirgood VC. Man and the Mediterranean Forest: A History of Resource Depletion. London: Academic Press; 1981. p. 194

[3] Massada AB, Kent R, Blank L, Perevolotsky A, Hadar L, Carmel Y. Automated segmentation of vegetation structure units in a Mediterranean landscape. International Journal of Remote Sensing. 2012;**33**(2):346-364

[4] Oliver CD, Larson BC, editors. Forest Stand Dynamics. Update ed. New York: John Wiley & Sons, Inc; 1996. p. 544

[5] O'Hara KL. Multiaged Silviculture Managing for Complex Forest Stand Structures. Oxford: Oxford University Press; 2014. p. 213

[6] Keeley J, Bond W, Bradstock R, Pausas J, Rundel P. Fire in Mediterranean Ecosystems: Ecology, Evolution and Management. Cambridge: Cambridge University Press; 2012. p. 515

[7] Alday JG, Santana VM, Lee H, Allen KA, Marrs RH. Aboveground biomass accumulation patterns in moorlands after prescribed burning and low-intensity grazing. Perspectives in Plant Ecology, Evolution and Systematics. 2015;**17**:388-396

[8] Odion DC, Moritz MA, DellaSala DA. Alternative community states maintained by fire in the Klamath Mountains, USA. Journal of Ecology. 2010;**98**:96-105

[9] Santana VM, Baeza MJ, Marrs RH, Vallejo VR. Old-field secondary succession in SE Spain: Can fire divert it? Plant Ecology. 2010;**211**:337-349

[10] Bowman DM, Murphy BP, Boer MM, Bradstock RA, Cary GJ, Cochrane MA, Fensham RJ, Krawchuk MA, Price OF, Williams RJ. Forest fire management, climate change, and the risk of catastrophic carbon losses. Frontiers in Ecology and the Environment. 2013;**11**:66-67

[11] San-Miguel-Ayanz J, Moreno JM, Camia A. Analysis of large fires in European Mediterranean landscapes: Lessons learned and perspectives. Forest Ecology and Management. 2013;**294**:11-22

[12] Alexandrian D, Esnault F, Calabri G. Forest fires in the Mediterranean area. Unasylva. 2000;**197**(50):35-41

[13] Guiomar N, Godinho S, Fernandes PM, Machado R, Neves N, Fernandes JP. Wildfire patterns and landscape changes in Mediterranean oak woodlands. Science of the Total Environment. 2005;**536**:338-352

[14] Naveh Z. The evolutionary significance of fire in the Mediterranean region. Vegetatio. 1975;**21**:199-208

[15] Daniau AL, d'Errico F, Goni MFS. Testing the hypothesis of fire use for ecosystem management by Neanderthal and upper Palaeolithic modern human populations. PLoS One. 2010;**5**(2):e9157. DOI: 10.1371/journal.pone.0009157

[16] Amatulli G, Camia A, San-Miguel-Ayanz J. Estimating future burned areas under changing climate in the EU-Mediterranean countries. Science of the Total Environment. 2013;**450-451**:209-222

[17] Bedia J, Herrera S, Gutiérrez JM, Zavala G, Urbieta IR, Moreno JM. Sensitivity of fire weather index to different reanalysis products in the Iberian Peninsula. Natural Hazards and Earth System Sciences. 2012;**12**:699-708

[18] Moriondo M, Good P, Durao R, Bindi M, Giannakopoulos C, Corte-Real J. Potential impact of climate change on fire risk in the Mediterranean area. Clinical Research. 2006;**31**:85-95

[19] Pausas JP. Changes in fire and climate in the eastern Iberian Peninsula (Mediterranean Basin). Climate Change. 2004;**63**:337-350. DOI: 10.1023/B:CLIM.0000018508.94901.9c

[20] Dale VH, Joyce LA, McNulty S, Neilson RP, Ayres MP, Flannigan MD, Hanson PJ. Climate change and forest disturbances. Bioscience. 2001;**51**(9):723-734

[21] Torn MS, Fried JS. Predicting the impacts of global warming on wildland fire. Climate Change. 1992;**21**:257-274. DOI: 10.1007/BF00139726

[22] Nunes AN, Lourenco L, Meira ACM. Exploring spatial patterns and drivers of forest fires in Portugal (1980-2014). Science of the Total Environment. 2016;**573**:1190-1202

[23] Modugno S, Balzter H, Cole B, Borrelli P. Mapping regional patterns of large forest fires in Wildland–Urban Interface areas in Europe. Journal of Environmental Management. 2016;**172**:112-126

[24] Stamou Z, Xystrakis F, Koutsias N. The role of fire as a long-term landscape modifier: Evidence from long-term fire observations (1922-2000) in Greece. Applied Geography. 2016;**74**:47-55

[25] Pausas JG, Keeley J. Abrupt climate-independent fire regime changes. Ecosystems. 2014;**17**:1109-1120

[26] Pausas JG, Keeley J. Evolutionary ecology of resprouting and seeding in fireprone ecosystems. New Phytologist. 2014;**204**(1):55-65

[27] Ganteaume A, Camia A, Jappiot M, San-Miguel-Ayanz J, Long-Founel M, Corinne L. A review of the main driving factors of forest fire ignition over Europe. Environmental Management. 2013;**51**:651-662

[28] Oliveira S, Oehler F, San-Miguel-Ayanz J, Camia A, Pereira JMC. Modeling spatial patterns of fire occurrence in Mediterranean Europe using multiple regression and random forest. Forest Ecology and Management. 2012;**275**:117-129

[29] Pausas JG, Fernández-Muñoz S. Fire regime changes in the Western Mediterranean Basin: From fuel-limited to drought-driven fire regime. Climatic Change. 2012;**110**:215-222

[30] Moreira F, Viedma O, Arianoutsou M, Curt T, Koutsias N, Rigolot E, Barbati A, Corona P, Vaz P, Xanthopoulos G, Mouillot F. Landscape-wildfire interactions in Southern Europe: Implications for landscape management. Journal of Environmental Management. 2011;**92**:2389-2402

[31] Catry FX, Rego F, Moreira F, Fernandes PM, Pausas JG. Post-fire tree mortality in mixed forests of Central Portugal. Forest Ecology and Management. 2010;**260**:1184-1192

[32] Wittenberg L, Malkinson D. Spatio-temporal perspectives of forest fires regimes in a maturing Mediterranean mixed pine landscape. European Journal of Forest Research. 2009;**128**:297-304

[33] Badia A, Saur D, Cerdan R, Llurdes JC. Causality and management of forest fires in Mediterranean environments: An example from Catalonia. Environmental Hazards. 2002;**4**(1):23-32

[34] Giovannini G, Vallejo R, Lucchesi S, Bautista S, Ciompi S, Llovet J. Effects of land use and eventual fire on soil erodibility in dry Mediterranean conditions. Forest Ecology and Management. 2001;**147**:15-23

[35] Pausas JG, Vallejo VR. The role of fire in European Mediterranean ecosystems. In: Chuvieco E, editor. Remote Sensing of Large Wildfires in the European Mediterranean Basin. Berlin: Springer; 1999. pp. 3-16

[36] Eugenio Gosalbo M, Lloret F. Fire recurrence effects on the structure and composition of Mediterranean *Pinus halepensis* communities in Catalonia (Northeast Iberian Peninsula). Ecoscience. 2004;**11**(4):446-454

[37] Ne'eman G, Lahav H, Izhaki I. Recovery of vegetation in a natural East Mediterranean pine forest on Mount Carmel, Israel as affected by management strategies. Forest Ecology and Management. 1995;**75**:17-26

[38] Naveh Z. Fire in the Mediterranean region. A landscape ecological perspective. In: Goldammer JG, Jenkins MJ, editors. Fire in Ecosystem Dynamics: Mediterranean and Northern Perspectives. The Hague: SPB Academic Publishing; 1990. pp. 1-20

[39] Naveh Z. Effects of fire in the Mediterranean region. In: Kozlowski TT, Ahlgren CE, editors. Fire and Ecosystems. Physiological Ecology: A Series of Monographs Texts and Treatises. New York: Academic Press; 1974. p. 423

[40] Díaz-Delgado R, Lloret F, Pons X, Terradas J. Satellite evidence of decreasing resilience in Mediterranean plant communities after recurrent wildfires. Ecology. 2002;**83**:2293-2303

[41] Sá-Torres JPC. Patterns and drivers of wildfire occurrence and post-fire vegetation resilience across scales in Portugal [thesis]. Porto: University of Porto; 2013

[42] Kazanis D, Arianoutsou M. Long-term post-fire vegetation dynamics in *Pinus halepensis* forests of Central Greece: A functional group approach. Plant Ecology. 2004;**171**:101-121

[43] Lee SW, Lee MB, Lee YG, Won MS, Kim JJ, Hong S. Relationship between landscape structure and burn severity at the landscape and class levels in Samchuck, South Korea. Forest Ecology and Management. 2009;**258**:1594-1604

[44] Silva JS, Vaz P, Moreira F, Catry F, Rego FC. Wildfires as a major driver of landscape dynamics in three fire-prone areas of Portugal. Landscape and Urban Planning. 2011;**101**:349-358

[45] Handmer J, Proudley B. The economics of interface wildfires. USDA Forest Service, Pacific Southwest Research Station, General Technical Report PSWGTR-208. Albany: USDA; 2008. pp. 627-637

[46] Schaffhauser A, Curt T, Tatoni T. Fire-vegetation interplay in amosaic structure of *Quercus suber* woodlands and Mediterranean maquis under recurrent fires. Forest Ecology and Management. 2011;**262**:730-738

[47] Baeza MJ, De Luis M, Raventos J, Escarré A. Factors influencing fire behaviour in shrublands of different stand ages and the implications for using prescribed burning to reduce wildfire risk. Journal of Environmental Management. 2002;**65**:199-208

[48] Curt T, Adra W, Borgniet L. Fire-driven oak regeneration in French Mediterranean ecosystems. Forest Ecology and Management. 2009;**258**:2127-2135

[49] Pausas GJ, Carbo E, Caturla RN, Gil JM, Vallejo R. Post fire revegetation patterns in the Eastern Iberian Peninsula. Acta Oecologica. 1999;**20**(5):499-508

[50] Jain T, Pilliod D, Graham R. Tongue-tied. Wildfire. 2004;**4**:22-26

[51] Lentile LB, Holden ZA, Smith AMS, Falkowski MJ, Hudak AT, Morgan P, Lewis SA, Gessler PE, Benson NC. Remote sensing techniques to assess active fire characteristics and post-fire effects. International Journal of Wildland Fire. 2006;**15**:319-345

[52] Bastos A, Gouveia CM, DaCamara CC, Trigo RM. Modelling post-fire vegetation recovery in Portugal. Biogeosciences. 2011;**8**:3593-3607

[53] Röder A, Hill J, Duguy B, Alloza JA, Vallejo R. Using long time series of Landsat data to monitor fire events and post-fire dynamics and identify driving factors. A case study in the Ayora region (Eastern Spain). Remote Sensing of Environment. 2008;**112**:259-273

[54] Smithwick EAH, Turner MG, Mack MC, Chapin FS. Postfire soil N cycling in Northern conifer forests affected by severe, stand-replacing wildfires. Ecosystems. 2005;**8**:163-181

[55] Neary DG, Klopatek CC, DeBano LF, Ffolliott PF. Fire effects on belowground sustainability: A review and synthesis. Forest Ecology and Management. 1999;**122**:51-71

[56] Strauss D, Bednar L, Mees R. Do one percent of the forest fires cause ninety-nine percent of the damage? Forest Science. 1989;**35**:319-328

[57] Beverly J, Martell D. Characterizing extreme fire and weather events in the Boreal Shield Ecozone of Ontario. Agricultural and Forest Meteorology. 2005;**133**:5-16

[58] Pyne SJ, Andrews PL, Laven RD. Introduction to Wildland Fire Science. New York: Wiley; 1996. p. 769

[59] Alexander ME. Help with making crown fire hazard assessments. In: Fischer WC, Arno SF, editors. Protecting people and homes from wildfire in the interior west. Proceedings of the symposium and workshop, October 6-8, 1988; Missoula, USA. USDA Forest Service, Intermountain Research Station, General Technical Report INT-251. Missoula: USDA Forest Service; 1988. pp. 147-156

[60] Alvarez A, Gracia M, RetanaJ. Fuel types and crown fire potential in *Pinus halepensis* forests. European Journal of Forest Research. 2012;**131**:463-474

[61] Raymond CL, Peterson DL. Fuel treatments alter the effects of wildfire in a mixed-evergreen forest, Oregon, USA. Canadian Journal of Forest Research. 2005;**35**:2981-2995

[62] Cruz MG, Alexander ME, Wakimoto RH. Modeling the likelihood of crown fire occurrence in conifer forest stands. Forest Science. 2004;**50**:640-658

[63] van Wagner CE. Conditions of the start and spread of crown fires. Canadian Journal of Forest Research. 1977;**7**:23-34

[64] Duguy B, Vallejo R. Land-use and fire history effects on post-fire vegetation dynamics in Eastern Spain. Journal of Vegetation Science. 2008;**19**:97-108

[65] Arca B, Duce P, Laconi M, Pellizzaro G, Salis M, Spano D. Evaluation of FARSITE simulator in Mediterranean maquis. International Journal of Wildland Fire. 2007;**16**:563-572

[66] Sandberg DV, Riccardi CL, Schaaf MD. Fire potential rating for wildland fuelbeds using the fuel characteristic classification system. Canadian Journal of Forest Research. 2007;**37**:2456-2463

[67] Keeley J, Fotheringham CJ, Baer-Keeley M. Determinants of postfire recovery and succession in Mediterranean-climate shrublands of California. Ecological Applications. 2005;**15**:1515-1534

[68] Hély C, Bergeron Y, Flannigan MD. Effects of stand composition on fire hazard in mixed-wood Canadian boreal forest. Journal of Vegetation Science. 2000;**11**:813-824

[69] Fernandes PM, Rigolot E. The fire ecology and management of maritime pine (*Pinus pinaster* Ait.). Forest Ecology and Management. 2007;**241**:1-13

[70] McRae R, Sharples J. A conceptual framework for assessing the risk posed by extreme bushfires. The Australian Journal of Emergency Management. 2011;**26**(2):47-53

[71] Williams J, Hamilton L. The Mega-Fire Phenomenon: Toward a More Effective Management Model. A Concept Paper. Washington: The Brookings Institution Center for Public Policy Education; 2005. Available from: http://www.wildfirelessons.net [Accessed: 1 February 2017]

[72] Turner MG, Romme WH, Gardner RH. Pre-fire heterogeneity, fire severity, and early post-fire plant reestablishment in subalpine forests of Yellowstone National Park, Wyoming. International Journal of Wildland Fire. 1999;**9**:21-36

[73] Fernandes PM, Pacheco AP, Almeida R, Claro J. The role of fire-suppression force in limiting the spread of extremely large forest fires in Portugal. European Journal of Forest Research. 2016;**135**:253-262. DOI: 10.1007/s10342-015-0933-8

[74] Tedim F, Remelgado R, Borges C, Carvalho S, Martins J. Exploring the occurrence of mega-fires in Portugal. Forest Ecology and Management. 2013;**294**:86-96

[75] Bond WJ, Keeley JE. Fire as a global "herbivore": The ecology and evolution of flammable ecosystems. Trends in Ecology & Evolution. 2005;**20**:387-394

[76] Carríon JS, van Geel B. Fine-resolution Upper Weichselian and Holocene palynological record from Navarrés (Valencia, Spain) and a discussion about factors of Mediterranean forest succession. Review of Palaeobotany and Palynology. 1999;**106**:209-236

[77] Baeza MJ, Valdecantos A, Aloloza JA, Vallejo VR. Human disturbance and environmental factors as drivers of long-term post-fire regeneration patterns in Mediterranean forests. Journal of Vegetation Science. 2007;**18**:243-252

[78] Retana J, Espelta JM, Habrouk A, Ordonez JL, de Sola-Morales F. Regeneration patterns of three Mediterranean pines and forest changes after a large wildfire in Northeastern Spain. Ecoscience. 2002;**9**:89-97

[79] Bradstock RA, Hammill KA, Collins L, Price O. Effects of weather, fuel and terrain on fire severity in topographically diverse landscapes of Southeastern Australia. Landscape Ecology. 2010;**25**(4):607-619

[80] Lloret F, Marí G. A comparison of medieval and the current fire regimes in managed pine forests of Catalonia (NE Spain). Forest Ecology and Management. 2001;**141**:155-163

[81] Mouillot F, Rambal S, Joffre R. Simulating climate change impacts on fire frequency and vegetation dynamics in a Mediterranean-type ecosystem. Global Change Biology. 2002;**8**:423-437

[82] Pal JS, Giorgi F, Bi X. Consistency of recent European summer precipitation trends and extremes with future regional climate projections. Geophysical Research Letters. 2004;**31**:L13202

[83] Maracchi G, Sirotenko O, Bindi M. Impacts of present and future climate variability on agriculture and forestry in the temperate regions: Europe. Climatic Change. 2005;**70**:117-135

[84] Rodrigo A, Retana J, Pico FX. Direct regeneration is not the only response of Mediterranean forests to large fires. Ecology. 2004;**85**:716-729

[85] Lavorel S, Canadell J, Rambal S, Terradas J. Mediterranean terrestrial ecosystems: Research priorities on global change effects. Global Ecology & Biogeography Letters. 1998;**7**:157-166

[86] Chauchard S, Carcaillet C, Guibal F. Patterns of land-use abandonment control tree-recruitment and forest dynamics in Mediterranean mountains. Ecosystems. 2007;**10**: 936-948

[87] Cheddadi R, Guiot J, Jolly D. The Mediterranean vegetation: What if the atmospheric CO_2 increased? Landscape Ecology. 2001;**16**:667-675

[88] González-De Vega S, De Las Heras J, Moya D. Resilience of Mediterranean terrestrial ecosystems and fire severity in semiarid areas: Responses of Aleppo pine forests in the short, mid and long term. Science of the Total Environment. 2016;**573**:1171-1177

[89] Capitanio R, Carcaillet C. Post-fire Mediterranean vegetation dynamics and diversity: A discussion of succession models. Forest Ecology and Management. 2008;**255**:431-439

[90] Guo Q. Early post-fire succession in California chaparral: Changes in diversity, density, cover and biomass. Ecological Research. 2001;**16**:471-485

[91] Herranz JM, Martínez-Sánchez JJ, Marín A, Ferrandis P. Post-fire regeneration of *Pinus halepensis* Miller in a semi-arid area in Albacete province (South-Eastern Spain). Ecoscience. 1997;**4**:86-90

[92] Thanos CA, Daskakakou EN, Nikolaidou S. Early post-fire regeneration of a *Pinus halepensis* forest on Mount Parnis, Greece. Journal of Vegetation Science. 1996;**7**:273-280

[93] Moore PF. Forest landscape restoration after fires. In: Stephanie M, Vallauri D, Dudley N, editors. Forest Restoration in Landscapes, Beyond Planting Trees. Cidade: Springer; 2005. p. 427

[94] Santana VM, Alday JG, Baeza MJ. Effects of fire regime shift in Mediterranean Basin ecosystems: Changes in soil seed bank composition among functional types. Plant Ecology. 2014;**215**:555-566

[95] De Las Heras J, Martínez-Sánchez JJ, González-Ochoa AI, Ferrandis P, Herranz JM. Establishment of *Pinus halepensis* Mill. saplings following fire: Effects of competition with shrub species. Acta Oecologica. 2002;**23**:91-97

[96] Wittenberg L, Malkinson D, Beeri O, Halutzy A, Tessler N. Spatial and temporal patterns of vegetation recovery following sequences of forest fire in a Mediterranean landscape Mt. Carmel, Israel. Catena. 2007;**71**:76-83. DOI: 10.1016/j.catena.2006.10.007

[97] Saïd S. Floristic and life form diversity in post-pasture succession on a Mediterranean island (Corsica). Plant Ecology. 2001;**162**:67-76

[98] Gracia M, Retana J, Roig P. Mid-term successional patterns after fire of mixed pine-oak forests in NE Spain. Acta Oecologica. 2002;**23**:405-411

[99] Tessler N, Wittenberg L, Greenbaum N. Vegetation cover and species richness after recurrent forest fires in the Eastern Mediterranean ecosystem of Mount Carmel, Israel. Science of the Total Environment. 2016;**572**:1395-1402

[100] Tessler N, Sapir Y, Wittenberg L, Greenbaum N. Recovery of Mediterranean vegetation after recurrent forest fires: Insight from the 2010 forest fire on Mount Carmel, Israel. Land Degradation & Development. 2015;**27**:1424-1431. DOI: 10.1002/ldr.2419

[101] Lloret F, Pausas J, Vila M. Responses of Mediterranean plant species to different fire frequency in Garraf Natural Park (Catalonia, Spain): Field observations and modeling predictions. Plant Ecology. 2003;**167**:223-235. Available from: http://www.jstor.org/stable/20146446

[102] Zedler PH, Gautier CR, McMaster GS. Vegetation change in response to extreme events: The effect of a short interval between fires in California chaparral and coastal scrub. Ecology. 1983;**64**:809-818. Available from: https://www.jstor.org/stable/1937204?seq=9#page_scan_tab_contents

[103] Eugenio Gosalbo M, Lloret F, Alcañiz JM. Regional patterns of fire recurrence effects on calcareous soils of Mediterranean *Pinus halepensis* communities. Forest Ecology and Management. 2006;**221**:313-318

[104] Pausas JG. Simulating Mediterranean landscape pattern and vegetation dynamics under different fire regimes. Plant Ecology. 2006;**187**:249-259

[105] Vilà-Cabrera A, Saura-Mas S, Lloret F. Effects of fire frequency on species composition in a Mediterranean shrubland. Ecoscience. 2008;**15**:519-528

[106] Hinojosa MB, Parra A, Laudicina VA, Moreno JM. Post-fire functionality and microbial community structure in a Mediterranean shrubland subjected to experimental drought. Science of the Total Environment. 2016;**573**:1178-1189

[107] Caon L, Vallejo VR, Coen RJ, Geissen V. Effects of wildfire on soil nutrients in Mediterranean ecosystems. Earth-Science Reviews. 2014;**139**:47-58

[108] Certini G. Effects of fire on properties of forest soils: A review. Oecologia. 2005;**143**:1-10

[109] Lane PN, Sheridan GJ, Noske PJ, Sherwin CB. Phosphorus and nitrogen exports from SE Australian forests following wildfire. Journal of Hydrology. 2008;**361**:186-198

[110] Shakesby RA, Doerr SH. Wildfire as a hydrological and geomorphological agent. Earth-Science Reviews. 2006;**74**:269-307

[111] Thomas AD, Walsh RP, Shakesby RA. Nutrient losses in eroded sediment after fire in eucalyptus and pine forests in the wet Mediterranean environment of Northern Portugal. Catena. 1999;**36**:283-302

[112] Karhu K, Dannenmann M, Kitzler B, Díaz-Pinés E, Tejedor J, Ramírez DA, Parra A, Resco V, Moreno JM, Rubio A, Guiaraes-Povoas L, Zechmeister-Boltenstern S, Butterbach-Bahl K, Ambus P. Fire increases the risk of higher soil N_2O emissions from Mediterranean macchia ecosystems. Soil Biology & Biochemistry. 2015;**82**:44-51

[113] Parra A, Ramírez DA, Resco V, Velasco A, Moreno JM. Modifying rainfall patterns in a Mediterranean shrubland: System design, plant responses and experimental burning. International Journal of Biometeorology. 2012;**56**:1033-1043

[114] González-Pérez JA, González-Vila FJ, Almendros G, Knicker H. The effect of fire on soil organic matter—a review. Environment International. 2004;**30**:855-870

[115] Hedo J, Lucas-Borja ME, Wic C, Andrés-Abellán M, De Las Heras J. Soil microbiological properties and enzymatic activities of long-term post-fire recovery in dry and semiarid Aleppo pine (*Pinus halepensis* M.) forest stands. Solid Earth. 2015;**6**:243-252

[116] Goberna M, Garcia C, Insam H, Hernández MT, Verdú M. Burning fire-prone Mediterranean shrublands: Immediate changes in soil microbial community structure and ecosystem functions. Microbial Ecology. 2012;**64**:242-255

[117] Vargas R, Collins SL, Thomey ML, Johnson JE, Brown RF, Natvig DO, Friggens MT. Precipitation variability and fire influence the temporal dynamics of soil CO_2 efflux in an arid grassland. Global Change Biology. 2012;**18**:1401-1411

[118] Schimel J, Balser TC, Wallenstein M. Microbial stress-response physiology and its implications for ecosystem function. Ecology. 2007;**88**:1386-1394

[119] Schoenholtz SH, Miegroet H, Burger JA. A review of chemical and physical properties as indicators of forest soil quality: Challenges and opportunities. Forest Ecology and Management. 2000;**138**:335-356

[120] Jenny H. The Soil Resource. Origin and Behaviour. Ecological Studies. Vol. 37. Berlin: Springer-Verlag; 1983. p. 377

[121] Kohnke H, Bertrand AR. Soil Conservation. New York: McGraw-Hill Book Company; 1959. p. 298

[122] Shakesby RA, Coelho COA, Ferreira AD, Terry JP, Walsh RPD. Wildfire impacts on soil erosion and hydrology in wet Mediterranean forest, Portugal. International Journal of Wildland Fire. 1993;**3**:95-110

[123] Inbar M, Tamir M, Wittenberg L. Runoff and erosion processes after a forest fire in Mount Carmel, a Mediterranean area. Geomorphology. 1998;**24**(1):17-33

[124] Cerda A. Changes in overland flow and infiltration after a rangeland fire in a Mediterranean scrubland. Hydrological Processes. 1998;**12**:1031-1042

[125] Campo J, Andreu V, Gimeno-Garcia E, Gonzalez O, Rubio JL. Occurrence of soil erosion after repeated experimental fires in a Mediterranean environment. Geomorphology. 2006;**82**:376-387

[126] Cerda A, Doerr SH. The influence of vegetation recovery on soil hydrology and erodibility following fire: An eleven year investigation. International Journal of Wildland Fire. 2005;**14**(4):423-437

[127] Malvar MC, Martins MA, Nunes JP, Robichaud PR, Keizer JJ. Assessing the role of pre-fire ground preparation operations and soil water repellency in post-fire runoff and inter-rill erosion by repeated rainfall simulation experiments in Portuguese eucalypt plantations. Catena. 2013;**108**:69-83

[128] Fernández C, Veja JA, Fonturbel T, Pérez-Gorostiaga P, Jiménez E, Madrigal J. Effects of wildfire salvage logging and slash treatments on soil degradation. Land Degradation & Development. 2007;**18**:591-607

[129] Gonzalez M, Augusto L, Gallet-Budynek A, Xue J, Yauschew-Raguenes N, Guyon D, Trichet P, Delerue F, Niollet S, Andreasson F, Achat DL, Bakker MR. Contribution of understory species to total ecosystem aboveground and belowground biomass in temperate *Pinus pinaster* Ait. forests. Forest Ecology and Management. 2013;**289**:38-47

[130] Wagenbrenner JW, MacDonald LH, Coats RN, Robichaud PR, Brown RE. Effects of post-fire salvage logging and a skid trail treatment on ground cover, soils, and sediment production in the interior Western United States. Forest Ecology and Management. 2015;**335**:176-193

[131] Boucher D, Gauthier S, Noël J, Greene DF, Bergeron Y. Salvage logging affects early post-fire tree composition in Canadian boreal forest. Forest Ecology and Management. 2014;**325**:118-127

[132] Maia P, Pausas JG, Vasques A, Keizer JJ. Fire severity as a key factor in post-fire regeneration of *Pinus pinaster* (Ait.) in Central Portugal. Annals of Forest Science. 2012;**69**:489-498

[133] Fernández C, Vega JA, Fonturbel T, Jiménez E, Pérez-Gorostiaga P. Effects of wildfire, salvage logging and slash manipulation on *Pinus pinaster* Ait. recruitment in Orense (NW Spain). Forest Ecology and Management. 2008;**255**:1294-1304

[134] Santana VM, González-Pelayo O, Maia PAA, Varela ME, Valdecantos A, Vallejo VR, Keizer JJ. Effects of fire recurrence and different salvage logging techniques on carbon storage in *Pinus pinaster* forests from Northern Portugal. European Journal of Forest Research. 2016;**135**:1107-1117

[135] Nunes L, Lopes D, Rego FC, Gower ST. Aboveground biomass and net primary production of pine, oak and mixed pine–oak forests on the Vila Real district, Portugal. Forest Ecology and Management. 2013;**305**:38-47

[136] Porté A, Trichet P, Bert D, Loustau D. Allometric relationships for branch and tree woody biomass of maritime pine (*Pinus pinaster* Ait.). Forest Ecology and Management. 2002;**158**:71-83

[137] Martí-Roura M, Casals P, Romanyà J. Temporal changes in soil organic C under Mediterranean shrublands and grasslands: Impact of fire and drought. Plant Soil. 2011;**338**:289-300

[138] Kaipainen T, Liski J, Pussinen A, Karjalainen T. Managing carbon sinks by changing rotation length in European forests. Environmental Science & Policy. 2004;**7**:205-219

[139] Shakesby RA. Post-wildfire soil erosion in the Mediterranean: Review and future research directions. Earth-Science Review. 2011;**105**:71-100

[140] Hosseini M, Keizer JJ, González-Pelayo O, Prats SA, Ritsema C, Geissen V. Effect of fire frequency on runoff, soil erosion, and loss of organic matter at the micro-plot scale in North-Central Portugal. Geoderma. 2016;**269**:126-137

[141] Martins MA, Machado AI, Serpa D, Prats SA, Faria SR, Varela ME, González-Pelayo O, Keizer JJ. Runoff and inter-rill erosion in a maritime pine and a eucalypt plantation following wildfire and terracing in North-Central Portugal. Journal of Hydrology and Hydromechanics. 2013;**61**:261-268

[142] Prats SA, Malvar MC, Vieira DCS, MacDonald L, Keizer JJ. Effectiveness of hydromulching to reduce runoff and erosion in a recently burnt pine plantation in Central Portugal. Land Degradation & Development. 2013;**27**:1319-1333

[143] Prats SA, MacDonald LH, Monteiro M, Ferreira AJ, Coelho CO, Keizer JJ. Effectiveness of forest residue mulching in reducing post-fire runoff and erosion in a pine and a eucalypt plantation in North-Central Portugal. Geoderma. 2012;**191**:115-124

[144] Loepfe L, Martinez-Vilalta J, Piñol J. Management alternatives to offset climate change effects on Mediterranean fire regimes in NE Spain. Climate Change. 2012;**115**:693-707

[145] Regos A, Aquilué N, Retana J, Cáceres M, Brotons L. Using unplanned fires to help suppressing future large fires in Mediterranean forests. PLoS One. 2014;**9**(4):e94906. DOI: 10.1371/journal.pone.0094906

[146] Barros AMG, Pereira JMC. Wildfire selectivity for land cover type: Does size matter? PLoS One. 2014;**9**(1):e84760. DOI: 10.1371/journal.pone.008476

[147] Fernandes P. Combining forest structure data and fuel modelling to classify fire hazard in Portugal. Annals of Forest Science. 2009;**66**:1-9

[148] Gutiérrez E, Lozano S. Avoidable damage assessment of forest fires in European countries: An efficient frontier approach. European Journal of Forest Research. 2013;**132**:9-21

[149] Tansey K, Grégoire J-M, Stroppiana D, Sousa A, Silva J, Pereira JMC, Boschetti L, Maggi M, Brivio PA, Fraser R, Flasse S, Ershov D, Binaghi E, Graetz D, Peduzzi P. Vegetation burning in the year 2000: Global burned area estimates from SPOT VEGETATION data. Journal of Geophysical Research Atmospheres. 2003;**109**:D14S03. DOI: 10.1029/2003JD003598

[150] Tansey K, Grégoire J-M, Binaghi E, Boschetti L, Brivio PA, Ershov D, Flasse S, Fraser R, Graetz D, Maggi M, Peduzzi P, Pereira JMC, Silva J, Sousa A, Stroppiana D. A global inventory of burned areas at 1 km resolution for the year 2000 derived from Spot Vegetation data. Climate Change. 2004;**67**(2):345-377

[151] Sousa AMO, Pereira JMC, Silva JMN. Evaluating the performance of multitemporal image compositing algorithms for burned area analysis. International Journal of Remote Sensing. 2003;**24**(6):1219-1236

[152] Pereira JMC, Sá ACL, Sousa AMO, Silva JMN, Santos TN, Carreiras JMB. Spectral characterisation and discrimination of burnt areas. In: Chuvieco E, editor. Remote Sensing of Large Wildfires in the European Mediterranean Basin. Berlin: Springer-Verlag; 1999. pp. 123-138

[153] Pereira JMC, Sousa AMO, Sá ACL. Regional-scale burnt area mapping in Southern Europe using NOAA-AVHRR 1km data. In: Chuvieco E, editor. Remote Sensing of Large Wildfires in the European Mediterranean Basin. Berlin: Springer-Verlag; 1999. pp. 139-155

[154] Gitas I, Mitri G, Veraverbeke S, Polychronaki A: Advances in remote sensing of post-fire vegetation recovery monitoring – A review. In: Fatoyinbo, L, editor. Remote sensing of biomass – principles and applications. Rijeka: InTech; 2012. pp. 144-170

[155] McRoberts R, Tomppo E, Naesset E. Advanced and emerging issues on national forest inventories. Scandinavian Journal of Forest Research. 2010;**25**:368-381

[156] Bolton DK, Coops NC, Wulder MA. Characterizing residual structure and forest recovery following high-severity fire in the western boreal of Canada using Landsat time-series and airborne lidar data. Remote Sensing of Environment. 2015;**163**:48-60

[157] Vila G, Barbosa P. Post-fire vegetation regrowth detection in the Deiva Marina region (Liguria-Italy) using Landsat TM and ETM+ data. Ecological Modelling. 2010;**221**:75-84

[158] Fernandez-Manso A, Quintano C, Roberts DA. Burn severity influence on post-fire vegetation cover resilience from Landsat MESMA fraction images time series in Mediterranean forest ecosystems. Remote Sensing of Environment. 2016;**184**:112-123

[159] Dubovyk O, Landmann T, Erasmus BFN, Tewes A, Schellberg J. Monitoring vegetation dynamics with medium resolution MODIS-EVI time series at sub-regional scale in southern Africa. International Journal of Applied Earth Observation and Geoinformation. 2015;**8**:175-183

[160] Meng R, Dennison PE, Huang C, Moritz MA, D'Antonio C. Effects of fire severity and post-fire climate on short-term vegetation recovery of mixed-conifer and red fir forests in the Sierra Nevada Mountains of California. Remote Sensing of Environment. 2015;**171**:311-325

[161] Chen W, Moriya K, Sakai T, Koyama L, Cao C. Monitoring of post-fire forest recovery under different restoration modes based on time series Landsat data. European Journal of Remote Sensing. 2014;**47**:153-168

[162] Chu T, Guo X. Remote sensing techniques in monitoring post-fire effects and patterns of forest recovery in boreal forest regions: A review. Remote Sensing. 2014;**6**:470-520

[163] Di Mauro B, Fava F, Busetto L, Crosta GF, Colombo R. Post-fire resilience in the Alpine region estimated from MODIS satellite multispectral data. International Journal of Applied Earth Observation and Geoinformation. 2014;**32**:163-172

[164] Lanorte A, Lasaponara R, Lovallo M, Telesca L. Fisher–Shannon information plane analysis of SPOT/VEGETATION normalized difference vegetation index (NDVI) time series to characterize vegetation recovery after fire disturbance. International Journal of Applied Earth Observation and Geoinformation. 2014;**26**:441-446

[165] Goetz SJ, Fiske GJ, Bunn AG. Using satellite time-series data sets to analyze fire disturbance and forest recovery across Canada. Remote Sensing of Environment. 2006;**101**:352-365

[166] Riaño D, Chuvieco E, Ustin S, Zomer R, Dennison P, Roberts D, Salas J. Assessment of vegetation regeneration after fire through multitemporal analysis of AVIRIS images in the Santa Monica Mountains. Remote Sensing of Environment. 2002;**79**:60-71

[167] Ireland G, Petropoulos GP. Exploring the relationships between post-fire vegetation regeneration dynamics, topography and burn severity: A case study from the Montane Cordillera Ecozones of Western Canada. Applied Geography. 2015;**56**:232-248

[168] Petropoulos GP, Griffiths HM, Kalivas DP. Quantifying spatial and temporal vegetation recovery dynamics following a wildfire event in a Mediterranean landscape using EO data and GIS. Applied Geography. 2014;**50**:120-131

[169] Sever L, Leach J, Bren L. Remote sensing of post-fire vegetation recovery; a study using Landsat 5 TM imagery and NDVI in North-East Victoria. Journal of Spatial Science. 2012;**57**:75-191

[170] White J, Ryan K, Key C, Running S. Remote sensing of forest fire severity and vegetation recovery. International Journal of Wildland Fire. 1996;**6**:125-136

[171] Quintano C, Fernández-Manso A, Shimabukuro YE, Pereira G. Spectral unmixing: A review. International Journal of Remote Sensing. 2012;**33**:5307-5340

[172] EFFIS. European Forest Fire Information System. 2017. Available from: http://forest.jrc.ec.europa.eu/effis/applications/data-and-services/ [Accessed: 3 January 2017]

[173] ICNF. Defesa da Floresta Contra Incêndios [Defense of Forest Against Fires]. Instituto da Conservação da Natureza e das Florestas [in Portuguese]. 2017. Available from: http://www.icnf.pt/portal/florestas/dfci [Accessed: 3 January 2017]

[174] Marques S, Borges JG, Garcia-Gonzalo J, Moreira F, Carreiras JMB, Oliveira MM, Cantarinha A, Botequim B, Pereira JMC. Characterization of wildfires in Portugal. European Journal of Forest Research. 2011;**130**:775-784

[175] Parente J, Pereira MG, Tonini M. Space-time clustering analysis of wildfires: The influence of dataset characteristics, fire prevention policy decisions, weather and climate. Science of the Total Environment. 2016;**559**:151-165

[176] Eichhorn MP, Paris P, Herzog F, Incoll LD, Liagre F, Mantzanas K, Mayus M, Moreno G, Papanastasis VP, Pilbeam DJ, Pisanelli A, Dupraz C. Silvoarable systems in Europe—past, present and future prospects. Agroforestry Systems. 2006;**67**:29-50

[177] Jose S, Gillespie AR, Pallardy SG. Interspecific interactions in temperate agroforestry. Agroforestry Systems. 2004;**61**:237-255

[178] Godinho-Ferreira P, Azevedo A, Rego F. Carta da Tipologia Florestal de Portugal Continental [Map of the forest typology of continental Portugal]. Silva Lusitana. 2005;**13**(1):1-34 [in Portuguese]

[179] Gonçalves AC, Dias SS, Ferreira AG, Ribeiro NA: Landscape diversity patterns in Alentejo forest area. In: Schnabel S, Ferreira AG, editors. Sustainability of Agrosilvopastoral Systems. Advances in GeoEcology. Reiskirchen: Catena Verlag; 2004; Vol. 37. 2004. pp. 255-261

[180] Rodrigues M, San Miguel J, Oliveira S, Moreira F, Camia A. An insight into spatial–temporal trends of fire ignitions and burned areas in the European Mediterranean countries. Journal of Earth Science and Engineering. 2013;**3**:497-505

[181] Nunes AN. Regional variability and driving forces behind forest fires in Portugal: An overview of the last three decades (1980-2009). Applied Geography. 2012;**34**:576-586

[182] Carmo M, Moreira F, Casimiro P, Vaz P. Land use and topography influences on wildfire occurrence in Northern Portugal. Landscape and Urban Planning. 2001;**100**:169-176. DOI: 10.1016/j.landurbplan.2010.11.017

[183] Bajocco S, Ricotta C. Evidence of selective burning in Sardinia (Italy): Which land cover classes do wildfires prefer? Landscape Ecology. 2008;**23**:241-248

[184] Sebastián-López A, Salvador-Civil R, Gonzalo-Jiménez J, San Miguel-Ayanz J. Integration of socio-economic and environmental variables for modelling long-term fire danger in Southern Europe. European Journal of Forest Research. 2008;**127**:149-163

[185] Nunes MCS, Vasconcelos MJ, Pereira JMC, Dasgupta N, Alldredge RJ, Rego FC. Land cover type and fire in Portugal: Do fires burn land cover selectively? Landscape Ecology. 2005;**20**:661-673

[186] IPMA. Normais Climatológicas [Climate Normals]. Instituto Português do Mar e da atmosfera [in Portuguese]. 2017. Available from: http://www.ipma.pt/pt/oclima/normais.clima/ [Accessed: 24 March 2017]

[187] Mutanga O, Adam E, Cho MA. High density biomass estimation for wetland vegetation using WorldView-2 imagery and random forest regression algorithm. International Journal of Applied Earth Observation and Geoinformation. 2012;**18**:399-406

[188] Rouse JW, Haas RH, Schell JA, Deering DW. Monitoring vegetation systems in the Great Plains with ERTS. In: 3rd ERTS Symposium, NASA SP-351. Vol. I. Washington DC: NASA; 1973. pp. 309-317

[189] van Leeuwen W. Monitoring the effects of forest restoration treatments on post-fire vegetation recovery with MODIS multitemporal data. Sensors. 2008;**8**:2017-2042

Chapter 14

A Common Approach to Foster Prevention and Recovery of Forest Fires in Mediterranean Europe

Sandra Oliveira, Giovanni Laneve, Lorenzo Fusilli,
Georgios Eftychidis, Adélia Nunes,
Luciano Lourenço and Ana Sebastián-López

Additional information is available at the end of the chapter

http://dx.doi.org/10.5772/intechopen.68948

Abstract

Most countries of Mediterranean Europe are strongly affected by forest fires, with major socio-economic and environmental impacts that can spread over several regions and countries. A transnational approach allows creating synergies regarding resource sharing and problem-solving strategies. The access to high quality and up-to-date information is critical to improve fire hazard mitigation measures and promote comparable appraisals between different regions. Several collaborative initiatives have been implemented in Europe to foster research and service development, focusing on common issues amongst countries. The PREFER project was one of these initiatives, with the purpose of contributing to protect human communities and forests from fire hazard, by providing cartographic products through the implementation of a new systematic framework. The participation of end users, such as civil protection organizations and forest services, covering the Euro-Mediterranean region, was crucial to ensure the operational application of the mapping products. Fuel classification, daily fire hazard indices, vulnerability assessment and damage severity levels were some of the mapping applications developed for several test areas in Mediterranean Europe. This chapter illustrates the potential enhancements for forest fire management offered by this framework, bearing in mind the benefits of applying shared and harmonized approaches for common issues.

Keywords: forest fires, Mediterranean Europe, common framework, collaboration, prevention, mitigation

1. Introduction

The European Mediterranean region is systematically affected by uncontrolled forest fires, which pervasively threaten most of the vegetated land and cause extensive environmental damage, with significant economic impacts [1–6]. In Mediterranean Europe, an average of ca. 60,000 fires burn over 400,000 ha of wildland and forest areas every year [7]. The countries in the southwestern part, namely Portugal, Spain, France, Italy and Greece, are the most affected by fires [8]. These countries share similar environmental conditions, which can explain such high fire incidence: the coincidence of the driest with the hottest season and the occurrence of wet and dry weather extremes throughout the year; the coexistence of urban settlements, infrastructure networks and vegetated areas (forest, agricultural and uncultivated areas) in a complex, dense and intimately interconnected patchwork; the diminished control on traditional practices involving fires as an instrument for land management and the changes in land use verified in recent decades [2, 9–16]. Portugal is the most affected country with regard to number of fires, whereas Spain generally records the highest burned area, despite annual variations (**Figure 1**).

Additionally, climate effects are expected to become even stronger in the upcoming decades, according to the most recent fire danger projections [17–20], highlighting the high sensitivity of the Mediterranean area to projected climatic change in terms of fire activity [21–24]. In this context, forest fires can pose serious threats to human communities and the environment. The investment in recent years in suppression policies and the deployment of more efficient and sophisticated techniques for detection and firefighting, despite the noteworthy improvements in early detection and extinction of fires, cannot counteract entirely the effects of repeated and simultaneous fire events or very high intensity fires [25–30]. Fostering a significant improvement in effectiveness and timeliness of prevention measures, in association with appropriate recovery actions and their proper integration in the overall fire management cycle, is, therefore, the most suitable way to mitigate efficiently the damages potentially caused by forest fires in the European Mediterranean region.

1.1. Common approaches for shared problems

At the European level, it is recognized that the prevention of natural and man-made disasters focusing on a common approach is more effective than separate national approaches [31], by linking actors and policies and promoting technical and knowledge development applicable to transnational situations. In the last 20 years, the European Commission and

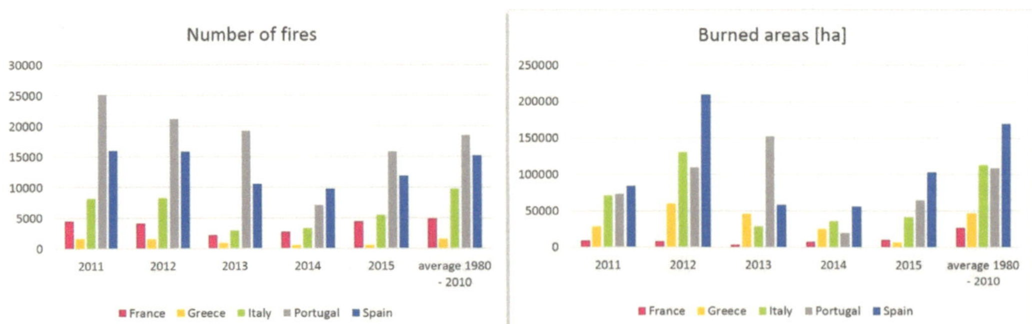

Figure 1. Distribution of the number of fires and burned areas (in ha) between 2011 and 2015 for the countries in Southern Europe most affected by forest fires, compared with the mean value of the previous 30 years.

the EU countries have worked together to establish common forest fire management and monitoring approaches, as is the case of the European Forest Fire Information System [32, 33], although geographic differences, diverse traditions, differing resources and interest with regard to fire issues have delayed this process [33]. Other initiatives have been implemented to promote collaboration in fire research in Europe, integrated in hazard and risk assessment studies at different levels, benefiting from synergies with regard to resource-sharing and problem-solving strategies for mutual concerns, such as the projects FIRE PARADOX [28], FIRESMART and MOVE [34].

This chapter presents the outcomes of one of these initiatives, the PREFER project (Space-based Information Support for Prevention and Recovery of Forest Fires Emergency in the MediteRranean Area) [35], which was developed to respond to the pragmatic need of protecting European forests and communities from fire hazard, capitalizing on the experience gained in the last decade by the participating institutions. The project was based on the development of a shared framework applicable to European Mediterranean countries, to deliver timely and high-quality cartographic products, suitable for decision-making at different levels within the fire management process.

1.2. Geographic coverage

The target geographic area of the PREFER project was composed of the European territories located in the Mediterranean area where fire occurrence is particularly relevant. To test and demonstrate the products and services developed, five pilot areas were selected based on the availability of data required to develop the products, the interest of end users, the biophysical and social conditions of these areas and their fire occurrence history (**Figure 2**).

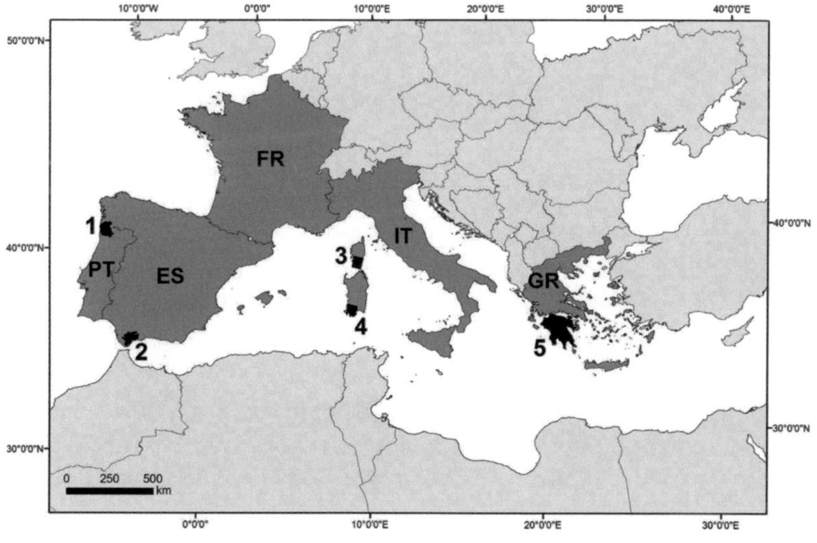

Figure 2. Geographic coverage of the PREFER project and location of the test areas in the participating countries. 1-Minho region, Portugal (PT); 2-Los Alcornocales, Andaluzia, Spain (ES); 3-SW Corsica, France (FR); 4-SW Sardinia, Italy (IT); 5-Peloponnese region, Greece (GR).

2. A common framework for forest fires in Mediterranean Europe: the PREFER project

The PREFER project ("Space-based Information Support for Prevention and REcovery of Forest Fires Emergency in the MediteRranean Area") was developed under the framework of the EU service Copernicus, between 2012 and 2015. PREFER was configured as an industry-driven/science-controlled initiative and the consortium was composed of eight partners from five most fire-affected countries in Southern Europe, among which research institutes and industries of technological development and security (**Table 1**). The main purpose of PREFER was to set up a common information communication technology (ICT) infrastructure to provide mapping tools and services, adapted to the needs of end users interested in different stages of forest fire management. To attain this goal, the project was driven by three main conditions as follows:

(1) The development of a common mapping framework regarding fire prevention and recovery, applicable to European Mediterranean countries.

(2) The creation of a service available at the operational level and useful for multiple users from different sectors and countries.

(3) The timely delivery of easily accessible cartographic tools based on harmonized, high-quality and up-to-date information.

2.1. A common framework for forest fire management

Although the peril of forest fires is a shared issue among the countries of the European Mediterranean region, there are differences in the way each country reacts, which are mirrored in the administrative structure and operational organization of each country or region. In most countries, specific regulatory conditions exist at the national level with regard to forest fire management. France, Greece, Italy, Portugal and Spain have corpus of laws with specific norms punishing fire crimes, and their regulations foresee administrative penalties towards burned surfaces or unmaintained vegetation. Specific regulations concerning the preparedness and prevention phase exist for all these countries, but this is not the case for the recovery phase (e.g. Italy). As such, fire prevention plans are foreseen by law and operational procedures include risk analysis; however, standard operational procedures of forest fire management are not established by any national legal framework. Fire management structures vary among countries, from the dispersed governmental organization of Spain and the regional autonomies [36] to the provincial structure of Italy and to the more centralized systems of France, Greece and Portugal [37, 38]. At the governmental level, resources and responsibilities on forest fire management can be spread over different departments (e.g. Portugal). Overall, preventive actions and recovery strategies are assigned to various actors and institutions that differ among countries, crossing local, regional or national management level. Among these, fire services, forest management and environmental protection services, civil protection agencies, environmental

http://dx.doi.org/10.5772/intechopen.68948

Country	Consortium partners		Participating users
France	University of Strasbourg	UNISTRA SERTIT	• Service Départemental d'Incendie et de Secours de la Haute Corse • Office National des Fôrets, Direction Generale Corse
Greece	Center for Security Studies	KEMEA	• General Secretariat for Civil Protection
	SATWAYS Ltd, Satcom & Telematics	SATW	• Aristotle University of Thessaloniki • InterBalkan Environment Center (i-BEC) • National Centre for Scientific Research DEMOKRITOS
Italy	Università Degli Studi di Roma "La Sapienza"	DIAEE	• Corpo Forestale e di Vigilanza Ambientale, Sardinia
	CGS SPA Compagnia Generale per lo Spazio	GCS	• Centro Nazionale di Meteorologia e Climatologia Aeronautica
	Intelligence for Environment & Security	IESC	• Fire Brigades
Portugal	University of Coimbra	UCO	• Autoridade Nacional de Proteção Civil (ANPC) • Instituto de Conservação da Natureza e das Florestas (ICNF) • SEPNA, Guarda Nacional Republicana
Spain	GMV, Aerospace and Defence SA Unipersonal	GMV	• Agencia de Medio Ambiente y Agua de Andalucía

Table 1. Countries, consortium partners and end users who participated in the PREFER project.

organizations, volunteer associations and armed forces assume different tasks, although with common interests.

Under these circumstances, the involvement of multiple actors and institutions from the participating countries was essential to develop a framework applicable at the transboundary level, with systematic data collection and analysis, but flexible enough to allow for specific adjustments according to each country's characteristics and needs.

2.2. Creation of a service available at the operational level and useful for multiple users

Concerning forest fire emergency, the civil protection authority is usually the main organization dealing with the strategic planning and coordination of resources. However, they are not involved in operational activity at the same level in all countries. Owing to the obvious differences among countries, the PREFER project devoted a specific activity to involve a core group of users in the project activity. These end users represented a variety of organizations with a legal mandate for managing forest fires in their respective region or country, between the most affected countries in southern Mediterranean Europe. Among these, civil protection authorities, regional and national forest services, fire brigades and environmental protection agencies were involved (**Table 1**). The application of the cartographic tools at the operational level required the testing and demonstration of the resulting products in view of current implemented procedures. The end users provided feedback along the different stages of project development and helped defining a harmonized set of requirements to ensure the operational applicability of the products and tools developed.

2.3. Delivery of timely and easily accessible cartographic tools based on harmonized information

The access to high-quality and up-to-date information is a key issue to improve the effectiveness of fire damage mitigation, grounded on preventive measures and suitable recovery actions. The PREFER project made use of advanced technologies to ensure the consistent collection and analysis of data and the systematic delivery of mapping tools. Considering the relevance of transnational approaches within the European forest fire context and the need to obtain harmonized information to enable equivalent appraisals between regions and countries, the project design was based on the exploitation of remote sensing observations, in order to: (i) provide systematic and repeatable data and support the consistent creation, update and delivery of products and (ii) harmonize information used to manage forest fires and promote transnational cooperation across Mediterranean Europe.

2.3.1. Satellite remote sensing

Remote sensing offers useful tools for fire monitoring and damage assessment to support fire management in a cost-effective way. In comparison to other methods of information gathering, satellite remote sensing provides some advantages, such as (i) large area coverage; (ii) frequent and repetitive coverage of the area of interest, even if this area is remote or difficult to access; (iii) easy data acquisition at different scales and resolutions; (iv) provides spatially continuous data, avoiding thus the need for interpolation which always degrades the original information to some extent; (v) one single image can be interpreted for different purposes and applications; (vi) quantitative measurements of ground features are obtained using radiometrically calibrated sensors and (vii) enables the implementation of a systematic approach and allows for monitoring of dynamic processes.

Currently, several remote sensing satellites provide imagery suitable for forest fire research and fire monitoring operations; from land cover/use maps to burned area and fire danger estimations, nowadays users can access several global, pan-European products for use in the field of fire preparedness and post-fire vegetation recovery. Several initiatives also exist to generate Earth observation (EO) data according to predetermined standards and to optimize the access of the users' community to geoinformation products derived from remote platforms; among these are as follows: the Global Fire Monitoring Center (GFMC, http://www.fire.uni-freiburg de), the fire mapping and monitoring theme of the GOFC/GOLD (Global Observations of Forest and Land Cover Dynamics, http://gofc-fire.umd.edu), the CEOS (Committee on Earth Observation Satellites) and several missions of NASA (National Aeronautics and Space Administration, http://www.nasa.gov).

The PREFER project was particularly associated with the European Programme Copernicus (http://www.copernicus.eu/), formerly known as GMES (Global Monitoring for Environment and Security). The goal of Copernicus is to develop operational information services on a global scale in support of environment and security policy needs, to provide the user community with accurate, timely and easily accessible information collected from Earth observing satellites and *in-situ* sensors. The thematic areas more relevant for PREFER were the land and the emergency management, and the project aimed to ensure complementarity of its products with those delivered by the other Copernicus services. Therefore, the potential synergies between different services, initiatives and current space missions were integrated in the design of the project. As well, the future EO space missions, particularly Sentinel [39], were examined to ensure the long-term sustainability of the project's services, the availability of cost-effective remote sensing data, and to identify potential gaps of future missions with respect to the users' requirements.

The information retrieved from satellite images was used in various ways, according to the type of cartographic product developed, the needs and requirements of end users and the local characteristics of forest fires. In the participating countries, no specific legal framework exists regarding the use of space-based information in the domain of forest monitoring and forest fire management. This is why most countries do not retain a central Earth Observation database and they do not use official forest fuel type or other maps provided at the national level and based on EO data. The exception is found in Spain, where the situation differs between autonomous regions.

3. Service portfolio and cartographic tools

The approach implemented enabled the creation of a regional service for the systematic provision and update of operational products suitable for use in different European Mediterranean countries. The PREFER portfolio included services providing information concerning the preparedness/prevention phase and recovery/reconstruction phase (**Table 2**). Each of these services focused on a particular aspect of the forest fire cycle, following the next specific criteria:

- The products developed were based on the exploitation of data from the Copernicus space infrastructure, taking advantage of the readily available space-borne observation data;
- The procedure of product development had to optimize the integration of different data types from a variety of sources, such as earth observation, digital terrain models, socio-economic data, meteorological data and *in-situ* data;
- The procedure developed had to be applicable in different countries of the Mediterranean region, in a systematic and sustainable way;
- The products had to be distributed by an interoperable service provision infrastructure (based on OGC/INSPIRE) that could allow easy access to the information.

Product	Description	Resolution	Refresh rate	Input EO data current (and future)
Preparedness/prevention service phase				
Fuel map	Classification map of forest fuel complexes	5–30 m	Yearly	OLI/Landsat8, Rapid-Eye
Fuel reduction map	Identification of the areas where the prevention procedure based on fuel reduction is advisable	5–20 m	Yearly	OLI/Landsat8 (Sentinel-2)
Prescribed fire map	Spatio-temporal map of the areas where prescribed fire would be useful and safe to apply	250 m	Daily	OLI/Landsat8 (Sentinel-2)
Daily fire hazard map	Fire danger index, indicating the proneness of a vegetated area to support a fire	250 m	Daily	MODIS/Terra & Aqua (Sentinel-3)
Seasonal fire hazard map	High resolution hazard index	250 m	Bi-weekly	MODIS/Terra & Aqua
Vulnerability map & economic value	Relative measure of the maximum potential for loss in the case of fire. Estimation of economic losses	100 m	Yearly	(OLI/Landsat8)
Seasonal risk map	The probability of occurrence of a fire event that can cause losses	100 m	Bi-weekly	MODIS/Terra & Aqua
Recovery/reconstruction service phase				
Post-fire vegetation recovery map	Identification of areas previously damaged by fire event where regrowth of vegetation took place	Min 1 ha	Every 16 days	OLI/Landsat8 (Sentinel-2)
Damage severity map	Degree of damage based on vegetation status after a fire event	30 m, BA >10 ha	On request	OLI/Landsat8 (Sentinel-2)
3D Fire vegetation volume loss map	2D and 3D fire impact monitoring maps, to identify areas with increased risk of flash flooding and debris and to highlight areas prone to soil erosion	10 m	On request	SPOT, Pleiades, Sentinel-1
Soil erosion susceptibility map		30 m	On request	SPOT, Pléiades, Sentinel-1, OLI/ Landsat8
Burned scar map HR	Burned scar perimeters, at scale 1/10,000–1/50,000	Min. 1 ha	Every 16 days	OLI/Landsat8, Cosmo-SkyMed, Sentinel-1, (2)
Burned scar map VHR optical	Burned scar perimeters at cadastral scale (1/1.000–1/4.000)	Min. 0.25 ha	On request	Ikonos, Quickbird, Worldview & other

Table 2. Service portfolio and brief description of each cartographic product.

The information based on satellite images, combined with data from other sources according to the type of product developed, can improve forest management with regard to fire prevention and the prompt assessment of damages for planning recovery activities at a transboundary level. To illustrate this goal, details on some of the products and mapping tools developed, as well as the results obtained and their usage in view of fire management activities in the different countries, are presented. PREFER products aimed at enhancing forest fire management across Mediterranean Europe, and helping to meet the overall objective of improving the preparedness and responsiveness of national authorities in relation to wildfires, reducing the vulnerability of people, property and the environment to the adverse impacts of fire hazard.

3.1. Preparedness/prevention service phase

In Mediterranean Europe, over 90% of fires are due to human activities, either intentional or accidental, and therefore are mainly unpredictable and require strong awareness actions. Notwithstanding, particular environmental and biophysical conditions, such as vegetation, meteorological/climatic features and topography, influence fire propagation and the severity of damages caused by a fire [4, 6, 40–43]. The cartographic tools developed for this phase were, therefore, inspired on the possibility to measure these biophysical conditions and to foresee the behaviour of a fire in case an ignition occurs, as well as on the estimation of the potential damages a fire could cause, with the overall purpose of identifying suitable prevention and hazard mitigation strategies.

3.1.1. Fuel map

Forest fuel types can be defined as an identifiable association of fuel elements of distinctive species, form, size arrangement and continuity, which will exhibit particular fire behaviour under defined burning (meteorological) conditions [44]. Fuels vary broadly in spatial distribution, type, physical characteristics, load (amount) available for combustion, their contribution to fire potential and effect on fire behaviour.

The fuel type mapping method implemented in PREFER combined Geographic Information Systems (GIS) methods with remote sensing technology. GIS techniques were applied to integrate different static vegetation-related geospatial layers, while remote sensing aimed at capturing those dynamic aspects of vegetation, which relate to its phenology and thus require more frequent monitoring. Evolving from the results of previous projects, namely FUELMAP and ArcFUEL, the method applied within the PREFER project meant a step further, since it allowed for a more realistic updating that accounts for season, the clear-cuttings, reforestation and afforestation labours, as well as for urban sprawl, and burned areas. Additionally, in comparison with previous approaches in which the fuel load was a fixed number derived from standard fuel classifications (for example, see Ref. [45]), in PREFER the fuel load was included through a dynamic proxy, the fraction of vegetation cover (FVC) [46–48] computed as a function of the normalized difference vegetation index (NDVI). The forest fuel type product depicted forest fuel complexes in some test areas and classified them according to the assessed capacity of these complexes to support fire occurrence and contribute to fire potential. The fuel types were presented in categories, according to specific characteristics of the vegetation, such as density, arrangement and physical parameters related to fire behaviour (such as potential flame length and rate of spread) and then input into the following phases of the processing chain, that is, fire risk and behaviour modelling on the one hand and prescribed fires and vulnerability modelling, on the other hand. Adaptations at national and subnational levels were carried out when needed, using the best available local data sources to increase the fuel map precision while keeping its compatibility with the pan-European scale.

Figure 3 shows the fuel maps produced for the test sites in Italy (Sardinia) and Spain (Andalusia). The distribution of fuel types in these sites is very different; in Sardinia, non-wildland fuels (growing areas and other non-wild vegetation) and shrubs predominate and forests are interspersed with shrub lands dispersed by the NW and SE sides of the test site. Conversely in Cádiz, dense and open forests occupy a larger area, which can be a result of the classification as a natural protected area (Natural Park).

Fuel maps constitute a very important tool of fire management and are used by multiple institutions for allocating resources, planning prevention activities and developing fuel management programs. These mapping tools are also useful for civil protection authorities, since fire propagation modelling, based on fuel classification, serves to design evacuation plans for endangered population. A European fuel classification scheme enables equivalent evaluation between countries and regions, promotes the strategic implementation of pan-European approaches and a more efficient use of resources and effective preparedness.

3.1.2. Fuel reduction map

The reduction of fuel load accumulation is a common practice [49–52] to decrease hazard levels, through the use of several prevention practices such as: (i) manual or mechanical modification and removal of natural fuels, including cutting, crushing or stacking; (ii) other treatments such as application of herbicides, introduction of biological controls or pasturage; (iii) prescribed burning technique, with the deliberate use of fire in a given area and well-defined environmental conditions.

This product was found on the need to plan suitable programs for reducing fuel loads to reduce fire danger. The fuel reduction map was based on the identification of the territorial units with a higher probability of being affected by a fire and on the factors that favour

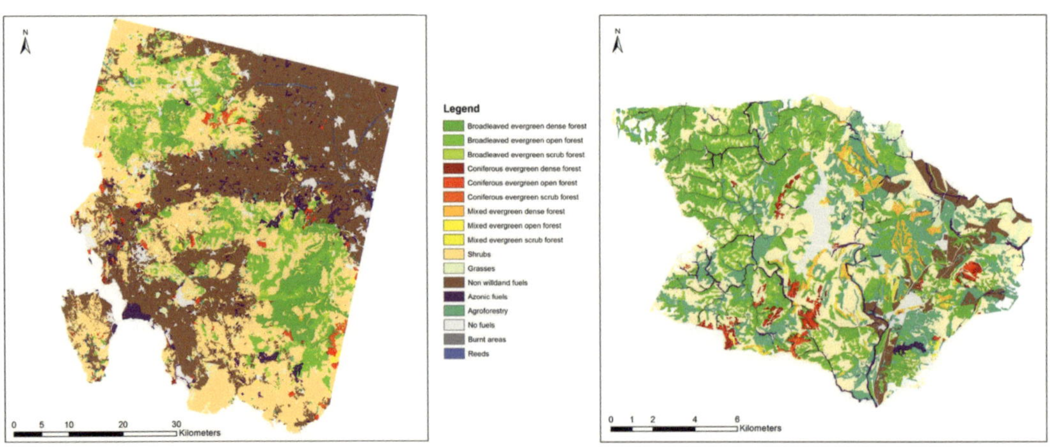

Figure 3. Fuel maps created for the test sites of Italy (left) and Spain (right).

fire spread, with the subsequent computation of a fire propagation probability (FPP). The methodology takes into account geospatial elements, representing the territorial factors that may affect the onset of fires and their propagation. Four types of factors were considered: (i) topographic (altitude, slope, aspect); (ii) environmental (fuel type, climatic conditions); (iii) land use (urban areas, agricultural areas, roads, natural parks, forests) and (iv) fire incidence (burned areas of, at least, the last 5 years).

The combination of all these parameters provided a type of landscape classification, where each class represents a particular combination of these parameters which were defined as homogeneous territorial classes (HTC). The FPP was then obtained for each HTC, following the formula:

$$FFP = \frac{\sum_{k=1}^{N_i} Cb_{ik}}{TC_i^{\text{tot}}} \tag{1}$$

where Cb_{ik} is the burned surface of the k_i-th fire within the HTC$_i$-th (Ha), T is the time range (8 years), N_i is the number of fires in the time range T in the type$_i$-th, C_i^{tot} is the area of the HTC$_i$-th.

The FPP values were reclassified in five classes expressing the geospatial hazard (GSH), which indicates the priority level for fuel load reduction (**Table 3**). The final output is a cartographic product (**Figure 4**) showing the areas where fuel load reduction measures should be carried out in order to mitigate potential fire effects.

3.1.3. *Daily fire hazard index map*

The daily fire hazard map provided a medium spatial resolution fire danger index, through a dimensionless number indicating the proneness of a vegetated area to burn or to support a fire. This product was based on the evidence of a strong relationship between fire and the fuel characteristics (vegetation type, density and humidity content), topographic features (slope, altitude and solar aspect angle) and meteorological conditions (rainfall, wind direction and speed, air humidity, surface and air temperature) [43, 53–56]. The resulting maps represent a normalized index with values from low (0) to very high hazard (100), subsequently classified in six hazard levels (**Figure 5**). Every day, three maps are produced for the required area, showing the spatial distribution of fire hazard levels for the present day and the two following days.

The daily hazard levels are strongly dependent on weather and fuel conditions, which are highly variable on a daily basis. This dynamic index is a valuable tool also for allocating fire suppression and emergency resources for a short timeframe, during the main fire season (usually summertime) in Mediterranean countries. The computation of this product has been active beyond the duration of the project and it has been applied in additional areas, by request of local users (**Figure 6**), which confirms its applicability within the Mediterranean region.

3.1.4. *Seasonal fire hazard map*

The strong influence of anthropic factors on forest fire distribution in the context of Mediterranean Europe requires the integration of the human factor in fire hazard definition, besides the

FPP values	Geospatial hazard	FR classes
>0.1	5	Very high
0.01–0.1	4	High
0.01–0.001	3	Moderate
0.001–0.0001	2	Low
<0.0001	1	Very low
0	0	Null

Table 3. Geospatial hazard and fuel reduction (FR) classes according to fire propagation probability (FPP) values.

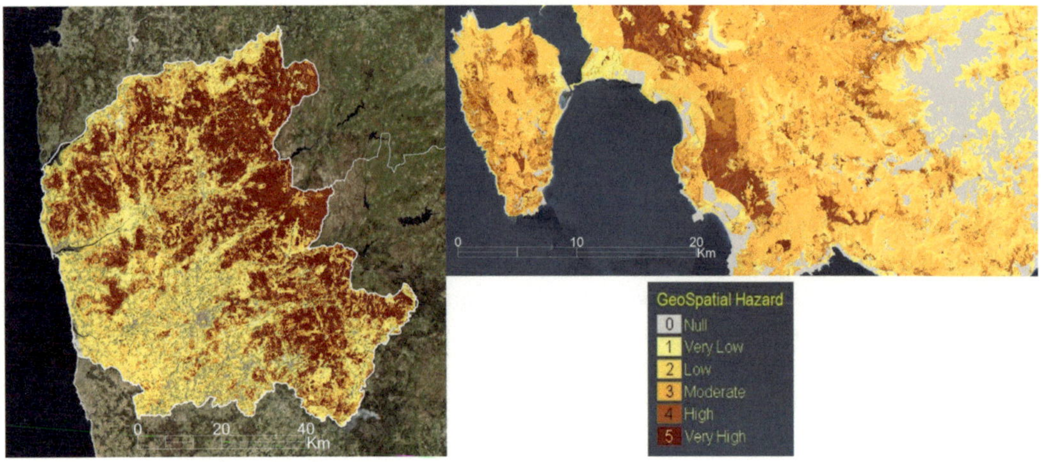

Figure 4. Fuel reduction maps computed for the test sites of Minho region, Portugal and SW Sardinia, Italy, in 2015. Darker areas are expected to be more prone to fire. The map can be computed for areas with different sizes and at several scales.

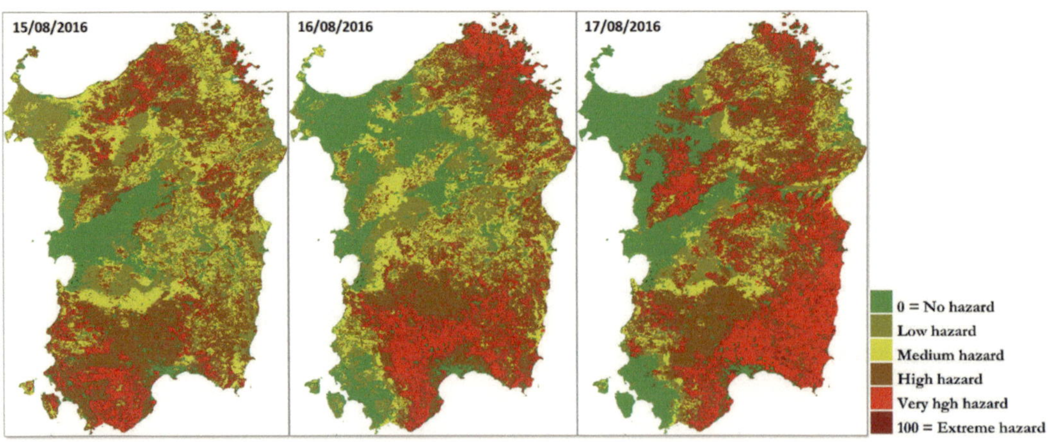

Figure 5. Daily hazard index maps computed for the Sardinia region in August 2016. The maps produced on the August 15, 2016 depict the situation for that day and provide the prediction of fire hazard levels for the next 2 days (August 16–17, 2016).

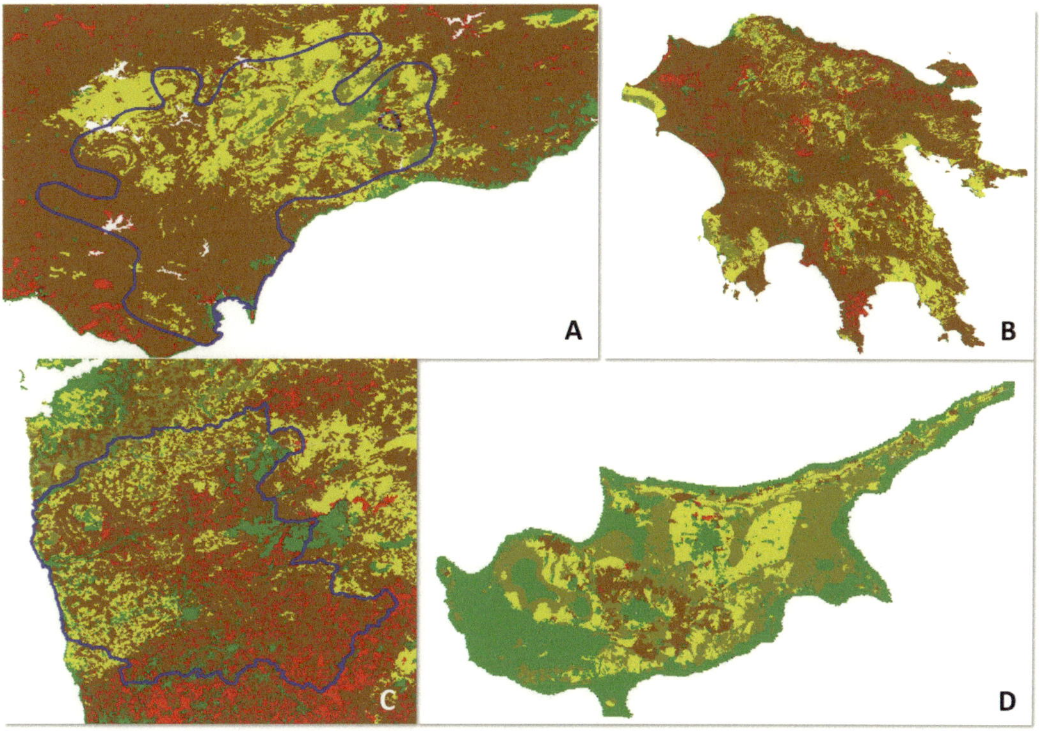

Figure 6. Daily hazard index maps computed for other test sites for the August 15, 2016, at different scales: (A) Spain, (B) Greece, (C) Portugal, (D) Cyprus island, requested later by the local Civil Protection.

biophysical and variable weather conditions. The seasonal fire hazard product had the objective to combine both human and meteorological factors to create a new index able to represent these different aspects of fire hazard, based on the integration of the following information: (i) natural factors that take into account the morphological characteristics distinguishable in static (such as slope, aspect and climatic zone) and dynamic data, related to vegetation stress (e.g. normalized difference vegetation index, NDVI), meteorological data, updated fuel map with burned areas and daily fire hazard index averaged over 15 days; (ii) human factors, considering the accessibility (urban areas and roads), the cultural factors (fire occurrences for the last 5–10 years, representing the security of a place) and the seasonal factors (current season fire trend with respect to the 5 previous years). This approach is based on the computation of homogeneous territorial classes according to the probability that a forest fire may occur, useful to manage prevention actions and firefighting activities and adjust the distribution of human resources and other means. Based on statistics of the previous years, in combination with the other included factors, the Portuguese test area had the highest proportion of spatial units classified as high and very high hazard, with a decreasing trend between June and September (**Figure 7**).

3.1.5. Vulnerability map

In a general sense, vulnerability means the potential for loss, which can affect different types of elements, either biophysical or anthropogenic [34, 57–59]. The vulnerability approach applied in PREFER was comprehensive, resulting from the combination of three different components:

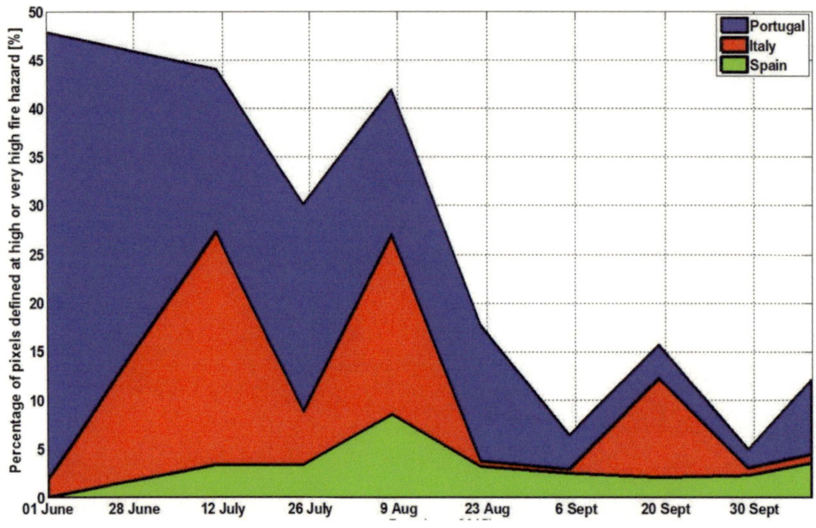

Figure 7. Variation of the area (% of pixels) classified at high or very high fire hazard during the 2015 summer season in three test areas.

exposure, sensitivity and coping capacity. Exposure represented the presence of people, property, systems or other elements in hazard zones that are, for that reason, subject to potential losses. Sensitivity characterized the conditions that influence the predisposition of the exposed elements to suffer a certain level and extension of damages. Coping capacity represented the circumstances that could reduce or, instead, amplify the ability of the elements to respond and recover from the impacts of a hazard, being related to the level of resilience of a community. Each of these components was first computed separately as an intermediate index, resulting from the combination of normalized variables that reflected multiple dimensions (socio-demographic, environmental, economic and institutional conditions). The vulnerability map resulted from the aggregation of the three intermediate components and represented a relative measure of the potential for loss, identifying the areas with higher likelihood to suffer losses in case a forest fire occurs.

The spatial patterns of forest fire vulnerability levels were uneven in the test sites (**Figure 8**). In Portugal, the northeastern (NE) part of Minho region showed higher vulnerability, due to the cumulative presence of population, infrastructures and fire-prone fuels, higher sensitivity derived from ageing population in that area and the presence of an internationally recognized protected area (Peneda-Gerês). Lower vulnerability levels largely coincide with the more urbanized areas, where fuel is nearly absent and institutional resources, such as fire stations, are concentrated. In the Italian test site, the lowest vulnerability levels are mainly concentrated in the NE part and seem to be closely related to fuel patterns; non-wildland fuels (which include agricultural land and are classified as less sensitive to fire) occupy about 37% of the test site and mostly occur in the northeastern side. Despite its complexity and challenging interpretation, vulnerability assessment is a relevant tool for fire prevention. In this case, the stepwise approach applied, with the creation of intermediate indices and the possibility to provide cartographic tools to end users for all the variables and components integrated,

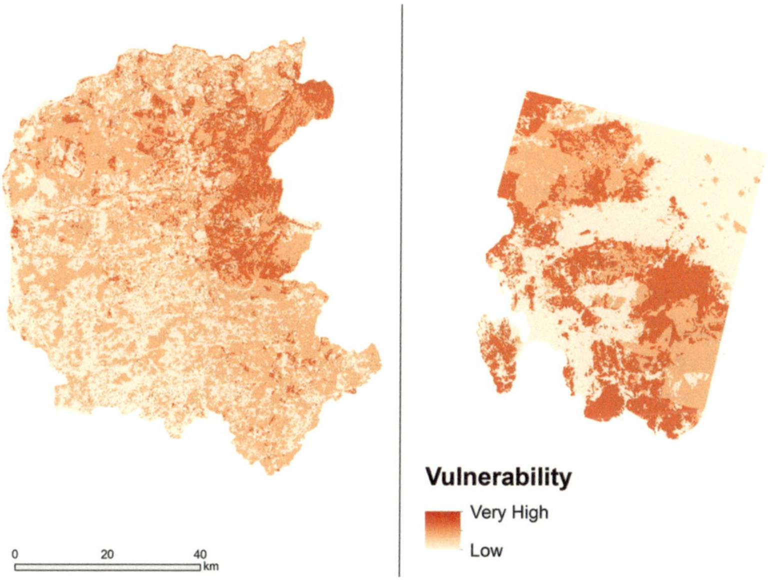

Figure 8. Vulnerability maps for the test sites of Portugal (left) and Italy (right).

facilitated the interpretation of the outputs and fostered their use in established procedures in the different countries. A common approach for vulnerability assessment enables comparable evaluations of exposed assets and coping capacity levels and promotes the implementation of mutual approaches to deal with lack of resilience.

3.2. Recovery/reconstruction service phase

The recovery phase products are strongly interconnected with each other. High spatial resolution (HR) burned perimeters are used to identify the areas where damage severity and post-fire vegetation recovery are evaluated. Post-event products are meant to improve the planning of the post-event intervention (recovery) by providing information on the level of damage caused by fires for vegetation and soil erosion, among other consequences. These products were computed based mainly on Landsat8 images [60, 61], acquired before and after the fire event, although the processing chain has been designed to use Sentinel-2 images when they become systematically available, ensuring the sustainability and further improvement of the cartographic tools at the disposal of multiple users.

3.2.1. Post-fire vegetation recovery map

The methodology for detecting and mapping burnt areas was based on the use of invariant or pseudo-invariant components of an image, called 'PRs' or permanent reflectors, to separate the variations which are due to intrinsic reflectance of the vegetation, from those caused by other external factors, such as sensor type, acquisition parameters or atmospheric conditions [61]. Preliminary detailed analysis of large sets of optical data was carried out to identify the pseudo-invariant targets to be used as PR, for burn scars mapping on the geographic area of interest, considering pre- and post-fire spectral signature changes in diverse land cover classes versus the associated PR reflectance spectra, as a function of time elapsed after fire. This process was necessary to define appropriate thresholding criteria for burn scar detection and mapping from HR optical data. The procedure was defined taking into consideration its possible application to any kind of satellite image having the required spectral channels (RED, near infrared (NIR) and short wave infrared (SWIR)). Hitherto, the mapping of burn scars was carried out on specific corine land cover (CLC) classes, namely permanent crops, agro-forestry areas, forests, scrub and/or herbaceous vegetation and sparsely vegetated areas, allowing to obtain a statistical analysis of fire impacts regarding affected land cover at the end of the fire season. **Table 4** shows the results of the burnt scars (BSs) for several test sites between 2014 and 2015. These areas showed different patterns of affected land cover. In France and Portugal test sites, scrub and/or herbaceous vegetation types were the most affected, whereas in Greece and Italy permanent crops were dominant in the burned perimeters. These differences between test areas can also be associated with firefighting resources efficiency, weather conditions or fire ignition causes, a relation that should be further explored.

3.2.2. Damage severity map

Maps with the level of damage in vegetation due to the fire were systematically computed on each burned area larger than 10 ha. The procedure downloaded automatically the burned areas for the test site (obtained within the PREFER project), retrieved two Landsat8 images, from before and after the fire, and computed three different indices: damage severity index (DSI), burn severity index (BSI) and differenced normalized burn ratio (dNBR) [60, 62, 63]. Validation was carried out with very high resolution pre-event images (RapidEye) and post-event field campaigns. The level of damage, as defined here, takes into account the biomass available in the burned pixel and high damage means that a long period is necessary for regeneration of the vegetation affected by the fire, for example, a wooded area completely burned that would require

Test sites		**FR**	**GR**	**IT**	**PT**
Temp. interval		**September 14, 2014–October 19, 2015**	**October 1, 2014–September 18, 2015**	**September 14, 2014–September 17, 2015**	**September 27, 2014–September 30, 2015**
Total N. of BS		62	92	58	1465
Total BS area (ha)		488	737	366	21,616
N. BS < 3 ha (%)		48	**77**	**71**	49
N. BS > 3 ha (%)		52	23	29	51
CLC of the BS (%)	22	29	**64**	**41**	4
	244	0	0	2	<0.2
	31	26	5	5	17
	32	**42**	29	52	**71**
	333	3	2	0	7

Note: Bold values highlight the highest values for the the percentage of fires smaller than 3 ha (77% in Greece and 71% in Italy) and the prevalent type of vegetation burned in each area.

Table 4. Results of the burnt scars (BS) for several test sites between 2014 and 2015.

several years to recover. On the contrary, the regeneration period of grasslands and shrublands is between 1 and 3 years, thus the damage severity on this type of vegetation is always lower. Based on this evaluation regarding damage level by vegetation type, the three indices gave rather different results [62] and the DSI seemed to provide better estimations, since a lower number of burned pixels in grasslands was considered high or very high damage (**Figure 9**).

In 2015, this procedure was applied to 173 burned perimeters in the test area of Portugal (**Figure 10**), showing higher damage levels in the north. Damage severity maps allow evaluating the

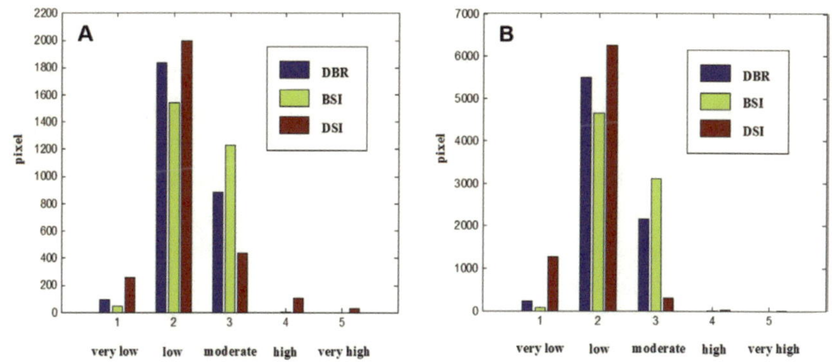

Figure 9. Distribution of the damage severity levels of burned (A) grassland areas and (B) woodland areas in Portugal (>10 ha).

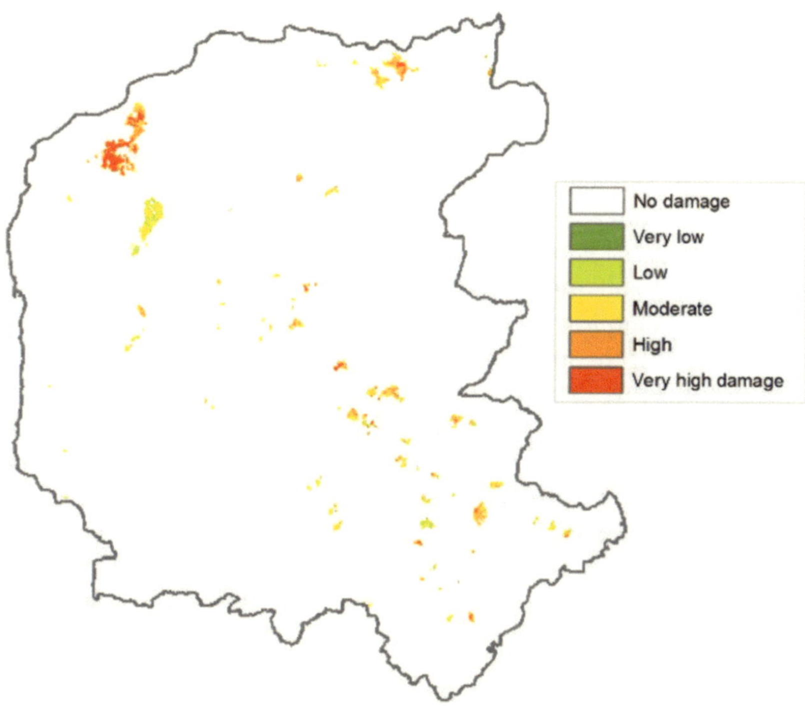

Figure 10. Distribution of the damage severity levels in Minho region, Portugal, in 2015.

incidence of fires at local, regional or national scales and help identifying areas where artificial plantation should be considered for reforestation of burned areas.

4. Concluding remarks

The forest fire context in Mediterranean Europe is characterized by a high incidence of fires causing substantial damages, with strong seasonal variability. The efficient mitigation of fire damages requires additional investment in the early phases of the fire management cycle, such as planning and resource allocation, since further improvements in firefighting and suppression above a certain threshold are associated with unsustainable costs. Despite regional and national differences, the European Mediterranean countries share common issues, for which equivalent appraisals are beneficial in view of facilitating knowledge-transfer and synergy at the transnational level. Fostering significant improvements in effectiveness and timeliness of prevention measures remains the most cost-effective strategy, which can be enhanced by sharing scientific knowledge, operational expertise and technical resources among countries with common challenges and environmental issues.

The PREFER project contributed to this direction, as an international initiative focused on the development of services for supporting fire management tasks, driven by cutting-edge technologies and state-of-the-art scientific results. The project has set up a regional online service, able to process and distribute to end-users spatial information to support forest fire management and intended to stimulate further coordination between countries to promote cooperation and operational deployment.

Relying upon an improvement in quality, quantity, scale and timeliness of mapping, and being complementary to services provided by other institutions, the project searched for efficiency improvements in preventing ignitions and supporting the planning of response and recovery strategies of burned areas. A common framework for product development was applied, ensuring comparability, sharing of best practices and reinforcing cooperation between institutions and countries, without precluding the possibility for adjustments on the procedures, by introducing further regional knowledge and level of detail, depending on the conditions of the area, data availability and user needs. Moreover, the framework was designed to allow for the integration of new or improved satellite images and other data when they become systematically available, ensuring the sustainability and further upgrading of the cartographic tools. The availability of a synoptic view of fire management procedures and the provision of a pan-European portfolio of products, based on similar requirements and usable by multiple users, are important tools to provide quantitative criteria to help prioritising the management of resources. An important aftermath of the project lies on the possibility to apply the knowledge and technical resources developed to other areas in the Mediterranean region, fostering collaboration between institutions and improving the transnational cooperation to increase resilience of the population and the environment to forest fires by providing valuable tools, at the disposal of the authorities in charge of forest fire management in the Mediterranean region.

http://dx.doi.org/10.5772/intechopen.68948

Acknowledgements

This work was developed under the framework of the PREFER project ("Space-based information support for prevention and recovery of forest fire emergency in Mediterranean areas"), financed by the European Community's Seventh Framework Programme ([FP7/2007-2013]) under G.A. nr. 312931.

Author details

Sandra Oliveira[1]*, Giovanni Laneve[2], Lorenzo Fusilli[2], Georgios Eftychidis[3], Adélia Nunes[4], Luciano Lourenço[4] and Ana Sebastián-López[5]

*Address all correspondence to: sisoliveira@gmail.com

1 Institute of Geography and Spatial Planning, University of Lisbon (Universidade de Lisboa), Cidade Universitária, Lisbon, Portugal

2 School of Aerospace Engineering, University of Rome 'La Sapienza', Rome, Italy

3 KEMEA, Center for Security Studies, Athens, Greece

4 Department of Geography and Tourism, CEGOT, University of Coimbra, Coimbra, Portugal

5 GMV Aerospace & Defence SAU, Isaac Newton, Tres Cantos, Madrid, Spain

References

[1] Bowman DMJS, Balch JK, Artaxo P, Bond WJ, Carlson JM, Cochrane MA, et al. Fire in the earth system. Science. 2009;**324**(5926):481-484. Available from: http://www.sciencemag.org/cgi/doi/10.1126/science.1163886

[2] Moreira F, Viedma O, Arianoutsou M, Curt T, Koutsias N, Rigolot E, et al. Landscape–wildfire interactions in southern Europe: Implications for landscape management. Journal of Environmental Management. 2011;**92**(10):23892402. Available from: http://www.sciencedirect.com/science/article/pii/S0301479711002258

[3] Moritz MA, Batllori E, Bradstock RA, Gill AM, Handmer J, Hessburg PF, et al. Learning to coexist with wildfire. Nature. 2014;**515**(7525):58-66. Available from: http://www.nature.com/doifinder/10.1038/nature13946

[4] Ganteaume A, Camia A, Jappiot M, San-Miguel-Ayanz J, Long-Fournel M, Lampin C. A review of the main driving factors of forest fire ignition over Europe. Environmental Management. 2013;**51**(3):651-662. Available from: http://link.springer.com/10.1007/s00267-012-9961-z

[5] San-Miguel-Ayanz J, Moreno JM, Camia A. Analysis of large fires in European Mediterranean landscapes: Lessons learned and perspectives. Forest Ecology and Management. 2013;**294**:11-22. Available from: http://linkinghub.elsevier.com/retrieve/pii/S0378112712006561

[6] Turco M, Bedia J, Di Liberto F, Fiorucci P, von Hardenberg J, Koutsias N, et al. Decreasing fires in Mediterranean Europe. PLoS One. 2016;**11**(3):e0150663. Available from: http://dx.plos.org/10.1371/journal.pone.0150663

[7] Schmuck G, San-Miguel-Ayanz J, Camia A, Durrant T, Boca R, Libertà G, et al. Forest Fires in Europe, Middle East and North Africa 2013 [Internet]. Luxembourg: Publications Office of the European Union, 2014. ISBN 978-92-79-39627-4 (pdf). doi:10.2788/99870 (online)

[8] Jesús San-Miguel-Ayanz, Tracy Durrant, Roberto Boca, Giorgio Libertà, Francesco Boccacci, Margherita Di Leo, Jorge López Pérez, Ernst Schulte; Forest Fires in Europe, Middle East and North Africa 2015; European Union, EUR 28158 EN; doi:10.2788/914

[9] Badia A, Serra P, Modugno S. Identifying dynamics of fire ignition probabilities in two representative Mediterranean wildland-urban interface areas. Applied Geography. 2011; **31**(3):930-940. Available from: http://linkinghub.elsevier.com/retrieve/pii/S0143622811000178

[10] Leone V, Koutsias N, Martínez J, Vega-García C, Allgöwer B, Lovreglio R. The human factor in fire danger assessment. Chuvieco, E. (Ed.) Wildland Fire Danger Estimation and Mapping. The Role of Remote Sensing Data, World Scientific Publishing, Singapore, 2003. pp. 143-196. (Series in Remote Sensing; Volume 4).

[11] Lampin-Maillet C, Long-Fournel M, Ganteaume A, Jappiot M, Ferrier JP. Land cover analysis in wildland–urban interfaces according to wildfire risk: A case study in the South of France. Forest Ecology and Management. 2011;**261**(12):2200-2213. Available from: http://linkinghub.elsevier.com/retrieve/pii/S0378112710006924

[12] Pausas JG. Changes in fire and climate in the eastern Iberian Peninsula (Mediterranean basin). Climatic Change. 2004;**63**(3):337-350

[13] San-Miguel-Ayanz J, Rodrigues M, Oliveira SS, Pacheco CK, Moreira F, Duguy B, et al. Land cover change and fire regime in the European Mediterranean Region. In: Moreira F, Arianoutsou M, Corona P, De las Heras J, editors. Post-Fire Management and Restoration of Southern European Forests. Dordrecht, Netherlands: Springer; 2012. pp. 21-43. Available from: http://www.springerlink.com/index/10.1007/978-94-007-2208-8_2

[14] Oliveira S, Pereira JMC, San-Miguel-Ayanz J, Lourenço L. Exploring the spatial patterns of fire density in Southern Europe using geographically weighted regression. Applied Geography. 2014;**51**:143-157. Available from: http://linkinghub.elsevier.com/retrieve/pii/S0143622814000630

[15] Nunes, A, Lourenço L Castro Meira AC "Exploring spatial patterns and drivers of forest fires in Portugal (1980-2014)." Science of the Total Environment. 573(**2016**):1190-1202.

[16] Modugno S, Balzter H, Cole B, Borrelli P. Mapping regional patterns of large forest fires in wildland-urban interface areas in Europe. Journal of Environmental Management. 2016;**172**:112-126

[17] Bedia J, Herrera S, San Martín D, Koutsias N, Gutiérrez JM. Robust projections of fire weather index in the Mediterranean using statistical downscaling. Climatic Change. 2013;**120**:229-247

[18] Bedia J, Herrera S, Gutiérrez JM. Assessing the predictability of fire occurrence and area burned across phytoclimatic regions in Spain. Natural Hazards and Earth System Sciences. 2014;**14**:53-66

[19] Turco M, Llasat M-C, von Hardenberg J, Provenzale A. Climate change impacts on wildfires in a Mediterranean environment. Climatic Change. 2014;**125**:369-380

[20] Jolly WM, Cochrane M a, Freeborn PH, Holden ZA, Brown TJ, Williamson GJ, et al. Climate-induced variations in global wildfire danger from 1979 to 2013. Nature Communications. 2015;**6**(7537):1-11. Available from: http://www.ncbi.nlm.nih.gov/pubmed/26172867

[21] Batllori E, Parisien MA, Krawchuk MA, Moritz MA. Climate change-induced shifts in fire for Mediterranean ecosystems. Global Ecology and Biogeography. 2013;**22**(10):1118-1129

[22] Bedia J, Herrera S, Gutiérrez JM, Benali A, Brands S, Mota B, et al. Global patterns in the sensitivity of burned area to fire-weather: Implications for climate change. Agricultural and Forest Meteorology. 2015;**214**:369-379

[23] Sousa PM, Trigo RM, Pereira MG, Bedia J, Gutiérrez JM. Different approaches to model future burnt area in the Iberian Peninsula. Agricultural and Forest Meteorology. 2015;**202**:11-25. Available from: http://dx.doi.org/10.1016/j.agrformet.2014.11.018

[24] Spies TA, White EM, Kline JD, Paige Fischer A, Ager A, Bailey J, et al. Examining fire-prone forest landscapes as coupled human and natural systems. Ecological Society. 2014; **19**(3):9

[25] Brotons L, Aquilué N, de Cáceres M, Fortin M-J, Fall A. How fire history, fire suppression practices and climate change affect wildfire regimes in Mediterranean landscapes. PLoS One. 2013;**8**(5):e62392. Available from: http://www.pubmedcentral.nih.gov/articlerender.fcgi?artid=3642200&tool=pmcentrez&rendertype=abstract

[26] Podur J, Wotton M. Will climate change overwhelm fire management capacity? Ecological Modelling. 2010;**221**(9):1301-1309. Available from: http://linkinghub.elsevier.com/retrieve/pii/S0304380010000529

[27] Tedim F, Leone V, Xanthopoulos G. A wildfire risk management concept based on a social-ecological approach in the European Union: Fire smart territory. International Journal of Disaster Risk Reduction. 2016;**18**:138-153. Available from: http://dx.doi.org/10.1016/j.ijdrr.2016.06.005

[28] Silva JS, Rego FC, Fernandes P, Rigolot E. Towards Integrated Fire Management: Outcomes of the European Project Fire Paradox. European Forest Institute; Finland. 2010

[29] Corona P, Ascoli D, Barbati A, Bovio G, Colangelo G, Elia M, et al. Integrated forest management to prevent wildfires under Mediterranean environments. Annals of Silvicultural Research . 2015;**39**(1):1-22. Available from: http://cra-journals.cineca.it/index.php/asr/article/view/946

[30] Fernandes PM, Pacheco AP, Almeida R, Claro J. The role of fire-suppression force in limiting the spread of extremely large forest fires in Portugal. European Journal of Agronomy. 2016;**135**(January):1-10

[31] Communication from the Commission to the European Parliament, the Council, the European Economic and Social Committee and the Committee of the Regions - A Community approach on the prevention of natural and man made disasters. 2009

[32] Jesús San-Miguel-Ayanz, Ernst Schulte, Guido Schmuck, Andrea Camia, Peter Strobl, Giorgio Liberta, Cristiano Giovando, Roberto Boca, Fernando Sedano, Pieter Kempeneers, Daniel McInerney, Ceri Withmore, Sandra Santos de Oliveira, Marcos Rodrigues, Tracy Durrant, Paolo Corti, Friderike Oehler, Lara Vilar and Giuseppe Amatulli (2012). Comprehensive Monitoring of Wildfires in Europe: The European Forest Fire Information System (EFFIS), Approaches to Managing Disaster - Assessing Hazards, Emergencies and Disaster Impacts, Prof. John Tiefenbacher (Ed.), InTech, DOI: 10.5772/28441. Available from: https://www.intechopen.com/books/approaches-to-managing-disaster-assessing-hazards-emergencies-and-disaster-impacts/comprehensive-monitoring-of-wildfires-in-europe-the-european-forest-fire-information-system-effis-

[33] San-Miguel-Ayanz J, Schulte E, Schmuck G, Camia A. The European forest fire information system in the context of environmental policies of the European Union. Forest Policy and Economics. 2013;**29**:19-25. Available from: http://linkinghub.elsevier.com/retrieve/pii/S138993411200127X

[34] Birkmann J, Cardona OD, Carreño ML, Barbat AH, Pelling M, Schneiderbauer S, et al. Framing vulnerability, risk and societal responses: The MOVE framework. Natural Hazards. 2013;**67**(2):193-211

[35] Laneve G, Bonis R De, Fusilli L, Ferrucci F, Sebástian-Lopéz A, Oliveira S, et al. PREFER FP7 project for the management of the pre- and post-fire phases: Presentation of the products. In: Advances in Forest Fire Research. 2014. pp. 903-911

[36] Vélez R. La defensa contra incendios forestales: fundamentos y experiencias. 2000

[37] Rego FC, Rigolot E, Fernandes P, Montiel C, Silva JS. Towards Integrated Fire Management. 2010

[38] Leventakis G. PREFER Legal framework document, Deliverable D21-2 [Internet]. 2013. Available from: http://www.prefer-copernicus.eu/index.php/downloads/view download/5-analysis-of-legal-framework/23-prefer-legal-framework-document

[39] European Space Agency (ESA). Sentinels: Space for Copernicus [Internet]. ESA BR-319. 2015. Available from: http://esamultimedia.esa.int/multimedia/publications/BR-319 (Accessed: 5 March 2017)

[40] Tedim F, Xanthopoulos G, Leone V. Forest fires in Europe: Facts and challenges. In: Schroder JF, Paton D, editors. Wildfire Hazards, Risks, and Disasters. Elsevier Inc. USA; 2015. pp. 77-99

[41] Leone V, Lovreglio R, Martín MP, Martínez J, Vilar L. Human factors of fire occurrence in the mediterranean. In: Chuvieco E, editor. Earth Observation of Wildland Fires in Mediterranean Ecosystems. Berlin, Heidelberg: Springer; 2009. pp. 149-170. Available from: http://www.springerlink.com/index/10.1007/978-3-642-01754-4_11

[42] Chas-Amil ML, Prestemon JP, McClean CJ, Touza J. Human-ignited wildfire patterns and responses to policy shifts. Applied Geography. 2015;**56**:164-176. Available from: http://dx.doi.org/10.1016/j.apgeog.2014.11.025

[43] Oliveira S, Oehler F, San-Miguel-Ayanz J, Camia A, Pereira JMC. Modeling spatial patterns of fire occurrence in Mediterranean Europe using multiple regression and random forest. Forest Ecology and Management. 2012;**275**:117-129. Available from: http://linkinghub.elsevier.com/retrieve/pii/S0378112712001272

[44] Merrill DF, Alexander ME. Glossary of Forest Fire Management Terms. 1987

[45] Sebastián-Lopéz A, San-Miguel Ayanz J, Burgan RE. Integration of satellite sensor data, fuel type maps and meteorological observations for evaluation of forest fire risk at the pan-European scale. International Journal of Remote Sensing. 2002;**23**(13):2713-2719

[46] Gutman G, Ignatov A. The derivation of the green vegetation fraction from NOAA/AVHRR data for use in numerical weather prediction models. International Journal of Remote Sensing. 1998;**19**(8):1533-1543

[47] Jiapaer G, Chen X, Bao A. A comparison of methods for estimating fractional vegetation cover in arid regions. Agricultural and Forest Meteorology. 2011;**151**(12):1698-1710. Available from: http://www.sciencedirect.com/science/article/pii/S0168192311002176

[48] Yue Y, Wang K, Zhang B, Jiao Q, Liu B, Zhang M. Remote sensing of fractional cover of vegetation and exposed bedrock for karst rocky desertification assessment. Procedia Environmental Sciences. 2012;**13**(2011):847-853. Available from: http://www.sciencedirect.com/science/article/pii/S1878029612000795

[49] Salis M, Laconi M, Ager AA, Alcasena FJ, Arca B, Lozano O, et al. Evaluating alternative fuel treatment strategies to reduce wildfire losses in a Mediterranean area. Forest Ecology and Management. 2016;**368**:207-221

[50] Oliveira TM, Barros AMG, Ager AA, Fernandes PM. Assessing the effect of a fuel break network to reduce burnt area and wildfire risk transmission. International Association of Wildland Fire. 2016;**25**(6):619-632

[51] Fernandes PM, Loureiro C, Guiomar N, Pezzatti GB, Manso FT, Lopes L. The dynamics and drivers of fuel and fire in the Portuguese public forest. Journal of Environmental Management. 2014;**146**:373-382. Available from: http://dx.doi.org/10.1016/j.jenvman.2014.07.049

[52] Kalies EL, Yocom Kent LL. Tamm review: Are fuel treatments effective at achieving ecological and social objectives? A systematic review. Forest Ecology and Management. 2016;**375**:84-95. Available from: http://dx.doi.org/10.1016/j.foreco.2016.05.021

[53] Moreira F, Vaz P, Catry FX, Silva JS. Regional variations in wildfire susceptibility of land-cover types in Portugal: Implications for landscape management to minimize fire hazard. International Journal of Wildland Fire. 2009;**18**(5):563-574. Available from: http://www.publish.csiro.au/paper/WF07098

[54] Koutsias N, Allgöwer B, Kalabokidis K, Mallinis G, Balatsos P, Goldammer JG. Fire occurrence zoning from local to global scale in the European Mediterranean basin: Implications for multi-scale fire management and policy. iForest - Biogeosciences for Forestry. 2015;**9**(12):195-204. Available from: http://www.sisef.it/iforest/contents/?id=ifor1513-008%5Cnhttp://www.sisef.it/iforest/?doi=ifor1513-008

[55] Verde JC, Zêzere JL. Assessment and validation of wildfire susceptibility and hazard in Portugal. Natural Hazards and Earth System Sciences. 2010;**10**(3):485-497

[56] Rodrigues M, San Miguel J, Oliveira S, Moreira F, Camia A. An insight into spatial-temporal trends of fire ignitions and burned areas in the European Mediterranean countries. Journal of Earth Sciences and Engineering. 2013;**3**:497-505

[57] Cutter SL, Barnes L, Berry M, Burton C, Evans E, Tate E, et al. A place-based model for understanding community resilience to natural disasters. Global Environmental Change. 2008;**18**(4):598-606

[58] Cutter SL. A ciência da vulnerabilidade: Modelos, métodos e indicadores. Rev Crit Cienc Sociais. 2011;**93**:59-69

[59] UNISDR. Terminology on Disaster Risk Reduction [Internet]. United Nations International Strategy for Disaster Reduction (UNISDR). 2009. Available from: www.unisdr.org/publications

[60] Roy DP, Boschetti L, Trigg SN. Remote sensing of fire severity: Assessing the performance of the normalized burn ratio. IEEE Geoscience and Remote Sensing Letters. 2006;**3**(1):112-116

[61] Hirn BR, Ferrucci F. MYME2: A multi-payload integrated procedure for the automated, high-resolution remote sensing of burn scars. International Geoscience and Remote Sensing Symposium. 2005;**8**(1):5721-5724

[62] Laneve G, de Bonis R, Fusilli L. Development of a vegetation damage severity index based on hyperspectral sensor data. In: 33rd EARSel Symposium; 3-6 June; Matera, Italy. 2013

[63] Laneve G, Fusilli L, Marzialetti P, De Bonis R, Bernini G, Tampellini L. Development and validation of fire damage-severity indices in the framework of the PREFER project. IEEE Journal of Selected Topics in Applied Earth Observations and Remote Sensing. 2016;**9**(6):2806-2817

Section 4

Cultural Ecology and Mediterranean Lifestyle - Within and Beyond

Chapter 15

Climate and Urban Morphology in the City of Barcelona: The Role of Vegetation

Gilkauris Rojas-Cortorreal, Francesc Navés Viñas,
Julio Peña, Jaime Roset and Carlos López-Ordóñez

Additional information is available at the end of the chapter

http://dx.doi.org/10.5772/intechopen.69125

Abstract

Urban morphology defines many characteristics of a city, such as its urban quality, density, and free spaces layout with its vegetation, microclimate, and thermal comfort. All these variables have in common the element of urban space, which is where users develop themselves, making it an element of value. As the city evolves, environmental conditions of the area also change. One example is the urban heat island (UHI) effect. The UHI is defined as the accumulation of heat produced by the urban surfaces emitted to the environment, affecting the urban structure and the thermal comfort of users. As discomfort increases, urban public spaces lose their appeal and may be turned into deserted areas. One of the reasons is the lack of vegetation, particularly arboreal species, which protect these spaces from the solar incidence. In this regard, the questions arise as to how urban morphology affects the incidence of solar radiation in the urban canyon, and how vegetation and its characteristics influence climatic comfort. To develop these concepts, three introductory sections are addressed: the city of Barcelona, Barcelona climate, and urban morphology. Where the main variables are solar radiation, urban canyon, orientation, and the role that vegetation plays as a protection.

Keywords: Mediterranean urbanism, urban heat island, vegetation, outdoor thermal comfort, radiation

1. Introduction

Urban morphology defines many characteristics of a city, such as its viability, microclimate, vegetation and thermal comfort. All these variables have in common the element of urban space, which is the space where users develop themselves, making it an element of value because it is the place where their travel and moments of leisure occur. As the city evolves, the characteristics change, leading to the birth of new variables, including the urban heat island (UHI). The UHI is defined as the accumulation of heat produced by urban surfaces emitting it to the environment, affecting the urban structure and the thermal comfort of users. This situation directly affects thermal comfort and recreation spaces.

Thermal comfort can be defined as the state in which a thermal balance exists. Thus, the definition of discomfort is the opposite state to this. Gradually, this urban thermal discomfort increases, turning urban areas that were previously widely used into deserted areas. One of the reasons is the lack of vegetation, particularly arboreal species, which protects these spaces from the solar incidence. For this purpose, one of the major interventions being implemented at the international level is the improvement of urban spaces, and the key element for this improvement is vegetation, since it provides great contributions in different climatic and urban aspects.

Questions arise as to how urban morphology affects the incidence of solar radiation in the urban canyon and how vegetation and its characteristics influence the microclimatic thermal comfort, with woodland being the main element of study, since it is one of the tools of greater dimension and use at the urban level. Previous studies [1–8] have validated that it is one of the greatest mitigating tools. The radiation, orientation and incidence in the urban canyon influence to a great extent how this urban space could behave that is why they are important variables of study [9–15].

To develop these concepts, four introductory sections will be developed: first, the Mediterranean, where concepts and data about the Mediterranean region and its cities will be developed; second, Barcelona city, which is the city of study; third, the climate of Barcelona and fourth, the urban morphology where the radiation and vegetation are dissected. The main variables will be solar radiation, urban canyon, orientation and the role played by vegetation as a protection and design tool, leading to relevant conclusions on the subject.

2. Methodology

The research methodology was based on analysis of the state of the technique of the urban morphology of Barcelona, which allowed evaluation of the main indicators. The method developed for this study combines energy simulations and *in situ* measurements of environmental conditions to evaluate the role of vegetation. Heliodon is a computer tool to evaluate and study the solar radiation over 3D urban models. It has been used to assess the impact of the urban canyon on the incidence of solar radiation in the streets of the city. Additionally, radiation and surface temperature measurements have been carried out. Through these simulations and measurements, calculations of the thermal comfort of these points of analysis were obtained to determine the quality of the space.

3. Mediterranean

A remarkable number of important cities converge in the Mediterranean region, such as Malaga, Valencia and Barcelona in Spain; Marseilles and Nice in France; Genoa, Rome, Naples and Bari in Italy; and Algiers, Oran and Tunisia in the south of the Mediterranean. Each of these cities has a different location in relation to the climate, soil and vegetation, as well as a different history and urban development. All these are relevant variables for the evaluation of a city. The climate and urban morphology will be the indicators developed in the climate and urban morphology.

Climate can be defined as the statistical description of the characteristics of the state of time over a period of a few months to millions of years [16]. The Mediterranean climate is characterised by temperate and rainy winters, autumns with torrential rains, rainy and temperate springs, and very hot and dry summers. Mean maximum temperature variations in different cities have been recorded during the period 1971–2000. The mean maximum temperature ranges reached 35°C in the city of Seville [16].

These climatic conditions depend on the location of the city and its urban conditions. For example, there are cities like Barcelona and Tarragona with a mean maximum temperature of up to 28°C in summer and a mean minimum temperature as low as 4°C in winter. Cities like Madrid record mean maximum temperatures of up to 31°C in summer and mean minimum temperatures as low as 2.6°C in winter. Cities such as Rome and Naples record mean maximum temperatures of up to 30°C in summer and mean minimum temperatures as low as 4°C in winter [16, 17]. However, cities such as Seville register mean maximum temperatures of up to 35°C in summer, being the highest in comparison to the previously mentioned cities, and mean minimum temperatures as low as 5°C in winter [16].

3.1. Vegetation

Vegetation is a living element that varies according to the conditions of a given environment. Studies have been conducted on how the minimum temperatures affect the frost resistance of a species [18, 19]. Species were classified in 10 zones of application of the vegetation according to resistance to these frosts (**Table 1**).

Classifications were performed to determine how the climatic zone influences the resistance of tree species. In this study, the areas of Valencia, Malaga, Alicante, Bari and Naples are classified as zone 10; the cities of Barcelona, Nice, Genoa and Rome belong to area 9. More inland cities like Madrid or Florence would be area 8.

It can be seen that these cities, due to their climatic conditions, have species resistant to medium frosts or, in the case of cities like Barcelona, not resistant to frost, indicating that climate is a factor that influences the contributions of a species in the urban environment. For this reason, when selecting a species for urban use, the climatic conditions and resistance of the tree species to this environment should be considered.

Zone	Resistance	
1–7	Hf	Strong frost resistance
8	Hm	Medium frost resistance
9	Hd	Resistance to weak frost
10	H	Not frost resistant
T	E	Not resistant to cooling

Table 1. Classifications of tree species according to frost resistance.

3.2. Radiation

Solar radiation is an essential element for the development of any kind of life on the planet [9–15]. It is both a factor and a climatic macro feature in which the direction of the incidence of radiation depends on the relative movements of the Earth and the Sun [20].

The type of radiation received from the Sun is distributed over a wide area of the electromagnetic spectrum. It is preferably in three types: short-wave (ultraviolet) radiation, visible radiations and long-wave (infrared) radiation. This set of radiation affects the outer limits of the Earth's atmosphere with an approximate irradiation value of 1400 W/m^2, which is called the solar constant [9–15, 20].

The cities mentioned above have different levels of incidence as far as solar radiation is concerned. The city of Seville, in the Autonomous Community of Andalusia in Spain, has the highest levels of average global irradiance, reaching 8.10 kWh/m^2 a day during July, although cities such as Malaga, Valencia and Barcelona have values very close to Seville, being maintained during the summer months in limits of 6.12–8.10 kWh/m^2 a day [21]. Cities like Genoa, Rome, Bari and Naples, to mention other cities of the Mediterranean, display lower values (although not low) of global irradiance compared to those reached in the cities of Spain, staying in a range from 5.22 to 6.86 kWh/m^2 a day during the same summer months (June, July and August) [22].

The study of the incidence of solar radiation in cities is of great importance in observing these values, since it is one of the parameters with the greatest impact on urban thermal comfort. The city of Barcelona presents an urban morphology in which the Eixample, a product of the Cerdà Plan [20], and its location and climatological conditions, is a case study of great interest in the subject to be developed in this chapter.

4. Barcelona: city

Barcelona is a large co-capital city of Spain like Madrid, belonging to the autonomous community of Catalonia. It is located on the shores of the Mediterranean Sea with the mountain chain of the Pyrenees and it is 160 km from France. The city is bounded on the north by the

http://dx.doi.org/10.5772/intechopen.69125

river Besos, to the south by the river Llobregat, to the east by the Mediterranean Sea and to the west by the Collserola mountain range.

Barcelona is a city with unique characteristics. This is due to its history and its transformation at both urban and economic levels, making it a city that values users, urban space and vegetation, resulting in an urban model that has come to be called the Barcelona model of urbanism. The climate, natural elements, orography, Collserola, rivers Besos and Llobregat, along with historical elements such as the walls and fortresses of Montjuic and Ciutadella largely determined the morphology of the city until the nineteenth century and the beginning of the twentieth century, when the old municipalities of Gracia, Sants and Sarria, among others, were added, and the expansion of Cerda was chosen as a new hygienic-rational means of growth and connection with these municipalities.

At the same time, the present form of the city with its natural elements, buildings, parks, gardens, squares and streets notably influences its own microclimate. In the last 40 years, the city has invested in elements of infrastructure, reorganisation of streets leading to an increase of vegetation, remodelling of sidewalks and urban spaces and creation of spaces open to users like rivers and beaches, among other endeavours. This has generated a modernised city where the main elements are the user and his or her needs. Such modernisation and investment of resources to improve urban elements have given the city an added value in large part by the urban landscape perceived by the user, since the equipment and conditioning of the urban spaces create a state of comfort linked to well-being and quality of life.

5. Barcelona: Mediterranean climate

Barcelona, with a latitude of 41° North, has a Mediterranean climate. It is characterised by seasons marked with air temperature with an annual average of 16° and global radiation in the horizontal plane of 174 W/m^2. The typical Mediterranean coastal climate consists of five types [18, 19]. The climate of the city of Barcelona is temperate, in which the average of the coldest month is less than 18°C and in summers 19°C, with defined dry seasons and humid periods during the year. Barcelona is part of the Mediterranean climate, registering an average temperature of 16.5°C and precipitation of 599 mm in October. This climate can be defined as dry and warm in summer and cool and humid in winter. It has moderate oscillation, thus creating sclerophyllous forests [18, 19]. These characteristics of marked stations become a challenge at both the level of design of buildings and the urban level, posing challenges to the designer to generate elements and tools necessary to create comfortable interior and exterior spaces.

6. Barcelona: urban morphology

Barcelona is a metropolis that comprises several distinctive urban tissues. The most recognisable is the Eixample, not only because of its urban design but also because of its remarkable extension. This urban area is linked to urban spaces such as streets, squares and walks that make it a city that invites its users to coexist in it, giving life and value to these spaces.

To achieve this well-being, there are climatic variables involved in its design, including air temperature, relative humidity, solar radiation and wind speed and direction, which are linked to key elements that are part of the environment: vegetation and building. For this reason, we will discuss the main indicators that influence this urban morphology and the thermal comfort of the solar radiation and vegetation.

6.1. Solar radiation and urban morphology

Solar radiation is a key variable that modifies the thermal contributions to the environment. Thus, its analysis and behaviour play an important role. Knowing its behaviour, we can determine what improvements could be made to protect users living in this urban environment.

The indicators involved this section are latitude, street orientation, urban canyon morphology (street width and height of the building), sky view factor (SVF), and the percentage of radiation that affect. We will evaluate each characteristic and how it influences the urban morphology of Barcelona. By evaluating these indicators, we can obtain solutions to improve these contributions that the environment receives through solar radiation, with vegetation as a key element.

As previously mentioned, latitude is a factor that directly influences the amount of radiation received, and this will be greater as we approach the equator. This is due not only to the inclination with which the Sun's rays fall on the surface of the Earth but also to the amount of atmosphere they must pass through [11–13]. In order to determine the direction and inclination of the radiation at each moment, we must know the relative position of the Sun and of the plane in question. Abacuses and solar charts graphically indicate the solar height (**Figure 1**).

Another important factor to consider in the study of the incidence of solar radiation is the orientation, which directly affects the access of solar rays in both the horizontal plane and the vertical plane [11]. Arnfield [23] developed different models of urban canyons, taking into account the aspect ratio (H/W) of Oke (street design and urban canopy Layer climate [24]) with different orientations. He concluded that the orientation of a street affects to a greater extent the amount of solar energy that affects the vertical planes, whereas the relation of aspect ratio (H/W) (**Figure 2**) directly influences the amount of radiation incident on the horizontal plane [11–13, 15].

The aspect ratio of an urban canyon is the geometric proportion between the height of the buildings adjacent to the street (vertical planes) and the width of the road (horizontal plane). This proportion is directly related to the SVF (**Figure 3**).

The SVF is the percentage of sky seen from a point located on any surface. This depends on the inclination and orientation of the surface. An SVF = 1 means a visible sky in its entirety, unobstructed. An SVF = 0 denotes a sky completely obstructed by obstacles (**Figure 4**) [11].

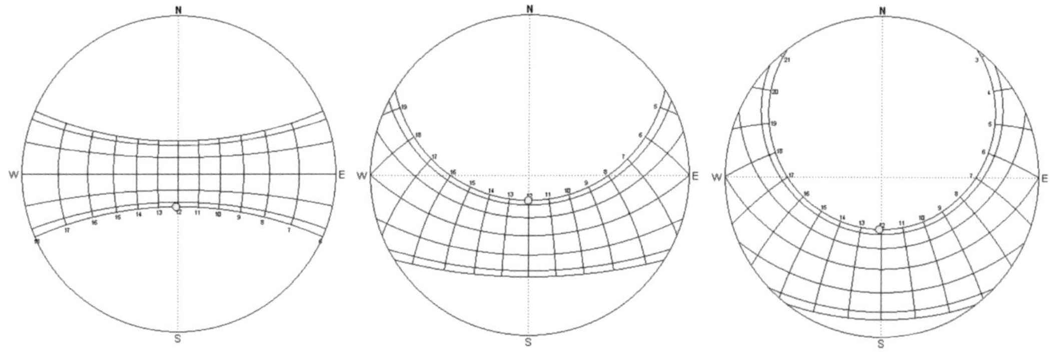

Figure 1. Stereographic solar charts, where one can observe the different solar trajectories throughout the day and the year. (Left and right): latitude 0°, latitude 41° N (Barcelona), latitude 51° N. Elaborated by the authors.

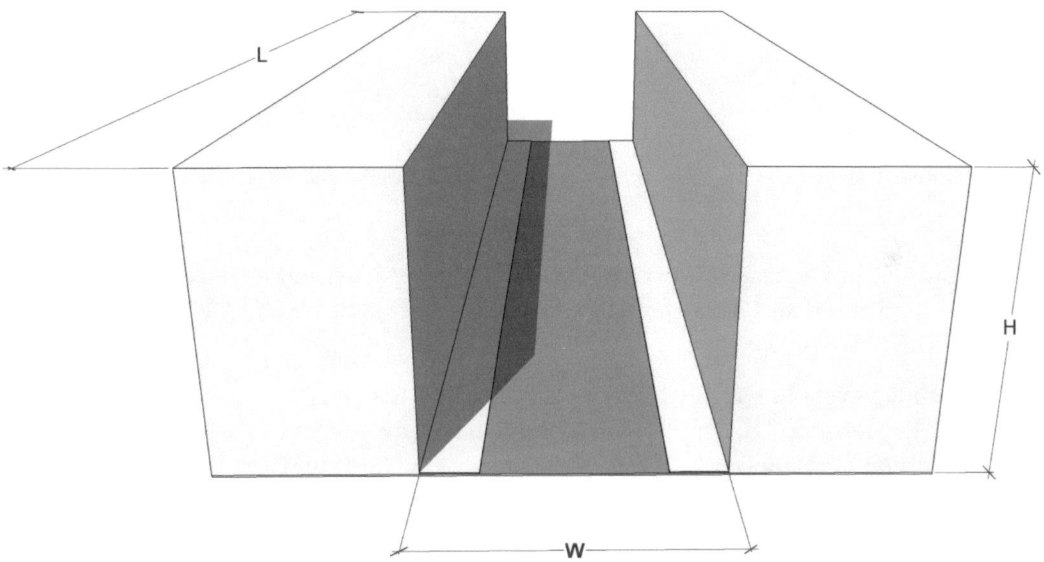

Figure 2. Urban canyons model, H/W aspect ratio. Elaborated by the authors.

The main climatic parameters that are affected by these two variants are the incidence of the solar radiation and the wind; however, as a consequence, the temperature and relative humidity are also affected [2].

The city of Barcelona, having a well-defined morphology (Eixample), exemplifies the energy performance of the urban canyon in terms of the incidence of radiation. Since it can be corrected through the proper use of vegetation, Avenida Diagonal, which crosses the Eixample district and is one of the main communication routes of the city, is taken as

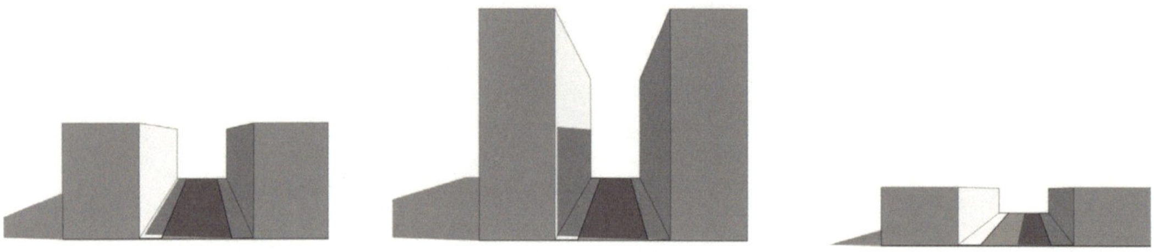

Figure 3. Schemes of different aspect ratios (H/W) of urban canyons (left and right): H/W = 1, H/W = 2 and H/W = 0.5. Elaborated by the authors.

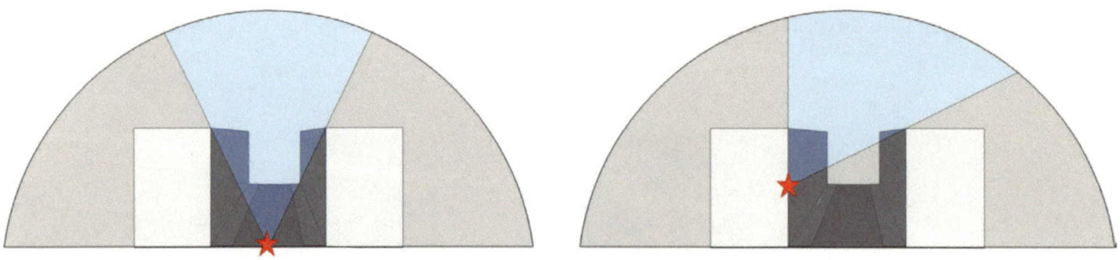

Figure 4. Schemes that show how the geometry and the plane of incidence affect the sky view factor. Elaborated by the authors.

a case study. It has a steady flow of pedestrians and vehicles and a continuous section of 50 m wide and about 10.5 km in length, with an orientation of WEST SOUTH WEST–EAST NORTH EAST (WSW - ENE).

For this study, Avenida Diagonal will be divided into two sections: the left section of the Ensanche (Esquerra de l'Eixample) would run from Plaza Macia to Passeig de Gracia and the right section of the Ensanche (Dreta de l'Eixample) would run from Passeig de Gracia to Passeig de Sant Joan. These sections differ mainly by the tree species they present and the height of their adjacent buildings.

The aspect ratio presented in the left section is 0.48 (24/50 m), while the right section is 0.42 (21/50 m). They are very similar, so urban vegetation will play a very important role in the amount of radiation incident on the horizontal plane, which presents notorious differences between the two cases. The left section presents species like the plane tree, canary palm tree and Mexican fan palm; the right section of the avenue presents trees like the hackberry tree, plane tree, holm oak and Judas tree (i.e. trees with denser foliage) (**Figure 5**).

The simulations and results of this case study were obtained using the software Heliodon2 [25] and were prepared for June 21 (i.e. at the beginning of summer), when temperatures are quite high. The right section has an SVF of 69.3% (placing the point at the centre of the street) and an aspect ratio of 0.42; the left section has an SVF of 62.1% and an aspect ratio of 0.48 [15].

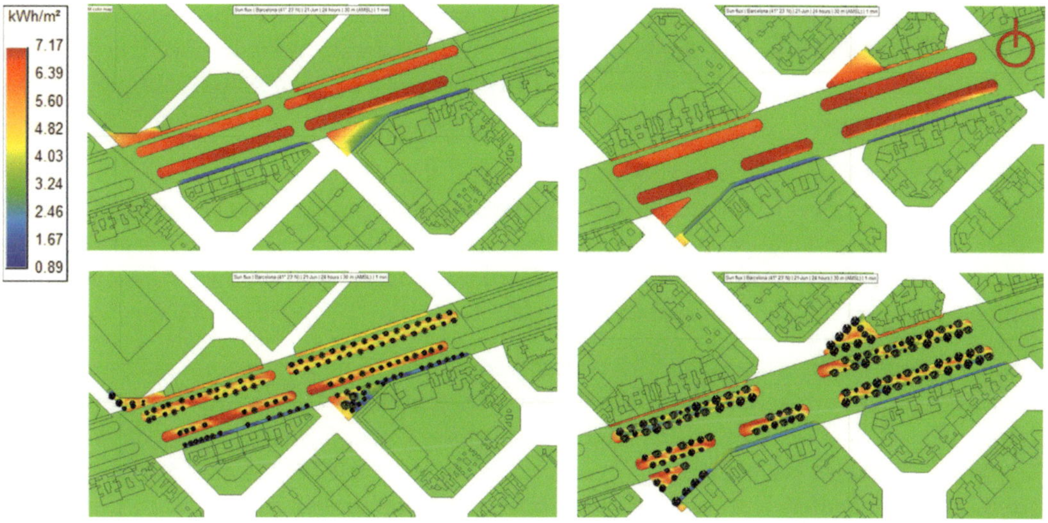

Figure 5. The images show the simulations of incident radiation in the left and right sections. Above: No vegetation is considered only the morphology. Below: Considering the existing vegetation. Authors´ elaboration, software Heliodon2.

Simulations show that the presence or absence of urban vegetation in the street can make a significant difference (**Figure 5**). In the case of the right section, the percentage of reduction in the incidence of solar radiation ranges from 11 to 51%; in the left section, the percentages are lower, falling within a range of reduction between 9 and 38%. (This difference of reduction percentages between the two sections of Avenida Diagonal is due to the different tree species with which it is counted. Although the left section receives a smaller amount of solar radiation due to its greater aspect ratio and the right section has better energy performance thanks to the correct use of vegetation.) [15].

6.2. Vegetation and urban morphology

The vegetation is a living element that plays a role in the beautification and protection of the urban canyon. Three levels of green are defined: creepy plants, shrubs and woodland. These three elements are the indicators that offer improvement to the urban environment at the microclimatic and psychological levels.

The level of ground cover plants is the surface of the ground covered by grass or turf. The role of these creepy plants is embellishment in the horizontal plane of green areas. At the level of comfort, these green horizontal zones absorb the solar radiation obtaining less contribution radiant temperature to the environment, contrary to the behaviour of the urban surfaces.

The shrub is classified as medium level, depending on the type of pruning can take different forms. The role of shrubs at the level of landscaping can be trajectory, barrier protection, among others. At the level of thermal comfort, shrubs create projected shade on the surfaces surrounding them, reducing the solar radiation reaching these urban surfaces.

The level of woodland is the species with the largest dimension and the most used at the urban level. Trees are used for the beautification of urban areas. At the level of thermal comfort is the protection of facade and pavement surfaces from solar radiation, diminishing the horizontal and vertical thermal contributions to the environment.

In this section, we will evaluate the contributions of these levels of vegetation to the urban space. Studies have been conducted on the behaviour and contributions of this vegetation at the urban level in the city of Barcelona on urban roads such as Passeig de Sant Joan, Avenida Diagonal, Carrer Bailen and Carrer Londres. The data obtained showed the levels of contributions provided by the vegetation in these spaces.

Passeig de Sant Joan has been a well-known and popular walk in the city for many years. It has been the centre of urban use for the inhabitants and tourists of the city due to its direct connection to historical elements of Barcelona, such as the Arc de Triomphe, Parc de la Citadel, Sagrada Familia and the beginning of connection between the Turons and the mountain of Collserola. It also offers a great variety of urban recreation spaces. This walk consists of the three levels of pasture, shrub and woodland in linear and perimeter layout. On the way to Passeig de Sant Joan is a variety of species, including the hackberry, China tree, Judas tree, plane tree, lime of large leaves and Tipuana.

The methodology used involved *in situ* measurements of the environmental parameters, High DynamicRange (HDR) photographs and field evaluation in the summer season at three specific times during the day. These measurements were made in the summer in July 2013. The results showed that people often live longer in recreational spaces that are protected by the projected shade of trees and shrubs.

The measurements showed a decrease of up to 23.6°C of the contributions of radiant temperature to the environment, thanks to the shade projected by the shrub and tree species. On the other hand, the areas covered by grasses dissipated less radiation to the environment due to the natural process of photosynthesis. This provided more comfortable urban spaces for visiting and recreation due to the decrease in radiant temperature.

Studies implemented in Carrer Casanova and Carrer Londres, urban roads in the expansion of Barcelona with different orientations SOUTH-EAST (SE) and NORTH-EAST (NE) but with similarities in the typology of urban canyon and vegetation, have evaluated the behaviour of the solar radiation and the contributions of the radiation in this type of urban canyon.

The methodology implemented in this study involved *in situ* measurements of environmental parameters, simulations of the behaviour of solar radiation in the Heliodon programme, evaluation of the SVF of the streets and photographic survey of the tree species. The tree species studied in this case were *Platanus x hispanic*a and *Celtis australis*. These measurements were made in summer at three specific times of day in the year 2015.

The results showed that Carrer Casanova has an SVF of 37.4% and Carrer Londres of 39.5%, determining that they are roads with a similar SVF. The hour of greatest solar incidence is when you can appreciate the contributions of the tree species to the environment. Similarly, due to the magnitude of the urban canyon, there was the effect where the shadow of the buildings managed to reduce the solar incidence on the road.

For the species *Celtis australis* at 13:00 hours, Carrer Casanova managed to decrease 852 W/m^2, which is reflected in a decrease of 27.1°C of radiant temperature to the environment. However, in Carrer Londres it managed to decrease 907 W/m^2, which is reflected in a decrease in the radiant temperature of 11°C. For the species *Platanus x hispanica* at 13:00 hours, Carrer Casanova managed to decrease 852 W/m^2, which is reflected in a 19°C decrease in radiant temperature to the environment. However, in Carrer Londres it managed to decrease 853 W/m^2, which is reflected in a decrease in the radiant temperature of 23°C.

The results show that both species provide similar contributions but at different levels, with *Celtis australis* contributing the most. This is due to the density of leaves, with *Platanus x hispanica* having medium density and *Celtis australis* high density. In the same way, it can be seen how the orientation influences the level of incident solar radiation. This proves that the density of the foliage of a species as well as the orientation of the urban canyon greatly influence the behaviour of the contributions to the environment.

Allowing to check that knowing and valuing urban spaces and how this influences the variable of a person's energy balance that is the balance between the flow of heat input that brings the environment to the body and the output of it.

Previous studies [1–3] have created a formula to determine the role of vegetation in this parameter. The formula consists of several variables that intervene in the immediate environment to determine this exchange and the contribution of the tree species. The calculation is

$$B = M + Ra \pm Conv - E - Re \tag{1}$$

B = energy balance, M = metabolism, Ra = radiation absorbed, $Conv$ = convection, E = evaporation and Re = radiation of long wave emitted by the body.

Studies have developed a methodology to determine the foliage of an arboreal species and included in this calculation, allowing to know at a higher average the contributions that each species provides. The methodology implemented in this study involved *in situ* measurements of the environmental parameters in two cities of different climates, selection of different tree species for the study and realisation of infrared photographs and photographic shoots for evaluation of the foliage. This methodology has been applied to two types of climates in the Mediterranean (Barcelona, Spain) and warm, humid weather (Santo Domingo and Dominican Republic).

A number of arboreal species were evaluated, determining the foliage of the tree species of urban use. In order to determine the lushness, the measurements were made *in situ* in summer and winter in deciduous and perennial species to learn the contributions that the species provide in their lack of foliage. After the data were obtained, calculations were made of the energy balance of a person exposed under the foliage of each tree species.

Technical data were obtained to determine the contributions that this species provided in the improvement of the thermal comfort in these urban spaces. After obtaining these results, we evaluated the behaviour of these urban spaces with the lack of these species. The results reflect the importance of the use of vegetation in urban spaces, since if these species did not have the state of thermal comfort of a person it would be stressful [26–32].

A difference of protection of the tree species could be observed in this study. The results on a typical summer day at 11:00 a.m. solar time differ by 7°C in the case of a light species and 8°C in the case of a medium and dense species (**Figure 6**). These values are reflected in the users in a state of heat in the case of light species and a state of comfort in the medium and dense species. In lack of these species although its foliage is different it is reflected in the users in a hot state (**Figures 6** and **7**).

This shows that no matter the type of foliage of the species at the urban level, their contributions will always be positive. That depending the climate the foliage in the urban design is important, since many factors can vary from the thermal comfort of the users. The same analyses were carried out in several tree species of Barcelona and Santo Domingo to determine in quantitative and qualitative ways, the contributions they can offer to the immediate environment and its users. Technical data sheets were compiled to gather the relevant information for urban design, architects or landscapers, providing a tool for the correct choice of species. The obtained results show how the variables of urban canyon and the typology of arboreal species include in the environmental conditions and in the users, turning them into variables of great relief in the design.

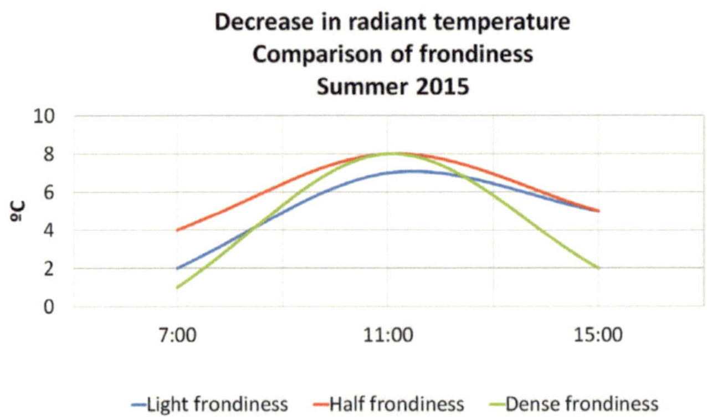

Figure 6. Comparison of frondiness of tree behaviour in radiant temperature.

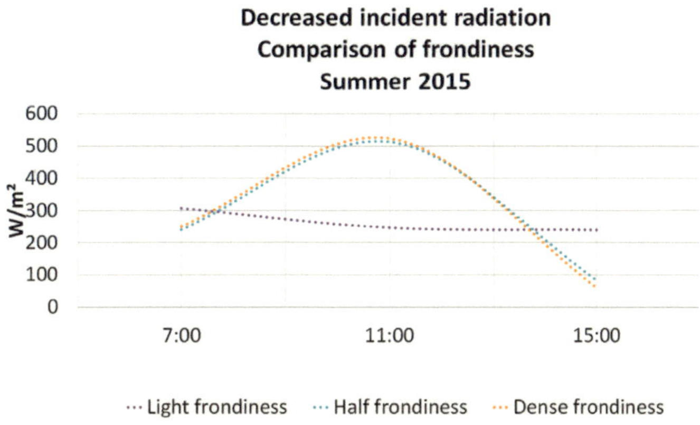

Figure 7. Comparison of frondiness of tree behaviour in incident radiation.

7. Conclusion

In this study, evaluations of incident solar radiation in Barcelona streets have shown which façades and time of year are more favourable or unfavourable. It is clear that the aspect with the greatest impact on the incidence or lack of incidence of solar radiation over the horizontal surface of the urban canyon is its morphology. As can be seen in Barcelona's Avenida Diagonal, this situation can be corrected by the right use of urban vegetation.

The typologies of vegetation of two cities were evaluated to learn their contributions in two different climates: a Mediterranean climate and a warm and humid climate. Vegetation is one of the best tools to improve the climatic conditions of urban spaces. Quantitative and qualitative results were obtained from the influence of vegetation on the thermal comfort of a person in the urban morphology, showing that vegetation is one of the most suitable tools for urban use.

According to the results, vegetation can be considered a useful tool to improve the climatic conditions of urban spaces, adding quality and value to the city. Evaluation of each tree species provides relevant knowledge about its environmental performance, which can be used in urban design as a tool of high value for designers.

Acknowledgements

The authors wish to thank the Ministry of Science and Technology Education of the Dominican Republic and its Minister Ligia Amada Melo; Dr Arq Helena Coch and her research team from the Department of Architecture, Energy and Environment; Msc Arq Curro Peña Ariza; Msc Arq Badia Masoud; Msc Arq Elena García and Lic Elaynne Rojas.

Author details

Gilkauris Rojas-Cortorreal[1*], Francesc Navés Viñas[2], Julio Peña[3], Jaime Roset[4] and Carlos López-Ordóñez[1]

*Address all correspondence to: gilkaurisrojas@gmail.com

1 Department of Architectural Technology, Polytechnic University of Catalonia, Barcelona, Spain

2 Department of Structure, Polytechnic University of Catalonia, Barcelona, Spain

3 Department of Architecture, Autonomous University of Santo Domingo, Santo Domingo, Dominican Republic

4 Department of Physics, Polytechnic University of Catalonia, Barcelona, Spain

References

[1] Ochoa de la Torre JM. La vegetación como instrumento para el control microclimatico [thesis]. Barcelona, Spain: Universidad Politecnica de Catalunya; 1999. p. 600

[2] Ochoa de la Torre JM. Ciudad, Vegetación e Impacto Climático. 1st ed. Barcelona, Spain: Erasmus; 2009. p. 450

[3] Ochoa de la Torre JM. Scudo G. Spazi verdi urbani. Napoli, Italy: Sistemi Editoriali; 2003

[4] Abreu L. V., Labaki L. C., Matzarakis A. Reduction of mean radiant temperature by cluster of trees in urban and architectural planning in tropical climates the case of Campinas, Brazil. Passive Low Energy Architecture (PLEA). 2012. Volumen 1. Page: 1-6.

[5] Arboit M.E.. Permeabilidad del arbolado urbano a la radiación solar: Estudio de dos especies representativas en entornos urbanos de baja densidad del área Metropolitana de Mendoza, Argentina. Hábitat sustentable. 2013; Volumen 3, No. 2, Page 3-18.

[6] Armson D, Stringer P, Ennos AR. The effect of tree shade and grass on surface and globe temperatures in an urban area. Urban Forestry & Urban Greening. 2012;**11**(3):245-255

[7] Echave Martinez C. Ver des de la ciudad. Potencial de habitabilidad térmica en espacios urbanos [thesis]. Barcelona, Spain: Universitat Politecnica de Catalunya; 2007. Available from: http://issuu.com/cynthiaechave/docs/ver-des_de_la_ciudad_cynthia_echave

[8] Correa E. N., Ruiz M. A., Catón M. A. Morfología forestal y confort térmico en ¨ciudades oasis¨ de zonas áridas. Ambiente construido. 2010; Volumen 10, No. 4, Page 119-137.

[9] Salvati A, Coch H, Cecere C. Which climate for each urban context? A preliminary comparative study on urban climate prediction and measurement in different districts in Rome and Barcelona. In: Word Renewable Energy Congress XIII; 3-8 August; London, UK. 2014. DOI: 10.13140/RG.2.1.4777.3207

[10] Salvati A, Coch H, Cecere C. Urban morphology and energy performance: The direct and indirect contribution in Mediterranean climate. In: PLEA: 9-11 September 2015; Bologna. DOI: 10.13140/RG.2.2.19036.59526

[11] López-Ordóñez C, Roset J. Análisis de la radiación en las calles de Barcelona y de Hermosillo [thesis]. Barcelona, Spain: Universitat Politécnica de Cataluña; 2013

[12] Garcia-Nevado E. Toldo urbano: Posibilidades de reducción de la demanda de refrigeración [thesis]. Barcelona, Spain: Universitat Politecnica de Catalunya; 2013. Available from: https://mastersuniversitaris.upc.edu/aem/informacion-academica/2022-tesinas-finales-de-master/2012-13/TESINAGarcaElena.pdf

[13] Garcia-Nevado E, Pages-Ramon A, Coch H. Solar access assessment in dense urban environments: The effect of intersections in an urban canyon. Energies. 2016; volumen 9, No. 10, Page 1-12 . DOI: 10.3390/en9100796

[14] Wang Y, Akbari H. Effect of sky view factor on outdoor temperature and comfort in Montreal. Environmental Engineering Science. 2014;**31**(6):272-287. DOI: 10.1089/ees.2013.0430

[15] López-Ordóñez CF, Roset Calzada J, Rojas-Cortorreal G. Análisis de la radiación solar directa en las calles de Barcelona, en base a la relación entre su morfología y vegetación. Architecture, City and Environment. 2017. Volume 11, No. 34

[16] Ministerio de medio ambiente y medio rural y marino, Agencia estatal de meteorología de España. Iberian Climate Atlas. Closas-Orcoyen S. L. ed. Spain; 2000. p. 80

[17] Temperature Weather. Temperature Weather [Internet]. 2017. Available from: http://www.temperatureweather.com/mediterr/tiempo/es-tiempo-en-italia-rome.htm [Accessed: 15 February 2017]

[18] Navés Viñas F. Arquitectura del paisaje natural. Barcelona, Spain: Omega; 2005

[19] Navés Viñas F, Pujol Solanich J, Argimon de Vilardaga X, Sampere Montlló L. El árbol en Jardinería y Paisaje. Barcelona, Spain: Omega; 1995

[20] Serra Florensa R, Coch H. Arquitectura y energía natural. Barcelona, Spain: Universitat Politecnica de Catalunya; 2009. p. 395

[21] Sancho Ávila JM, Riesgo Martín J, Sánchez de Cos Escuin MC, Montero Cadalso J, López Bartolomé M. Atlas de Radiación solar en España utilizando datos del SAF de clima de EUMETSAT. Barcelona, Spain: Agencia estatal de meteorología y ministerio de agricultura, alimentacion y medio ambiente; 2012. p. 162

[22] Petrarca S, Cogliani E, Spinelli F. La radiazione solare globale al suolo in Italia. Stampa 2000. p. 193

[23] Arnfield AJ. Street design and urban canyon solar access. Energy and Buildings. 1990; **14**(2):117-131

[24] Oke TR. Street design and urban canopy layer climate. Energy and Buildings. 1988; **11**(1-3):103-113

[25] Beckers B. Heliodon [Internet]. 2010. Available from: http://www.heliodon.net/ [Accessed: 05 May 2016]

[26] Rojas-Cortorreal G. Cuantificación de la mejora de las condiciones ambientales producida por el arbolado urbano. Caso de estudio Barcelona, España y Santo Domingo, República Dominicana [thesis]. Barcelona, Spain: Universitat Politecnica de Catalunya; 2016. Available from: http://futur.upc.edu/19361653

[27] Rojas-Cortorreal G, Roset J, Navés Viñas F, López-Ordoñez C. The effect of comfort heat produced by tree species Celtis australis (Almez) in the public space. Architecture, City and Environment . 2016; Volumen 11, No. 32. Page 187-202 DOI: 10.5821/ace.11.32. 4839

[28] Rojas-Cortorreal G, Roset J, Navés Viñas F. La vegetación en el confort micro climático. Comparación de especies del clima mediterráneo de Barcelona, España. Architecture, City and Environment. 2014;**10**(29):59-84. DOI: 10.5821/ace.10.29.3589

[29] Rojas-Cortorreal G, Roset J, Navés Viñas F, López-Ordoñez C, Vidmar J. Evaluation of vegetation in urban space Barcelona base model and proposed to Dominican Republic. In: Borg RP, Gauci P, Spiteri Staines C, editors. Europe and the Mediterranean: Towards a Sustainable Built Environment; 16-18 March 2016; Malta. Malta: SBE Malta—Sustainable Built Environment; 2016. pp. 195-206

[30] Rojas-Cortorreal G, Roset J, Navés Viñas F. La vegetación tropical en el confort climático urbano. Aplicado a Santo Domingo, República Dominicana en comparación con el clima mediterráneo, Barcelona, España [thesis]. Barcelona, Spain: Universitat Politecnica de Catalunya; 2013. Available from: http://hdl.handle.net/2099.1/20638

[31] Rojas-Cortorreal G, Roset J, Navés Viñas F. Contributions to micro climate comfort of tree species Platanus X hispanica. In: JIPI, editor. 4ª Jornada de Investigadores Predoctorales Interdisciplinaria (4ª JIPI); 8 March 2016; Barcelona. Barcelona: JIPI; 2016. pp. A-9. DOI: 10.13140/RG.2.1.2786.0248

[32] Masoud B, Beckers B, Coch H. Sky access versus shading for pedestrian comfort in the hot tropical climate of Jeddah. In: FICUP, editor. First International Conference on Urban Physics (FICUP); 26/09/2016; Quito, Ecuador. Quito:2016

Chapter 16

Wheat: A Crop in the Bottom of the Mediterranean Diet Pyramid

Conxita Royo, Jose Miguel Soriano and
Fanny Alvaro

Additional information is available at the end of the chapter

http://dx.doi.org/10.5772/intechopen.69184

Abstract

Wheat currently provides 18% of the daily intake of calories and 20% of proteins for humans. Since its domestication in the Fertile Crescent, wheat has been the basic staple food of the major civilizations of Europe, West Asia and North Africa. The wheat-growing area within the Mediterranean Basin represents 27% of the arable land, and the region represents 60% of the world's growing area for durum wheat, the species used for pasta manufacturing. Many changes have occurred from the low-productive plants cultivated in prehistoric times to the modern varieties that are now grown, which offer high productivity and quality standards. During the migration process of ancient forms of wheat from the east to the west of the Mediterranean Basin, both natural and human selections occurred, resulting in the development of local landraces characterized by their huge genetic diversity and their documented resilience to abiotic stresses. Wheat breeding activities conducted in the Mediterranean Basin during the twentieth century resulted in large genetic gains in yield and quality. New wheat varieties to be grown in the Mediterranean Basin will need to be resilient to climate change because more frequent episodes of higher temperatures and water scarcity are to be expected.

Keywords: wheat origin, wheat domestication, landrace, population structure, core collection, adaptation, productivity, grain quality, wheat breeding, genetic gains, yield, yield components

1. Introduction

Wheat is currently the most widespread crop. It is grown on about 219 million hectares all over the world (**Figure 1**) and is the basic staple food of mankind, providing humans with 18% of their daily intake of calories and 20% of their protein (http://faostat.fao.org/). The

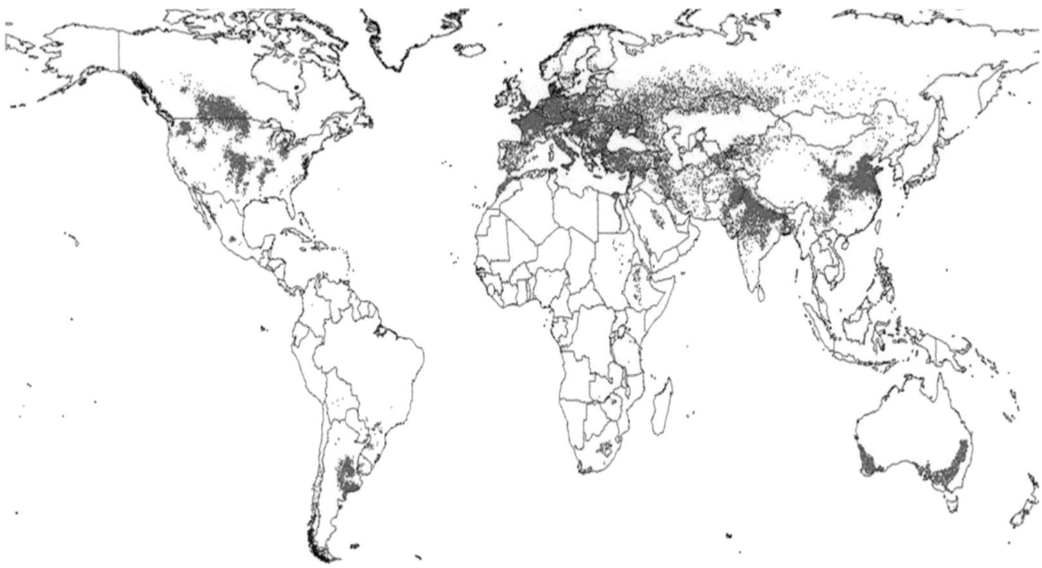

Figure 1. Global wheat distribution. Each point represents 20,000 t of grain production (modified from CIMMYT).

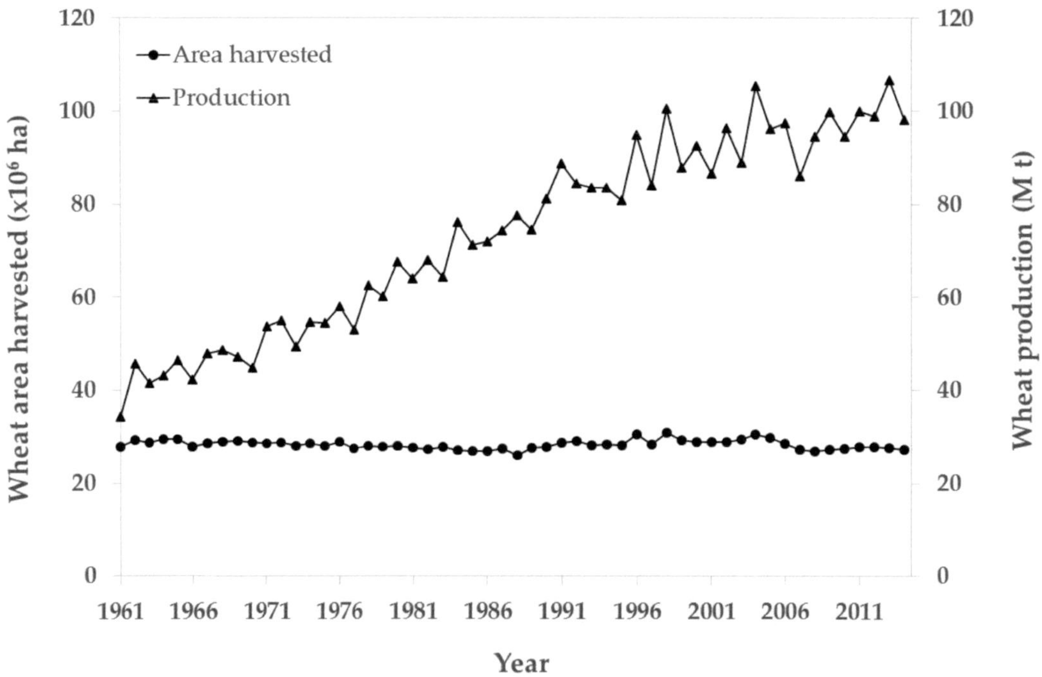

Figure 2. Total wheat harvested area and wheat production in coastal countries to the Mediterranean Sea from 1961 until 2014 (own elaboration from FAOSTAT data; www.fao.org/faostat).

wheat-growing area within the Mediterranean Basin represents 27% of the arable land (**Figure 2**), and the region represents 60% of the world's growing area for durum wheat, the species used for pasta manufacturing. The Mediterranean dietary traditions have often been related to health benefits and the prevention of cardiovascular disease [1]. The Mediterranean diet is the heritage of millennia of exchanges between people, cultures and foods of all countries around the Mediterranean Basin, and during the twentieth century, it has been the basis of food habits in all countries of the region, originally based on Mediterranean agricultural and rural models [2]. Cereals, and mostly wheat in the form of bread, pasta or couscous, form the base of the pyramid and are daily included as part of the main meals [3].

There are several species of the wheat genus (*Triticum*). The most widespread is common or bread wheat (*Triticum aestivum* L.), which occupies 94% of the total area cultivated with wheat, is mainly used for manufacturing bread and biscuits. Durum wheat (*Triticum turgidum* L. var. *durum*) is grown on about 13 million hectares, about 60% of them located in the Mediterranean Basin, where it is considered a typical crop. Durum wheat is mostly used for pasta making, but it is also the raw material for producing other traditional goods of Mediterranean countries such as flat breads, couscous and bulgur. The genetic differences between the two species lie in the number of chromosomes, as durum wheat is a tetraploid species (with 28 chromosomes in four sets), while bread wheat is a hexaploid species (with 42 chromosomes in 6 sets).

2. The origin of wheat and its spreading through the Mediterranean Basin

Wheat was one of the first domesticated food crops and its history is that of humanity. The domestication of wheat and the beginning of agriculture go hand in hand. Kislev [4] classified the data of wheat husbandry into three major phases: (*i*) the agro-technical revolution, which occurred within a still hunting-gathering society during the Natufian period (Epipalaeolithic, 13,000–10,300 BP); (*ii*) the domestication revolution (Pre-Pottery Neolithic 10,300–7500 BP); and (*iii*) the expansion of agriculture, mostly during the Ceramic or Pottery Neolithic (7500–6200 BP).

The crucial separation of individuals of the *Triticeae* tribe that resulted in the different cereal species (wheat, barley, rye, etc.) is believed to have occurred during the Pleistocene, a glacial epoch. The major climate changes that started on the eastern Mediterranean coast about 15,000 BP replaced the original cold and arid conditions with warmer and moister ones, thus allowing the expansion of grasses [5, 6]. Palaeobotanical investigations and other indications show that, from about 11,500 BP, the climate of the eastern Mediterranean region (Levant) became dry and cold [7] and the large variations in rainfall and temperature between years and seasons forced vegetation to make important changes in order to adapt to the new environmental conditions. It is believed that at that time, self-fertilization (autogamy) increased as a mechanism of reproductive assurance [8]. The growth habit of vegetation became annual and seed dormancy augmented, allowing seeds to overcome periods of harsh environmental conditions by remaining in the soil until conditions were suitable for germination [6]. It was probably during this time that, somewhere along the Fertile Crescent, the hunter-gatherers who were accustomed to collecting grains of wild cereals, fruits and roots of other plants started to cultivate grasses [9]. It has been suggested that the Natufian tribe, who lived around Mount Carmel in present-day Israel and showed advanced preadaptive traits, or the dwellers in the Karacadağ Mountains in southeast Turkey may have been the first [4, 10]. This assumption is supported by genetic studies demonstrating that in this area, wild einkorn grass (a diploid ancestor species of wheat) contains an identical genetic fingerprint to modern domesticated wheat. As women had the primary responsibilities for plant gathering in hunter-gatherer societies, it is believed that they probably planted the first seeds. Einkorn (diploid), emmer (tetraploid) and spelt (hexaploid) are among the earliest cultivated wheats and are commonly referred to as 'ancient' wheats [11].

The first signs of cultivation of the so-called emmer (an awned wild wheat) correspond to the Pre-Pottery Neolithic A period (10,300–9500 BP), at the end of which all basic agricultural practices had already been established [6]. The transformation of some wild cultivated forms into domesticated wheats proceeded very rapidly from this stage. From the Karacadağ Mountains, emmer spread first northward and then southward. There is a general agreement that domestication occurred at the beginning of the Pre-Pottery Neolithic B (9500–7500 BP) [7, 12], when the spontaneous crosses between grasses that led to the appearance of bread wheat probably took place. Plant domestication was driven by humans' need to secure the greatest possible amount of food with the least possible labour.

Wheat domestication involved major morphological, physiological and adaptive changes in plants, most of them induced by humans. One of the clearest examples of the contribution of humans to domestication was the transformation of the spike axis from brittle to tough. In wheat ancestors, the spike became brittle at maturity, falling apart into small pieces (spikelets) containing the seeds, which were spread by wind and animals as an essential mechanism of propagation and survival. However, a small number of plants (those carrying a recessive allele conferring axis robustness) tended to develop robust spike axes, and this caused the seeds to remain together in the spike at ripeness without falling down. This feature was very beneficial for humans, as it allowed them to harvest complete spikes at ripening instead of unripe spikes. It is likely that seed collectors gradually increased the proportion of tough-axis spikes gathered, thus unconsciously favouring the tough-axis genes in the harvested grains, which led to a suppression of the brittle axis in domesticated wheats [13]. Thus, due to the loss of the seed-dispersal mechanism, wheat started to depend on humans for survival. Other important changes that occurred during domestication were the reduction of grain self-protection (due to the loss of the leaf-like glumes that covered each seed), which made the grains free-threshing, and the loss of seed dormancy, which favoured a uniform and rapid seed germination.

The establishment of agriculture in the Levant and the neighbouring regions was a very gradual evolutionary process that took place over a period of several hundred years [7, 12]. Studies conducted today to imitate different harvest techniques of wild wheats grown in a dense stand suggest that at that time, it was possible to obtain about 0.5–1 kg of pure grain per hour or 300–700 kg of grain per hectare or even more [14, 15]. This significant improvement led to a substantial population growth.

In the Ceramic or Pottery Neolithic, the wheat culture spreads from the western flank of the Fertile Crescent to southeast Europe through Transcaucasia, reaching the Balkan Peninsula and Greece in about 8000 years BP. Primitive wheat was transported by ships along the coast of the Mediterranean Sea to Italy and Spain (7000 BP) [5, 16] and south of Gibraltar. Two possible ways have been proposed for the introduction of durum wheat into the Iberian Peninsula: North Africa and south-eastern Europe [6, 17]. Wheat reached Egypt through Israel and Jordan [5].

After arriving in a given territory, wheat underwent a progressive adaptation to the varying conditions of the new area and gradually established new strategies for yield formation, which likely conferred adaptive advantages under the new environmental conditions [18]. During the dispersal of wheat along the Mediterranean Basin, the farmers took their habits wherever they went, not just sowing, reaping and threshing but also other well-established technologies such as baking and fermenting. This process of migration and natural and human selection resulted in the establishment of a wide diversity of local landraces specifically adapted to different agro-ecological zones. These dynamic populations with distinct identities are considered to be genetically more diverse than currently cultivated varieties (**Figure 3**); they show local adaptation and are associated with traditional farming systems [19].

The Mediterranean Basin comprises countries between 27° and 47°N and between 10°W and 37°E, including three continents with a coastline of 46,000 km. In this region, wheat is grown in a range of environmental conditions varying from favourable to dry land areas. In the Mediterranean

Figure 3. Variability in spike morphology in durum wheat Mediterranean landraces.

climate, most rain falls in autumn and winter, and a water deficit appears in spring, resulting in moderate stress for wheat around anthesis that increases in severity throughout the grain-filling period. However, the climatic conditions of the north and the south of the Mediterranean Basin have great differences. While the north has temperate and cold climates (classes C and D, respectively, according to the Koppen climate classification), the south has a dry climate (class B according to the same classification) [20]. Scientific evidence has shown that contrasting adaptation strategies occurred during the spread of wheat over the north (via Turkey, Greece and Italy) and south (via North Africa) of the Mediterranean Basin. The different climates prevalent in the zones of adaptation may have induced gradual changes in crop phenology and in the strategies used by wheat to form its yield during its dispersal from the east to the west of the Mediterranean Basin [21]. Royo et al. [18] demonstrated that the number of days to heading and flowering of traditional durum wheat varieties (landraces) increased from the warmest and driest zone of the Mediterranean Basin to the coldest and wettest one. Durum wheat landraces collected in the north of the Mediterranean Basin have been found to have more stems per unit area, more biomass, a higher proportion of biomass and leaf area allocated on tillers at flowering and heavier grains than those collected in the south [18, 21, 22]. African landraces of diploid, tetraploid and hexaploid wheats have been reported to carry genes for tolerance to physical environmental stresses [23] and are therefore better adapted to drought environments than those of northern countries [22]. Durum wheat landraces from southern Mediterranean countries allocate more biomass to the main stem, produce more grains per spike and per unit area and have higher harvest index than those from northern countries [21, 22]. These differences mean that grain yield of landraces collected in northern Mediterranean countries is mainly related to variations in grain weight, whereas grain yield of landraces collected in the drier and warmer southern countries is mainly related to the number of spikes per unit area [21].

3. Mediterranean wheat landraces

During the first few decades of the twentieth century, the wheat varieties grown in the Mediterranean were the so-called traditional varieties or landraces. Landraces resulted jointly from the evolution of wheat during its dispersal to new territories and from the role of humans in selecting large spikes for planting the next generation after the advent of agriculture. Landraces are heterogeneous in their shape because they are populations formed by sets of plants with different genetic constitutions. Their huge genetic diversity makes them a particularly important group of genetic resources. Knowledge of the genetic diversity and population structure of landraces is essential for their conservation and efficient use in breeding programmes. The first diversity studies used phenotypic markers of morphological and physiological traits, but DNA-based markers are currently extensively used as they are not affected by the environment and can be detected in all tissues at all developmental stages. Among them, microsatellites or simple sequence repeats (SSRs) have proven to be very useful for evaluating the genetic diversity and population structure of Mediterranean wheat collections [17, 24–29].

A number of studies have been conducted by our team to assess the genetic diversity and population structure of Mediterranean durum wheat germplasm. In all cases, the genetic structure of the landrace populations proved to be associated with the geographical origin of accessions. The study of Moragues et al. [17], which used a set of 63 durum wheat landraces from 12 Mediterranean countries, grouped the accessions in two clusters: (*i*) landraces from north and east of the Mediterranean Basin and (*ii*) landraces from North Africa and the Iberian Peninsula. These results support the hypothesis of two dispersal patterns of durum wheat in the Mediterranean Basin previously proposed by MacKey [6], one through the north side and one through the south side.

A recent study by Soriano et al. [29] classified a collection of 152 durum wheat landraces and old varieties from 21 Mediterranean countries into four subpopulations that showed an eastern-western geographical pattern (**Figure 4**): eastern Mediterranean, eastern Balkans and Turkey, western Balkans and Egypt and western Mediterranean. The genetic diversity found by Soriano et al. [29] was lower in the eastern Mediterranean group, indicating that the diversity of wheat increased during the dispersal from its area of domestication to the western Mediterranean Basin.

A study was carried out by Ruiz et al. [26] on a collection of 190 durum wheat Spanish landraces. The results showed that the diversity and agro-morphological traits were correlated with geographic and climatic features. The distribution of the collection in nine clusters was largely determined by the three subspecies, *dicoccon*, *turgidum* and *durum*, which were present on it, with an east-west geographic structure for *dicoccon* and a northeast-southwest structure

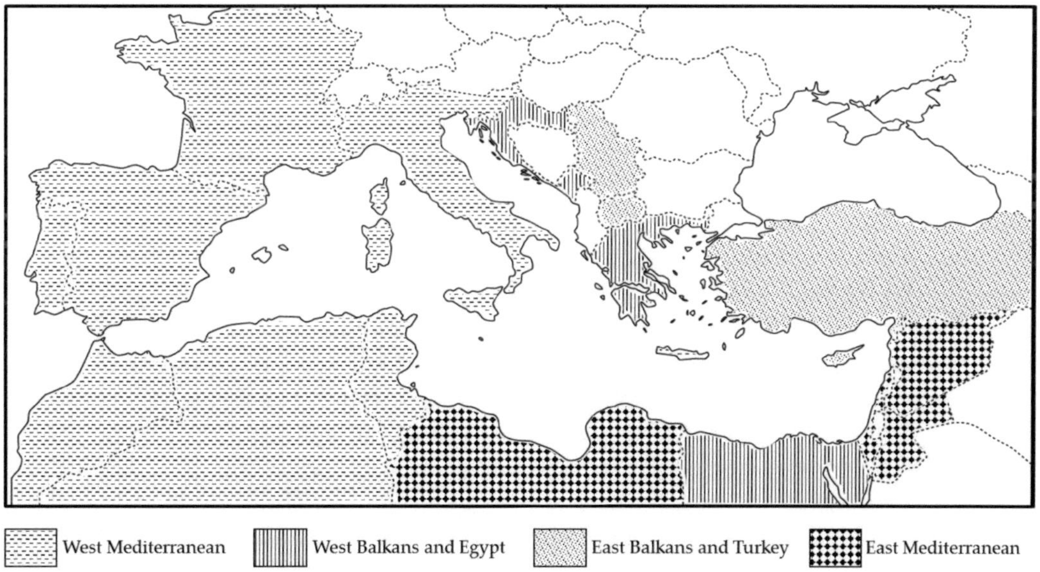

Figure 4. Genetic structure of the Mediterranean durum wheat landraces and old varieties reported by Soriano et al. [29] according to their countries of origin.

for *turgidum*. The results of the phylogenetic study showed that some *durum* accessions were more related to ssp. *turgidum* from northern Spain, while others were more related to *durum* wheats from North Africa [26]. These results also support the hypothesis of two possible forms of introduction of durum wheat in the Iberian Peninsula, South-East Europe and North Africa, as suggested by Moragues et al. [17].

During the last century, a large number of wheat germplasm accessions have been collected by Mediterranean gene banks. Western Mediterranean countries (Portugal, Spain, southern France, Morocco, Algeria and Tunisia) have been identified as one of the principal regions for collecting tetraploid wheats due to the variability gathered by local germplasm [23]. As the management of the whole collections is costly and inefficient, if the collection shows a significant level of redundancy, core collections consisting of a limited set of accessions that maximize the genetic variation contained in the whole collection with a minimum of repetitiveness [30] have been created. The bread wheat worldwide core collection formed by Balfourier et al. [31] from 3942 cultivars included 372 accessions, 149 of which came from 18 countries around the Mediterranean Basin. The accessions were grouped using molecular markers according to their geographical distribution—western and southern Europe, the eastern Mediterranean Basin, North Africa, Turkey, the Balkans and finally France—which was grouped with cultivars from central and northern European countries. The Spanish durum wheat core collection created by Ruiz et al. [26] includes 94 accessions representative of a collection of 555 Spanish landraces and old cultivars and contains a wide range of genotypes adapted to Mediterranean environments.

Mediterranean landraces have a good adaptation to Mediterranean environments. They can be considered as likely sources of putatively lost variability and may provide favourable alleles for the improvement of commercially valuable traits, especially in breeding for suboptimal environments. However, their plant height, general lateness and low harvest index limit the attainment of high yields, particularly when they are grown in intensive agricultural systems. A study by our team conducted on 154 durum landraces from 20 Mediterranean countries revealed that landraces from western Mediterranean countries had greater grain-filling rates and heavier grains than those of eastern ones [32]. The contribution of landraces in wheat breeding programmes also seems possible in terms of end-product quality, given the high level of polymorphism of key quality genes and the large genetic diversity found for quality traits between and within landrace populations [33–35]. It has been reported that specific Mediterranean durum wheat landraces may be used as sources of quality improvement for grain protein content, gluten strength, grain weight, test weight and general quality [32]. In order to identify durum wheat landraces as potential parents in breeding programmes for gluten strength enhancement, Nazco et al. [35, 36] analysed the allelic composition at five glutenin loci on a collection of 155 durum wheat landraces and old varieties from 21 Mediterranean countries and 18 representative modern cultivars. The results indicated that landraces with outstanding gluten strength were more frequent in eastern than in western Mediterranean countries. Only 9 different allelic combinations were identified in modern cultivars, but 126 in the landraces, 3 of them new with a positive effect on gluten strength [36]. Twelve banding patterns positively affecting gluten strength were identified in the landraces [35].

4. From local landraces to modern Mediterranean wheat cultivars

The pioneer breeders or entrepreneurial Mediterranean agriculturalists started selecting from within landrace populations (sometimes from foreign countries) the plants with the most favourable characteristics in terms of vigour, phenological adaptation, spike length and yield in order to produce superior lines. This pure-line selection did not entail the development of new genotypes as the improvement was only achieved by identifying and isolating the best lines already existing within the original landrace. This methodology was used by Nazareno Strampelli in Italy to release the durum wheat cultivar 'Senatore Cappelli' in 1915 from the Algerian population 'Jean Retifah' [37], by Enrique Sánchez-Monge Parellada in Spain to release the barley variety 'Albacete' from a selection within a local population, the bread wheat 'Aragón 03' selected by Manuel Gadea from a selection within the local variety 'Catalán de Monte' and the durum wheat varieties 'Andalucía 344' and 'Jerez 36' obtained in Spain by Juan Bautista Camacho from 'Manchón de Alcalá la Real' and 'Raspinegro de Alcolea', respectively [38].

The first organized wheat programme in France was implemented by the Vilmorin family in the eighteenth century [39]. When Mendel's laws were rediscovered at the beginning of the twentieth century plant breeding was established as a science, making crosses between varieties or breeding lines selected in the previous phase of breeding. The Italian breeder Nazareno Strampelli, considered the local promotor of the Mendelian findings, started to make crosses around 1900 [40]. In parallel, breeders started to interchange germplasm and to use foreign varieties or lines developed by their colleagues in other countries, for crossing with their best types. Farmer breeding was also encouraged by the collection and distribution of wheat seed from all over the world. In the 1850s, the harvest index (the partitioning of photosynthates between the grain and the vegetative plant) of most wheats was about 0.3 or less [41].

From around 1940, breeding programmes based on scientific findings were created in a number of countries. One of the most famous was the Rockefeller-Mexico programme, led by Dr. Norman Borlaug, which started in 1945, which used germplasm from different origins in his crosses. The particularities of Mexico allowed Borlaug's programme to grow two cropping seasons every year, thus speeding the breeding process. Using Norin 10, a Japanese variety, and the cross Norin/Brevor as a parental, Borlaug obtained 'semi-dwarf wheats' that yielded far better than the taller wheats grown in most parts of the world at that time. The incorporation of the dwarfing genes designated as *Rht-B1b* (formerly *Rht1*) and *Rht-D1b* (formerly *Rht2*) resulted in an increased earliness, a reduction in plant height and lodging without significant decreases in total plant dry weight and a larger allocation of resources in grains, thus improving the harvest index [41–44]. In addition, the incorporation of photoperiod insensitivity in the wheats developed in Mexico, as consequence of the shuttle breeding between contrasting environments, allowed them to be adapted to a wide range of environments all around the world, showing good adaptation to a number of environmental conditions. Semi-dwarf varieties have the capacity to redistribute the plant weight so as to allocate a higher percentage of it to the grain than in unimproved varieties. As a result, plant height decreased (**Figure 5**), but harvest index increased.

Figure 5. Field plots of a durum wheat landrace (left) and a semi-dwarf improved variety (right).

The semi-dwarf varieties developed in Mexico were rapidly adopted by Mediterranean countries, leading to the progressive abandonment of the cultivation of landraces. The adoption of improved semi-dwarf varieties was accompanied by the intensification of management practices to allow the semi-dwarf wheats to express their potentiality. The progress achieved for grain yield until 1982 was the result of combining improved varieties with appropriate crop management strategies. Sowing densities, application of fertilizers (particularly nitrogen), irrigation and the use of pesticides to control weeds and diseases increased resulting in yield rises in many countries. International Maize and Wheat Improvement Center (CIMMYT) was formally launched in 1966, and Norman Borlaug was honoured with the Nobel Peace prize in 1970 for his contribution to the Green Revolution.

Advances in yield during the twentieth century in Mediterranean countries due to variety improvement have been widely reported in the literature for bread wheat [43–45] and durum wheat ([42, 46] and references therein). The role played by the variety is generally ascertained by growing historical series of cultivars in a common environment. Following this approach, our group quantified yield increases during the last century in 35.1 kg ha^{-1} y^{-1}, or 0.88% y^{-1} in relative terms, for bread wheat in Spain [43] and in 16.9 and 23.6 kg ha^{-1} y^{-1} (0.51 and 0.72% y^{-1} in relative terms) for durum wheat in Italy and Spain, respectively [47]. For both species, the modification of plant architecture by the introduction of dwarfing alleles played a major role in the achieved gains [42, 43], by reducing plant height and reallocating photosynthates in reproductive organs of plants with more grains per spike and improved tillering capacity [42, 43]. In Spain, changes of harvest index during the twentieth century have been estimated by our team to be from 0.25 to 0.40 in bread wheat [44] and from 0.36 to 0.44 in durum wheat [42, 48].

However, the GA insensitivity conferred by the dwarfing genes *Rht-B1b* and *Rht-D1b* resulted in grains of similar weight.

The effect of dwarfism was greater on the root system than on aerial biomass, reducing the aerial biomass of each plant at anthesis by 7.6% and the root by 28.1% [49]. However, despite their reduced root biomass, modern cultivars are more responsive in terms of yield and number of grains per spike to environments with high water input after anthesis [50]. Breeding also improved adaptation to Mediterranean conditions by reducing cycle length to flowering [44, 51], thus benefiting grain setting [46] and improving photosynthesis during grain filling [44] in environments characterized by terminal stresses. The adaptation pattern of bread wheat changed towards varieties with a wider adaptation to variable environmental conditions and spring types that performed better than landraces in environments with high temperatures before heading [52]. However, our results also evidenced a slowdown in bread wheat yield increases since 1970 [43]. Breeding activities improved the overall processing quality of wheat for making bread and pasta. Although most modern cultivars have less grain protein content than traditional varieties, breeding activities during the twentieth century in Mediterranean countries resulted in an improvement of global grain quality in both bread and durum wheat [44, 53].

Comprehensive information about the history of wheat breeding in Mediterranean countries may be found in Bonjean et al. [39, 54] and Royo et al. [55].

Although the Green Revolution was critical for raising wheat production enough to mitigate the effect of rapid demographic growth, it affected the natural habitat of wheat. Landraces and pure-line cultivars obtained through mass selection from them during the first decades of the twentieth century were widely grown until the late 1960s, but due to the massive introduction of the homogeneous and more productive semi-dwarf cultivars released since the Green Revolution, they practically disappeared from farmer's fields. Particularly, in the domestication area of wheat and the Mediterranean regions, which are the reservoir of the greatest genetic variability of the species, wild relatives and landraces were displaced by improved varieties. In consequence, the variability present on farmer fields due to the cultivation of old unimproved varieties (landraces) gave way to the genetic uniformity of the most productive modern cultivars. The decrease in cultivar diversity and the loss of the natural variation present in landraces increased the genetic vulnerability of wheat crops [23] and led to a loss of the diversity exploitable by plant breeders, the so-called genetic erosion. Among the factors that have been reported to contribute to the narrow genetic background underlying successful modern wheat varieties, the reduced number of ancestors and the relatively small number of varieties cultivated at present are among the most significant [56, 57].

Another important change derived from the advent of modern cereal culture was the requirement of field uniformity, which led to the planting of large extensions of a single variety or a small number of varieties, managed under similar cultural practices. This homogeneity is very convenient from the industrial and commercial viewpoints as it allows sets of tons of wheat grains with similar quality characteristics to be obtained. However, it most likely increases the vulnerability to diseases as the pressure exerted by large extensions of uniform varieties pushes the races of fungal species to mutate, and to very rapidly overcome the genetic resistance of cultivars [58].

5. Future prospects

Several global and regional climate models suggest that the Mediterranean Basin might be an especially vulnerable region to global change [59]. A pronounced decrease in precipitation and warming are projected, particularly in the summer season, with large interannual variability leading to a greater occurrence of extremely high-temperature events [59, 60]. The impact of agricultural practices on climate change has redesigned the breeding paradigm. While past yield improvements relied on the development of improved varieties that needed the intensification of agricultural practices to maximize yields, the new released varieties have to be able to produce with the minimum environmental impact, that is, they must fit into the concept of 'sustainable agricultural ecosystems'. This entails their genetic adaptation to environmental conditions, making it unnecessary to modify the environment through the use of non-sustainable practices to cover the variety requirements, as was the case in the past. This is a huge challenge for breeders, as wheat breeding today largely depends on the incorporation in improved varieties of adaptive traits for specific environments.

Given that most traits useful for improving the adaptation of modern cultivars to abiotic or biotic constraints cannot be found in modern cultivars, in many cases the enlargement of the genetic variability has to be sought in local landraces and close-related species. The high genetic diversity of landraces buffers them against spatial and temporal variability and upgrades the resilience to abiotic and biotic stresses in comparison with modern varieties [61, 62]. The essential role of landraces as likely sources of highly beneficial untapped diversity has led them to be considered essential for food security because they are potential providers of new favourable genes to be incorporated into modern cultivars. However, as the genetic variation contained in them is usually unknown, the effective use of landraces in breeding programmes will make necessary to evaluate the existing diversity in the gene pool and to characterize the available accessions [62]. Detecting the presence of variants of potential interest for breeding purposes in landraces may be particularly useful in situations of breeding for suboptimal environments.

Among the set of wheat landraces, the ones coming from the Mediterranean Basin are considered to hold the largest genetic variability within the species as shown by the genetic variability found in Portuguese [63] and Spanish [34] wheats. Mediterranean wheat landraces are considered as a potential genetic resource of drought resistance, frost tolerance and biotic and abiotic stresses in general. In addition, as mentioned above, an increase in the available genetic variation through the use of landraces in breeding programmes seems possible in terms of yield component enhancement and end-product quality.

The enormously expanding potential of recently developed technologies offers opportunities for improvement of plant traits and agricultural management that were inconceivable few decades ago. Genomics offers new opportunities to dissect traits of quantitative inheritance and chromosomal regions whose allelic variation may be statistically associated with a specific trait. During the last few decades, several types of molecular markers have been used for wheat genetic studies, providing effective genotyping but resulting costly and time-consuming due to the low number of markers to be screened in a single reaction. In the last few years, the advances in next-generation sequencing (NGS) technologies has reduced the costs of DNA sequencing to the point that genotyping based on sequence data is now feasible for high-diversity and large-genome species. New high-throughput platforms have been developed in bread wheat, such as single nucleotide polymorphism (SNP) arrays [64, 65] and genotyping by sequencing (GBS) platforms, e.g., DArT-seq, developed by Diversity Arrays Technology Pty Ltd (Canberra, Australia). In early 2017 the bread wheat genome sequence was released providing wheat researchers with a new resource to identify the most influential genes that are important to wheat adaptation, stress response, pest resistance and improved yield. Recent scientific breakthroughs in genome-editing technologies, such as the clustered regularly interspaced short palindromic repeat (CRISPR), have opened new avenues for accelerating basic research and plant breeding by providing the means to modify genomes rapidly in a precise and predictable manner [66]. With recent biotechnology developments, advances in statistics, precision agriculture and information technologies such as geographic information system (GIS), remote sensing and the exploitation of big data among other new tools will hopefully help to meet the challenges of breeders and agronomists in the next few decades.

Acknowledgements

The authors thank INIA, MINECO (projects AGL2012-37217 and AGL2015-65351-R) and IRTA for supporting wheat research programmes in Spain. JM Soriano is employed by the INIA-DOC program funded by the European Social Fund with the contribution of the CERCA Programme (Generalitat de Catalunya).

Author details

Conxita Royo*, Jose Miguel Soriano and Fanny Alvaro

*Address all correspondence to: conxita.royo@irta.es

IRTA (Institute for Food and Agricultural Research and Technology), Field Crops Programme, Lleida, Spain

References

[1] Estruch R, Ros E, Salas-Salvadó J, Covas MI, Corella D, Arós F, Gómez-Gracia E, Ruiz-Gutiérrez V, Fiol M, Lapetra J, Lamuela-Raventos R, Serra-Majem Ll, Pintó X, Basora J, Muñoz MA, Sorlí JV, Martínez JA, Martínez-González MA. Primary prevention of cardiovascular disease with a Mediterranean diet. The New England Journal of Medicine. 2013;**368**:1279-1290. DOI: 10.1056/NEJMoa1200303

[2] Bach-Faig A, Berry EM, Lairon D, Reguant J, Trichopoulou A, Dernini S, Medina FX, Battino M, Belahsen R, Miranda G, Serra-Majem Ll. Mediterranean diet pyramid today.

Science and cultural updates. Public Health Nutrition. 2011;**14**:2274-2284. DOI: 10.1017/S1368980011002515

[3] Willett WC, Sacks F, Trichopoulou A, Drescher G, Ferro-Luzzi A, Helsing E, Trichopoulos D. Mediterranean diet pyramid: A cultural model for healthy eating. The American Journal of Clinical Nutrition. 1995;**61**:14025-14065

[4] Kislev ME. Emergence of wheat agriculture. Paléorient. 1984;**10**:61-70. DOI: 10.3406/paleo.1984.940

[5] Feldman M. Origin of cultivated wheat. In: Bonjean P, Angus WJ editors. The World Wheat Book. A History of Wheat Breeding. Paris: Tec & Doc/Intercept Ltd; 2001. pp. 3-56

[6] MacKey J. Wheat, its concept, evolution and taxonomy. In: Royo C, Nachit M, Di Fonzo N, Araus JL, Pfeiffer WH, Slafer GA, editors. Durum Wheat Breeding: Current Approaches and Future Strategies. New York: Haworth Press; 2005. pp. 3-62

[7] Hillman G. Late pleistocene changes in wild plant-food available to hunter-gatherers of the northern Fertile Crescent: Possible preludes to cereal cultivation. In: Harris DR, editor. The Origins and Spread of Agriculture and Pastoralism in Eurasia. London: UCL Press and Smithsonian Institution Press; 1996. pp. 159-203

[8] Takebayashi N, Morrell PL. Is self-fertilization an evolutionary dead end? Revisiting an old hypothesis with genetic theories and a macroevolutionary approach. American Journal of Botany. 2001;**88**:1143-1150

[9] Harlan JR. The early history of wheat: Earliest trace to the sack of Rome. In: Evans LT, Peacock WJ, editors. Wheat Science – Today and Tomorrow. Cambridge: Cambridge University Press; 1981. pp. 1-19

[10] Heun M, Schäfer-Pregl R, Klawan D, Castagna R, Accerbi M, Borghi B, and Salamini F. Site of einkorn wheat domestication identified by DNA fingerprinting. Science. 1997;**278**:1312-1314. DOI: 10.1126/science.278.5341.1312

[11] Abdel-Aal ESM, Sosulski FW, Hucl P. Origins, characteristics, and potentials of ancient wheats. Cereal Foods World. 1998;**43**:708-715

[12] Kisley ME. The agricultural situation in the Middle East in the VIIth millennium BP. In: Anderson PC, editor. Préhistoire de l'agriculture; nouvelles approches expérimentales et ethnographiques. Paris: Edition du CNRS, Monographie du Centre recherches Archéologiques; 1992. 6. pp. 87-93

[13] Helbaek H. Domestication of food plants in the old world. Science. 1959;**130**:365-372. DOI: 10.1126/science.130.3372.365

[14] Harlan JR. A wild wheat harvest in Turkey. Archaeology. 1967;**20**:197-201

[15] Harlan JR. Wild grass-seed harvesting and implications for domestication. In: Anderson PC, editor. Préhistoire de l'agri-culture: nouvelles approches expérimentales et ethnographiques. Valbonne: Monographies du Centre de Recherches Archéologiques; 1990. pp. 21-27

[16] Zohary D, Hopf M. Domestication of Plants in the Old World. 3rd ed. New York: Oxford University Press Inc.; 2000. pp. 239-240. DOI: 10.2307/23325662

[17] Moragues M, Moralejo MA, Sorrells ME, Royo C. Dispersal of durum wheat landraces across the Mediterranean basin assessed by AFLPs and microsatellites. Genetic Resources and Crop Evolution. 2007;**54**:1133-1144. DOI: 10.1007/s10722-006-9005-8

[18] Royo C, Nazco R, Villegas D. The climate of the zone of origin of Mediterranean durum wheat (*Triticum durum* Desf.) landraces affects their agronomic performance. Genetic Resources and Crop Evolution. 2014;**61**:1345-1358. DOI: 10.1007/s10722-014-0116-3

[19] Camacho Villa TC, Maxted N, Scholten MA, Ford-Lloyd BV. Defining and identifying crop landraces. Plant Genetic Resources. 2005;**3**:373-384. DOI: 10.1079/PGR200591

[20] Leemans R, Cramer W. The IIASA Database for Mean Monthly Values of Temperature, Precipitation and Cloudiness on a Global Terrestrial Grid. Luxemburg: International Institute of Applied Systems Analyses; 1991. RR-91-18

[21] Moragues M, García del Moral LF, Moralejo M, Royo C. Yield formation strategies of durum wheat landraces with distinct pattern of dispersal within the Mediterranean basin: I. Yield components. Field Crops Research. 2006;**95**:194-205. DOI: 10.1016/j.fcr.2005.02.009

[22] Moragues M, García del Moral LF, Moralejo M, Royo C. Yield formation strategies of durum wheat landraces with distinct pattern of dispersal within the Mediterranean basin: II. Biomass production and allocation. Field Crops Research. 2006;**95**:182-193. DOI: 10.1016/j.fcr.2005.02.008

[23] Skovmand B, Warburton ML, Sullivan SN, Lage J. Managing and collecting genetic resources. In: Royo C, Nachit MN, Di Fonzo N, Araus JL, Pfeiffer WH, Slafer GA, editors. Durum Wheat Breeding: Current Approaches and Future Strategies. New York: Haworth Press; 2005. pp. 142-163

[24] Royo C, Maccaferri M, Álvaro F, Moragues M, Sanguineti MC, Tuberosa R, Maalouf F, García del Moral LF, Demontis A, Rhouma S, Nachit M, Nserallah N, Villegas D. Understanding the relationships between genetic and phenotypic structures of a collection of elite durum wheat accessions. Field Crops Research. 2010;**119**:91-105. DOI: 10.1016/j.fcr.2010.06.020

[25] Oliveira HR, Campana MG, Jones H, Hunt HV, Leigh F, Redhouse D, Lister DL, Jone MK. Tetraploid wheat landraces in the Mediterranean Basin: Taxonomy, evolution and genetic diversity. PLoS One. 2012;**7**:e37063. DOI: 10.1371/journal.pone.0037063

[26] Ruiz M, Giraldo P, Royo C, Villegas D, Aranzana MJ, Carrillo JM. Diversity and genetic structure of a collection of Spanish durum wheat landraces. Crop Science. 2012;**52**:1-14. DOI: 10.2135/cropsci2012.02.0081

[27] Laidò G, Mangini G, Taranto F, Gadaleta A, Blanco A, et al. Genetic diversity and population structure of tetraploid wheats (*Triticum turgidum* L.) estimated by SSR, DArT and pedigree data. PLoS One. 2013;**8**:e67280. DOI: 10.1371/journal.pone.0067280

[28] Sahri A, Chentoufi L, Arbaoui M, Ardisson M, Belqadi L, Birouk A, Roumet P, Muller MH. Towards a comprehensive characterization of durum wheat landraces in Moroccan traditional agrosystems: Analysing genetic diversity in the light of geography, farmers' taxonomy and tetraploid wheat domestication history. BMC Evolutionary Biology. 2014;**14**:264. DOI: 10.1186/s12862-014-0264-2

[29] Soriano JM, Villegas D, Aranzana MJ, García del Moral LF, Royo C. Genetic structure of modern durum wheat cultivars and Mediterranean landraces matches with their agronomic performance. PLoS One. 2016;**11**:e0160983. DOI: 10.1371/journal.pone.0160983

[30] Frankel OH. Genetic perspectives of germplasm conservation. In: Arber W, Llimensee K, Peacock WJ, Starlinger P, editors. Genetic Manipulation: Impact on Man and Society. Cambridge: Cambridge University Press; 1984. pp. 161-170

[31] Balfourier F, Roussel V, Strelchenko P, Exbrayat-Vinson F, Sourdille P, Boutet G, Koenig J, Ravel C, Mitrofanova O, Beckert M, Charmet G. A worldwide bread wheat core collection arrayed in a 384-well plate. Theoretical and Applied Genetics. 2007;**114**:1265-1275. DOI: 10.1007/s00122-007-0517-1

[32] Nazco R, Villegas D, Ammar K, Peña RJ, Moragues M, Royo C. Can Mediterranean durum wheat landraces contribute to improved grain quality attributes in modern cultivars? Euphytica. 2012;**185**:1-17. DOI: 10.1007/s10681-011-0588-6

[33] Carrillo JM, Vázquez JF, Orellana J. Relationship between gluten strength and glutenin proteins in durum wheat cultivars. Plant Breeding. 1990;**104**:325-333. DOI: 10.1111/j.1439-0523.1990.tb00443.x

[34] Moragues M, Zarco-Hernández J, Moralejo MA, Royo C. Genetic diversity of glutenin protein subunits composition in durum wheat landraces [*Triticum turgidum* ssp. *turgidum* convar. *durum* (Desf.) MacKey] from the Mediterranean Basin. Genetic Resources and Crop Evolution. 2006;**53**:993-1002. DOI: 10.1007/s10722-004-7367-3

[35] Nazco R, Peña RJ, Ammar K, Villegas D, Crossa J, Royo C. Durum wheat Mediterranean landraces as sources of variability for allelic combinations at *Glu-1/Glu-3* loci affecting gluten strength and pasta cooking quality. Genetic Resources and Crop Evolution. 2014;**61**:1219-1236. DOI: 10.1007/s10722-014-0104-7

[36] Nazco R, Peña RJ, Ammar K, Villegas D, Crossa J, Royo C. Durum wheat (*Triticum durum* Desf.) Mediterranean landraces as sources of variability for allelic combinations at *Glu1/Glu-3* loci affecting gluten strength and pasta cooking quality. Genetic Resources and Crop Evolution. 2014;**61**:1219-1236. DOI: 10.1017/S0021859613000117

[37] Di Fonzo N, Ravaglia S, DeAmbrogio E, Blanco A, Troccoli A. Durum wheat improvement in Italy. In: Royo C, Nachi MN, Di Fonzo N, Araus JL, Pfeiffer WH, Slafer GA, editors. Durum Wheat Breeding: Current Approaches and Future Strategies. New York: Ed. Haworth Press; 2005. pp. 825-881

[38] Royo C, Briceño-Félix GA. Spanish wheat pool. In: Bojean AP, Angus WJ, van Ginkel M, editors. The World Wheat Book, Volume 2. A History of Wheat Breeding. Lavoisier; 2011. pp. 121-154. DOI: 10.1016/j.jcs.2011.10.001

[39] Bonjean AP, Doussinault G, Stragliati J. French wheat pool. In: Bonjean P, Angus WJ, editors. The World Wheat Book. A History of Wheat Breeding. Paris: Tec & Doc/Intercept Ltd; 2001. pp. 126-165

[40] Borghi B. Italian wheat pool. In: Bonjean P, Angus WJ, editors. The World Wheat Book. A History of Wheat Breeding. Paris: Tec & Doc/Intercept Ltd; 2001. pp. 288-309

[41] Sinclair TR. Historical changes in harvest index and crop nitrogen accumulation. Crop Science. 1998;**38**:638-643. DOI: 10.2135/cropsci1998.0011183X003800030002x

[42] Royo C, Álvaro F, Martos V, Ramdani A, Isidro J, Villegas D, García Del Moral LF. Genetic changes in durum wheat yield components and associated traits in Italian and Spanish varieties during the 20th century. Euphytica. 2007;**155**:259-270. DOI: 10.1007/s10681-006-9327-9

[43] Sanchez-Garcia M, Royo C, Aparicio N, Martín-Sánchez JA, Alvaro F. Genetic improvement of bread wheat yield and associated traits in Spain during the 20th century. The Journal of Agricultural Science. 2013;**151**:105-118. DOI: 10.1017/S0021859612000330

[44] Sanchez-Garcia M, Álvaro F, Peremarti A, Trevaskis B, Martín-Sánchez JA, Royo C. Breeding effects on dry matter accumulation and partitioning in Spanish bread wheat during the 20th century. Euphytica. 2014;**203**:321-336. DOI: 10.1007/s10681-014-1268-0

[45] Sanchez-Garcia M, Álvaro F, Peremarti A, Martín-Sánchez JA, Royo C. Changes in bread-making quality attributes of bread wheat varieties cultivated in Spain during the 20th century. European Journal of Agronomy. 2015;**63**:79-88. DOI: 10.1016/j.eja.2014.11.006

[46] Álvaro F, Isidro J, Villegas D, García del Moral LF, Royo C. Old and modern durum wheat varieties from Italy and Spain differ in main spike components. Field Crops Research. 2008;**106**:86-93. DOI: 10.1016/j.fcr.2007.11.003

[47] Royo C, Martos V, Ramdami A, Villegas D, Rharrabti Y, Garcia del Moral LF. Changes in yield and carbon isotope discrimination of Italian and Spanish durum wheat during the 20th century. Agronomy Journal. 2008;**100**:352-360. DOI: 10.2134/agronj2007.0060

[48] Álvaro F, Isidro J, Villegas D, García del Moral LF, Royo C. Breeding effects on grain filling, biomass partitioning, and remobilization in Mediterranean durum wheat. Agronomy Journal. 2008;**100**:361-370. DOI: 10.2134/agronj2007.0075

[49] Subira J, Ammar K, Álvaro F, García del Moral LF, Dreisigacker S, Royo C. Changes in durum wheat root and aerial biomass caused by the introduction of the *Rht-B1b* dwarfing allele and their effects on yield formation. Plant and Soil. 2016;**403**:291-304. DOI: 10.1007/s11104-015-2781-1

[50] Subira J, Álvaro F, García del Moral LF, Royo C. Breeding effects on the cultivar×environment interaction of durum wheat yield. European Journal of Agronomy. 2015;**68**:78-88. DOI: 10.1016/j.eja.2015.04.009

[51] Isidro J, Alvaro F, Royo C, Villegas D, Miralles DJ, García del Moral LF. Changes in duration of developmental phases of durum wheat caused by breeding in Spain and Italy during the 20th century and its impact on yield. Annals of Botany. 2011;**107**:1355-1366. DOI: 10.1093/aob/mcr063

[52] Sanchez-Garcia M, Álvaro F, Martín-Sánchez JA, Sillero JC, Escribano J, Royo C. Breeding effects on the genotype×environment interaction for yield of bread wheat grown in Spain during the 20th century. Field Crops Research. 2012;**126**:79-86. DOI: 10.1016/j.fcr.2011.10.001

[53] Subira J, Peña RJ, Álvaro F, Ammar K, Ramdani A, Royo C. Breeding progress in the pasta-making quality of durum wheat cultivars released in Italy and Spain during the 20th century. Crop and Pasture Science. 2014;**65**:16-26. DOI: 10.1071/CP13238

[54] Bonjean AP, Angus W, van Ginkel M, editors. 2011. The World Wheat Book. A History of Wheat Breeding. Vol. 2. Paris: Tec & Doc/Intercept Ltd; 2001. p. 1200

[55] Royo C, Nachit MN, Di Fonzo N, Araus JL, Pfeiffer WH, Slafer GA, editors. Durum Wheat Breeding: Current Approaches and Future Strategies. New York: Ed. Haworth Press; 2005. p. 1112

[56] Soleimani VD, Baum BR, Johnson DA. AFLP and pedigree-based genetic diversity estimates in modern cultivars of durum wheat [*Triticum turgidum* L. subsp. *durum* (Desf.) Husn.]. Theoretical and Applied Genetics. 2002;**104**:350-357. DOI: 10.1007/s001220100714

[57] Maccaferri M, Sanguineti MC, Donini P, Porceddu E, Tuberosa R. A retrospective analysis of genetic diversity in durum wheat elite germplasm based on microsatellite analysis: A case of study. In: Royo C, Nachit MN, Di Fonzo N, Araus JL, Pfeiffer WH, Slafer GA, editors. Durum Wheat Breeding: Current Approaches and Future Strategies. New York: Ed. Haworth Press; 2005 pp. 99-142

[58] Hovmøller MS, Justesen AF. Appearance of atypical *Puccinia striiformis* sp. *tritici* phenotypes in north-western Europe. Australian Journal of Agricultural Research. 2007;**58**:518-524. DOI: 10.1071/AR06146

[59] Giorgi F, Lionello P. Climate change projections for the Mediterranean region. Global and Planetary Change. 2008;**63**:90-104. DOI: 10.1016/j.gloplacha.2007.09.005

[60] Gibelin AL, Déqué M. Anthropogenic climate change over the Mediterranean region simulated by a global variable resolution model. Climate Dynamics. 2003;**20**:327. DOI: 10.1007/s00382-002-0277-1

[61] Lin BB. Resilience in agriculture through crop diversification: Adaptive management for environmental change. Bioscience. 2011;**61**:183-193. DOI: 10.1525/bio.2011.61.3.4

[62] Lopes MS, El-Basyoni I, Baenziger S, Singh S, Royo C, Ozbek K, Aktas H, Ozer E, Ozdemir F, Manickavelu A, Ban T, Vikram P, Saint Pierre C. Exploiting genetic diversity from landraces in wheat breeding for adaptation to climate change. Journal of Experimental Botany. 2015;**66**:3477-3486. DOI: 10.1093/jxb/erv122

[63] Carvalho A, Lima-Brito J, Maças B, Guedes-Pinto H. Genetic diversity and variation among botanical varieties of old Portuguese wheat cultivars revealed by ISSR assays. Biochemical Genetics. 2009;**47**:276-294. DOI: 10.1007/s10528-009-9227-5

[64] Cavanagh, et al. Genome-wide comparative diversity uncovers multiple targets of selection for improvement in hexaploid wheat landraces and cultivars. PNAS. 2013;**110**:8057-8062. DOI: 10.1073/pnas.1217133110

[65] Wang, et al. Characterization of polyploid wheat genomic diversity using a high-density 90 000 single nucleotide polymorphism array. Plant Biotechnology Journal. 2014;**12**:787-796. DOI: 10.1111/pbi.12183

[66] Bortesi L, Fischer R. The CRISPR/Cas9 system for plant genome editing and beyond. Biotechnology Advances. 2015;**33**:41-52. DOI: 10.1016/j.biotechadv.2014.12.006

Chapter 17

Mediterranean Diet beyond the Mediterranean Basin: Chronic Disease Prevention and Treatment

Guadalupe Echeverría, Catalina Dussaillant, Emma McGee, Inés Urquiaga, Nicolás Velasco and Attilio Rigotti

Additional information is available at the end of the chapter

http://dx.doi.org/10.5772/intechopen.68937

Abstract

The Mediterranean diet (MedDiet) is considered one of the healthiest dietary patterns. Current scientific evidence supports that this dietary pattern is associated with lower prevalence and incidence of a number of chronic diseases, such as cardiovascular disease, diabetes, cancer, and age-related cognitive decline as well as reduced overall mortality. The Mediterranean diet includes a wide variety of foods that are eaten in moderation and enjoyed in a positive social environment. It is characterized by a high intake of fruits and vegetables, whole grains, legumes, nuts, fish and seafood, white meats, olive oil, herbs, and spices paired with moderate consumption of fermented dairy products and wine and low intake of red meat, butter, and sugar. The generic term "Mediterranean diet" was coined in the *Seven Countries Study* led by Ancel Keys in the 1950s. Yet, in spite of its name, this dietary pattern and its benefits are not confined exclusively to the Mediterranean Basin. Among other world regions, Central Chile exhibits climate, agriculture, and culinary traditions similar to various Mediterranean countries. It is therefore essential to increase awareness about the Mediterranean-like richness of both produce and culinary culture beyond the Mediterranean Basin. Active promotion of this dietary pattern may offer health benefits and improve the quality of life in many populations worldwide.

Keywords: Mediterranean diet, human health, chronic disease prevention, Chile, lifestyle

1. Introduction

Noncommunicable diseases (NCDs), also known as chronic diseases, are the leading cause of death worldwide, representing an increasing healthcare burden and a growing concern for national and international public health agencies [1]. Currently, most NCD deaths occur in low- and middle-income countries rather than in developed regions. The rise of NCDs—obesity, diabetes, cardiovascular diseases, chronic lung disease, and cancer—has been driven by predisposing lifestyle-related risk factors, including physical inactivity, unhealthy diet, and tobacco use. Notably, diets rich in saturated fats, polyunsaturated omega-6 fatty acids, trans-fatty acids, and refined carbohydrates—along with a low supply of antioxidants and fiber—are harmful to human health [2].

At the population level, there is a consensus that lifestyle improvement is the most cost-effective approach that can be used in preventive and therapeutic programs aimed at stopping the ongoing NCD epidemic and avoiding excessive medicalization of NCD management through pharmacological or surgical interventions. Among these lifestyle improvement measures, promotion of a healthy diet plays a critical role. Indeed, healthy dietary patterns can modify the population risk profile for chronic diseases, and the association between high-quality diet, better health, and less disease is widely accepted [3].

A large amount of scientific evidence supports the claim that the Mediterranean diet (MedDiet) is beneficial for human health and disease prevention [4, 5]. However, this dietary pattern and its benefits are not exclusively confined to the Mediterranean Basin. Among others, Central Chile is one region that exhibits climate, agriculture, and culinary traditions similar to various Mediterranean countries. In this book chapter, we (1) review the origin of the MedDiet concept; (2) analyze evidence supporting its positive impact on human health; (3) assess geographical, ecological, and agricultural evidence indicating that Central Chile may be considered a Mediterranean-like region; (4) discuss traditional Chilean cuisine as related to Mediterranean diets; and (5) analyze pilot studies on dietary Mediterranization and its impact on biomarkers, risk factors, and disease conditions in Chilean adults.

2. The Mediterranean diet

The MedDiet is currently considered one of the healthiest dietary patterns in the world [4, 5]. MedDiet was born and popularized as a concept in the middle of the twentieth century as a consequence of major findings reported in the *Seven Countries Study*, led by Ancel Keys and colleagues in the 1950s, which compared the relationship between dietary habits and cardiovascular disease outcomes in 16 different cohorts from the USA, Japan, Finland, the Netherlands, former Yugoslavia, Italy, and Greece [6]. Comparative follow-up from 5 to 25 years confirmed a lower mortality from coronary heart disease and showed decreased overall mortality and longer life expectancy in association with dietary patterns found in Crete and Corfu (Greece), Montegiorgio (Italy), and Dalmatia (Croatia, formerly in Yugoslavia), despite the consumption of a high-fat diet due to significant intake of olive oil [7].

It is important to emphasize that there is no single MedDiet pattern. In fact, the Mediterranean Basin includes several countries that differ in their dietary intake. However, an overall

http://dx.doi.org/10.5772/intechopen.68937

Mediterranean dietary pattern can be defined in Spain, Southern France, Italy, and Greece—and even in North Africa and the Middle East—where diets are characterized by high intake of fruits and vegetables, whole grains, legumes, nuts, fish and seafood, white meats, olive oil, herbs, and spices; moderate consumption of fermented dairy products and wine; and low intake of red meat, whole milk, and butter as well as sugar and sugary foods. From a nutritional perspective, the MedDiet is low in saturated fats, refined carbohydrates, and terrestrial animal protein, is high in antioxidants, fiber, and monounsaturated fats; and exhibits an adequate omega-6/omega-3 fatty acid balance. As indicated by this description, the MedDiet includes a wide variety of foods that are combined in a specific consumption frequency (**Table 1**) to prepare appealing and tasty meals rather than restricting specific foods or nutrients.

The MedDiet includes a wide range of foods rich in bioactive components, which are responsible for the dietary pattern's beneficial health effects [8–12]. Among these components, the most important are antioxidants, fiber, and phytosterols from vegetables, fruits, virgin olive oil, and wine; monounsaturated fatty acids from olive oil; omega-3 fatty acids mainly from marine products and nuts; and probiotics provided by fermented foods such as cheese, yogurt, and others. Through multiple concurrent and synergistic mechanisms, the MedDiet generates diverse and favorable physiological and metabolic changes (e.g., improvement in antioxidant defense, lipid profile, blood pressure, inflammation,

Food group	Consumption frequency
Vegetables	Daily, in abundant amount, three or more servings per day, raw or cooked
Fruits	Daily, in abundant amount, two or more servings per day
Olive oil	Daily, three to six tablespoons per day, as main fat source
Cereal grains, mainly bread and pastries	Daily, three to four times a day, in moderate amount
Legumes	At least three times a week
Nuts	At least three times a week
Dairy products, mainly fermented yogurts and cheeses	Daily, two to four servings per day
Eggs	One to four units per week
Fish and seafood	Two to four times per week
Poultry	Two to four times per week
Red and processed meats	One time per week or less
Wine	Daily, moderate consumption (one glass a day for women and two glasses a day for men), on a regular basis, mainly with meals
Spices and condiments	Daily, varied use in the preparation of meals

Table 1. Food components and consumption frequencies in a Mediterranean diet.

and coagulation together with modulation of gene expression) [8, 13–16] that ultimately improve health, lower the risk of chronic diseases, promote longevity, and enhance quality of life.

However, the MedDiet includes more than just food. The term diet is being derived from the Greek word *diaita* which has the meaning of the way of life/lifestyle. The MedDiet constitutes a set of skills, knowledge, and traditions stretching from the land to the table, including farming, fishing, conservation and processing, preparation, and—finally—consumption of food. Furthermore, this lifestyle also promotes social interaction, since familial and communal meals are the cornerstone of social customs and festive events. As a result, UNESCO recognized the MedDiet as Intangible Cultural Heritage in 2010 [17].

3. Mediterranean diet and human health/disease prevention

Increasing observational and interventional scientific evidence illustrates that MedDiet is beneficial for human health and disease prevention, leading to considerable interest in its application as a tool for attenuating the global impact of NCDs.

3.1. Observational studies

Several cross-sectional and prospective cohort studies have associated the MedDiet with lower prevalence and incidence of a number of chronic conditions, such as metabolic syndrome (MS), diabetes, coronary artery disease, cancer, and neurodegenerative diseases, as well as reductions in overall mortality.

MS is a set of clinical conditions that increases the risk of cardiovascular disease and other health complications. A large meta-analysis concluded that adherence to a Mediterranean-type diet is associated with a lower risk of MS, with a protective effect on all its components (hyperglycemia, high triglycerides, low HDL cholesterol, high blood pressure, and abdominal obesity) [18]. A high MedDiet score has also been correlated with a lower incidence of diabetes as well as improved metabolic control and decreased total mortality in diabetic patients [19–21]. Furthermore, meta-analysis of several large prospective cohort studies indicates that increased adherence to a MedDiet is significantly associated with reduced cardiovascular disease incidence and mortality [22, 23].

In regards to cancer, meta-analysis based on cumulative information showed that high adherence to MedDiet consumption was associated with lower incidence and mortality from malignant neoplasia [24]. More specifically, the lower overall risk of cancer found in Mediterranean countries is mainly due to a lower incidence of colon, endometrial, breast, and prostatic cancer [25].

Given the ongoing aging of the world population, the prevalence of neurodegenerative diseases has dramatically increased. In several observational studies, the MedDiet pattern has been linked with attenuation of mild cognitive impairment associated with aging as well as reduced incidence of Alzheimer's and Parkinson's diseases [26]. All these findings have been confirmed by meta-analytical approaches [22, 24].

http://dx.doi.org/10.5772/intechopen.68937

3.2. Interventional studies

To date, several randomized clinical trials have shown that Mediterranean-type diets are more effective than other diets in inducing clinically favorable long-term changes in cardiovascular risk factors and inflammatory markers [27]. This evidence strongly suggests that a MedDiet may have positive effects on clinical outcomes related to NCDs. Indeed, prospective interventional clinical studies—particularly the PREDIMED (Prevention with Mediterranean Diet) initiative in Spain [28]—have established that following a Mediterranean-type diet lowers the incidence of cardiovascular disease and attenuates the risk of diabetes, breast cancer, and age-related cognitive decline.

The first prospective intervention trial assessing MedDiet's impact on clinical outcomes was the *Lyon Diet Heart Study* (1988–1992). In this study, patients with a previous myocardial infarction (MI) were randomized to either α-linolenic fatty acid-enriched MedDiet or a prudent control diet [29]. The major finding was that, after a follow-up period of 27 months, subjects following the MedDiet had a significantly reduced recurrence of MI and other cardiac events as well as decreased overall mortality. An extended follow-up (46 months) confirmed these findings indicating the long-term effectiveness of the MedDiet in reducing coronary clinical events in secondary prevention [30].

More recently, the multicenter PREDIMED study was conducted in Spain, between the years 2003 and 2011. This study included men and women with high risk for NCDs but without previous cardiovascular events. Participants were randomized to MedDiet supplemented with extra virgin olive oil, MedDiet supplemented with nuts, or a low fat diet (control group) [28]. Both types of MedDiet intervention groups showed benefits on different cardiovascular risk factors such as dyslipidemia, insulin resistance, inflammation, and oxidation [28]. They also had a favorable effect on MS, improving its components and decreasing its overall prevalence via reversion with respect to baseline [31, 32]. After 4.8 years of follow-up, participants assigned to MedDiets—supplemented with either olive oil or nuts—exhibited a 30% reduction in cardiovascular event risk, mainly due to stroke, compared to the control group [33]. Additional analyses of this trial have revealed a significantly lower incidence of newly diagnosed cases of diabetes among nondiabetic subjects who followed a MedDiet as compared to the control group [34]. Finally, the PREDIMED study also demonstrated that the MedDiet reduces cognitive impairment associated with aging [35] as well as breast cancer incidence [36]. The PREDIMED study therefore provides strong scientific evidence regarding the beneficial role of MedDiet in the prevention of cardiovascular disease and other NCD-related conditions, validating the evidence derived from previous cross-sectional and longitudinal observational studies.

4. Central Chile: Mediterranean-like ecosystem and food production

The Mediterranean Basin is the largest Mediterranean ecosystem in the world. However, there are four more places in the world where Mediterranean ecosystems can be found:

California, South Africa, Southwest Australia, and Central Chile (**Figure 1**). Geographically, these Mediterranean ecosystems are localized on the western coasts of continental landmasses, at latitudes comprised between 30 and 45° north or south [37]. All of these regions are considered biodiversity hot spots with some of the world's best environments for agriculture and human habitation.

Cool, wet winters and hot, dry summers characterize the climate in these regions, and the vegetation is primarily evergreen or deciduous shrublands. Mediterranean agriculture is also defined by the cultivation of specific fruits, vegetables, and grains such as olives, grapes, garlic, lemons, oranges, tomatoes, chickpeas, eggplants, peppers, and fennel. In fact, the distribution of the primary olive oil and wine-producing regions of the world, two emblematic products from Mediterranean agriculture, is remarkably similar to that of the Mediterranean ecosystems. These ecosystems also determine livestock use, which includes mostly sheep and goats, leading to high availability of cheese and yogurt but low access to red meat and butter. These foods, together with fish and seafood harvested from Mediterranean coastal regions, are crucial components of MedDiets as indicated above.

The MedDiet is also a consequence of the global exchange of domesticated animal and plant species and gastronomic traditions of the Mediterranean region. Inhabitants of the Mediterranean regions devised culinary practices that rendered their cuisine attractive, healthy, and culturally important. However, the Mediterranean food basket is in fact the result of domesticated food species exchanged mainly between the Americas, Europe, North Africa, and Asia. Some

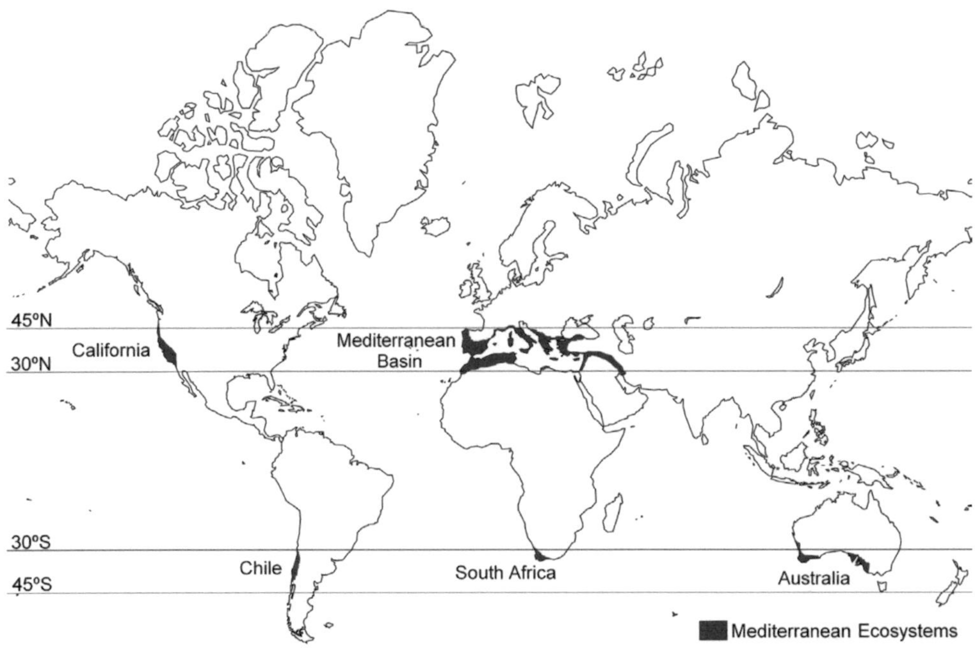

Figure 1. Mediterranean ecosystems of the world. There are five main Mediterranean ecosystems located between latitudes 30 and 45° of both hemispheres [50].

emblematic foods are native to the Americas, such as potatoes and tomatoes, which are from the Andean regions of Ecuador, Peru, and Chile. Common beans and green beans along with sweet and hot peppers are native to Mexico, Central America, Bolivia, and Peru. Corn is native to Mexico and the northern zone of Central America, broad beans and artichokes come from North Africa, chicken and rice come from China, and citrus comes from Southeast Asia, while garlic and onions are native to Central Asia. Meanwhile, eggplants come from India, and many spices come from Asia and other regions (**Figure 2**) [38].

Central Chile, a 1000 km narrow band (30–36°S) located west of the Andes mountains, also exhibits a Mediterranean-like environment, as previously reviewed in Ref. [39]. The ecological conditions of Central Chilean result in agricultural and livestock patterns similar to those generated in countries located in the Mediterranean Basin. In fact, data from the UN Food and Agriculture Organization indicated that Chilean food production, availability, and consumption at the end of the twentieth century were similar to that which was reported in Italy and Spain [40].

Figure 2. The Mediterranean diet includes foodstuffs derived from animal and plant species that originated in different regions of the world.

5. Mediterranean diet style in Chile

Traditional Chilean cuisine uses ingredients and cooking techniques that are very similar to those utilized in the Mediterranean culinary traditions of Spain, Italy, and Greece [37, 40]. Characteristic Chilean dishes, such as *charquicán* (stewed vegetables with minced meat), *porotos granados* (stewed vegetables, basil, and fresh beans), *porotos con riendas* (stewed vegetables, beans, and pasta), *cazuela de ave* (boiled chicken, potatoes, pumpkin pieces, green beans, carrots, parsley, rice, and corn), and *tomaticán* (stewed onion, tomato, corn, and minced meat), are prepared on a flavored base that contains onions, oregano, cumin, black pepper, paprika, and sautéed garlic. This flavor base is similar to Spanish *sofrito*, Italian *soffritto*, or French *mirepoix*. Fish (such as conger eel in *caldillo de congrio*) or seafood (such as mussels, scallops, sea squid, barnacles, and other shellfish in *mariscal*) are often prepared in Mediterranean-type soups made with tomatoes, carrots, potatoes, onions, red and green peppers, garlic, oregano, chili peppers, and parsley. Traditionally, Chilean dishes are also served with fresh parsley and cilantro. Two popular salads are prepared with tomato, onion, and parsley or celery combined with avocado and walnuts. Salads are usually dressed with oil, salt, and lemon juice. Traditional *pebres* are prepared using finely chopped onion, garlic, tomato, parsley, chili, and cilantro and are dressed with vinegar, salt, and oil (**Figure 3**). Wine consumption is also an important part of Chilean culture.

Traditional Chilean cuisine thus exhibits key Mediterranean characteristics, and the promotion of MedDiet in Chile offers a great opportunity for the management of ongoing epidemiological transition and increasing incidence of risk conditions and NCDs in Chile [37, 41]. In fact,

Figure 3. Traditional dishes from Chile. Chilean recipes use ingredients and cooking techniques that are very similar to those utilized in culinary traditions of the Mediterranean region. Upper row: *charquicán*, *porotos granados*, and *porotos con riendas*; middle row: *cazuela de ave*, *tomaticán*, and *caldillo de congrio*; lower row: *mariscal*, *pebre*, Chilean salad, and salad with celery, avocado, and nuts.

data from 2010 to 2015 collected through *Programa Aliméntate Sano* (www.alimentatesano.cl), an open-access online platform created and led by our institution to provide general lifestyle (including the MedDiet) and health information and resources to the Chilean population, indicate that Chilean adults who exhibit higher adherence to MedDiet (based on a locally developed and validated score) [42] have a lower prevalence of overweight, obesity, and MS [43]. Furthermore, this cross-sectional study suggests that even small improvements in adherence to MedDiet may be associated with a significant decrease in obesity/overweight and MS, though this claim requires further testing.

In spite of the importance of MedDiet in Chile, current economical and epidemiological transitions as well as ongoing lifestyle changes are negatively impacting the use of traditional cuisine and dietary habits in the Chilean population [37, 40, 41]. Information obtained from the *Programa Aliméntate Sano* database (n = 53,366 subjects) demonstrate that only 10% of the sampled population complies with a Mediterranean-type diet (2) with higher adherence in women, at older ages, and at higher educational levels. These latter demographic trends are comparable to those obtained with similar MedDiet indexes applied in other Mediterranean and non-Mediterranean populations.

6. Dietary Mediterranization studies in Chile

In Chile, some studies have already been conducted to evaluate the impact of the MedDiet on laboratory and clinical parameters.

In one study conducted in 1998, individuals subjected to a 3-month MedDiet intervention showed greater antioxidant capacity and less oxidative damage [44], better fatty acid profile [45], hemostasis [46], and endothelial function [47] compared with those following a Western diet. In a second more recent study (2014), a calorie-restricted MedDiet resulted in decreased serum levels of advanced glycation end products in premenopausal women with overweight or obesity [48].

Finally, dietary Mediterranization and its effect on the prevalence of MS and its components were evaluated at a workplace. During this intervention, a Mediterranization of the participants' diet was observed (measured by MedDiet score [42]) in association with a significant improvement of MS components as well as MS itself by the end of the study [49].

7. Concluding remarks

The MedDiet has been shown, with ample scientific solvency, through both observational and interventional studies, numerous benefits in the prevention, and treatment of high-risk conditions and/or NCDs; these benefits are most likely explained by the MedDiet's favorable composition including a wide variety of important bioactive compounds and nutrients.

With regard to our country and based on the ecosystem, agricultural conditions, and culinary traditions of Central Chile, the MedDiet may play a key role in public policies for health promotion and disease prevention. It also offers enormous expectations—beyond healthcare—for well-being improvement, economic and social development, and global quality of life advancement in the Chilean population and society.

Acknowledgements

We would like to acknowledge the leading role of the late Dr. Federico Leighton in promoting and studying the application of the Mediterranean diet in Chile. Research reported by our group in this article has been supported by grants from Fundación Alimenta and Fundación Banmédica (Chile).

Author details

Guadalupe Echeverría[1], Catalina Dussaillant[1], Emma McGee[1], Inés Urquiaga[1], Nicolás Velasco[2] and Attilio Rigotti[1,2*]

*Address all correspondence to: arigotti@med.puc.cl

1 Center for Molecular Nutrition and Chronic Diseases, School of Medicine, Pontificia Universidad Católica de Chile, Santiago, Chile

2 Department of Nutrition, Diabetes and Metabolism, School of Medicine, Pontificia Universidad Católica de Chile, Santiago, Chile

References

[1] WHO. Attaining the Nine Global Noncommunicable Diseases Targets; A Shared Responsibility: Switzerland; 2014

[2] Willett WC, Stampfer MJ. Current evidence on healthy eating. Annual Review of Public Health. 2013;**34**:77-95

[3] Mozaffarian D, Appel LJ, Van Horn L. Components of a cardioprotective diet: New insights. Circulation. 2011;**123**(24):2870-2891

[4] Sofi F, Macchi C, Abbate R, Gensini GF, Casini A. Mediterranean diet and health. Biofactors. 2013;**39**(4):335-342

[5] Dussaillant C, Echeverria G, Urquiaga I, Velasco N, Rigotti A. Current evidence on health benefits of the mediterranean diet. Revista Médica de Chile. 2016;**144**(8):1044-1052

[6] Keys A. Coronary heart disease in seven countries. Circulation. 1970;**41**(4 Suppl):1-211

[7] Keys A. Mediterranean diet and public health: Personal reflections. American Journal of Clinical Nutrition. 1995;**61**(6 Suppl):1321S–1323S

[8] Urquiaga I, Echeverría G, Dussaillant C, Rigotti A. [Origin, components and mechanisms of action of the Mediterranean diet]. Revista Médica de Chile. 2017; **145**(1):85-95.

[9] Simopoulos AP. The Mediterranean diets: What is so special about the diet of Greece? The scientific evidence. Journal of Nutrition. 2001;**131**(11 Suppl):3065S–3073S

[10] Saura-Calixto F, Goni I. Definition of the Mediterranean diet based on bioactive compounds. Critical Reviews in Food Science and Nutrition. 2009;**49**(2):145-152

[11] Bullo M, Casas R, Portillo MP, Basora J, Estruch R, Garcia-Arellano A, et al. Dietary glycemic index/load and peripheral adipokines and inflammatory markers in elderly subjects at high cardiovascular risk. Nutrition, Metabolism, and Cardiovascular Diseases. 2013;**23**(5):443-450

[12] Konstantinidou V, Covas MI, Sola R, Fito M. Up-to date knowledge on the in vivo transcriptomic effect of the Mediterranean diet in humans. Molecular Nutrition & Food Research. 2013;**57**(5):772-783

[13] Schwingshackl L, Hoffmann G. Mediterranean dietary pattern, inflammation and endothelial function: A systematic review and meta-analysis of intervention trials. Nutrition, Metabolism, and Cardiovascular Diseases. 2014;**24**(9):929-939

[14] Damasceno NR, Sala-Vila A, Cofan M, Perez-Heras AM, Fito M, Ruiz-Gutierrez V, et al. Mediterranean diet supplemented with nuts reduces waist circumference and shifts lipoprotein subfractions to a less atherogenic pattern in subjects at high cardiovascular risk. Atherosclerosis. 2013;**230**(2):347-353

[15] Estruch R, Martinez-Gonzalez MA, Corella D, Salas-Salvado J, Ruiz-Gutierrez V, Covas MI, et al. Effects of a Mediterranean-style diet on cardiovascular risk factors: A randomized trial. Annals of Internal Medicine. 2006;**145**(1):1-11

[16] Fito M, Guxens M, Corella D, Saez G, Estruch R, de la Torre R, et al. Effect of a traditional Mediterranean diet on lipoprotein oxidation: A randomized controlled trial. Archives of Internal Medicine. 2007;**167**(11):1195-1203

[17] UNESCO. Representative Lists of the Intangible Cultural Heritage of Humanity; 2010 [cited April 2011]. Available from: http://www.unesco.org/culture/ich/en/RL/00394

[18] Kastorini CM, Milionis HJ, Esposito K, Giugliano D, Goudevenos JA, Panagiotakos DB. The effect of Mediterranean diet on metabolic syndrome and its components: A meta-analysis of 50 studies and 534,906 individuals. Journal of the American College of Cardiology. 2011;**57**(11):1299-1313

[19] Martinez-Gonzalez MA, de la Fuente-Arrillaga C, Nunez-Cordoba JM, Basterra-Gortari FJ, Beunza JJ, Vazquez Z, et al. Adherence to Mediterranean diet and risk of developing diabetes: Prospective cohort study. British Medical Journal. 2008;**336**(7657):1348-1351

[20] Esposito K, Maiorino MI, Di Palo C, Giugliano D, Campanian Postprandial Hyperglycemia Study G. Adherence to a Mediterranean diet and glycaemic control in type 2 diabetes mellitus. Diabetic Medicine. 2009;**26**(9):900-907

[21] Bonaccio M, Di Castelnuovo A, Costanzo S, Persichillo M, De Curtis A, Donati MB, et al. Adherence to the traditional Mediterranean diet and mortality in subjects with diabetes. Prospective results from the MOLI-SANI study. European Journal of Preventive Cardiology. 2016; **23**(4):400-7

[22] Sofi F, Abbate R, Gensini GF, Casini A. Accruing evidence on benefits of adherence to the Mediterranean diet on health: An updated systematic review and meta-analysis. American Journal of Clinical Nutrition. 2010;**92**(5):1189-1196

[23] Sofi F, Macchi C, Abbate R, Gensini GF, Casini A. Mediterranean diet and health status: An updated meta-analysis and a proposal for a literature-based adherence score. Public Health Nutrition. 2014;**17**(12):2769-2782

[24] Sofi F, Cesari F, Abbate R, Gensini GF, Casini A. Adherence to Mediterranean diet and health status: Meta-analysis. British Medical Journal. 2008;**337**:a1344

[25] Trichopoulou A, Lagiou P, Kuper H, Trichopoulos D. Cancer and Mediterranean dietary traditions. Cancer Epidemiology, Biomarkers and Prevention. 2000;**9**(9):869-873

[26] Sofi F, Macchi C, Casini A. Mediterranean diet and minimizing neurodegeneration. Current Nutrition Reports. 2013;**2**:75-80

[27] Serra-Majem L, Roman B, Estruch R. Scientific evidence of interventions using the Mediterranean diet: A systematic review. Nutrition Reviews. 2006;**64**(2 Pt 2):S27–S47

[28] Martinez-Gonzalez MA, Salas-Salvado J, Estruch R, Corella DD, Fito M, Ros E, et al. Benefits of the Mediterranean Diet: Insights from the PREDIMED Study. Progress in Cardiovascular Diseases. 2015; **58**(1):50-60

[29] de Lorgeril M, Renaud S, Mamelle N, Salen P, Martin JL, Monjaud I, et al. Mediterranean alpha-linolenic acid-rich diet in secondary prevention of coronary heart disease. Lancet. 1994;**343**(8911):1454-1459

[30] de Lorgeril M, Salen P, Martin JL, Monjaud I, Delaye J, Mamelle N. Mediterranean diet, traditional risk factors, and the rate of cardiovascular complications after myocardial infarction: Final report of the Lyon Diet Heart Study. Circulation. 1999;**99**(6):779-785

[31] Salas-Salvado J, Fernandez-Ballart J, Ros E, Martinez-Gonzalez MA, Fito M, Estruch R, et al. Effect of a Mediterranean diet supplemented with nuts on metabolic syndrome status: One-year results of the PREDIMED randomized trial. Archives of Internal Medicine. 2008;**168**(22):2449-2458

[32] Babio N, Toledo E, Estruch R, Ros E, Martinez-Gonzalez MA, Castaner O, et al. Mediterranean diets and metabolic syndrome status in the PREDIMED randomized trial. Canadian Medical Association Journal. 2014;**186**(17):E649-E657

[33] Estruch R, Ros E, Salas-Salvado J, Covas MI, Corella D, Aros F, et al. Primary prevention of cardiovascular disease with a Mediterranean diet. New England Journal of Medicine. 2013;**368**(14):1279-1290

[34] Salas-Salvado J, Bullo M, Babio N, Martinez-Gonzalez MA, Ibarrola-Jurado N, Basora J, et al. Reduction in the incidence of type 2 diabetes with the Mediterranean diet: Results of the PREDIMED-Reus nutrition intervention randomized trial. Diabetes Care. 2011;**34**(1):14-19

[35] Valls-Pedret C, Sala-Vila A, Serra-Mir M, Corella D, de la Torre R, Martinez-Gonzalez MA, et al. Mediterranean diet and age-related cognitive decline: A randomized clinical trial. Journal of the American Medical Association Internal Medicine. 2015;**175**(7):1094-1103

[36] Toledo E, Salas-Salvado J, Donat-Vargas C, Buil-Cosiales P, Estruch R, Ros E, et al. Mediterranean diet and invasive breast cancer risk among women at high cardiovascular risk in the PREDIMED trial: A randomized clinical trial. Journal of the American Medical Association Internal Medicine. 2015;**175**(11):1752-60

[37] Urquiaga I, Echeverria G, Polic G, Castillo O, Liberona Y, Rozowski J, et al. Mediterranean food and diets, global resource for the control of metabolic syndrome and chronic diseases. World Review of Nutrition and Dietetics. 2008;**98**:150-173

[38] Leighton F, Urquiaga, I. The Mediterranean diets: Nutrition and gastronomy. In: Smith J, Charter E, editors. Functional Food Product and Development. Canada: Wiley-Blackwell; 2010. pp. 322-343

[39] Armesto JJ, Arroyo MTK, Hinojosa, LF. The Mediterranean Environment of Central Chile. New York: Oxford University Press; 2007

[40] Rozowski J, Castillo O. Is the Chilean diet a Mediterranean-type diet? Biological Research. 2004;**37**(2):313-319

[41] Albala C, Vio F, Kain J, Uauy R. Nutrition transition in Latin America: The case of Chile. Nutrition Reviews. 2001;**59**(6):170-176

[42] Echeverría G, Urquiaga I, Concha MJ, Dussaillant C, Villarroel L, Velasco N, et al. [Validation of self-applicable questionnaire for a mediterranean dietary index in Chile]. Revista Médica de Chile. 2016 Dec;**144**(12):1531-1543

[43] Echeverría G, McGee E, Urquiaga I, Jimenez P, D'Acuña S, Villarroel L, et al. Inverse associations between a locally validated Mediterranean Diet Index, Obesity/Overweight, and Metabolic Syndrome: Findings from a self-selecting sample of Chilean adults. Manuscript in preparation. 2017

[44] Urquiaga I, Strobel P, Perez D, Martinez C, Cuevas A, Castillo O, et al. Mediterranean diet and red wine protect against oxidative damage in young volunteers. Atherosclerosis. 2010;**211**(2):694-699

[45] Urquiaga I, Guasch V, Marshall G, San Martin A, Castillo O, Rozowski J, et al. Effect of Mediterranean and Occidental diets, and red wine, on plasma fatty acids in humans. An intervention study. Biological Research. 2004;**37**(2):253-261

[46] Mezzano D, Leighton F, Martinez C, Marshall G, Cuevas A, Castillo O, et al. Complementary effects of Mediterranean diet and moderate red wine intake on haemostatic cardiovascular risk factors. European Journal of Clinical Nutrition. 2001;**55**(6): 444-451